Representation and Invariance
of Scientific Structures

Representation and Invariance of Scientific Structures

Patrick Suppes

CSLI PUBLICATIONS — Center for the Study of Language and Information, Stanford, California

Copyright © 2002
CSLI Publications
Center for the Study of Language and Information
Leland Stanford Junior University
Printed in the United States
06 05 04 03 02   5 4 3 2 1

Library of Congress Cataloging-in-Publication Data

Suppes, Patrick, 1922–
Representation and invariance of scientific structures /
Patrick Suppes.
  p.   cm. – (CSLI lecture notes ; no. 130)
Includes bibliographical references and index.
ISBN 1-57586-333-2 (cloth : alk. paper)

1. Science–Philosophy.
I. Title. II. Series.
Q175 .S93945 2001
501—dc21 2001035637
CIP

∞ The acid-free paper used in this book meets the minimum requirements of the American National Standard for Information Sciences—Permanence of Paper for Printed Library Materials, ANSI Z39.48-1984.

CSLI was founded early in 1983 by researchers from Stanford University, SRI International, and Xerox PARC to further research and development of integrated theories of language, information, and computation. CSLI headquarters and CSLI Publications are located on the campus of Stanford University.

CSLI Publications reports new developments in the study of language, information, and computation. In addition to lecture notes, our publications include monographs, working papers, revised dissertations, and conference proceedings. Our aim is to make new results, ideas, and approaches available as quickly as possible. Please visit our web site at
http://cslipublications.stanford.edu/
for comments on this and other titles, as well as for changes and corrections by the author and publisher.

To My Children
Patricia, Deborah, John,
Alexandra and Michael

Contents

Preface   xiii

1 Introduction   1
    1.1   General Viewpoint   1
    1.2   What Is a Scientific Theory?   2
              The traditional sketch.   2
              Models versus empirical interpretations of theories.   3
              Intrinsic-versus-extrinsic characterization of theories.   5
              Coordinating definitions and the hierarchy of theories.   7
              Instrumental view of theories.   8
    1.3   Plan of the Book   10
    1.4   How To Read This Book   14

2 Axiomatic Definition of Theories   17
    2.1   Meaning of *Model* in Science   17
              Comments on quotations.   20
    2.2   Theories with Standard Formalization   24
              Example: ordinal measurement.   25
              Axiomatically built theories.   26
              Difficulties of scientific formalization.   27
              Useful example of formalization   28
    2.3   Theories Defined by Set-theoretical Predicates   30
              Example: theory of groups.   31
              Meaning of 'set-theoretical predicate'.   32
              Set theory and the sciences.   33
              Basic structures.   33
              Reservations about set theory.   34
    2.4   Historical Perspective on the Axiomatic Method   35
              Before Euclid.   35

Euclid.   36
Archimedes.   37
Euclid's Optics.   40
Ptolemy's Almagest.   41
Jordanus de Nemore.   43
Newton.   44
Modern geometry.   45
Hilbert and Frege.   47
Physics.   48

3 Theory of Isomorphic Representation   51

3.1 Kinds of Representation   51
Definitions as representations.   53

3.2 Isomorphism of Models   54

3.3 Representation Theorems   57
Homomorphism of models.   58
Embedding of models.   62

3.4 Representation of Elementary Measurement Structures   63
Extensive measurement.   63
Difference measurement.   66
Bisection measurement.   67
Conjoint measurement.   69
Proofs of Theorems 2–4   70

3.5 Machine Representation of Partial Recursive Functions   74
Unlimited register machines (URM).   76
Partial recursive functions over an arbitrary finite alphabet.   80

3.6 Philosophical Views of Mental Representations   81
Aristotle.   81
Descartes.   83
Hume.   83
Kant.   86
James.   88
Special case of images.   92
Psychological views of imagery.   93

4 Invariance   97

4.1 Invariance, Symmetry and Meaning   97
Meaning.   102
Objective meaning in physics.   103

4.2 Invariance of Qualitative Visual Perceptions   105
Oriented physical space.   106

	4.3	Invariance in Theories of Measurement 110

- 4.3 Invariance in Theories of Measurement   110
  - Second fundamental problem of measurement: invariance theorem.   112
  - Classification of scales of measurement.   114
- 4.4 Why the Fundamental Equations of Physical Theories Are Not Invariant   120
  - Beyond symmetry.   122
  - Covariants.   122
- 4.5 Entropy as a Complete Invariant in Ergodic Theory   123
  - Isomorphism of ergodic processes.   125

5 Representations of Probability   129

- 5.1 The Formal Theory   130
  - Primitive notions.   130
  - Language of events.   132
  - Algebras of events.   133
  - Axioms of probability.   134
  - Discrete probability densities.   136
  - Conditional probability.   138
  - Independence.   144
  - Random variables.   146
  - Joint distributions.   153
  - Modal aspects of probability.   154
  - Probabilistic invariance.   155
- 5.2 Classical Definition of Probability   157
  - Laplace.   159
  - Classical paradoxes.   163
  - Historical note on Laplace's principles 3–10.   166
- 5.3 Relative-frequency Theory for Infinite Random Sequences   167
  - Von Mises.   171
  - Church.   173
- 5.4 Random Finite Sequences   178
  - Kolmogorov complexity.   179
  - Universal probability.   182
  - Relative frequencies as estimates of probability.   183
- 5.5 Logical Theory of Probability   184
  - Keynes.   184
  - Jeffreys.   185
  - Carnap's confirmation theory.   190
  - Hintikka's two-parameter theory.   198
  - Kyburg.   200
  - Model-theoretic approach.   200
  - Chuaqui.   200

5.6 Propensity Representations of Probability   202
  Propensity to decay.   203
  Discrete qualitative densities.   210
  Propensity to respond.   211
  Propensity for heads.   214
  Propensity for randomness in motion of three bodies.   218
  Some further remarks on propensity.   220
5.7 Theory of Subjective Probability   225
  De Finetti's qualitative axioms.   226
  General qualitative axioms.   230
  Qualitative conditional probability.   234
  Historical background on qualitative axioms.   238
  De Finetti's representation theorem.   240
  Defense of objective priors.   241
  General issues.   242
  Decisions and the measurement of subjective probability.   245
  Inexact measurement of belief: upper and lower probabilities.   248
5.8 Epilogue: Pragmatism about Probability   256
  Early statistical mechanics.   256
  Quantum mechanics.   257
  Pragmatism in physics.   261
  Statistical practice.   262

6 Representations of Space and Time   265

6.1 Geometric Preliminaries   266
6.2 Classical Space-time   269
  Historical remarks.   272
6.3 Axioms for Special Relativity   275
  Historical remarks.   278
  Later qualitative axiomatic approaches.   281
6.4 How to Decide if Visual Space is Euclidean   282
  The hierarchy of geometries.   287
6.5 The Nature of Visual Space: Experimental and Philosophical Answers   288
6.6 Partial Axioms for the Foley and Wagner Experiments   297
6.7 Three Conceptual Problems About Visual Space   300
  Contextual geometry.   300
  Distance perception and motion.   301
  Objects of visual space.   302
6.8 Finitism in Geometry   303
  Quantifier-free axioms and constructions.   305

Affine Axioms. 306
Theorems. 308
Analytic representation theorem. 309
Analytic invariance theorem. 310

7 Representations in Mechanics 313
7.1 Classical Particle Mechanics 313
Assumed mathematical concepts. 313
Space-time structure. 318
Primitive notions. 319
The axioms. 320
Two theorems—one on determinism. 323
Momentum and angular momentum. 325
Laws of conservation. 327
7.2 Representation Theorems for Hidden Variables in Quantum Mechanics 332
Factorization. 333
Locality. 335
GHZ-type experiments. 338
Second-order Gaussian theorems. 342
7.3 Weak and Strong Reversibility of Causal Processes 343
Weak reversibility. 344
Strong reversibility. 346
Ehrenfest model. 348
Deterministic systems. 349

8 Representations of Language 353
8.1 Hierarchy of Formal Languages 354
Types of grammars. 356
Normal forms. 357
Operations on languages. 359
Unsolvable problems. 360
Natural-language applications. 361
8.2 Representation Theorems for Grammars 361
Finite automata. 361
Languages accepted by finite automata. 364
Regular grammars and finite automata. 367
Remark on the empty sequence. 371
Pushdown automata and context-free languages. 371
Turing machines and linear bounded automata. 373
8.3 Stimulus-response Representation of Finite Automata 374
Stimulus-response theory. 377

Representation of finite automata. 380

Response to criticisms. 387

Another misconception: restriction to finite automata. 394

Axioms for register learning models. 397

Role of hierarchies and more determinate reinforcement. 401

8.4 Representation of Linear Models of Learning by Stimulus-sampling Models 403

Modification of general axioms. 404

Preliminary theorems. 405

Theorems involving the sequence ω_n. 411

Limit assumptions. 414

8.5 Robotic Machine Learning of Comprehension Grammars for Ten Languages 419

Problem of denotation. 420

Background cognitive and perceptual assumptions. 421

Internal language. 424

General learning axioms. 425

Specialization of certain axioms and initial conditions. 429

The Corpora. 432

Empirical results. 433

Grammatical rules. 438

Related work and unsolved problems. 441

8.6 Language and the Brain. 442

Some historical background. 442

Observing the brain's activity. 444

Methods of data analysis. 446

Three experimental results. 450

Criticisms of results and response. 453

Computation of extreme statistics. 456

Analysis of earlier studies. 458

Other pairs in the first experiment with 48 sentences. 461

Test of a timing hypothesis for the experiment with 100 sentences. 461

Censoring data in the visual-image experiment. 463

8.7 Epilogue: Representation and Reduction in Science 465

Summary Table of Representation and Invariance Theorems by Chapter 471

References 475

Author Index 503

Index 511

Preface

This book has a long history of development. My earliest preliminary edition, entitled *Set-theoretical Structures in Science,* with temporary binding and cover, dates from 1962, but I know that even earlier drafts were produced for the lectures I gave in the basic undergraduate course in the philosophy of science at Stanford, in the 1950s. Course notes materials were developed rather quickly that conceptually followed the final chapter on the set-theoretical foundations of the axiomatic method of my *Introduction to Logic*, first published in 1957.

I can remember being asked on several occasions in those early years, But what is the general point of having set-theoretical structures play a central role in the philosophy of science? At first, I answered with a stress on the many general intellectual virtues of the axiomatic method apparent since the appearance long ago of Euclid's *Elements*. But I gradually came to see there was a better answer, of more philosophical interest. This was that such structures provided the right settings for investigating problems of representation and invariance in any systematic part of past or present science. Of course, this answer prompts another question. Why are representation and invariance important in the foundations or philosophy of science? In a sense, it is the purpose of this entire book to provide an answer. But some standard examples can be helpful, even without any serious details.

One of the great intellectual triumphs of the nineteenth century was the mechanical explanation of such familiar concepts as temperature and pressure by their representation simply in terms of the motion of particles. An equally great triumph, one more disturbing on several counts, was the realization at the beginning of the twentieth century that the separate invariant properties of space and time, standard in classical physics, even if often only implicitly recognized, must be replaced by the space-time of Einstein's special relativity and its new invariants.

But physics is not the only source of such examples. Sophisticated analysis of the nature of representation in perception is to be found already in Plato and Aristotle, and is still alive and well in the controversies of contemporary psychology and philosophy.

So I offer no apologies for my emphasis on representation and invariance, reflected in the revised title of this book. Central topics in foundational studies of different sorts through the ages easily fall under this heading. Certainly it is not the whole of philosophy of science, but it is a major part.

The first four chapters offer a general introduction to the concepts of representation and invariance. The last four provide applications to four areas of thought, important in the philosophy of science, but important because they are, more generally, of scientific significance. They are the nature of probability, concepts of space and time, both physical and psychological, representations in classical and quantum mechanics, and, finally, representations of language examined from several different angles. I include at the end of the book a summary table of the representation and invariance theorems stated and discussed in various chapters. Many are not proved and represent famous results in the mathematical and scientific literature. Those that are proved usually, but not always, represent some aspect of my own work, and most of the proofs given are elementary from a mathematical standpoint. As will be evident to any persistent reader, analysis and clarification of concepts, not formal proofs, are the main focus.

Although I have been devoted to formal methods in the philosophy of science, there are two other approaches to which I am nearly as faithful, and consequently they have had a big influence. One is concern for empirical details. This is reflected in the many experiments, especially in psychology, I consider at various points. Moreover, the content of this book does not adequately reflect the large number of experiments I have conducted myself, almost always in conjunction with colleagues, in many areas of psychology. I originally intended to write a long final chapter on set-theoretical representations of data as a necessary, but still desirable, abstraction of the complicated activity of conducting experiments. With the rapid increase in computer power in the last decade or so, the analysis of data in many parts of science is being radically transformed. I hope to extend the present work in that direction, perhaps in a new chapter to be made available on the internet, which will surely become the medium for the majority of detailed scientific publications in the next decade or so. The next-to-last section of the last chapter, 8.6, exemplifies what I have in mind.

The other approach close to my heart is concern for the historical background and development of many different scientific ideas. There is a rich and attractive tradition in philosophy of being concerned with the historical development of concepts and theories. Analysis, even of an often rather sketchy kind, of the background of a new idea about probability, physical invariance, visual space, mental representation or nearly any other important scientific concept I almost always find enlightening and helpful. I hope that some readers will feel the same about my many historical excursions, which do not give anything like a fully detailed account of the evolution of a single concept.

The writing of a book like this over many years entails indebtedness for corrections, insights and suggestions on innumerable topics by more people than I can possibly thank explicitly. By now, many are gone and many others will have forgotten how they contributed to what is said here. I thank them one and all. I certainly do want to mention those from whom I still have relevant remarks, or with whom I have written a joint article, used and acknowledged at one or more points in this book.

Chapter 2. Dana Scott at the beginning, later Rolando Chuaqui, Newton da Costa, Francisco Doria, Jaakko Hintikka, Alfred Tarski, Paul Weingartner and Jules Vuillemin. Chapter 3. Dana Scott again, Kenneth Arrow, Dagfinn Follesdal, Duncan Luce and Jesús Mosterín. Chapter 4. Nancy Cartwright, Maria Luisa Dalla Chiara, Jan Drösler and Donald Ornstein. Chapter 5. In alphabetical order and over many years, David Blackwell, Thomas Cover, Persi Diaconnis, Jean Claude Falmagne, Jens Erik Fenstad, Terence Fine, Maria Carla Galavotti, Ian Hacking, Peter Hammond, Paul Holland, Joseph Keller, Duncan Luce, David Miller, Marcos Perreau-Guimaraes, Karl Popper, Roger Rosenkrantz, Henri Rouanet and Mario Zanotti. Chapter 6. Jan Drösler, Tarow Indow, Brent Mundy, Gary Oas, Victor Pambuccian, Fred Roberts and Herman Rubin. Chapter 7. Acacio de Barros, Arthur Fine, Ubaldo Garibaldi, Gary Oas, Adonai S. Sant' Anna and Mario Zanotti. Chapter 8. Theodore W. Anderson, Michael Böttner, William Estes, Bing Han, Lin Liang, Zhong-Lin Lu, Marcos Perreau-Guimaraes and Timothy Uy.

I have also benefited from the penetrating questions and skeptical comments of many generations of students, who read various chapters in courses and seminars over these many years. I mention especially a group of former graduate students of mine whose comments and corrections of mistakes were numerous to say the least: Colleen Crangle, Zoltan Domotor, Anne Fagot-Largeault, Paul Holland, Paul Humphreys, Christoph Lehner, Michael Levine, Brent Mundy, Frank Norman, Fred Roberts, Deborah Rosen, Roger Rosenkrantz, Joseph Sneed, Robert Titiev, Raimo Tuomela and Kenneth Wexler.

In the final throes of publication, I must thank Ben Escoto, who read the entire manuscript most carefully, looking for misprints and mistakes, and also Ann Gunderson, who has labored valiantly to produce the camera-ready copy for the printer. Claudia Arrighi has done as much as I have in organizing and checking the extensive references and constructing the subject index. Ernest Adams read the entire next-to-final draft and made many important suggestions for improvements in content and style, most of which I have been able to accomodate. Once again, as I have for many years, I benefited from his well thought-out criticisms.

I began this book as a young man. Well, at least I think of under 40 as being young, certainly now. I finish it in my tenth year of retirement, at the age of 80. As I look back over the pages, I can see many places that could still stand improvement and, perhaps above all, additional details and more careful perusal for errors. But I know it is time to stop, and so I do.

I dedicate this book to my five children. I have been working on it during their entire lives, except for the oldest, Patricia, but she was a young child when I began. I also express my gratitude to my wife Christine for her patience and tolerance of my many years of intermittent effort, now, at last, at an end.

PATRICK SUPPES

Stanford, California
March, 2002.

1

Introduction

1.1 General Viewpoint

There is no simple or direct way to characterize the philosophy of science. Even the much narrower and more well-defined discipline of formal logic is not really subject to an exact definition. Individual philosophers have widely varying conceptions of the philosophy of science, and there is less agreement about what are the most important topics. All the same, there is a rather wide consensus about certain topics like causality, induction, probability and the structure of theories. In this book I approach these and related topics with certain formal methods and try to show how we can use these methods to make clear distinctions that can easily be lost at the level of general discourse. On the other hand, it is my aim not to let formal matters get out of hand. I have tried to follow a path that will not lose the reader in the underbrush of purely technical problems.

It might be thought that the emphasis on formal methods follows from a desire to emphasize a discussion of particular well-developed theories in science which are usually given a mathematical formulation. Undoubtedly this consideration has weighed to some extent, but it is not the most important. From my own point of view, there are two reasons of a more fundamental character for emphasizing the role of formal methods in a systematic discussion of the philosophy of science. One is the desirability for any large-scale discussion of having a fixed frame of reference, or a fixed general method, that may be used to organize and criticize the variety of doctrines at hand. Formal, set-theoretical methods provide such a general framework for discussion of the systematic problems of the philosophy of science. Such methods are more appropriate here than in the foundations of mathematics, because the foundations of set theory are themselves a central subject of investigation in the foundations of mathematics. It seems a wise division of labor to separate problems in the foundations of mathematics from problems in the foundations of science. As far as I can see, most problems of central importance to the philosophy of science can be discussed in full detail by accepting something like a standard formulation of set theory, without questioning the foundations of mathematics. In the discussion of problems of the philosophy of science, I identify formal methods with set-theoretical methods. The reasons for this identification I pursue later in more detail, but I do wish to make the point that I do not have any dogmatic commitment to the ultimate character of this identification, or even to the ultimate character of a set-theoretical approach. It will be clear from what I have to say in Chapter 8 about behavioristic and neural theories of language that my own conception of an ultimately satisfactory theory of these phenomena will fall outside the standard set-theoretical approach.

However, a virtue of the set-theoretical approach is that we may easily meet a frequent criticism of the artificial languages often introduced in the philosophy of science–namely, that such languages are not powerful enough to express most scientific results. The set-theoretical devices and framework we use are powerful enough easily to express any of the systematic results in any branch of empirical science or in the general logic of science.

Another reason for advocating formal methods in the philosophy of science is the conviction that both the commonsense treatment and the artificial-language treatment of problems of evidence are inadequate. Both these approaches give a much too simplified account of the extraordinarily complex and technically involved practical problems of assessing evidence in the empirical sciences. Various parts of the long fifth chapter on probability provide examples of the many subtle issues involved.

Before turning in the second chapter to a detailed consideration of set-theoretical methods, it will be useful in this introductory chapter to have an informal discussion of scientific theories. This discussion is intended to adumbrate many of the issues that are examined more thoroughly later.

1.2 What Is a Scientific Theory?

Often when we ask what is a so-and-so, we expect a clear and definite answer. If, for example, someone asks me what is a rational number, I may give the simple and precise answer that a rational number is the ratio of two integers. There are other kinds of simple questions for which a precise answer can be given, but for which ordinarily a rather vague answer is given and accepted. Someone reads about nectarines in a book but has never seen a nectarine, or possibly has seen nectarines but is not familiar with their English name. He may ask me, 'What is a nectarine?' and I would probably reply, 'a smooth-skinned sort of peach'. Certainly, this is not a very exact answer, but if my questioner knows what peaches are, it may come close to being satisfactory. The question, 'What is a scientific theory?', fits neither one of these patterns. Scientific theories are not like rational numbers or nectarines. Certainly they are not like nectarines, for they are not physical objects. They are like rational numbers in not being physical objects, but they are totally unlike rational numbers in that scientific theories cannot be defined simply or directly in terms of other nonphysical, abstract objects.

Good examples of related questions are provided by the familiar inquiries, 'What is physics?', 'What is psychology?', 'What is science?'. To none of these questions do we expect a simple and precise answer. On the other hand, many interesting remarks can be made about the sort of thing physics or psychology is. I hope to show that this is also true of scientific theories.

The traditional sketch. The traditional sketch of scientific theories–and I emphasize the word 'sketch'–runs something like the following. A scientific theory consists of two parts. One part is an abstract logical calculus, which includes the vocabulary of logic and the primitive symbols of the theory. The logical structure of the theory is fixed by stating the axioms or postulates of the theory in terms of its primitive symbols. For many theories the primitive symbols are thought of as theoretical terms like 'electron' or 'particle', which cannot be related in any simple way to observable phenomena.

The second part of the theory is a set of rules that assign an empirical content to the logical calculus by providing what are usually called 'coordinating definitions' or 'empirical interpretations' for at least some of the primitive and defined symbols of the calculus. It is always emphasized that the first part alone is not sufficient to define a scientific theory; for without a systematic specification of the intended empirical interpretation of the theory, it is not possible in any sense to evaluate the theory as a part of science, although it can be studied simply as a piece of pure mathematics.

The most striking thing about this characterization is its highly schematic nature. Concerning the first part of a theory, there are virtually no substantive examples of a theory actually worked out as a logical calculus in the writings of philosophers of science. Much hand waving is indulged in to demonstrate that working out the logical calculus is simple in principle and only a matter of tedious detail, but concrete evidence is seldom given. The sketch of the second part of a theory, that is, the coordinating definitions or empirical interpretations of some of the terms, is also highly schematic. A common defense of the relatively vague schema offered is that the variety of different empirical interpretations, for example, the many different methods of measuring mass, makes a precise characterization difficult. Moreover, as we move from the precisely formulated theory to the very loose and elliptical sort of experimental language used by almost all scientists, it is difficult to impose a definite pattern on the rules of empirical interpretation.

The view I want to support is not that this standard sketch is flatly wrong, but rather that it is far too simple. Its very sketchiness makes it possible to omit both important properties of theories and significant distinctions that may be introduced between theories.

Models versus empirical interpretations of theories. To begin with, there has been a strong tendency on the part of many philosophers to speak of the first part of a theory as a logical calculus purely in syntactical terms. The coordinating definitions provided in the second part do not in the sense of modern logic provide an adequate semantics for the formal calculus. Quite apart from questions about direct empirical observations, it is pertinent and natural from a logical standpoint to talk about the models of the theory. These models are abstract, nonlinguistic entities, often remote in their conception from empirical observations. So, apart from logic, someone might well ask what the concept of a model can add to the familiar discussions of empirical interpretation of theories.

I think it is true to say that most philosophers find it easier to talk about theories than about models of theories. The reasons for this are several, but perhaps the most important two are the following. In the first place, philosophers' examples of theories are usually quite simple in character, and therefore, are easy to discuss in a straightforward linguistic manner. In the second place, the introduction of models of a theory inevitably introduces a stronger mathematical element into the discussion. It is a natural thing to talk about theories as linguistic entities, that is, to speak explicitly of the precisely defined set of sentences of the theory and the like, when the theories are given in what is called standard formalization. Theories are ordinarily said to have a standard formalization when they are formulated within first-order logic. Roughly speaking, first-order logic is just the logic of sentential connectives and predicates holding for one type of object. Unfortunately, when a theory assumes more than first-order logic, it is neither natural nor simple to formalize it in this fashion. For example, if in axiomatizing geometry we want to define lines as certain *sets* of points, we must work within a framework that already includes the ideas of set theory. To be sure, it is theoretically possible to axiomatize simultaneously geometry and the relevant portions of set theory, but this is awkward and unduly laborious. Theories of more complicated structure, like quantum mechanics, classical thermodynamics, or a modern quantitative version of learning theory, need to use not only general ideas of set theory but also many results concerning the real numbers. Formalization of such theories in first-order logic is utterly impractical. Theories of this sort are very similar to the theories mainly studied in pure mathematics in their degree of complexity. In such contexts it is very much simpler to assert things about models of the theory rather than to talk directly and explicitly about the sentences of the theory. Perhaps the main reason for this is that the notion of a sentence of the theory is not well defined, when the theory is not given in standard formalization.

I would like to give just two examples in which the notion of model enters in a natural and explicit way in discussing scientific theories. The first example is concerned with the nature of measurement. The primary aim of a given theory of measurement is to show in a precise fashion how to pass from qualitative observations to the quantitative assertions needed for more elaborate theoretical stages of science. An analysis of how this passage from the qualitative to the quantitative may be accomplished is provided by axiomatizing appropriate algebras of experimentally realizable operations and relations. Given an axiomatized theory of measurement of some empirical quantity such as mass, distance or force, the mathematical task is to prove a representation theorem for *models* of the theory which establishes, roughly speaking, that any empirical model is isomorphic to some numerical model of the theory.[1] The existence of this isomorphism between models justifies the application of numbers to things. We cannot literally take a number in our hands and apply it to a physical object. What we can do is show that the structure of a set of phenomena under certain empirical operations is the same as the structure of some set of numbers under arithmetical operations and relations. The definition of isomorphism of models in the given context makes the intuitive idea of same structure precise. The great significance of finding such an isomorphism of models is that we may then use all our familiar knowledge of computational methods, as applied to the

---

[1] The concept of a representation theorem is developed in Chapter 3. Also of importance in this context is to recognize that the concept of an empirical model used here is itself an abstraction from most of the empirical details of the actual empirical process of measurement. The function of the empirical model is to organize in a systematic way the *results* of the measurement procedures used. I comment further on this point in the subsection below on coordinating definitions and the hierarchy of theories.

1.2 What Is a Scientific Theory?

arithmetical model, to infer facts about the isomorphic empirical model. It is extremely awkward and tedious to give a linguistic formulation of this central notion of an empirical model of a theory of measurement being isomorphic to a numerical model. But in model-theoretic terms, the notion is simple, and in fact, is a direct application of the very general notion of isomorphic representation used throughout all domains of pure mathematics, as is shown in Chapter 3.

The second example of the use of models concerns the discussion of reductionism in the philosophy of science. Many of the problems formulated in connection with the question of reducing one science to another may be formulated as a series of problems using the notion of a representation theorem for the models of a theory. For instance, for many people the thesis that psychology may be reduced to physiology would be appropriately established by showing that for any model of a psychological theory, it is possible to construct an isomorphic model within some physiological theory. The absence at the present time of any deep unitary theory either within psychology or physiology makes present attempts to settle such a question of reductionism rather hopeless. The classical example from physics is the reduction of thermodynamics to statistical mechanics. Although this reduction is usually not stated in absolutely satisfactory form from a logical standpoint, there is no doubt that it is substantially correct, and it represents one of the great triumphs of classical physics.

One substantive reduction theorem is proved in Chapter 8 for two closely related theories of learning. Even this conceptually rather simple case requires extensive technical argument. The difficulties of providing a mathematically acceptable reduction of thermodynamics to statistical mechanics are formidable and well recognized in the foundational literature (see, e.g., Khinchin (1949), Ruelle (1969)).

Intrinsic-versus-extrinsic characterization of theories. Quite apart from the two applications just mentioned of the concept of a model of a theory, we may bring this concept to bear directly on the question of characterizing a scientific theory. The contrast I wish to draw is between intrinsic and extrinsic characterization. The formulation of a theory as a logical calculus or, to put it in terms that I prefer, as a theory with a standard formalization, gives an intrinsic characterization, but this is certainly not the only approach. For instance, a natural question to ask within the context of logic is whether a certain theory *can* be axiomatized with standard formalization, that is, within first-order logic. In order to formulate such a question precisely, it is necessary to have some extrinsic way of characterizing the theory. One of the simplest ways of providing such an extrinsic characterization is simply to define the intended class of models of the theory. To ask if we can axiomatize the theory is then just to ask if we can state a set of axioms such that the models of these axioms are precisely the models in the defined class.

As a very simple example of a theory formulated both extrinsically and intrinsically, consider the extrinsic formulation of the theory of simple orderings that are isomorphic to a set of real numbers under the familiar less-than relation. That is, consider the class of all binary relations isomorphic to some fragment of the less-than relation for the real numbers. The extrinsic characterization of a theory usually follows the sort given for these orderings, namely, we designate a particular model of the theory (in this case, the numerical less-than relation) and then characterize the entire class of models of the theory in relation to this distinguished model. The problem of intrinsic characterization is now to formulate a set of axioms that will characterize this class of models without referring to the relation between models, but only to the intrinsic properties of any one model. With the present case the solution is relatively simple, although even it is not naturally formulated within first-order logic.[2]

A casual inspection of scientific theories suggests that the usual formulations are intrinsic rather than extrinsic in character, and therefore, that the question of extrinsic formulations usually arises only in pure mathematics. This seems to be a happy result, for our philosophical intuition is surely that an intrinsic characterization is in general preferable to an extrinsic one.

However, the problem of intrinsic axiomatization of a scientific theory is more complicated and considerably more subtle than this remark would indicate. Fortunately, it is precisely by explicit consideration of the class of models of the theory that the problem can be put in proper perspective and formulated in a fashion that makes possible consideration of its exact solution. At this point I sketch just one simple example. The axioms for classical particle mechanics are ordinarily stated in such a way that a coordinate system, as a frame of reference, is tacitly assumed. One effect of this is that relationships deducible from the axioms are not necessarily invariant with respect to Galilean transformations. We can view the tacit assumption of a frame of reference as an extrinsic aspect of the familiar characterizations of the theory. From the standpoint of the models of the theory the difficulty in the standard axiomatizations of mechanics is that a large number of formally distinct models may be used to express the same mechanical facts. Each of these different models represents the tacit choice of a different frame of reference, but all the models representing the same mechanical facts are related by Galilean transformations. It is thus fair to say that in this instance the difference between models related by Galilean transformations does not have any theoretical significance, and it may be regarded as a defect of the axioms that these trivially distinct models exist. It is important to realize that this point about models related by Galilean transformations is not the kind of point usually made under the heading of empirical interpretations of the theory. It is a conceptual point that just as properly belongs to the theoretical side of physics. I have introduced this example here in order to provide a simple instance of how the explicit consideration of models can lead to a more subtle discussion of the nature of a scientific theory. It is certainly possible from a philosophical standpoint to maintain that particle mechanics as a scientific theory should be expressed only in terms of Galilean invariant relationships, and that the customary formulations are defective in this respect. These matters are discussed in some detail in Chapter 6; the more general theory of invariance is developed in Chapter 4.

---

[2]The intrinsic axioms are just those for a simple ordering plus the axiom that the ordering must contain in its domain a countable subset, dense with respect to the ordering in question. The formal details of this example are given in Chapter 3.

1.2 WHAT IS A SCIENTIFIC THEORY?

Coordinating definitions and the hierarchy of theories. I turn now to the second part of theories mentioned above. In the discussion that has just preceded we have been using the word 'theory' to refer only to the first part of theories, that is, to the axiomatization of the theory, or the expression of the theory, as a logical calculus; but as I emphasized at the beginning, the necessity of providing empirical interpretations of a theory is just as important as the development of the formal side of the theory. My central point on this aspect of theories is that the story is much more complicated than the familiar remarks about coordinating definitions and empirical interpretations of theories would indicate. The kind of coordinating definitions often described by philosophers have their place in popular philosophical expositions of theories, but in the actual practice of testing scientific theories more elaborate and more sophisticated formal machinery for relating a theory to data is required. The concrete experience that scientists label an experiment cannot itself be connected to a theory in any complete sense. That experience must be put through a conceptual grinder that in many cases is excessively coarse. What emerges are the experimental data in canonical form. These canonical data constitute a model of the results of the experiment, and direct coordinating definitions are provided for this model rather than for a model of the theory. It is also characteristic that the model of the experimental results is of a relatively different logical type from that of any model of the theory. It is common for the models of a theory to contain continuous functions or infinite sequences, but for the model of the experimental results to be highly discrete and finitistic in character.

The assessment of the relation between the model of the experimental results and some designated model of the theory is a characteristic fundamental problem of modern statistical methodology. What is important about this methodology for present purposes is that, in the first place, it is itself formal and theoretical in nature; and, second, a typical function of this methodology has been to develop an elaborate theory of experimentation that intercedes between any fundamental scientific theory and raw experimental experience. My only point here is to make explicit the existence of this hierarchy and to point out that there is no simple procedure for giving coordinating definitions of a theory. It is even a bowdlerization of the facts to say that coordinating definitions are given to establish the proper connections between models of the theory and models of the experimental results, in the sense of the canonical form of the data just mentioned. The elaborate methods, for example, for estimating theoretical parameters in the model of the theory from models of the experimental results are not adequately covered by a reference to coordinating definitions.[3]

If someone asks 'What is a scientific theory?' it seems to me there is no simple response to be given. Are we to include as part of the theory the well-worked-out statistical methodology for testing the theory? If we are to take seriously the standard claims that the coordinating definitions are part of the theory, then it would seem inevitable that we

---

[3]It would be desirable also to develop models of the experimental *procedures*, not just the results. A really detailed move in this direction would necessarily use psychophysical and related psychological concepts to describe what experimental scientists actually do in their laboratories. This important foundational topic is not developed here and it has little systematic development in the literature of the philosophy of science.

must also include in a more detailed description of theories a methodology for designing experiments, estimating parameters and testing the goodness of fit of the models of the theory. It does not seem to me important to give precise definitions of the form: X is a scientific theory if, and only if, so-and-so. What is important is to recognize that the existence of a hierarchy of theories arising from the methodology of experimentation for testing the fundamental theory is an essential ingredient of any sophisticated scientific discipline.

In the chapters that follow, the important topic of statistical methodology for testing theories is not systematically developed. Chapter 5 on interpretations or representations of probability, the longest in the book, is a detailed prolegomena to the analysis of statistical methods. I have written about these matters in several earlier publications,[4] and there is a fairly detailed discussion of experiments on the nature of visual space in Chapter 6, as well as data on brain-wave representations of words and sentences in the final section of Chapter 8, with empirical application of the concept of an extreme statistics.

Instrumental view of theories. I have not yet mentioned one view of scientific theories which is undoubtedly of considerable importance; this is the view that theories are to be looked at from an instrumental viewpoint. The most important function of a theory, according to this view, is not to organize or assert statements that are true or false but to furnish material principles of inference that may be used in inferring one set of facts from another. Thus, in the familiar syllogism 'all men are mortal; Socrates is a man; therefore, Socrates is mortal', the major premise 'all men are mortal', according to this instrumental viewpoint, is converted into a principle of inference. And the syllogism now has only the minor premise 'Socrates is a man'. From a logical standpoint it is clear that this is a fairly trivial move, and the question naturally arises if there is anything more substantial to be said about the instrumental viewpoint. Probably the most interesting argument for claiming that there is more than a verbal difference between these two ways of looking at theories or laws is the argument that when theories are regarded as principles of inference rather than as major premises, we are no longer concerned directly to establish their truth or falsity but to evaluate their usefulness in inferring new statements of fact. No genuinely original formal notions have arisen out of these philosophical discussions which can displace the classical semantical notions of truth and validity. To talk, for instance, about laws having different jobs than statements of fact is trivial unless some systematic semantical notions are introduced to replace the standard analysis.

From another direction there has been one concerted serious effort to provide a formal framework for the evaluation of theories which replaces the classical concept of truth. What I have in mind is modern statistical decision theory. It is typical of statistical decision theory to talk about actions rather than statements. Once the focus is shifted from statements to actions it seems quite natural to replace the concept of truth by that of expected loss or risk. It is appropriate to ask if a statement is true but does

---

[4]Suppes and Atkinson 1960, Ch. 2; Suppes 1962, 1970b, 1973a, 1974b, 1979, 1983, 1988; Suppes and Zanotti 1996.

1.2 WHAT IS A SCIENTIFIC THEORY?

not make much sense to ask if it is risky. On the other hand, it is reasonable to ask how risky is an action but not to ask if it is true. It is apparent that statistical decision theory, when taken literally, projects a more radical instrumental view of theories than does the view already sketched. Theories are not regarded even as principles of inference but as methods of organizing evidence to decide which one of several actions to take. When theories are regarded as principles of inference, it is a straightforward matter to return to the classical view and to connect a theory as a principle of inference with the concept of a theory as a true major premise in an argument. The connection between the classical view and the view of theories as instruments leading to the taking of an action is certainly more remote and indirect.

Although many examples of applications of the ideas of statistical decision theory have been worked out in the literature on the foundations of statistics, these examples in no case deal with complicated scientific theories. Again, it is fair to say that when we want to talk about the evaluation of a sophisticated scientific theory, disciplines like statistical decision theory have not yet offered any genuine alternative to the semantical notions of truth and validity. In fact, even a casual inspection of the literature of statistical decision theory shows that in spite of the instrumental orientation of the fundamental ideas, formal development of the theory is wholly dependent on the standard semantical notions and in no sense replaces them. What I mean by this is that in concentrating on the taking of an action as the terminal state of an inquiry, the decision theorists have found it necessary to use standard semantical notions in describing evidence, their own theory, and so forth. For instance, I cannot recall a single discussion by decision theorists in which particular observation statements are treated in terms of utility rather than in terms of their truth or falsity.

It seems apparent that statistical decision theory does not at the present time offer a genuinely coherent or deeply original new view of scientific theories. Perhaps future developments of decision theory will proceed in this direction. Be that as it may, there is one still more radical instrumental view that I would like to discuss as the final point in this introduction. As I have already noted, it is characteristic of many instrumental analyses to distinguish the status of theories from the status of particular assertions of fact. It is the point of a more radical instrumental, behavioristic view of the use of language to challenge this distinction, and to look at the entire use of language, including the statement of theories as well as of particular matters of fact, from a behavioristic viewpoint. According to this view of the matter, all uses of language are to be analyzed with strong emphasis on the language users. It is claimed that the semantical analysis of modern logic gives a very inadequate account even of the cognitive uses of language, because it does not explicitly consider the production and reception of linguistic stimuli by speakers, writers, listeners and readers. It is plain that for the behaviorist an ultimately meaningful answer to the question "What is a scientific theory?" cannot be given in terms of the kinds of concepts considered earlier. An adequate and complete answer can only be given in terms of an explicit and detailed consideration of both the producers and consumers of the theory. There is much that is attractive in this behavioristic way of looking at theories or language in general. What it lacks at present, however, is sufficient scientific depth and definiteness to serve as a genuine alternative to the approach of modern logic and mathematics. Moreover, much of the language of models and theories discussed earlier is surely so approximately correct that any behavioristic revision of our way of looking at theories must yield the ordinary talk about models and theories as a first approximation. It is a matter for the future to see whether or not the behaviorist's approach will deepen our understanding of the nature of scientific theories. Some new directions are considered at the end of Chapter 8, the final section of the book, for moving from behavioristic psychology to cognitive neuroscience as the proper framework for extending the analysis of language to its brain-wave representations.

A different aspect of the instrumental view of science is its affinity with pragmatism, especially in the radical sense of taking seriously only what is useful for some other purpose, preferably a practical one. This pragmatic attitude toward theories of probability is examined in some detail in the final section of Chapter 5, which, as already remarked, is entirely focused on representations of probability.

1.3 Plan of the Book

Of all the remarkable intellectual achievements of the ancient Greeks, perhaps the most outstanding is their explicit development of the axiomatic method of analysis. Euclid's *Elements*, written about 300 BC, has probably been the most influential work in the history of science. Every educated person knows the name of Euclid and in a rough way what he did—that he expounded geometry in a systematic manner from a set of axioms and geometrical postulates.

The purpose of Chapter 2 is to give a detailed development of modern conceptions of the axiomatic method. The deviation from Euclid is not extreme, but there are many formal details that are different. The central point of this chapter on modern methods of axiomatization is to indicate how any branch of mathematics or any scientific theory may be axiomatized within set theory. The concentration here is on scientific theories, not on parts of mathematics, although simple mathematical examples are used to illustrate key conceptual points. The axiomatization of a scientific theory within set theory is an important initial step in making its structure both exact and explicit. Once such an axiomatization is provided, it is possible to ask the kind of structural questions characteristic of modern mathematics. For instance, when are two models of a theory isomorphic, that is, when do they have exactly the same structure? Above all, Chapter 2 is focused on showing how, within a set-theoretical framework, to axiomatize a theory is, as I have put it for many years, to *define a set-theoretical predicate*.

After the exposition of the axiomatic method in Chapter 2, the next two chapters are devoted to problems of representation, the central topic of Chapter 3, and problems of invariance, the focus of Chapter 4. Here I only want to say briefly what each of these chapters is about.

The first important distinction is that in talking about representations for a given theory we are not talking about the theory itself but about models of the theory. When the special situation obtains that any two models for a theory are isomorphic, then the theory is said to be categorical. In standard formulations, the theory of real numbers is such an example. Almost no scientific theories, on the other hand, are meant to be categorical in character. When a theory is not categorical, an important problem is to discover if an interesting subset of models for the theory may be found such that any model is isomorphic to some member of this subset. To find such a distinguished subset of models for a theory and show that it has the property indicated, is to prove a *representation theorem* for the theory. The purpose of Chapter 3 is to provide a rather wide-ranging discussion of the concept of representation, and initially not be restricted to just models of theories. The general concept of representation is now very popular in philosophy and, consequently, many different delineations of what is meant by representation are to be found in the literature. I do not try to cover this wide range of possibilities in any detail, but concentrate on why I think the set-theoretical notion of isomorphic representation just described in an informal way is the most important one.

Chapter 4 is devoted to the more subtle concept of invariance. Many mathematical examples of invariance can be given, but, among scientific theories, the two principle areas in which invariance theorems are prominent are theories of measurement and physical theories. Here is a simple measurement example. If we have a theory of the fundamental measurement of mass, for example, then any empirical structure satisfying the axioms of the theory will be isomorphic, or something close to isomorphic, to a numerical model expressing numerical measurements for the physical objects in the empirical structure. Now, as we all know, these measurements are ordinarily expressed with a particular unit of mass or weight, for example, grams, kilograms or pounds. The point of the analysis of invariance is to show that any other numerical model related to the first numerical model by a change in the unit of measurement, which is arbitrary and not a reflection of anything in nature, is still a satisfactory numerical model from a structural standpoint. This seems simple enough, but, as the variety of theories of measurement is expanded upon, the questions become considerably more intricate.

Another way to think about invariance is in terms of symmetry. Symmetry and invariance go hand in hand, as I show at the beginning of Chapter 4. The shape of a square is invariant under rotations of 90° around the center of the square, because of the obvious properties of symmetry of a square. A rectangle that is not a square does not possess such symmetry, but invariance only under rotations of 180°.

The second sort of invariance I mentioned is that of particular physical theories. Perhaps the most familiar and important example is the concept of invariance associated with the special theory of relativity. As physicists are inclined to formulate the matter, a modern physical theory dealing with physical phenomena approaching the speed of light in terms of their motion must be relativistically invariant. I shall not try to say exactly what this means at this point, but, in addition to the general exposition in Chapter 4, there is a rather thorough discussion of such matters in Chapter 6.

It is worth noting that many philosophical discussions of representation do not seem to take any notice of the problems of invariance. But in scientific theories of a definite kind, for which representation theorems can be proved, the notion of invariance is recognized as of critical importance. One of the purposes of Chapter 4 is to try to explain why this is so.

In Chapter 5, I turn to perhaps the central philosophical topic in the general methodology of science, namely, the nature of probability. Probability does not enter into many scientific theories. The most important examples are the theories of classical physics, ranging from Newton's theory of particle mechanics through the theory of fluid mechanics, Cauchy's theory of heat and on to Maxwell's theory of electromagetic phenomena. But probability was important already in the eighteenth and nineteenth centuries when it was recognized that errors of measurement somehow had to be taken account of in the testing of such theories. Fundamental memoirs on the probabilistic analysis of errors in observation were written by Simpson, Lagrange, Laplace, Gauss and others. Ever since that time and, especially starting roughly at the beginning of the twentieth century, probability and statistics have been a major component of all detailed discussions of scientific methodology.

In spite of the recognized importance of probability in the methodology of science, there has been no wide agreement on the nature of probability. The purpose of Chapter 5 is to give a leisurely analysis of the most prominent views that have been held. To give this discussion focus, as much as possible, I formulate for each of the characteristic views, from Laplace's conception of probability to Kolmogorov's complexity view, some kind of representation theorem. Philosophical discussion often accompanies the formal presentation of much of the material and accounts for the length of the chapter, the longest in the book.

The importance of probability, not only in the methodology, but also in the theoretical formulation, of science has come to be widely recognized. It is now an integral and fundamental concept in the very formulation of theories in many domains of science. The most important of these in the twentieth century were statistical mechanics and quantum mechanics, certain fundamental features of which are discussed in Chapter 7.

Because some interested philosophical readers will not be familiar with the standard formal theory of probability developed from the classic axiomatization given by Kolmogorov in the 1930s, I include a detailed and elementary exposition at the beginning of Chapter 5, before turning to any of the variety of representations or interpretations of the nature of probability.

The last three chapters deal with representation and invariance problems in the special sciences. Chapters 6 and 7 focus on physics and psychology. The final chapter, Chapter 8, focuses on linguistics, psychology and, at the very end, neuroscience.

The particular focus of Chapter 6 is representations of space and time. The first part deals with classical space-time and the second, relativistic space-time (in the sense of special relativity), the main results concern invariance; the representation is more or less assumed. This is because of the great importance of questions of invariance in classical and relativistic physics. Sections 6.4–6.7 move to questions of invariance in visual space and, more generally, problems of invariance in perception, so that here, the focus is psychological rather than physical. The questions addressed in these sections are more particular, they are less well-established as timeless matters, and they reflect rather particular interests of my own. Another reason for such an extended treatment is that it is certainly the case that philosophers of science are ordinarily more familiar with the topics discussed in classical and relativistic physics than they are with the experiments and concepts used in the analysis of visual space. The last section is on finitism in geometry, a topic that has its roots in ancient Greek geometry, and is related

1.3 PLAN OF THE BOOK

to finitism in the applications and foundations of mathematics more generally.

Chapter 7 turns to representations in mechanics. I originally intended for this chapter to focus entirely on one of my earliest interests, the foundations of classical particle mechanics. I enjoyed reworking those old foundations to give here a somewhat different formulation of the axioms. But as I reflected on the nature of representations in physics, I could not resist adding sections covering more recent topics. I have in mind here, above all, the extensive work by many people on the problem of hidden variables in quantum mechanics. I summarized several years ago with two physicists, Acacio de Barros and Gary Oas, a number of theorems and counterexamples about hidden variables. It seemed appropriate to include a version of that work, because there are many different representation theorems involved in the formulation of the results. I have also included still more recent work by de Barros and me. The subtle problems of distant entanglement and nonlocality associated with the nonexistence of local hidden variables present, in my opinion, the most perplexing philosophical puzzle of current quantum mechanical developments. I also include a final section on problems of reversibility in causal processes. The approach is written to include both deterministic and stochastic processes. The probabilistic concepts introduced in Chapter 5 are extended to cover temporal processes, especially dynamic ones. I think the distinction I introduce between weak and strong reversibility, although it is a distinction sometimes used in the literature on stochastic processes, does not seem to have had the emphasis in philosophical discussions of reversibility it should have. In any case, thinking through these problems led again to useful results about representation and, particularly, invariance of causal processes of all sorts.

The final chapter, Chapter 8, is a kind of donnybrook of results on representations of language accumulated over many years. It reflects the intersection of my own interests in psychology and linguistics. The chapter begins with a review of the beautiful classic results of Noam Chomsky, Stephen Kleene and others on the mutual representation theorems holding between grammars of a given type and automata or computers of a given strength. These results, now more than 40 years old, will undoubtedly be part of the permanent literature on such matters for a very long time. The organization of concepts and theorems comes from the lectures I gave in a course on the theory of automata I taught for many years in the 1960s and 1970s. After the sections on these results, I turn to more particular work of my own on representations of automata in terms of stimulus-response models. It is useful, I think, to see how to get from very simple psychological ideas that originated in classical behaviorism, but need mathematical expression, to representations of automata and, therefore, equivalent grammars. The following section is entirely focused on proving a representation theorem for a behavioral learning model having concepts only of response and reinforcement. Given just these two concepts, the probability of a response depends on the entire history of past reinforcements. To truncate this strong dependence on the past, models of a theory that is extended to stimuli and their conditioning (or associations) are used to provide means of representation that is Markov in nature, i.e., dependence on the past is limited to the previous trial. The next section (8.5) goes on to some detailed work I have done in recent years on machine learning of natural language. Here, the basic axioms are carefully crafted to express reasonable psychological ideas, but quite explicit ones, on how a machine (computer or robot) can learn fragments of natural language through association, abstraction and generalization. Finally, in the last section I include some of my most recent work on brain-wave representations of words and sentences. The combination of the empirical and the conceptual, in providing a tentative representation of words as processed in the brain, is a good place to end. It does not have the timeless quality of many of the results I have analyzed in earlier pages, but it is an opportunity I could not resist to exhibit ideas of representation and invariance at work in a current scientific framework.

1.4 How To Read This Book

As is evident from my description in the preceding section of the chapters of this book, the level of detail and technical difficulty vary a great deal. So, I try to provide in this section some sort of reading guide. It is written with several different kinds of readers in mind, not just philosophers of science, but also scientists of different disciplines interested in foundational questions. The historical background of many concepts and theories is sketched. It satisfies my own taste to know something about the predecessors of a given theory, even when the level of detail is far short of providing a full history.

It is my intention that Chapters 1–4 provide a reasonably elementary introduction to the main topics of the book, especially those of representation and invariance. There are two exceptions to this claim. First, the machine representation of partial recursive functions, i.e., the computable functions, in Section 3.5 is quite technical in style, even though in principle everything is explained. The general representation result is used at various later points in the book, but the proof of it is not needed later. Second, the final section of Chapter 4 on the role of entropy in ergodic theory requires some specific background to fully appreciate, but this example of invariance is so beautiful and stunning that I could not resist including it. I also emphasize that none of the central theorems are proved; they are only stated and explained with some care. The reader can omit this section, 4.5, without any loss of continuity. Another point about Chapters 1–4 is that what few proofs there are, with the exception of Section 3.5, have been confined to footnotes, or, in the case of Section 3.4, placed at the end of the section.

As I have said already, I count probability as perhaps the single most important concept in the philosophy of science. But here I have a problem of exposition for the general reader. Probability is not discussed at all in any detail in Chapters 1–4, with the exception of the final section of Chapter 4. So how should a general reader interested in an introduction to the foundations of probability tackle Chapter 5? First, someone not familiar with modern formal concepts of probability, such as the central one of random variable, should read the rather detailed introduction to the formal theory in Section 5.1. Those already familiar with the formal theory should skip this section. The real problem is what to do about the rest of the sections, several of which include a lot of technical detail, especially the long section (5.6) on propensity representations of probability and the equally long next section (5.7) on subjective views of probability. I suggest that a reader who wants a not-too-detailed overview read the nontechnical parts of each of the sections after the first one. Proofs and the like can easily be omitted. Another strategy useful for some readers will be to scan sections that cover views of probability that are less familiar, in order to decide if they are worth a more careful look.

In the case of the last three chapters, many topics are covered. Fortunately, most of the sections are nearly independent of each other. For example, someone interested in perceptual and psychological concepts could read Sections 6.4–6.7 without reading Sections 6.2 and 6.3 on classical and relativistic space-time, and vice versa for someone interested primarily in the foundations of physics.

All sections of Chapter 7 are relevant to physics, although the last section (7.3) on reversibility is of broader interest and applies to processes much studied in biology and the social sciences. On the other hand, no section of Chapter 8 is really relevant to central topics in the foundations of physics. The focus throughout is on language, but as much, or perhaps more, on psychological rather than linguistic questions.

I repeat here what I said in the preface about the Summary Table of Theorems at the end of the book. This table provides a list, by chapter, of the representation and invariance theorems explicitly stated, but not necessarily proved. The table is the best place to find quickly the location of a specific theorem.

As in love and warfare, brief glances are important in philosophical and scientific matters. None of us have time to look at all the details that interest us. We get new ideas from surprising places, and often through associations we are scarcely conscious of, while superficially perusing unfamiliar material. I would be pleased if this happens to some readers of this book.

2

Axiomatic Definition of Theories

This chapter begins with a detailed examination of some of the meanings of the word *model* which may be inferred from its use.[1] The second section gives a brief overview of the formalization of theories within first-order logic. Examples of models of such theories are considered. The third section, the core of the chapter, develops the axiomatic characterization of scientific theories defined by set-theoretical predicates. This approach to the foundations of theories is then related to the older history of the axiomatic method in Section 4. Substantive examples of axiomatized theories are to be found in later chapters.

2.1 Meaning of *Model* in Science

The use of the word *model* is not restricted to scientific contexts. It is used on all sorts of ordinary occasions. Everyone is familiar with the idea of a physical model of a building, a ship, or an airplane. Indeed, such models are frequently bought as gifts for young children, especially model airplanes and cars. (This usage also has a wide technical application in engineering.) It is also part of ordinary discourse to refer to model armies, model governments, model regulations, and so forth, where a certain design standard, for example, has been satisfied. In many cases, such talk of model government, for example, does not have reference to any actual government but to a set of specifications that it is felt an ideal government should satisfy. Still a third usage is to use an actual object as an exemplar and to refer to it as a model object of a given kind. This is well illustrated in the following quotation from Joyce's *Ulysses* (1934, p. 183):

> —The schoolmen were schoolboys first, Stephen said superpolitely. Aristotle was once Plato's schoolboy.
>
> —And has remained so, one should hope, John Eglinton sedately said. One can see him, a *model* schoolboy with his diploma under his arm.

In most scientific contexts, the use of *model* in the sense of an exemplar seldom occurs. But in the sense of a design specification abstracted from the full details, we come close to the usage that seems to be dominant. How to explain this in a more formal and mathematical way is the subject of the next section. However, in turning to a number

---

[1] The analysis given draws heavily on Suppes (1960a).

of quotations from scientific contexts, I begin with a formal and mathematical one.[2]

> A possible realization in which all valid sentences of a theory T are satisfied is called a *model* of T.
>
> (Tarski 1953, p. 11)

Quotations from the physical sciences.

> In the fields of spectroscopy and atomic structure, similar departures from classical physics took place. There had been accumulated an overwhelming mass of evidence showing the atom to consist of a heavy, positively charged nucleus surrounded by negative, particle-like electrons. According to Coulomb's law of attraction between electric charges, such a system will collapse at once unless the electrons revolve about the nucleus. But a revolving charge will, by virtue of its acceleration, emit radiation. A mechanism for the emission of light is thereby at once provided.
>
> However, this mechanism is completely at odds with experimental data. The two major difficulties are easily seen. First, the atom in which the electrons revolve continually should emit light all the time. Experimentally, however, the atom radiates only when it is in a special, 'excited' condition. Second, it is impossible by means of this *model* to account for the occurrence of spectral lines of a single frequency (more correctly, of a narrow range of frequencies). The radiating electron of our *model* would lose energy; as a result it would no longer be able to maintain itself at the initial distance from the nucleus, but fall in toward the attracting center, changing its frequency of revolution as it falls. Its orbit would be a spiral ending in the nucleus. By electrodynamic theory, the frequency of the radiation emitted by a revolving charge is the same as the frequency of revolution, and since the latter changes, the former should also change. Thus, our *model* is incapable of explaining the sharpness of spectral lines.
>
> (Lindsay and Margenau 1936, pp. 390-391)

> The author [Gibbs] considers his task not as one of establishing physical theories directly, but as one of constructing statistic-mechanical *models* which have some analogies in thermodynamics and some other parts of physics; hence he does not hesitate to introduce some very special hypotheses of a statistical character. (Khinchin 1949, p. 4)

> The modern quantum theory as associated with the names of de Broglie, Schrödinger, and Dirac, which of course operates with continuous functions, has overcome this difficulty by means of a daring interpretation, first given in a clear form by Max Born: —the space functions which appear in the equations make no claim to be a mathematical *model* of atomic objects. These functions are only supposed to determine in a mathematical way the probabilities of encountering those objects in a particular place or in a particular state of motion, if we make a measurement. This conception is logically unexceptionable, and has led to important successes. But unfortunately it forces us to employ a continuum of which the number of dimensions is not that of previous physics, namely 4, but which has dimensions increasing without limit as the number of the particles constituting the system under consideration increases. I cannot help confessing that I myself accord to this interpretation no more than a transitory significance. I still believe in the possibility of giving a *model* of reality, a theory, that is to say, which shall represent events themselves and not merely the probability of their occurrence. On the other hand, it seems to me certain that we have to give up the notion of an absolute localization of the particles in a theoretical *model*. This seems to me to be the correct theoretical interpretation of

---

[2] In all of the quotations in this section, I have italicized the word 'model'.

Heisenberg's indeterminacy relation. And yet a theory may perfectly well exist, which is in a genuine sense an atomistic one (and not merely on the basis of a particular interpretation), in which there is no localizing of the particles in a mathematical *model*.

(Einstein 1934, pp. 168-169)

Quotations from the biological sciences.

There are two important ways in which a chemist can learn about the three-dimensional structure of a substance. One is to apply a physical tool, which depends on a property of matter, to provide information on the relative spatial positions of the atoms in the molecules. The most powerful technique developed to do this is X-ray diffraction. The other approach is *model* building. A *model* of the molecule in question is constructed using scale *models* of the atoms present in it, with accurate values of the bond angles and bond distances between these atoms. The basic information of these parameters must, of course, come from physical measurements, such as X-ray diffraction studies on crystals, a technique known as X-ray crystallography. Thus, our understanding of the actual structure of DNA, and how it became known, must start with a consideration of X-rays and how they interact with matter. (Portugal and Cohen 1977, p. 204)

To begin to study the effects of duplication and dispersion of loci on such simple networks, I ignored questions of recombination, inversion, deletion, translocation, and point mutations completely, and *modelled* dispersion by using transposition alone. I used a simple program which decided at random for the haploid chromosome set whether a duplication or transposition occurred at each iteration, over how large a range of loci, between which loci duplication occurred, and into which position transposition occurred.

Even with these enormous simplifications, the kinetics of this system is complex and scantily explored. Since to simplify the *model*, loci cannot be destroyed, the rate of formation of a locus depends upon the number of copies already in existence, and approximately stochastic exponential growth of each locus is expected. This is modified by the spatial range of duplication, which affords a positive correlation of duplication of neighboring loci, and further modified by the frequency ratio of duplication to transposition. The further assumption of first-order destruction of loci would decrease the exponential growth rates. However, the kinetics are not further discussed here, since the major purpose of this simple *model* is to examine the regulatory *architecture* after many instances of duplication and transposition have occurred. (Kaufmann 1982, p. 28)

Quotations from the social sciences.

Thus, the *model* of rational choice as built up from pair-wise comparisons does not seem to suit well the case of rational behavior in the described game situation. (Arrow 1951, p. 21)

In constructing the *model* we shall assume that each variable is some kind of average or aggregate for members of the group. For example, D might be measured by locating the opinions of group members on a scale, attaching numbers to scale positions and calculating the standard deviation of the members' opinions in terms of these numbers. Even the intervening variables, although not directly measured, can be thought of as averages of the values for individual members. (Simon 1957, p. 116)

This work on mathematical *models* for learning has not attempted to formalize any particular theoretical system of behavior; yet the influences of Guthrie and Hull are most noticeable. Compared with the older attempts at mathematical theorizing, the recent work has been more concerned with detailed analyses of data relevant to the *models* and with the design of experiments for directly testing quantitative predictions of the *models*. (Bush and Estes 1959, p. 3)

Quotation from mathematical statistics.

I shall describe . . . various criteria used in adopting a mathematical *model* of an observed stochastic process. . . . For example, consider the number of cars that have passed a given point by time t. The first hypothesis is a typical mathematical hypothesis, suggested by the facts and serving to simplify the mathematics. The hypothesis is that the stochastic process of the *model* has independent increments. . . . The next hypothesis, that of stationary increments, states that, if $s < t$, the distribution of $x(t) - x(s)$ depends only on the time interval length $t - s$. This hypothesis means that we cannot let time run through both slack and rush hours. Traffic intensity must be constant.

The next hypothesis is that events occur one at a time. This hypothesis is at least natural to a mathematician. Because of limited precision in measurements it means nothing to an observer. . . . The next hypothesis is a more quantitative kind, which also is natural to anyone who has seen Taylor's theorem. It is that the probability that at least one car should pass in a time interval of length h should be $ch + o(h)$. (Doob 1960, p. 27)

Quotation from applied mathematics.

To deny the concept of *infinity* is as unmathematical as it is un-American. Yet, it is precisely a form of such mathematical heresy upon which discrete *model* theory is built.

(Greenspan 1973, p. 1)

Comments on quotations. The first of these quotations is taken from mathematical logic, the next three from the physical sciences, the following two from biology, the next three from the social sciences, the next one from mathematical statistics, and the last from applied mathematics. These quotations do not by any means exhaust the variant uses that might easily be collected.

It may well be thought that it is impossible to put under one concept the several uses of the word *model* exhibited by these quotations. It would, I think, be too much to claim that the word *model* is being used in exactly the same sense in all of them. The quotation from Doob exhibits one very common tendency, namely, to confuse or to amalgamate what logicians would call the model and the theory of the model. It is very widespread practice in mathematical statistics and in the behavioral sciences to use the word *model* to mean the set of quantitative assumptions of the theory, that is, the set of sentences which in a precise treatment would be taken as axioms, or, if they are themselves not explicit enough, would constitute the intuitive basis for formulating a set of axioms. In this usage a model is a linguistic entity and is to be contrasted with the usage characterized by the definition from Tarski, according to which a model is a nonlinguistic entity in which a theory is satisfied.

We also should note a certain technical usage in econometrics of the word *model*. In this sense a model is a class of models in the sense of logicians, and what logicians call a model is called by econometricians a *structure*.

It does not seem to me that these are serious difficulties. I claim that the concept of model in the sense of Tarski may be used without distortion as a fundamental concept

2.1 MEANING OF *MODEL* IN SCIENCE

in all of the disciplines from which the above quotations are drawn. In this sense I would assert that (the meaning of) the concept of model is the same in mathematics and the empirical sciences. The difference between these disciplines is to be found in their use of the concept. In drawing this comparison between constancy of meaning and difference of use, the sometimes difficult semantical question of how one is to explain the meaning of a concept without referring to its use does not actually arise. When I speak of the meaning of the concept of a model I shall always be speaking in well-defined technical contexts. What I shall be claiming is that, given this technical meaning of the concept of a model, mathematicians ask a certain kind of question about models, while empirical scientists tend to ask another kind of question.

It will be instructive to defend this thesis about the concept of *model* by analyzing uses of the word in the above quotations. As already indicated, the quotation from Tarski represents a standard definition of *model* in mathematical logic. At this point, I shall not enter into a technical characterization of models, which will be done in the next section of this chapter. Roughly speaking, a possible realization of a theory is a set-theoretical entity of the appropriate logical type. For example, a possible realization of the theory of groups is any ordered couple whose first member is a nonempty set and whose second member is a binary operation on this set. A model is then just a possible realization in which the theory is satisfied. An important distinction we shall need is that a theory is a linguistic entity consisting of a set of sentences, while models are in general nonlinguistic entities in which the theory is satisfied. (Theories could be taken to be collections of propositions rather than of sentences. This would not affect the main point being made here, but it would change the approach of the next section.)

I think that the use of the notion of models in the quotation from Lindsay and Margenau could be recast in these terms in the following manner. The orbital theory of the atom is formulated as a theory. The question then arises, whether a possible realization of this theory, in terms of entities defined in close connection with experiments, actually constitutes a model of the theory; or, put in perhaps a simpler way, do models of an orbital theory correspond well to data obtained from physical experiments with atomic phenomena? It is true that many physicists want to think of a model of the orbital theory of the atom as being more than a certain kind of set-theoretical entity. They envisage it as a very concrete physical thing built on the analogy of the solar system. I think it is important to point out that there is no real incompatibility in these two viewpoints. To define a model formally as a set-theoretical entity, which is a certain kind of ordered tuple consisting of a set of objects and relations and operations on these objects, is not to rule out the kind of physical model which is appealing to physicists. The physical model may be simply taken to define the set of objects in the set-theoretical model. Because of the importance of this point it may be well to illustrate it in somewhat greater detail. I select as an example classical particle mechanics, which is discussed at length in Chapter 7. We may axiomatize classical particle mechanics in terms of the five primitive notions of a set P of particles, an interval T of real numbers corresponding to elapsed times, a position function s defined on the Cartesian product of the set of particles and the time interval, a mass function m defined on the set of particles, and a force function f defined on the Cartesian product of the set of particles, the time interval and the set of positive integers (the set of positive integers enters into the definition of the force function simply in order to provide a method of naming the forces). A possible realization of the axioms of classical particle mechanics would then be an ordered quintuple (P, T, s, m, f). A model of classical particle mechanics would be such an ordered quintuple. (In fact, a more complex analysis is actually given in Chapter 7.) It is simple enough to see how an actual physical model in the physicist's sense of classical particle mechanics is related to this set-theoretical sense of models. For example, in the case of the solar system we simply can take the set of particles to be the set of planetary bodies. Another slightly more abstract possibility is to take the set of particles to be the set of centers of mass of the planetary bodies. This generally exemplifies the situation. A set-theoretical model of a theory will have among its parts a basic set which will consist of the objects ordinarily thought to constitute the physical model being considered, and the given primitive relations and functions may be thought of the same way. In the preceding paragraphs we have used the phrases, 'set-theoretical model' and 'physical model'. There would seem to be no use in arguing about which use of the word *model* is primary or more appropriate in the empirical sense. My own contention is that the set-theoretical usage is the more fundamental. The highly physically minded or empirically minded scientists, who may disagree with this thesis and who believe that the notion of a physical model is the more important in a given branch of empirical science, may still agree with my systematic remarks.

An historical illustration of this point is Kelvin's and Maxwell's efforts to find a mechanical model of electromagnetic phenomena.[3] Without doubt they both thought of possible models in a literal physical sense, but it is not difficult to recast their published memoirs on this topic as a search for set-theoretical models of the theory of continuum mechanics which will account for observed electromagnetic phenomena. Moreover, it is really the formal parts of their memoirs which have had permanent value. Ultimately it is the mathematical theory of Maxwell which has proved to be permanently useful to later generations of scientists, not the physical image of an ether behaving like an elastic solid. However, the historical and even current importance of iconic models in many aspects of science is evident (cf. Harré 1999).

The third quotation is from Khinchin's book on statistical mechanics. The phrase 'the author' refers to Gibbs, whom Khinchin is discussing at this point. The use of the word *model* in this quotation is particularly sympathetic to the set-theoretical viewpoint, for Khinchin is claiming that Gibbs did not appeal directly to physical reality or attempt to establish true physical theories in his work on the foundations of statistical mechanics; rather, he tried to construct models or theories having partial analogies to the complicated empirical facts of thermodynamics and other branches of physics. Perhaps even more directly than in the case of Doob, this quotation illustrates the tendency toward a confusion of the logical type of theories and models; but again this does not create a difficulty. The work of both Gibbs and Khinchin can easily and directly be formulated so as to admit explicitly and exactly the distinction between theories and models in mathematical logic. The abstractness of Gibb's work in statistical mechanics furnishes

---

[3] As examples of extended mathematical analysis of natural phenomena these efforts are heroic. See Kelvin (1846) and Maxwell (1861-1862).

a particularly good example for applying the exact notion of model used by logicians, for there is no direct and immediate tendency to think of Gibb's statistical mechanical models as being *the* theory of the one physical universe.

Models in physics. I think the following observation is empirically sound concerning the use of the word *model* in physics. In old and established branches of physics, which correspond well with the empirical phenomena they attempt to explain, there is only a slight tendency ever to use the word *model*. The language of theory, experiment and common sense is blended into one realistic whole. Sentences of the theory are asserted as if they are the one way of describing the universe. Experimental results are described as if there were but one obvious language for describing them. Notions of common sense, perhaps refined here and there, are taken to be appropriately homogeneous with the physical theory. On the other hand, in those branches of physics which as yet give an inadequate account of the detailed physical phenomena with which they are concerned, there is a much more frequent use of the word *model*. The connotation of the use of the word is that the model is like a model of an airplane or ship. It simplifies drastically the true physical phenomena and accounts only for certain major aspects. Again, in such uses of the word it is to be emphasized that there is a constant interplay between the model as a physical or nonlinguistic object and the model as a theory. This is well exemplified in an interesting negative way in the quotation from Einstein, who laments the absence of an appropriate mathematical model of the motion of particles in quantum mechanics.

Models in biology. The two quotations from biology do not differ in any drastic way in the use of the word *model* from the physical concept of model as a definite simplification of the complex real phenomena. It seems likely that the complexity of the structure of biological entities will necessarily keep this kind of simplification in vogue for the indefinite future. (In case there is any possible uncertainty about the context of the second quotation from biology, that of Kauffman, it is dealing directly with modern theories of genetics.)

Models in the social sciences. The quotation from Arrow exemplifies a similar necessary tendency in the social sciences to follow a course of extreme simplification. Arrow refers to the *model* of rational choice, because the theory he has in mind does not give an adequate description of the phenomena with which it is concerned, but only provides a highly simplified schema. The same remarks apply fairly well to the quotation from Simon. In Simon we have an additional phenomenon exemplified which is very common in the social and behavioral sciences. A certain theory is stated in broad and general terms. Some qualitative experiments to test this theory are performed. Because of the success of these experiments, scientists interested in more quantitative and exact theories then turn to what is called the 'the construction of a model' for the original theory. In the language of logicians, it would be more appropriate to say that rather than constructing a model they are interested in constructing a quantitative theory to match the intuitive ideas of the original theory.

In the quotation from Bush and Estes and the one from Doob, there is introduced an important line of thought which is, in fact, very closely connected with the concept of model as used by logicians. I am thinking here of the notion of model in mathematical statistics, the extensive literature on estimating parameters in models and testing hypotheses about them. In a statistical discussion of the estimation of the parameters of a model it is usually a trivial task to convert the discussion into one where the usage of terms is in complete agreement with that of logicians. The detailed consideration of statistical questions almost requires the consideration of models as mathematical or set-theoretical rather than simple physical entities. The question, 'How well does the model fit the data?' is a natural one for statisticians and behavioral scientists. Only recently has it begun to be so for physicists, and it is still true that some of the experimental work in physics is reported in terms of a rather primitive brand of statistics.

In citing and analyzing these many quotations, I have tried to argue that the concept of model used by mathematical logicians is the basic and fundamental one needed for an exact statement of any branch of empirical science. To agree with this thesis it is not necessary to rule out or to deplore variant uses or variant concepts of model now abroad in the empirical sciences. As has been indicated, I myself am prepared to admit the significance and practical importance of the notion of physical model current in much discussion in physics and engineering. What I have tried to claim is that in the exact statement of the theory or in the exact analysis of data, the notion of model in the sense of logicians is the appropriate one.

Much of the remainder of this book will be concerned to show how the set-theoretical notion of model may be used to clarify and deepen discussions of general topics in the philosophy of science. The rest of this chapter, however, is mainly reserved for clarification of the notion of model itself. Thus far, what has been said is not very precise or detailed. I now turn to various more formal considerations.

2.2 Theories with Standard Formalization

A theory with standard formalization is one that is formalized within first-order predicate logic with identity. The usual logical apparatus of first-order logic is assumed, mainly variables ranging over the same set of elements and logical constants, in particular, the sentential connectives and the universal and existential quantifiers, as well as the identity symbol. Three kinds of nonlogical constants occur: the predicates or relation symbols, the operation symbols and the individual constants.

The expressions of the theory, i.e., finite sequences of symbols of the language of the theory, are divided into terms and formulas. Recursive definitions of each are ordinarily given, and in a moment we shall consider several simple examples. The simplest terms are variables or individual constants. New terms are built up by combining simpler terms with operation symbols in the appropriate fashion. Atomic formulas consist of a single predicate and the appropriate number of terms. Molecular formulas, i.e., compound formulas, are built from atomic formulas by means of sentential connectives and quantifiers. A general characterization of how all this is done is quite familiar from any discussion of such matters in books on logic. We shall not enter into details here. The intuitive idea can easily be best conveyed by examples. In considering these examples,

2.2 THEORIES WITH STANDARD FORMALIZATION

we shall not say much of an explicit nature about first-order logic itself, but assume some familiarity with it. In a theory with standard formalization we must first begin with a recursive definition of terms and formulas based on the primitive nonlogical constants of the theory. Secondly, we must say what formulas of the theory are taken as axioms.[4] (Appropriate logical rules of inference and logical axioms are assumed as available without further discussion.)

Example: ordinal measurement. As a first example of a theory with standard formalization, we may consider the simple theory of ordinal measurement. To begin with, we use one standard notation for logical symbols. The sentential connectives '&', '∨', '→', '↔', and '¬' have the usual meaning of 'and', 'or', 'if...then', 'if and only if', and 'not', respectively. I use '∀' for the universal and '∃' for the existential quantifier. Parentheses are used in the familiar way for punctuation and are also omitted in practice when no ambiguity of meaning will arise. Mention of the identity symbol '=' and variables 'x', 'y', 'z',... completes the description of the logical notation.

The language of the theory $\mathfrak{O}$ of ordinal measurement, considered as an example here, has a single primitive nonlogical constant, the two-place relation symbol '$\succeq$'. Because the language of $\mathfrak{O}$ has no operation symbols among its primitive nonlogical constants, the definition of the terms of $\mathfrak{O}$ is extremely simple.[5] The terms of $\mathfrak{O}$ are just the variables 'x', 'y', 'z', ... We use these variables to make up formulas of $\mathfrak{O}$. First, we define atomic formulas of $\mathfrak{O}$. An atomic formula of $\mathfrak{O}$ is a sequence of three symbols, namely, a variable, followed by '=' or by '$\succeq$', followed by a variable. Thus '$x = y$' and '$z \succeq z$' are examples of atomic formulas of $\mathfrak{O}$. In terms of atomic formulas we may then give what is called a recursive definition of formulas.

(a) Every atomic formula is a formula.

(b) If S is a formula, then ¬(S) is a formula.

(c) If R and S are formulas then (R) & (S), (R) ∨ (S), (R) → (S), and (R) ↔ (S) are formulas.

(d) If R is a formula and v is any variable then (∀v)(R) and (∃v)(R) are formulas.

(e) No expression is a formula unless its being so follows from the above rules.

This definition of formulas is called recursive because, given any finite sequence of symbols of the language of $\mathfrak{O}$, we can apply this definition to decide in a finite number of steps if the sequence is a formula of $\mathfrak{O}$. It should also be apparent that this definition will be precisely the same for all theories with standard formalization. What will change from theory to theory is the definition of terms and atomic formulas.

---

[4] More needs to be said about how we describe or pick out the axioms of a theory, and also how the concept of a valid sentence of the theory may replace the usual concept of an axiomatized theory for certain purposes, but these matters can, I think, be safely ignored here. For an excellent intuitive discussion the reader is referred to the opening pages of Tarski (1953).

[5] To simplify the exposition I say 'terms of $\mathfrak{O}$' rather than 'terms of the language of $\mathfrak{O}$', in this passage and later. Using '$\mathfrak{O}$' to refer both to the language and the theory should, in the context of this section, cause no confusion.

The two axioms of $\mathfrak{O}$ are just the two sentences:

1. $(\forall x)(\forall y)(\forall z)(x \succeq y \ \& \ y \succeq z \to x \succeq z)$
2. $(\forall x)(\forall y)(x \succeq y \lor y \succeq x)$.

Intuitively the first axiom says that the relation designated by $\succeq$ is transitive, and the second axiom that it is weakly connected (for a discussion of these and similar properties of relations, see Suppes (1957/1999, Ch. 10, and 1960b, Ch. 3)).

I next turn to the definition of models of the theory. To begin with, we define possible realizations of $\mathfrak{O}$, already mentioned in Sec. 2.1. Let A be a nonempty set and $\succeq$ a binary relation defined on A, i.e., $\succeq$ is a subset of the Cartesian product $A \times A$. Then the structure $\mathfrak{A} = (A, \succeq)$ is a *possible realization* of $\mathfrak{O}$. The set A, it may be noted, is called the *domain* or *universe* of the realization $\mathfrak{A}$. I shall assume it to be intuitively clear under what circumstances a sentence of $\mathfrak{O}$, i.e., a formula of $\mathfrak{O}$ without free variables, is true of a possible realization. (For an explicit definition of truth in this sense, see the classical work of Tarski (1935).) We then say that a *model* of $\mathfrak{O}$ is a possible realization of $\mathfrak{O}$ for which the axioms of $\mathfrak{O}$ are true. For example, let

$$A = \{1, 2\}$$

and

$$\succeq\, = \{(1,1), (2,2), (1,2)\}.$$

Then $\mathfrak{A} = (A, \succeq)$ is a possible realization of $\mathfrak{O}$, just on the basis of its set-theoretical structure, but it is also easy to check that $\mathfrak{A}$ is also a model of $\mathfrak{O}$, because the relation $\succeq$ is transitive and weakly connected in the set A. On the other hand, let

$$A' = \{1, 2, 3\}$$

and

$$\succeq'\, = \{(1,2), (2,3)\}.$$

Then $\mathfrak{A}' = (A', \succeq')$ is a possible realization of $\mathfrak{O}$, but it is not a model of $\mathfrak{O}$, because the relation $\succeq'$ is neither transitive nor weakly connected in the set A'.

Axiomatically built theories. I began this chapter with Tarski's definition of a model of a theory as any possible realization of the theory in which all valid sentences of the theory are satisfied. Just above, I stated a variant of this formulation by requiring that the axioms of the theory be satisfied in the possible realization. When the set of valid sentences of a theory is defined as a set of given axioms of the theory together with their logical consequences, then the theory is said to be *axiomatically built*. Generally speaking, the analysis in this book shall be restricted to theories that are built axiomatically, but within the framework laid down in the next section rather than in this one. As we shall see, this is not a severe restriction of any theory with a well-defined scientific content.

Granted then that the models of the ordinal theory of measurement are just the possible realizations satisfying the two axioms of the theory, we may ask other sorts of questions about the theory $\mathfrak{O}$.

The kinds of questions we may ask naturally fall into certain classes. One class of question is concerned with relations between models of the theory. Investigation of

2.2 THEORIES WITH STANDARD FORMALIZATION

such structural questions is widespread in mathematics, and, as we shall see, does not require that the theory $\mathfrak{O}$ be given a standard formalization. Another class of questions does depend on this formalization. For example, metamathematical questions of decidability—is there a mechanical decision procedure for asserting whether or not a formula of the theory is a valid sentence of the theory?

A third class of questions concerns empirical interpretations and tests of the theory, which shall not be considered in much detail in this book, but suffice it to say that a standard formalization of a theory is almost never required to discuss questions of this class.

Difficulties of scientific formalization. I have written as though the decision to give a standard formalization of a theory hinged entirely upon the issue of subsequent usefulness of the formalization. Unfortunately the decision is not this simple. A major point I want to make is that a simple standard formalization of most theories in the empirical sciences is not possible. The source of the difficulty is easy to describe. Almost all systematic scientific theories of any interest or power assume a great deal of mathematics as part of their formal background. There is no simple or elegant way to include this mathematical background in a standard formalization that assumes only the apparatus of elementary logic. This single point has been responsible for the lack of contact between much of the discussion of the structure of scientific theories by philosophers of science and the standard scientific discussions of these theories. (For support of this view, see van Fraassen (1980).)

Because the point under discussion furnishes one of the important arguments for adopting the set-theoretical approach outlined in the next section, it will be desirable to look at one or two examples in some detail.

Suppose we want to give a standard formalization of elementary probability theory. On the one hand, if we follow the standard approach, we need axioms about sets to make sense even of talk about the joint occurrence of two events, for the events are represented as sets and their joint occurrence as their set-theoretical intersection. On the other hand, we need axioms about the real numbers as well, for we shall want to talk about the numerical probabilities of events. Finally, after stating a group of axioms on sets, and another group on the real numbers, we are in a position to state the axioms that belong just to probability theory as it is usually conceived. In this welter of axioms, those special to probability can easily be lost sight of. More important, it is senseless and uninteresting continually to repeat these general axioms on sets and on numbers whenever we consider formalizing a scientific theory. No one does it, and for good reason.

A second important example is mechanics. In this case we need to have available for the formulation of interesting problems not only the elementary theory of real numbers but much of standard mathematical analysis, particularly of the sort relevant to the solution of differential equations. Again, as a *tour de force* we can use any one of several approaches to bring this additional body of mathematical concepts within a standard formalization of mechanics, but the result is too awkward and ungainly.

Unfortunately, because a standard formalization of a scientific theory is usually not practical, some philosophers of science and scientists seem to think that any rigorous approach is out of the question. To show that this is not the case is the main objective of the next section.

Useful example of formalization.[6] Before turning to that task I would like to consider one example of how treatment of a theory in terms of its standard formalization can have interesting consequences within the philosophy of science. The problem I have in mind is one of finite axiomatizability of a particular theory of measurement. To set the scene we may briefly return to the ordinal theory $\mathfrak{O}$. We need not prove here the intuitively obvious fact that for any finite model $\mathfrak{A} = (A, \succeq)$ of the theory $\mathfrak{O}$ (i.e., the set A is finite), there exists a real-valued function f defined on A such that for every x and y in A

$$x \succeq y \text{ if and only if } f(x) \geq f(y). \tag{1}$$

In algebraic terms, any finite model of $\mathfrak{O}$ is homomorphic to a numerical model, and it is this simple theorem that is the basis for calling the theory $\mathfrak{O}$ a theory of measurement. (The theorem is not true, of course, for arbitrary models of $\mathfrak{O}$, a point discussed in more detail below.)

For numerous applications in psychology it is natural to pass from the ordinal theory to the theory in which *differences* between the ordinal positions of objects are also ordered. Thus we ask what axioms must be imposed on a quaternary relation D and the finite set A on which D is defined, such that there is a real-valued function f with the property that for every x, y, z and w in A[7]

$$xyDzw \text{ if and only if } f(x) - f(y) \geq f(z) - f(w). \tag{2}$$

(For extensive discussion of applications of the quaternary difference relation D in psychological theories of preference and measurement, see Suppes and Winet 1955; Davidson, Suppes and Siegel 1957; Luce and Suppes 1965; Suppes and Zinnes 1963; and Krantz et al. 1971.) We can use equivalence (2) to obtain an immediate extrinsic axiomatization of the theory of this difference relation, which it is natural to call the theory of *hyperordinal* measurement. This is, of course, just a way of making a well-defined problem out of the question of intrinsic axiomatization, as discussed in Chapter 1. We want elementary axioms in the technical sense of a standard formalization of the theory $\mathcal{H}$ of hypordinal measurement such that a finite realization $\mathfrak{A} = (A, D)$ of the theory $\mathcal{H}$ is a model of $\mathcal{H}$ if, and only if, there exists a real-valued function f on A such that (2) holds. Still other restrictions need to be imposed to make the problem interesting. If we are willing to accept a countable infinity of elementary axioms a positive solution may be found by enumerating for each n all possible isomorphism types, but this sort of axiomatization gives little insight into the general structural characteristics that must be possessed by any model of the theory $\mathcal{H}$. So we turn to the possibility of a finite axiomatization. Unfortunately, questions of finite axiomatizability are difficult to settle. In the present case, however, we want to catch with our axioms *all* finite models of the theory; so, it is reasonable to ask for axioms all of which are universal sentences, i.e., sentences all of whose quantifiers stand in front and have the

---

[6]This rather technical subsection may be omitted without any loss of continuity.
[7]In Chapter 3 I replace 'D' by the general weak inequality symbol '$\succeq$'.

remainder of the sentences as their scopes. The point is that such sentences are closed under submodels. This means that if a sentence is true of a realization $\mathfrak{A} = (A, D)$, then it is also true of $\mathcal{B} = (B, D')$ where B is a subset of A and D' is the relation D restricted to the set B. Such closure under submodels is characteristic of universal sentences, and also such a closure property is required if we intend to axiomatize all finite models of the theory.

As it turns out, the structure of the theory $\mathcal{H}$ is quite complex when expressed in terms of elementary universal sentences, for Dana Scott and I (1958) were able to prove that no finite set of universal sentences could provide an adequate set of axioms for $\mathcal{H}$. The method of proof consists of constructing a sequence of finite possible realizations of $\mathcal{H}$ with increasingly large domains and with each realization having the property that the deletion of any element in the domain of the realization yields a model of $\mathcal{H}$, but the full realization itself is not such a model. It follows, from a general axiomatizability criterion of Vaught (1954), that when such a sequence of realizations can be found, the theory cannot be axiomatized by a finite set of universal sentences, or, what comes to the same thing, by a single universal sentence.[8]

Negative results of this sort are particularly dependent on working within the framework of standard formalizations. In fact, few if any workable criteria exist for axiomatizability of theories outside the framework of standard formalization. For a theory like $\mathcal{H}$, significant light is shed on its structure by the sort of negative result just described, but for the reasons stated earlier few scientific theories have the simplicity of theories $\mathfrak{O}$ and $\mathcal{H}$. In fact, it is not easy to find other empirically significant examples of theories of comparable simplicity outside the area of measurement.

An elegant axiomatization of $\mathcal{H}$ in terms of a schema that stands for an infinite collection of elementary axioms has been given by Scott (1964).[9] However, his schema, expressed in terms of all possible permutations of the elements of the domain of a model of $\mathcal{H}$, does not provide the kind of intrinsic feeling for the structure of the theory we get from axioms like those of transitivity or connectivity. Moreover, the negative result about finite axiomatizability in terms of universal sentences shows that the search for a simple, intuitive intrinsic characterization of $\mathcal{H}$ is probably hopeless.

Even the characterization of the ordinal theory $\mathfrak{O}$ as a theory of measurement is not possible in first-order predicate logic, i.e., in a standard formalization, with $\succeq$ as the only nonlogical primitive constant, once we try to capture the infinite as well as the finite models. The difficulty centers around the fact that if a set of first-order axioms has one infinite model then it has models of unbounded cardinalities, and it easily follows from this fact that, for the infinite models of any additional axioms we impose, we cannot prove the existence of a real-valued function f satisfying (1). A minor complication arises from the fact that models of arbitrarily large cardinality might still be homomorphic to a numerical model, but this can be got around fairly directly, and

---

[8]Tait (1959) has shown that it is not possible to just consider universal sentences, even when closure under submodels is required, in establishing general results about finite axiomatizability of finite models. More general results on finite axiomatizability without restriction to universal sentences are too complicated to discuss here. For presentation of some previously unpublished results of Lindstrom that $\mathcal{H}$ is not finitely axiomatizable, see Luce et al. (1990).

[9]A related result of Scott's for probability is stated as Theorem 3 of Section 5.7.

could have been avoided in the first place by postulating that $\succeq$ is an antisymmetric relation, i.e., by adding the axiom

$$(\forall x)(\forall y)(x \succeq y \ \& \ y \succeq x \rightarrow x = y).$$

Once this axiom is included any homomorphism must also be an isomorphism, and of course, no isomorphic numerical model can be found for a model of cardinality greater than that of the continuum. The necessary and sufficient condition to impose on infinite models of $\mathfrak{O}$ to guarantee they are homomorphic to numerical models is not a first-order condition, i.e., is not formulated with the language of the theory $\mathfrak{O}$, but it does have a natural expression in the set-theoretical framework discussed in the next section, and is explicitly considered in Chapter 3.

2.3 Theories Defined by Set-theoretical Predicates[10]

Although a standard formalization of most empirically significant scientific theories is not a feasible undertaking for the reasons set forth in the preceding section, there is an approach to an axiomatic formalization of such theories that is quite precise and satisfies all the standards of rigor of modern mathematics. From a formal standpoint the essence of this approach is to add axioms of set theory to the framework of elementary logic, and then to axiomatize scientific theories within this set-theoretical framework. From the standpoint of the topics in the remainder of this book, it is not important what particular variant of set theory we pick for an axiomatic foundation. From an operational standpoint what we shall be doing could best be described as operating within naive set theory. The only important distinction between axiomatic and naive set theory is that in axiomatic set theory one continually must consider questions about the existence of sets, since it is precisely around such questions that problems of paradox and inconsistency arise. In naive set theory one proceeds as if there were no paradoxes and blithely assumes that all sets exist in the intended intuitive fashion. Although questions about the existence of sets shall not be considered explicitly here, it is a simple and routine matter to embed everything that is said in the remainder of this book within some well-defined axiomatic set theory, for example, the Zermelo-Fraenkel version described and worked out in detail in several books, including my own (Suppes 1960b). For some reservations about this point see the last part of this section.

My slogan for many years has been 'to axiomatize a theory is to define a set-theoretical predicate'. Perhaps the most important point of confusion to clear up about this slogan is the intimate relation between axiomatization and definition. We may begin by considering some purely mathematical examples of axiomatized theories. As we saw in the last section, we can treat simple theories as purely autonomous disciplines by giving them a formalization, which in a systematic sense will require only elementary logic and no other mathematics as a background. However, this is scarcely the standard way of doing mathematics among contemporary mathematicians, for the reason that has also been made clear, namely, the awkwardness of handling theories of any complexity within this restricted framework.

---

[10]The general viewpoint of this section is the same as that set forth many years ago in the last chapter of Suppes (1957/1999). For more general discussion of it and related viewpoints, see Suppe (1974) and van Fraassen (1980).

2.3 Theories Defined by Set-theoretical Predicates

Example: theory of groups. Let us first consider the axioms for a group, which are now often discussed in high-school mathematics texts. A standard formulation is the following:

A1. $x \circ (y \circ z) = (x \circ y) \circ z$.

A2. $x \circ e = x$.

A3. $x \circ x^{-1} = e$.

Here $\circ$ is the binary operation of the group, e is the identity element and $^{-1}$ is the inverse operation. The difficulty with these axioms, taken in isolation, is that one does not quite understand how they are related to other axioms of other theories, or exactly how they are related to mathematical objects themselves. These uncertainties are easily cleared up by recognizing that in essence the axioms are part of a definition, namely, the definition of the predicate 'is a group'. The axioms are the most important part of the definition of this predicate, because they tell us the most important properties that must be possessed by a mathematical object which satisfies the predicate 'is a group', or in other words, by a mathematical object that is a group. In order to make the axioms a part of a proper definition, one or two technical matters have to be settled, and it is one of the main functions of our set-theoretical framework to provide the methods for settling such questions. It is clear from the axioms that in some sense a group must be an object that has a binary operation, an identity element and an inverse operation. How exactly are we to talk about such an object? One suggestion is to formulate the definition in the following fashion:

A nonempty set A is a group with respect to the binary operation $\circ$, the identity element e, and the inverse operation $^{-1}$ if and only if for every x, y and z in A the three axioms given above are satisfied.

The first thing to note about this definition is that it is not a definition of a one-place predicate, but actually of a four-place predicate. This point is somewhat masked by the way in which the definition was just formulated. If we gave it an exact formulation within set theory, the definition would make it clear that not only the letter 'A' but also the operation symbols and the symbol for the identity element are variables that take as values arbitrary objects. Thus a formally exact definition, along the lines of the above but within set theory, would be the following.

A is a group with respect to $\circ$, e, and $^{-1}$ if and only if A is a nonempty set, $\circ$ is a binary operation on A, e is an element of A, and $^{-1}$ is a unary operation on A, such that for every x, y, and z in A, the three axioms given above are satisfied.

The real difficulty with this last version is that it is natural to want to talk about groups as objects. But in this version we talk instead about a four-place predicate as defined above, i.e., we talk about a set and three other objects simultaneously. Our next step is to take care of this difficulty.

$\mathfrak{A}$ is a group if and only if there exists a nonempty set A, a binary operation $\circ$ on A, an element e of A and an inverse operation $^{-1}$ on A such that $\mathfrak{A} = (A, \circ, e, ^{-1})$ and for every x, y, and z in A the three axioms given above are satisfied.

The point of this definition is to make the predicate 'is a group' a one-place predicate, and thus to introduce talk about groups as definite mathematical objects. As can be seen from this definition, a group is a certain kind of ordered quadruple. This characterization of the set-theoretical structure of a group answers the general question of what kind of object a group is. In ordinary mathematical practice, it is common to define the kind of entities in question, for example, ordered quadruples, which correspond to possible realizations in the sense defined in the preceding section, and then not to repeat in successive definitions the set-theoretical structure itself. For example, we might well define *algebras* which are ordered quadruples $\mathfrak{A} = (A, \circ, e, ^{-1})$, where $\circ$ is a binary operation on A, e is an element of A, and $^{-1}$ is a unary operation on A. Having introduced this notion of algebra we could then shorten our definition of a group to the following sort of standard format.

DEFINITION 1. *An algebra* $\mathfrak{A} = (A, \circ, e, ^{-1})$ *is a* group *if and only if for every* x, y, *and* z *in* A *the three axioms given above are satisfied.*

The format exemplified by this last definition is the one that we shall most often use in subsequent parts of this book. For those who prefer their definitions straight, it is a simple and routine matter to convert the sort of statement just used into a completely explicit formal definition in the style of the preceding definition.

To make the variety of possibilities more evident, we can also take algebras to be structures consisting just of a nonempty set and a binary operation on the set. Groups are then defined slightly differently.

DEFINITION 2. *An algebra* $\mathfrak{A} = (A, \circ)$ *is a* group *if and only if for every* x, y, *and* z *in* A

Axiom 1. $x \circ (y \circ z) = (x \circ y) \circ z$.

Axiom 2. There is a w *in* A *such that*
$$x = y \circ w.$$

Axiom 3. There is a w *in* A *such that*
$$x = w \circ y.$$

Meaning of 'set-theoretical predicate'. Because I have expounded these matters of definition in great detail elsewhere (Suppes 1957/1999, Chapters 8 and 12), I shall not go into them further here, except for one or two general remarks. In the first place, it may be well to say something more about the slogan 'to axiomatize a theory is to define a set-theoretical predicate'. It may not be entirely clear what is meant by the phrase 'set-theoretical predicate'. Such a predicate is simply a predicate that can be defined within set theory in a completely formal way. For a version of set theory based only on the primitive predicate of membership, '∈' in the usual notation, this means that ultimately any set-theoretical predicate can be defined solely in terms of membership. Any standard mathematical notions will be used freely in the definition of such a predicate, for we shall assume that these standard notions have already been fully developed and formalized. Whenever a previous set-theoretical formalization of a scientific theory is assumed, this shall be explicitly so indicated. In actual fact this will not be the case for many of the examples considered in this book.

Set theory and the sciences. This last remark suggests there may be some systematic difference between a set-theoretical definition embodying concepts of pure mathematics and one involving concepts of some particular science. I do not mean to suggest anything of the sort. It is one of the theses of this book that there is no theoretical way of drawing a sharp distinction between a piece of pure mathematics and a piece of theoretical science. The set-theoretical definitions of the theory of mechanics, the theory of thermodynamics, and a theory of learning, to give three rather disparate examples, are on all fours with the definitions of the purely mathematical theories of groups, rings, fields, etc. From a philosophical standpoint there is no sharp distinction between pure and applied mathematics, in spite of much talk to the contrary. The continuity between pure and applied mathematics, or between mathematics and science, will be illustrated here by many examples drawn from both domains.

Basic structures. The viewpoint of this section, which provides a general viewpoint for this book, could be formulated as saying that set-theoretical predicates are used to define classes of structures. The class of structures consists of those structures that satisfy the predicate. When the approach is set-theoretical rather than syntactic or formal in the sense of logic, the kind of approach described is similar to the well-known approach of Bourbaki to mathematics. (*Bourbaki* is the pseudonym for a group of mathematicians who have jointly written a many-volume treatise covering large parts of mathematics.) Someone familiar with the Bourbaki approach would regard what has been said thus far as unsatisfactory, insofar as the approach I am describing is meant to be like the Bourbaki approach to mathematics but with that approach applied to the empirical sciences. The source of the criticism is easy to state. Bourbaki asserts in a number of places, but especially in the general article (1950), that it is not at all a question of simply giving independent axiomatizations in the sense of characterizing structures for different parts of mathematics. What is much more important about the Bourbaki program is the identification of basic structures (Bourbaki calls them mother structures). Given this identification, the focus is then to explore with care the way the various basic structures are used in many different parts of mathematics. Simple examples of basic structures in the sense of Bourbaki would be groups as defined earlier, ordering relations or order structures, and also topological structures.

There is still another particular point of importance. In various places Bourbaki discusses the methods for generating the most familiar structures. Da Costa and Chuaqui (1988) have given an analysis in terms of Cartesian products and power sets of a given basic set which will cover a large number of structures defined in the present book. In this case the approach is to show how a restricted number of basic operations on a given set of objects is sufficient for generating the structures of interest.

It seems to me that the approach of da Costa and Chuaqui for generating structures is a fruitful one and can be used for theoretical study of many different scientific disciplines. On the other hand, it seems premature to hope for anything like the kind of identification of basic structures that Bourbaki has considered important for mathematics.

There is also a sense of structuralism that has been discussed in relation to the framework of ideas set forth in this section. Here I refer to the work of Sneed (1971), Stegmüller (1976, 1979), Moulines (1975, 1976), Moulines and Sneed (1979), and Balzer (1978), together with a number of other articles or books by these authors. The literature is now large and it will not be possible to survey it in detail. There are some important points raised by the structuralist viewpoint represented by Sneed, Stegmüller, and their colleagues. The general idea, beginning with Sneed's book (1971), is to start from the framework of characterizing theories by set-theoretical predicates and using the sort of axiomatization discussed in this chapter, but then to go on to the much more difficult problem of characterizing the concepts of a theory that are theoretical as opposed to empirical in nature.

Reservations about set theory. Total adherence to a reduction of all entities to sets, or sets and individuals, is not really crucial to the viewpoint adopted here. It is also natural and sometimes easier to use creative definitions of identity to make available new abstract entities that are not necessarily sets. Such definitions are called *creative* because new propositions can be proved with their introduction. Already in my book on set theory (1960b) I departed from pure set theory both by introducing individuals and using Tarski's axiom for cardinal numbers. Two sets A and B have the same cardinal number if and only if they are equipollent, i.e., can be put in one-one correspondence. This is a creative definition.[11] In symbols:

$$K(A) = K(B) \text{ iff } A \approx B.$$

When this creative definition is added to Zermelo-Fraenkel set theory it is undecidable whether or not cardinal numbers are sets. An even more elementary example is the creative definition of ordered pairs:

$$(x, y) = (u, v) \text{ iff } x = u \ \& \ y = v. \tag{1}$$

To me this approach to ordered pairs seems the most natural and obvious, in spite of the historical interest in the definition of ordered pairs as certain sets by Kuratowski (1921) and Wiener (1914). As new objects are created by such definitions as (1) we can add them to our universe as new individuals, or, in a category of new abstract objects, but in any case as potential members of sets. This way of looking at abstract objects will not satisfy those who want to create all such objects at the beginning in one fixed set of axioms, but I find the process of 'creating as needed' more appealing. Above all, abstract objects defined by creative definitions of identity carry no excess baggage of irrelevant set-theoretical properties. For example, Kuratowski's definition

$$(x, y) = \{\{x\}, \{x, y\}\}$$

generates an asymmetry between the sets to which x and y belong. This does not seem an intuitive part of our concept of ordered pair. I find this even more true of the various set-theoretical reductions of the concept of natural number, and therefore prefer Tarski's axiom given above as a creative definition.

---

[11] For a detailed but elementary discussion of creative definitions, see Suppes (1957/1999, Ch. 8).

Nonetheless, this reservation will not play an important role in the sequel. I have sketched it here to show that a pluralistic attitude toward the concept of structure can be taken. The modern mathematical theory of categories provides other arguments of a different sort. A very accessible introduction to these ideas is to be found in MacLane and Birkhoff (1967).

2.4 Historical Perspective on the Axiomatic Method

Before concluding this chapter, it will perhaps be of interest to examine, even if not in full detail, the history of the development of the axiomatic method.

Before Euclid. That the axiomatic method goes back at least to ancient Greek mathematics is a familiar historical fact. However, the story of this development in the century or two prior to Euclid and its relation to earlier Babylonian and Egyptian mathematics is not wholly understood by scholars. The axiomatic method, as we think of it as crystallized in Euclid's *Elements*, seems to have originated in the fourth century BC, or possibly somewhat earlier, primarily at the hands of Eudoxus. The traditional stories that name Thales as the father of geometry are poorly supported by the existing evidence, and certainly it seems mistaken to attribute to Thales the sophisticated axiomatic method exhibited in Euclid, and already partly evident in Plato's *Meno*. It is also important to keep the relatively modest role of Plato in appropriate perspective (Neugebauer 1957, pp. 151-152). An excellent detailed discussion of the historical development of Euclid's *Elements* in Greek mathematics is given by Knorr (1975),[12] but before turning to this work, something needs to be said about Aristotle.

In various places Aristotle discusses the first principles of mathematics, but the most explicit and detailed discussions are to be found in the *Posterior Analytics* (72a14-24, and 76a31-77a4). According to Aristotle, a demonstrative science must start from indemonstrable principles. The impossibility of having a demonstration of everything is especially emphasized in various passages in the *Metaphysics* (997a5-8, 1005b11-18, and 1006a5-8). Of these indemonstrable principles, Aristotle says in the *Posterior Analytics* that some are common to all sciences. These are the axioms ($\alpha\xi\iota\acute{\omega}\mu\alpha\tau\alpha$). Other principles are special to some particular science.

The standard examples of axioms for Aristotle are the principle that 'if equals be subtracted from equals, the remainders are equal', and the logical principle 'of two contradictories one must be true'. Aristotle called axioms by other names, for example 'common (things)' ($\tau\alpha$ $\kappa o\iota\nu\alpha$), or 'common opinions' ($\kappa o\iota\nu\alpha\iota$ $\delta o\xi\alpha\iota$).

---

[12] A more general analysis of the controversial history of the pre-Euclidean development of the axiomatic method in Greek mathematics may be found in Knorr (1986). I say 'controversial history' because the extant evidence is so fragmentary. Some more comments on this history, with some emphasis on applications are to be found in Section 6.8, the last section of Chapter 6. Knorr is more balanced than Neugebauer in his evaluation. Knorr's view is that there was a serious interaction between the geometers and philosophers in Plato's Academy. Various passages in Plato's *Republic* (510C–511C, 527A, 528B–D) criticize the rigor of the geometers. These philosophical views about the foundations probably had an impact on Eudoxus and other geometers in their development of formal methods. But, and this is the other side of the coin, philosophers like Plato had little impact on the many technical developments of geometry occurring at the same time. There was no foundational crisis stopping these ever deeper technical results.

As for the principles special to a science, Aristotle divided them in two: first, hypotheses which assert existence, for example, the existence of the genus or subject matter, which is magnitude in the case of geometry. The second kind of thesis is definition. Thus, for Aristotle the first principles of a science are divided into axioms, hypotheses, and definitions. It is worth noting that although Aristotle gives a rather thorough general discussion of these ideas, nowhere does he present an extensive and detailed example of a discipline with its principles so formulated. He does have the merit of giving the first extant discussion of first principles in a general and systematic fashion.

Euclid. Admitting that not much is to be said about precursors of Euclid, let us turn now to a somewhat more detailed discussion of the axiomatic method as reflected in his *Elements*. Euclid begins Book I with a list of 23 definitions, 5 postulates, and 5 axioms or common notions (Heath translation, 2nd ed., 1926).[13] The standard of rigor followed by Euclid in deriving the 48 propositions of Book I is remarkably high. On the other hand, as might be expected, his approach is not in complete agreement with modern conceptions of the axiomatic method. He does not seem to see clearly that the axiomatic development of geometry must begin with some primitive ideas which are not themselves defined in terms of others. Perhaps it is fair to say that in this respect he confuses formal or axiomatic questions with problems concerning the application of geometry. For example, Definition 1 of Book I asserts that a point is that which has no parts. Definition 2 asserts that a line is length without breadth, but subsequent proofs do not really appeal to either of these definitions. What is closer to the case is this. The concepts of point and line are in practice primitive for Euclid and are characterized intuitively but not defined mathematically by the explicit and tacit assumptions he makes.

Another distinction not carried over into modern discussions is Euclid's distinction between postulates and axioms. Proclus, in his famous commentary on Euclid (about 450 AD), has this to say about the distinction:

> They differ from one another in the same way as theorems are also distinguished from problems. For, as in theorems we propose to see and determine what follows on the premises, while in problems we are told to find and do something, in like manner in the *axioms* such things are assumed as are manifest of themselves and easily apprehended by our untaught notions, while in the postulates we assume such things as are easy to find and effect (our understanding suffering no strain in their assumption), and we require no complication of machinery. . . . Both must have the characteristic of being simple and readily grasped, I mean both the postulate and the axiom; but the postulate bids us contrive and find some subject-matter ($υλψ$) to exhibit a property simple and easily grasped, while the axiom bids us assert some essential attribute which is self-evident to the learner . . .

---

[13] Interestingly enough, Euclid does not actually use the term 'axioms' but only 'common notions' ($κóιναδ$), although the meaning is the same as Aristotle's meaning of axioms or common things. On the other hand, Euclid replaces Aristotle's terminology of hypotheses by postulates ($αιτουμενα$). In spite of the great influence of Euclid, this terminology is not fixed in Greek mathematics but continues to change; for example, the Greek terms for postulates, hypotheses, and assumptions are all to be found in various works of Archimedes. But in all these cases, essentially the same thing is meant as was meant by Aristotle in the discussion of hypotheses or by Euclid in the introduction of postulates.

Thus it is the geometer who knows that all right angles are equal and how to produce in a straight line any limited straight line, whereas it is a common notion that things which are equal to the same thing are also equal to one another, and it is employed by the arithmetician and any scientific person who adapts the general statement to his own subject.[14]

It is perhaps the last distinction introduced, between postulates as peculiar to geometric subject matter and axioms as assumptions common to all investigation, which best represents the view adopted by Euclid himself. Although it is easy enough to pick out certain flaws in Euclid's *Elements,* and to emphasize differences between his conception of the axiomatic method and modern ones, the essential point remains that in basic conception and execution the axiomatic method as reflected in his *Elements* is extremely close to modern views.

It will perhaps be instructive to look at a few historical examples across the centuries. Because of the interest in the theoretical sciences in this book, I shall draw mainly upon examples which would ordinarily be associated with physics rather than with pure mathematics, even though the distinction between the two is not clear from a formal standpoint, as has already been emphasized. The discussion of these historical examples is meant to highlight two important points. The first is the existence of a continuous Euclidean-Archimedean mathematical tradition that reaches all the way to Newton and is almost completely independent of the Aristotelian tradition of analysis, which was dominant among philosophers from the Hellenistic age to the seventeenth century. The second point is that the standards of rigor in physics, although meant to imitate those in geometry, have never actually been as strict, starting with what is probably the first treatise on mathematical physics as we would define it, Archimedes' treatise on static equilibrium.[15]

Archimedes. We may begin by looking at Archimedes' postulates in Book I of his treatise *On the Equilibrium of Planes* (taken from Heath's edition of Archimedes' works, 1897, pp. 189–190).

I postulate the following:

1. Equal weights at equal distances are in equilibrium, and equal weights at unequal distances are not in equilibrium but incline towards the weight which is at the greater distance.

2. If, when weights at certain distances are in equilibrium, something be added to one of the weights, they are not in equilibrium but incline towards that weight to which the addition was made.

---

[14]The passages quoted are taken from T. L. Heath's introduction and commentary on Euclid's *Elements* (Heath 1926, pp. 122–123).

[15]But not the first work on mathematical physics. That signal honor belongs to the Babylonian astronomers, at least as early as the reign of Nabonassar (746-732 BC) and over the following several centuries, a fact now widely recognized. For recent analysis of the Babylonian astronomers' beginning of the historically first mathematical science, see Swerdlow (1998). Also to be emphasized is that the detailed mathematical computations about astronomical phenomena flourished only after several thousand years of preceding observations.

3. Similarly, if anything be taken away from one of the weights, they are not in equilibrium but incline towards the weight from which nothing was taken.

4. When equal and similar plane figures coincide if applied to one another, their centres of gravity similarly coincide.

5. In figures which are unequal but similar the centres of gravity will be similarly situated. By points similarly situated in relation to similar figures I mean points such that, if straight lines be drawn from them to the equal angles, they make equal angles with the corresponding sides.

6. If magnitudes at certain distances be in equilibrium, (other) magnitudes equal to them will also be in equilibrium at the same distances.

7. In any figure whose perimeter is concave in (one and) the same direction the centre of gravity must be within the figure.

Before examining in detail the character of these seven postulates, it may be useful to say something about the historical setting of this treatise of Archimedes. The exact date at which it was written is certainly not known. The usual, but not fully verified, dates for his life are 287 BC to 212 BC; the date of his death seems fairly reliable because according to several sources he was killed in 212 BC at the time of the Roman conquest of Syracuse. A good deal of Greek mathematics precedes the present treatise but it occupies a unique place in the long history of mathematical physics, because it is apparently the first systematic treatise to establish a close relationship between mathematics and mechanics. The postulates that Archimedes enunciates would now be regarded as postulates for that part of mechanics labeled *statics*.

There are a number of methodological comments to be made about Archimedes' axioms. Many of the difficulties that beset them are characteristic of the problems that have arisen throughout the centuries in attempting to give a finished formulation to the fundamental assumptions of various branches of physics. In terms of the viewpoint that has been developed thus far in this chapter, probably the most striking characteristic of Archimedes' axioms is their lack of completeness. The reader feels immediately that a mathematically complete account of the assumptions intended has not been given. It is fair to suppose that certain of the geometric notions are meant to be taken over directly from Greek geometry—for example, the notion of similarity introduced in Postulate 5—but this is not true of the concept of equal distance, of being at equilibrium, or in particular of the concept of center of gravity. Without question, elementary properties of the relation of two objects being in equilibrium are tacitly assumed, for example, that the relation is transitive and symmetric. The most obvious problem centers around the concept of the center of gravity of a plane figure. This concept is essential to the formulation of Postulates 4, 5, and 7, but it is quite evident on the other hand that these postulates in themselves do not provide a complete characterization of the concept. By this I mean that if we knew nothing about centers of gravity except what is stated in Postulates 4, 5, and 7, we would not be able to derive the theorems in which Archimedes is interested, and which he does derive. As Dijksterhius (1957) points out in his book on Archimedes, it is possible to argue that the concept of center of gravity is being taken over by Archimedes from more elementary discussions and thus really has the

2.4 HISTORICAL PERSPECTIVE ON THE AXIOMATIC METHOD 39

same status as the geometric concept of similarity in his treatise. On the face of it this argument seems sounder than that of Toeplitz and Stein (published in Stein (1930)), who propose that the postulates are to be taken as explicitly defining centers of gravity once the postulates are enlarged by the obvious and natural assumptions.

It is also clear that a standard formalization of Archimedes' theory, in the sense of the second section of this chapter, cannot be given in any simple or elegant way. It is possible to give a standard formalization of the part of the theory embodied in Postulates 1, 2, 3, and 6. (For an argument that this is the first axiomatic example of the theory of conjoint measurement, see Suppes 1980.[16]) Quite apart from the question of standard formalization, which we could hardly expect for a subject like statics, there are serious problems involved in giving a reconstruction in set-theoretical terms of Archimedes' postulates. In such a set-theoretical formulation, we can without difficulty use a geometric notion like similarity. If we take over from prior developments a definition of center of gravity, then it would seem that Postulate 4, for example, would simply be a theorem from these earlier developments and would not need separate statement. Put another way, under this treatment of the concept of center of gravity, no primitive notion of Archimedes' theory would appear in Proposition 4 and thus it would clearly be an eliminable postulate. The same remarks apply to Postulates 5 and 7. It would seem that Archimedes has constructed a sort of half-way house; his postulates do not give a complete characterization of center of gravity, but on the other hand, they cannot be said to depend upon a completely independent characterization of this concept.

These difficulties are the sort that have plagued foundational studies of physics and other empirical sciences from the very beginning. The case of geometry is deceptive. Geometry can be erected on a fairly small number of concepts, and even though Euclid's *Elements* do not represent a perfect axiomatic development, it is a straightforward matter today to provide a completely rigorous development. In Archimedes' treatise on statics, as in all cases of physical treatment of the empirical world, it is necessary to introduce more than geometric concepts. In spite of the elementary character of Postulates 1, 2, 3, and 6, a host of notions is introduced, which from a formal standpoint would have to be considered as primitive. The list is formidable considering the simplicity of the axioms: the concept of two objects being of equal weight; a relation of comparative distance (from a fulcrum); a binary relation for equilibrium, or at least a binary weak ordering relation in terms of which we may define both equilibrium and the relation of inclination; and finally a binary operation for combining objects, needed for the formulation of Postulates 2 and 3. The explicit analysis of all these concepts, many of which have a very well-defined intuitive meaning, undoubtedly seemed unnecessary to Archimedes as it has to most of his successors in mathematical physics. But this remark about intuitive meaning is far too simplistic. If this line of thought were to be pursued with any thoroughness, it would leave as an unexplained mystery the extraordinary Greek concern for rigor and the deductive structure of mathematics. It is perhaps reasonable to suppose that the treatise was written at a rather advanced level, and a good deal of prior analysis of the subject was assumed. To determine the extent to which this is true would require more textual analysis than is possible in the present

---

[16]Elementary axioms of conjoint measurement are given in Definition 9 of Section 3.4.

context.

Archimedes' treatise is in no sense an isolated case. Other advanced ancient treatises often state no axioms, but give at the beginning a list of definitions. Two examples are Apollonius' treatise *Conics*, famous for its mathematical rigor, and Archimedes' own short treatise *On Spirals*. A different, more physical example, but at the same level of rigor, is Diocles' *On Burning Mirrors*. What does seem correct is that these works all characteristically accepted without question the axiomatic framework and the level of deductive rigor set by Euclid's *Elements*. This broad acceptance of the *Elements* corresponds, to a surprising degree, to a similar twentieth-century acceptance of Zermelo-Fraenkel set theory with the axiom of choice (ZFC), or some minor variant of ZFC, as a proper framework for almost all of modern pure mathematics. An elementary exposition may be found in Suppes (1960b).

In closing the discussion of Archimedes' treatise, it is worth noting that Mach, in his famous treatise on mechanics, seems to be badly confused about Archimedes' proofs. The focus of Mach's discussion is Proposition 6 in the treatise which asserts that commensurable magnitudes are in equilibrium at distances reciprocally proportional to their weights, which is just Archimedes' formulation of the famous principle of the lever. Mach (1942, p. 20) is particularly exercised by the fact that 'the entire deduction (of this proposition) contains the proposition to be demonstrated, by assumption if not explicitly'. A central point of Mach's confusion seems to be a complete misunderstanding as to the nature of the application of mathematics to physics. Mach seems to have no real conception as to how mathematics is used to derive particular propositions from general assumptions, and what the relation of these general assumptions to the particular proposition is. He seems mistakenly to think that any such proposition as the one just quoted must somehow be established directly from experience. His sentiments on these matters are clearly expressed in the following passage:

> From the mere assumption of the equilibrium of equal weights at equal distances is derived the inverse proportionality of weight and lever arm! How is that possible? If we were unable philosophically and a priori to excogitate the simple fact of the dependence of equilibrium on weight and distance, but were obliged to go for *that* result to experience, in how much less degree shall we be able, by speculative methods, to discover the form of this dependence, the proportionality! (Mach 1942, p. 19)

Mach's mistaken views about the relation between mathematics and physics had a very unfortunate influence on discussions of the philosophy of physics throughout the twentieth century.

To show that Archimedes' lack of explicit axiomatic rigor was characteristic of other theoretical work in ancient science I discuss briefly two other ancient examples.

Euclid's Optics. It is important to emphasize that Euclid's *Optics* is really a theory of vision and not a treatise on physical optics. A large number of the propositions are concerned with vision from the standpoint of perspective in monocular vision. Indeed, Euclid's *Optics* could be characterized as a treatise on perspective within Euclidean geometry. The tone of Euclid's treatise can be seen from quoting the initial part, which consists of seven 'definitions':

2.4 Historical Perspective on the Axiomatic Method

1. Let it be assumed that lines drawn directly from the eye pass through a space of great extent;

2. and that the form of the space included within our vision is a cone, with its apex in the eye and its base at the limits of our vision;

3. and that those things upon which the vision falls are seen, and that those things upon which the vision does not fall are not seen;

4. and that those things seen within a larger angle appear larger, and those seen within a smaller angle appear smaller, and those seen within equal angles appear to be of the same size;

5. and that those things seen within the higher visual range appear higher, while those within the lower range appear lower;

6. and, similarly, that those seen within the visual range on the right appear on the right, while those within that on the left appear on the left;

7. but that things seen within several angles appear to be more clear.

(The translation is taken from that given by Burton in 1945.) The development of Euclid's *Optics* is mathematical in character, but it is not axiomatic in the same way that the *Elements* are. For example, Euclid later proves two propositions, 'to know how great is a given elevation when the sun is shining' and 'to know how great is a given elevation when the sun is not shining'. As would be expected, there is no serious introduction of the concept of the sun or of shining but they are treated in an informal, commonsense, physical way with the essential thing for the proof being rays from the sun falling upon the end of a line. Visual space is of course treated by Euclid as Euclidean in character.

It might be objected that there are similar formal failings in Euclid's *Elements*. But it does not take much reflection to recognize the very great difference between the introduction of many sorts of physical terms in these definitions from the *Optics* and the very restrained use of language to be found in the *Elements*. It seems to me that the formulation of fundamental assumptions in Euclid's *Optics* is very much in the spirit of what has come to be called, in our own time, physical axiomatics. There is no attempt at any sort of mathematical rigor but an effort to convey intuitively the underlying assumptions.

Ptolemy's Almagest. The third and most important example I cite is Ptolemy's *Almagest*. It is significant because it is the most important scientific treatise of ancient times and because it does not contain any pretense of an axiomatic treatment. Ptolemy does use mathematical argument, and indeed mathematical proof, with great facility, but he uses the mathematics in an applied way. He does not introduce explicit axioms about the motion of stellar bodies, but reduces the study of their motion to geometric propositions, including of course the important case of spherical trigonometry.

The following passage near the beginning of the *Almagest* illustrates very well the spirit in which Ptolemy brings in assumptions:

> The general preliminary discussion covers the following topics: the heaven is spherical in shape, and moves as a sphere; the earth too is sensibly spherical in shape, when taken as a whole; in position it lies in the middle of the heavens very much like its centre; in

size and distance it has the ratio of a point to the sphere of the fixed stars; and it has no motion from place to place.

(H10, Toomer translation, 1984, p. 38)

There then follows a longer and more detailed discussion of each of these matters, such as the proposition that the heavens move spherically. My point is that the discussion and the framework of discussion are very much in the spirit of what we think of as nonaxiomatic theoretical science today. There is not a hint of organizing these ideas in axiomatic fashion.

When Ptolemy gets down to details he has the following to say:

We are now about to begin the individual demonstrations, the first of which, we think, should be to determine the size of the arc between the aforementioned poles [of the ecliptic and equator] along the great circle drawn through them. But we see that it is first necessary to explain the method of determining chords: we shall demonstrate the whole topic geometrically once and for all.

(H31, Toomer translation, 1984)

The detailed calculations, then, on the size of chords inscribed in a circle emphasizes, above all, calculation and would make a modern physicist happy by its tone and results as well. This long and important analysis of computations is concluded with a numerical table of chords.

The thesis I am advancing is illustrated, in many ways even more strikingly, by the treatment of the motion of the moon in Book IV. Here Ptolemy is concerned with the kind of observations that are appropriate for a study of the moon's motion and especially with the methodology of how a variety of observations are to be rectified and put into a single coherent theory.

Various hypotheses introduced in later books, e.g., the hypothesis of the moon's double anomaly in Book V, are in the spirit of modern astronomy or physics, not axiomatic mathematics. Moreover, throughout the *Almagest,* Ptolemy's free and effective use of geometric theorems and proofs seems extraordinarily similar to the use of the differential and integral calculus and the theory of differential equations in a modern treatise on some area of mathematical physics.[17]

In this analysis of the use of axiomatic methods and their absence in explicit form from ancient theoretical sciences such as optics and astronomy, I have only partially entered into a discussion of the philosophical analysis of the status of axioms, postulates and hypotheses. There is a substantial ancient literature on these matters running from Plato to Proclus, who was quoted earlier. Perhaps the best and most serious extant

---

[17] With some reservations the same thing can be said about Euclid's fourth century BC treatise *Phaenomena* (Berggren and Thomas translation 1996) on spherical astronomy. Incidentally, Proposition 1 was famous in the ancient world as a serious mathematical proof from the astronomical assumptions made that, to quote, "The Earth is in the middle of the cosmos and occupies the position of centre with respect to the cosmos" (p. 52). Several hundred years later, about 130 AD, Galen, in his treatise *On the Doctrine of Hippocrates and Plato,* had this to say about this result,

Euclid showed in Theorem 1 of the *Phaenomena* in a few words that the Earth is in the midst of the cosmos, as a point or a centre, and the students trust the proof as if it were two and two is four. (Berggren and Thomas, p. 8).

2.4 HISTORICAL PERSPECTIVE ON THE AXIOMATIC METHOD 43

discussion is to be found in Aristotle's *Posterior Analytics.* Aristotle explains in a very clear and persuasive way how geometric proofs can be appropriately used in mechanics or optics (75b14ff). But just as Aristotle does not really have any developed examples from optics, mechanics or astronomy, so it seems to me that the interesting distinctions he makes do not help us understand any better the viewpoint of Euclid toward the definitions of his optics or the postulates of Archimedes about centers of gravity cited above.

Jordanus de Nemore. The kind of axiomatic approach used by Archimedes had a very strong influence on discussions of physics in the middle ages. Readers interested in a comprehensive history of mechanics in this period are referred to the classical volume of Clagett (1959). I consider here only one important example. Contrary to some popular opinions about the absence of any serious scientific and mathematical developments in medieval times, scholarship in the twentieth century, beginning with Duhem and since 1950 spearheaded by Marshall Clagett and his collaborators and colleagues, has demonstrated otherwise. The many medieval treatises now edited and translated by this group show clearly enough how much was going on. Moreover, in many cases the methods used were axiomatic in the spirit of Euclid and Archimedes. There is, in this collection, a large body of works on the theory of weights, an important sample of which is published in Moody and Clagett (1952). Probably the most important of these treatises on weight was the *Liber de ratione ponderis* by the thirteenth-century Parisian mathematician Jordanus de Nemore.

The postulates and especially the theorems of Jordanus make use of the classical concepts and theorems of Euclidean geometry—without explicitly saying so, in conformity with long-standing tradition. Here is an English translation of Jordanus' seven *suppositiones* or postulates (Moody and Clagett 1952, p. 175):

Postulates

1. The movement of every weight is toward the center (of the world), and its force is a power of tending downward and of resisting movement in the contrary direction.
2. That which is heavier descends more quickly.
3. It is heavier in descending, to the degree that its movement toward the center (of the world) is more direct.
4. It is heavier in position when in that position its path of descent is less oblique.
5. A more oblique descent is one which, in the same distance, partakes less of the vertical.
6. One weight is less heavy in position, than another, if it is caused to ascend by the descent of the other.
7. The position of equality is that of equality of angles to the vertical, or such that these are right angles, or such that the beam is parallel to the plane of the horizon.

I quote here only the second theorem of Part One, which is especially interesting because of its clear formulation (and proof) of the stable horizontal equilibrium of an equal-arm balance with equal weights suspended.

Theorem 2. When the beam of a balance of equal arms is in the horizontal position, then, if equal weights are suspended from its extremities, it will not leave the horizontal position, and if it is moved from the horizontal position, it will revert to it, but if unequal weights are suspended, the balance will fall on the side of the heavier weight until it reaches the vertical position. (p. 177)

Newton. I next look at the initial definitions and axioms of what has been probably the most important work in the history of modern science. The main historical point I want to emphasize is that Newton wrote his *Principia* in the Greek tradition of Euclid's *Elements*. The mathematical methods of attack he used, and even more the style of formulation, are very much in that tradition. In considering either Archimedes or Newton, as well as Euclid, it is important not to be put off by the verbal formulation of axioms and theorems. This again represents a well-established Greek mathematical tradition. Even though the mathematical treatises of these authors are formulated in verbal terms to a very large extent, there is a very sharp conceptual difference between their formulation and that to be found in Aristotle's *Physics* and in the medieval tradition that derives from Aristotle rather than from Archimedes and Euclid. Treatises written in the Aristotelian tradition are genuinely nonmathematical in character. Their basic assumptions do not have any natural mathematical formulation, and mathematical proofs of conclusions are almost nonexistent. From the standpoint of method Descartes's famous *Principles of Philosophy,* first published in 1644, is in the Aristotelian tradition, whereas Newton's *Principia,* first published in 1687, is in the tradition of Euclid and Archimedes.

It is often asserted that the mathematical methods of modern physics originated in the seventeenth century, and arose after a tortuous evolution from the Aristotelian tradition of the middle ages. The facts certainly do not seem to support such a thesis. There is a continuous mathematical tradition in the spirit of Euclid and Archimedes running down through the middle ages up to the seventeenth century. Certainly the continuity of this mathematical approach strongly influenced Galileo as well. The branch of physics that perhaps most clearly shows this continuity is astronomy.[18] The history of both observational and mathematical astronomy has an essentially unbroken line of development running back to the pre-Greek astronomers of Babylon, and has scarcely been affected in any period by Aristotelian modes of analysis, in spite of Ptolemy's acceptance of Aristotle's general cosmology.

Newton begins his *Principia* with the following eight definitions and three axioms or laws of motion (I have drawn upon the Cajori translation of 1946):

Definition I. The quantity of matter is the measure of the same, arising from its density and bulk conjointly.

Definition II. The quantity of motion is the measure of the same, arising from the velocity and quantity of matter conjointly.

Definition III. The *vis insita*, or innate force of matter, is a power of resisting, by which every body, as much as in it lies, continues in its present state, whether it be of rest, or of moving uniformly forwards in a right line.

---

[18] An important exception is Huygens' *Pendulum Clock* (1673/1986). It is easy to argue that the last two great original works of mathematical physics written in the traditional geometric style were this treatise of Huygens and Newton's *Principia*.

Definition IV. An impressed force is an action exerted upon a body, in order to change its state, either of rest, or of uniform motion in a right line.

Definition V. A centripetal force is that by which bodies are drawn or impelled, or any way tend, towards a point as to a centre.

Definition VI. The absolute quantity of a centripetal force is the measure of the same, proportional to the efficacy of the cause that propagates it from the centre, through the space round about.

Definition VII. The accelerative quantity of a centripetal force is the measure of the same, proportional to the velocity which it generates in a given time.

Definition VIII. The motive quantity of a centripetal force is the measure of the same, proportional to the motion which it generates in a given time.

Law I. Every body continues in its state of rest, or of uniform motion in a right line, unless it is compelled to change that state by forces impressed upon it.

Law II. The change of motion is proportional to the motive force impressed; and is made in the direction of the right line in which that force is impressed.

Law III. To every action there is always opposed an equal reaction: or, the mutual actions of two bodies upon each other are always equal, and directed to contrary parts.

It should be remarked at once that each of these definitions and laws is accompanied by additional comments and explanation. What is important from a methodological standpoint is how closely the initial organization of the *Principia* resembles Euclid's *Elements*. Just as in the case of Euclid, Newton begins with a series of definitions that attempt to define his key concepts in terms of more familiar notions. As has already been remarked, from a formal axiomatic standpoint, such definitions are out of place. They certainly do have a place, on the other hand, in giving the intended empirical meaning. However, the definitions play a more systematic role than that of simply providing such an empirical interpretation, for they relate in a systematic formal way various concepts occurring essentially in the formal statement of the theory. There is no clear separation of these two aspects of theory construction in Euclid or Newton.[19]

Modern geometry. The historical source of the modern axiomatic method was the intense scrutiny of the foundations of geometry in the nineteenth century. Undoubtedly the most important driving force behind this effort was the discovery and development of non-Euclidean geometry at the beginning of the nineteenth century by Bolyai, Lobachevski and Gauss. Serious logical deficiencies in Euclid's *Elements* were also given sustained attention. Early in the century, e.g., Poncelet (1822) introduced his rather unsatisfactorily formulated principle of continuity, which he could use to prove that if a straight line has one point inside and one point outside a circle then it must have two points of intersection with the circle—a result that could not be proved from Eu-

---

[19]The *Elements* and *Principia* do differ radically in that Newton included detailed astronomical data (Book III) and even some data on fluids relevant to fluid mechanics (Book II) rather than astronomy. In this respect the *Principia* is closer to Ptolemy's *Almagest* than the *Elements*. On the other hand, the strict axiomatic and mathematical development of Book I is much closer in spirit to the *Elements* than the *Almagest*.

clid's postulates. Later, this result was shown to follow from Dedekind's Postulate of Completeness (1872/1963) which had a satisfactory formulation:

> If all points of a straight line fall into two classes such that every point of the first class lies to the left of every point of the second class, there exists one and only one point which produces this division of all the points into two classes, this division of the straight line into two parts. (Dedekind 1872/1963, p. 11)

A more directly geometric and thus more intuitively appealing axiom that closed a gap in Euclid was Pasch's axiom (1882). This axiom asserts that if a line intersects one side of a triangle it must also intersect a second side.

It was Pasch, above all, who led the way to the modern formal conception of geometry in his influential book of 1882. (An excellent detailed discussion of these developments is to be found in Nagel (1939a). Here is what Pasch says about the meaning of geometric concepts (translation taken from Nagel's article):

> Indeed, if geometry is to be really deductive, the deduction must everywhere be independent of the *meaning* of geometric concepts, just as it must be independent of the diagrams; only the *relations* specified in the propositions and definitions employed may legitimately be taken into account. During the deduction it is useful and legitimate, but in *no* way necessary, to think of the meanings of the terms; in fact, if it is necessary to do so, the inadequacy of the proof is made manifest. If, however, a theorem is rigorously derived from a set of propositions—the *basic* set—the deduction has a value which goes beyond its original purpose. For if, on replacing the geometric terms in the basic set of propositions by certain other terms, true propositions are obtained, then corresponding replacements may be made in the theorem; in this way we obtain new theorems as consequences of the altered basic propositions without having to repeat the proof.
>
> (pp. 237-238)

Pasch's viewpoint is in all essential respects in agreement with the set-theoretical approach to axiomatic theories outlined in the preceding section.

Pasch's formal axiomatic approach was extensively commented on and extended by the Italian geometers Pieri and Veronese, as well as by Peano. But clearly the most influential work in this line of development was Hilbert's *Grundlagen der Geometrie*, first edition, 1897. Not only was Hilbert one of the most prominent European mathematicians at that time, his short treatise was written in a very clear, but elementary, way with a minimum use of mathematical notation. After Euclid it may be the most widely cited book in the history of geometry—at least in the twentieth century. But it is important to keep in mind that it was really Pasch, rather than Hilbert, who gave the first explicit formulation of the modern axiomatic approach to geometry, which has been so influential in the development of modern mathematics.

A superficial glance at the history of geometry in the nineteenth century raises a natural question about the development of the axiomatic method as it reached conceptual and logical maturity, especially in the work of Pasch and Hilbert. Why was projective geometry rather than non-Euclidean geometry not the basis of this axiomatic focus? From a modern standpoint it seems strange. For example, we get hyperbolic geometry, the most important non-Euclidean geometry, by changing only one axiom of a long list needed for Euclidean geometry in the meticulous axiomatization of Borsuk and

2.4 HISTORICAL PERSPECTIVE ON THE AXIOMATIC METHOD

Szmielew (1960). In contrast, any of the standard modern axiomatizations of projective geometry look, feel and are very different from the Euclidean ones. But, all the same, the answer to the question is simple. The early and even the flourishing period of development of projective geometry, associated with the work of Poncelet, Chasles, Steiner and von Staudt, stayed within the Euclidean analytic framework. They typically studied what were called the graphical properties of figures represented analytically. For example, Poncelet studied invariant properties which did not keep invariant magnitudes of distances or angles, but were preserved under projective transformations of their coordinates.[20]

Hilbert and Frege. These developments, however, have not been without their philosophical critics, especially Frege. Although Frege corresponded with Pasch, Frege's part of the correspondence is lost. In contrast, the surviving correspondence between Frege and Hilbert has been widely discussed. Here is the core of Frege's criticisms of Hilbert's *Grundlagen* expressed in a letter to Hilbert on December 27, 1899:

> The explanation of sections 1 and 3 are apparently of a very different kind, for here the meanings of the words 'point', 'line', 'between' are not given, but are assumed to be known in advance. At least it seems so. But it is also left unclear what you call a point. One first thinks of points in the sense of Euclidean geometry, a thought reinforced by the proposition that the axioms express fundamental facts of our intuition. But afterwards you think of a pair of numbers as a point. I have my doubts about the proposition that a precise and complete description of relations is given by the axioms of geometry (sect. 1) and that the concept 'between' is defined by axioms (sect. 3). Here the axioms are made to carry a burden that belongs to definitions. To me this seems to obliterate the dividing line between definitions and axioms in a dubious manner, and beside the old meaning of the word 'axiom', which comes out in the proposition that the axioms express fundamental facts of intuition, there emerges another meaning but one which I can no longer quite grasp.
>
> (Frege 1899/1980, pp. 35-36)

A little later in his long letter Frege remarks, "Thus axioms and theorems can never try to lay down the meaning of a sign or word that occurs in them, but it must already be laid down".

Hilbert responds with firm words of disagreement in a letter to Frege two days later, December 29, 1899. A central passage in the letter is the following:

> You write: 'I call axioms propositions . . . From the truth of the axioms it follows that they do not contradict one another.' I found it very interesting to read this very sentence in your letter, for as long as I have been thinking, writing and lecturing on these things, I have been saying the exact reverse: if the arbitrarily given axioms do not contradict one another with all their consequences, then they are true and the things defined by the axioms exist. This is for me the criterion of truth and existence.
>
> (Hilbert 1899/1980, pp. 39-40)

A way of focusing on the difference is Frege's objection to Hilbert's analytic model of the Euclidean plane—with points represented by pairs of numbers—to prove the

---

[20]This remark was added after a discussion of Poncelet with Jules Vuillemin.

consistency of the axioms. Frege cannot accept that this is an acceptable model of the axioms. He is far removed in spirit from the set-theoretical viewpoint I have adopted. The axioms of geometry are supposed to be true only of 'genuine' geometric objects.

In a later letter of January 6, 1900, Frege does concede that Hilbert's viewpoint could be seen as one of 'placing yourself in a higher position from which Euclidean geometry appears as a special case of a more comprehensive theoretical structure . . .' (Frege 1900/1980, p. 43). Frege makes this statement in commenting on Hilbert's proofs of independence of various of his axioms. The statement by Frege just quoted seems sympathetic to a set-theoretical approach, but Frege goes on to express his uncertainty about its correctness.

It also must be noted that Hilbert's own views are not as clearly put as might be desired, but his viewpoint and intuitions became those of most mathematicians working in the foundations of geometry. The absence of explicit discussion in this correspondence and elsewhere at the time of the need for representation theorems is considered in the next chapter, where this historical perspective is extended.

Subsequent to this correspondence, Frege published several articles on the foundations of geometry (1903a,b, 1906a,b,c) in which he spelled out in greater detail the ideas sketched in his letters to Hilbert. By ordinary standards, the articles are in fact prolix and far too detailed and repetitious. All the same, Frege does end up formulating more clearly than Hilbert, how Hilbert's axiomatic treatment of Euclidean geometry could be properly treated as a second-level formal theory. In the end, however, it is Pasch's and Hilbert's approach to the axiomatic method that has triumphed, not Frege's, as far as general use in mathematics is concerned.

Physics. There is not a clear separation between mathematics and mathematical physics in either Archimedes or Newton; this is to be regarded as a modern distinction. From the standpoint of the axiomatic method, what is interesting historically is that the mathematical physics of the seventeenth century is written as part of mathematical treatises, and in the same, mainly geometric, fashion. In the nearly three centuries since the publication of Newton's *Principia,* the divergence between the deductive methods used in mathematics and in mathematical physics has become quite pronounced. The direction of mathematics has been to move toward increasingly well-defined standards of rigor; the kind of set-theoretical approach outlined in the preceding section is now the approach characteristic of the great bulk of published mathematical research. Naturally, minor differences in style and verbal formulation of set-theoretical definitions are to be found; but in almost all cases reconstruction of the set-theoretical characterization is a routine affair.

This is far from being the case in physics, although even here a distinction exists between mathematical physics and theoretical physics. Mathematical physics is increasingly a subject done by mathematicians and a concern for questions of rigor and clarity of assumptions is evident. Theoretical physics, on the other hand, is done in a fashion that is very far from satisfying modern mathematical standards. None of the historically important papers of this century in the theory of relativity or in quantum mechanics were written in clearly delineated axiomatic fashion. (In referring to relativity theory, I

2.4 Historical Perspective on the Axiomatic Method

have in mind the early papers of Einstein and Lorentz, not the later mathematical work of Minkowski, Veblen, Robb, and others.[21]) Even von Neumann's book on quantum mechanics (1932) does not give an axiomatic development of quantum mechanics, but only of Hilbert space.

What has not been properly studied is the wonderful use of physical intuition by physicists to replace mathematical rigor. A fine example of this is the last sentence of Einstein's reply to Gödel's construction of periodic cosmological solutions to Einstein's gravitational equations permitting, in principle, 'a round trip on a rocket ship in a sufficiently wide curve ... to travel into any region of the past, present and future and back again ... ' (Gödel 1949, p. 560). Here is Einstein's reply, referring to Gödel's cosmological solutions: "It will be interesting to weigh whether these are not to be excluded on physical grounds" (Einstein 1949, p. 688).

Physics is focused on problem solving. When mathematical rigor gets in the way, it, like other concepts or techniques, is downplayed. This pragmatic attitude is discussed more generally in an earlier article of mine (Suppes 1998). See also the analysis of physicists' pragmatic attitudes toward the nature of probability in Section 5.8.

It is difficult to predict the general future of axiomatic methods in the empirical sciences. Economists use them extensively. In any case, axiomatic methods are now widely used in foundational investigations of particular sciences, as well as in the pursuit of certain general questions of methodology, especially those concerning probability, statistics and induction. The use of such methods permits us to bring to the philosophy of science the standards of rigor and clarity that are an accepted part of the mathematical sciences. A conservative prediction is that they will continue to be applied in foundational work throughout this century, even when surrounded by a context of informal philosophical or scientific analysis. Such a mixture of the formal and informal is to be found in all the chapters that follow. This is both desirable and necessary, in the sense that many significant ideas in the philosophy of science are not expressed in a way that makes them suitable to formulate in terms of systematic axioms. Experimental and statistical practices in all parts of science are the source of the largest numbers of such examples. But granting this,–and I consider it important to do so–, there remains much useful work to be done in clarifying the nature of theories, by examining as thoroughly as possible questions of representation and invariance of the structures satisfying the explicit theoretical axioms of different theories.

---

[21] For further relevant details about relativity theory, see Section 6.3.

3

Theory of Isomorphic Representation

A central topic in the philosophy of science is the analysis of the structure of scientific theories. Much of my own work has been concerned with this topic, but in a particular guise. The fundamental approach I have advocated for a good many years is the analysis of the structure of a theory in terms of the models of the theory. In a general way, the best insight into the structure of a complex theory is by seeking representation theorems for its models, for the syntactic structure of a complex theory ordinarily offers little insight into the nature of the theory. I develop that idea here in a general way, and expand upon things I have written earlier. I begin, in the first section, with some informal remarks about the nature of representations. The second section is devoted to the central concept of isomorphism of models of a theory. The third section focuses on some simple but significant representation theorems. The fourth section considers some elementary measurement examples of numerical representation. The fifth section concerns an important formal result, that of the representation of any partial recursive function by a program for an unlimited register machine or a universal Turing machine. The sixth section, partially historical in character, turns to philosophical views of mental representations and the extent to which they require some notion of isomorphism.

3.1 Kinds of Representation

A *representation* of something is an image, model, or reproduction of that thing. References to representations are familiar and frequent in ordinary discourse.[1] Some typical instances are these:

Sleep is a certain image and representation of death.

The Play is a representation of a world I once knew well.

---

[1] Other meanings of representation will not be analyzed here, even though a close affinity can be found for many of them, as in *The representation of the union approached management yesterday*. A familiar contrast between objective and subjective representations is nicely drawn in G. H. Lewes' review (1847) of Charlotte Brontë's novel *Jane Eyre* (1847/1971) "In her delineation of the country-houses and good society there is the ease and accuracy of one who has well known what she describes. ...This faculty for objective representation is also united to a strange power of subjective representation. We do not simply mean the power over the passions–the psychological intuition of the artist, but the power also of connecting external appearances with internal effects–of representing the psychological interpretation of material phenomena. This is shewn in many a fine description; but we select that of the punished child shut up in the old bedroom, because it exhibits at the same time the power we speak of, and the power before-mentioned of representing the material aspects of things."

It is the very representation of heaven on earth.

The representation of Achilles in the painting was marvelous.

This is a representation of the triumphal arch erected by Augustus.

An intuitive and visual representation of nuclear forces is not possible.

In some cases we can think of a representation as improving our understanding of the object represented. Many of us certainly understand the proportions of a building better—especially the layout of the interior—, after examining its architectural drawings.

The formal or mathematical theory of representation has as its primary goal such an enrichment of the understanding, although there are other goals of representation of nearly as great importance—for instance, the use of numerical representations of measurement procedures to make computations more efficient. Representation, in the formal sense to be developed here, has also been closely associated with reduction. An admirable goal accepted on almost all sides is to reduce the unknown to the known. Controversies arise when claims about reduction are ideological rather than scientific in character. It is not usually appreciated how involved and technical the actual reduction of one part of science—even a near neighbor—is to another.

Philosophical claims about the reduction—and thus representation—, of one kind of phenomenon or set of ideas by another are as old as philosophy itself. Here is Epicurus' reduction of everything to simple bodies, i.e., atoms, and space:

> Moreover, the universe is bodies and space: for that bodies exist, sense itself witnesses in the experience of all men, and in accordance with the evidence of sense we must of necessity judge of the imperceptible by reasoning, as I have already said. And if there were not that which we term void and place and intangible existence, bodies would have nowhere to exist and nothing through which to move, as they are seen to move. And besides these two nothing can even be thought of either by conception or on the analogy of things conceivable such as could be grasped as whole existences and not spoken of as the accidents or properties of such existences. Furthermore, among bodies some are compounds, and others those of which compounds are formed. And these latter are indivisible and unalterable. (Epicurus, Oates edition, p. 4)

This passage from Epicurus, written about 300 BC, is nearly duplicated in several places in Lucretius' long poem *De Rerum Natura* (Oates 1940, pp. 69–219) written about 250 years later. The reduction of all phenomena to the motion of atoms in the void was a central theme of ancient atomism, and the speculative development of the ideas were of significance for the scientific developments that occurred much later.

A claimed reduction much closer to the formal spirit promoted here and one of great importance in the history of ideas is Descartes' reduction of geometry to algebra. He puts the matter this way in the opening lines of his *La Geometrie* (1637/1954, p. 2):

> Any problem in geometry can easily be reduced to such terms that a knowledge of the lengths of certain straight lines is sufficient for its construction. Just as arithmetic consists of only four or five operations, namely, addition, subtraction, multiplication, division and the extraction of roots, which may be considered a kind of division, so in geometry, to find required lines it is merely necessary to add or subtract other lines; or else, taking one line which I shall call unity in order to relate it as closely as possible to numbers, and which arbitrarily, and having given two other lines, to find a fourth line which shall

be to one of the given lines as the other is to unity . . .

The difference between these two theses of reduction could hardly be greater in the degree to which they were carried out at the time of their conception. The ancient atomists could establish in a satisfactory scientific sense practically nothing about their reductive thesis. Descartes' detailed mathematical treatment constituted one of the most important conceptual breakthroughs of early modern mathematics. On the other hand, Descartes' attempted reduction of matter to nothing but extension in his *Principles of Philosophy* (1644) was in its way just as speculative as that of Epicurus or Lucretius.

I emphasize that these comparisons are not meant to encourage a reductionistic methodology that asserts we should only talk about reductions that can be fully carried out from a formal standpoint. Nothing could be further from the truth. As an unreconstructed pluralist, I am happy to assign a place of honor to speculation as well as results, especially in view of how difficult it is to establish specific results on reduction for any advanced parts of science. We just need to recognize speculation for what it is.

Definitions as representations. Before turning to isomorphism of models of a theory as the general formal approach to representation, it is important to consider explicitly the most significant way to think about the representation of concepts of a theory that is not in terms of its models. This approach is the useful and widespread practice of defining new concepts within a theory in terms of the primitive concepts of the theory.

To be more explicit, the first definition in a theory is a sentence of a certain form which establishes the meaning of a new symbol of the theory in terms of the primitive symbols of the theory. The second definition in a theory is a sentence of a certain form which establishes the meaning of a second new symbol of the theory in terms of the primitive symbols and the first defined symbol of the theory. And similarly for subsequent definitions. The point to be noted is that the definitions in a theory are introduced one at a time in some fixed sequence. Because of this fixed sequence we may always speak meaningfully of *preceding* definitions in the theory. Often it is convenient to adopt the viewpoint that any defined symbol must be defined in terms only of the primitive symbols of the theory. In this case there is no need to introduce definitions in some fixed sequence. However, the common mathematical practice is to use previously defined symbols in defining new symbols; and to give an exact account of this practice, a fixed sequence of definitions is needed.

From the standpoint of the logic of inference a definition in a theory is simply regarded as a new axiom or premise. But it is not intended that a definition shall strengthen the theory in any substantive way. The point of introducing a new symbol is to facilitate deductive investigation of the structure of the theory, but not to add to that structure. Two criteria which make more specific these intuitive ideas about the character of definitions are that (i) a defined symbol should always be eliminable from any formula of the theory, and (ii) a new definition does not permit the proof of relationships among the old symbols which were previously unprovable; that is, it does not function as a creative axiom.[2]

---

[2]See Suppes (1957/1999, Ch. 8) for a detailed exposition of the theory of definition. A related question, discussed in the same chapter, concerns the possible definability of one primitive concept of

A trivial but ubiquitous example of such representation by definition is the definition in terms of the weak inequality $\succeq$ of the strict order $\succ$, the equivalence $\approx$, and the reverse inequality $\preceq$.

The great successful example of representation by definition is the defining of all standard mathematical concepts within set theory, beginning with just the single primitive concept of set membership. (For this development, see Suppes (1960b).)

3.2 Isomorphism of Models

One of the most general and useful set-theoretical notions that may be applied to a theory is the concept of two models or structures of a theory being isomorphic. Roughly speaking, two models of a theory are isomorphic when they exhibit the same structure from the standpoint of the basic concepts of the theory. The point of the formal definition of isomorphism for a particular theory is to make this notion of *same structure* precise. It is to be emphasized, however, that the definition of isomorphism of models of a theory is not dependent on the detailed nature of the theory, but is in fact sufficiently independent often to be termed 'axiom free'. The use of the phrase 'axiom free' indicates that the definition of isomorphism depends only on the set-theoretical character of models of a theory. Thus two theories whose models have the same set-theoretical character, but whose substantive axioms are quite different, would use the same definition of isomorphism.[3]

These ideas may be made more definite by giving the definition of isomorphism for algebras as defined in Chapter 2. Here a structure $(A, \circ, e, ^{-1})$ is an *algebra* if A is a nonempty set, $\circ$ is a binary operation from $A \times A$ to A, e is an element of A, and $^{-1}$ is a unary operation from A to A.

DEFINITION 1. *An algebra* $\mathfrak{A} = (A, \circ, e, ^{-1})$ *is isomorphic to an algebra* $\mathfrak{A}' = (A', \circ', e', ^{-1'})$ *if and only if there is a function* f *such that*

(i) *the domain of* f *is* A *and the range of* f *is* A',

(ii) f *is a one-one function*,

(iii) *if* a *and* b *are in* A, *then* $f(a \circ b) = f(a) \circ' f(b)$,

(iv) *if* a *is in* A, *then* $f(a^{-1}) = f(a)^{-1'}$,

(v) $f(e) = e'$.

---

a theory in terms of the others. Padoa's principle (1902) provides a clear intuitive method for proving independence, i.e., nondefinability of a concept, in terms of the other concepts of the theory. The principle is simple: to prove the independence of a given concept, give two models of the axioms of the theory such that the given primitive concept is different in the two models, but the remaining primitive concepts are the same in both models.

[3]The etymological derivation of *isomorphism* is from the Greek meaning *same form*. As I argue in the final section of this chapter (3.6) the concept of form in Aristotle, and also in Plato, is very close to that of structure used here. It is very natural to say that two structures that are isomorphic have the same form. Moreover, the insistence on making clear exactly what structural part of one entity is isomorphic to that of some other entity corresponds rather closely to Aristotle's relative concept of form–the house is the form of the matter, the bricks out of which it is built, but a brick is the form of the matter out of which it is shaped. From a more technical and abstract viewpoint, we could say that, extensionally, the form of a structure, in the sense used here, is the class of all structures isomorphic to it, where we may want to restrict this class to some given domain, to avoid possible paradox.

3.2 Isomorphism of Models

When we ask ourselves whether or not two distinct objects have the same structure, we obviously ask relative to some set of concepts under which the objects fall. It is an easy matter to show that the relation of isomorphism just defined is an equivalence relation among algebras, i.e., it is reflexive, symmetric, and transitive. As a rather interesting example, we might consider two distinct but isomorphic groups which have application in the theory of measurement. Let one group be the additive group of integers. In this case, the set A is the set of all integers, the operation $\circ$ is the operation of addition, the identity element e is 0, and the inverse operation $^{-1}$ is the negative operation. As the second group, isomorphic to the first, consider the multiplicative group of all integer powers of 2. In this case, the set A' is the set of all numbers that are equal to 2 to some integer power, the operation $\circ'$ is the operation of multiplication, the identity element is the integer 1, and the inverse operation is the standard reciprocal operation, i.e., the inverse of x is $1/x$. To establish the isomorphism of the two groups $\mathfrak{A} = (A, +, 0, -)$ and $\mathfrak{A}' = (A', \cdot, 1, ^{-1})$, we may use the function f such that for every integer n in the set A

$$f(n) = 2^n.$$

Then it is easy to check that the range of f is A', that f is one-one, and

$$f(m \circ n) = f(m+n) = 2^{m+n} = 2^m \cdot 2^n = f(m) \cdot f(n) = f(m) \circ' f(n)$$

$$f(n^{-1}) = f(-n) = 2^{-n} = \frac{1}{2^n} = f(n)^{-1'},$$

and

$$f(0) = 2^0 = 1.$$

It should be apparent that the same isomorphism between additive and multiplicative groups is possible if we let the set of objects of the additive group be the set of all real numbers, positive or negative, and the set of objects of the multiplicative group be the set of all positive real numbers. From the standpoint of the theory of measurement, this isomorphism is of interest primarily because it means that there is no mathematical basis for choosing between additive and multiplicative representations. Standard discussions of extensive quantities, for example, those concerning the measurement of mass or distance, often do not emphasize that a multiplicative representation is as acceptable and correct as an additive representation. Because measurements of mass or distance are never negative, it may be thought that the remarks about groups do not apply precisely, for the additive groups considered all have negative numbers as elements of the group. The answer is that in considering the actual measurements of mass or distance, we restrict ourselves to the semigroup of positive elements of the additive group in question. However, the details of this point are not relevant here. Concerning the earlier remark that isomorphism or sameness of structure is relative only to a given set of concepts, note that the integers and the multiplicative group of powers of two differ in many number-theoretical properties.

As another simple example of a theory axiomatized by defining a set-theoretical predicate, we may consider the ordinal theory of measurement discussed in Chapter 2. Models of this theory are customarily called weak orderings and we shall use this terminology in defining the appropriate predicate. The set-theoretical structure of models of this theory is a nonempty set A and a binary relation R defined on this set. Let us call such a couple $\mathfrak{A} = (A, R)$ a *simple relation structure*. In the style of Definition 1 of Chapter 2, we then have the following.[4]

DEFINITION 2. *A simple relation structure $\mathfrak{A} = (A, R)$ is a weak ordering if and only if for every a, b, and c in A*

(i) *if aRb and bRc then aRc,*

(ii) *aRb or bRa.*

The definition of isomorphism of simple relation structures should be apparent, but for the sake of explicitness I give it anyway, and emphasize once again that the definition of isomorphism depends only on the set-theoretical structure of the simple relation structures and not on any of the substantive axioms imposed.

DEFINITION 3. *A simple relation structure $\mathfrak{A} = (A, R)$ is isomorphic to a simple relation structure $\mathfrak{A}' = (A', R')$ if and only if there is a function f such that*

(i) *the domain of f is A and the range of f is A',*

(ii) *f is a one-one function,*

(iii) *if a and b are in A then aRb if and only if $f(a) \, R' \, f(b)$.*

To illustrate this definition of isomorphism let us consider the question, "Are any two finite weak orderings with the same number of elements isomorphic?" Intuitively it seems clear that the answer should be negative, because in one of the weak orderings all the objects could stand in the relation R to each other and not so in the other. What is the counterexample with the smallest domain we can construct to show that such an isomorphism does not exist? It is clear at once that two one-element sets will not do, because within isomorphism there is only one weak ordering with a single element, namely, the ordering that makes that single element stand in the given relation R to itself. However, a counterexample can be found by adding one more element. In one of the weak orderings we can let R be the universal relation, i.e., $R = A \times A$, the Cartesian product of A with itself, and in the other, let R' be a 'minimal' relation satisfying the axioms for a weak ordering. More formally, let

$$A = \{1, 2\}$$
$$R = \{(1,1), (2,2), (1,2), (2,1)\}$$
$$A' = A$$
$$R' = \{(1,1), (2,2), (1,2)\}.$$

Then it is easily checked that $\mathfrak{A} = (A, R)$ and $\mathfrak{A}' = (A', R')$ are both weak orderings with domains of cardinality two, but $\mathfrak{A}$ cannot be isomorphic to $\mathfrak{A}'$. For suppose there were a function f establishing such an isomorphism. Then we would have

$$1 \, R \, 2 \text{ if and only if } f(1) \, R' \, f(2)$$

---

[4]For analysis of isomorphism I use the general binary relation symbol 'R' here, rather than the suggestive symbol '$\succeq$' introduced in Section 2 of Chapter 2. Later, where convenient, I also use '$\preceq$', for the converse of $\succeq$.

and
$$2 \; R \; 1 \text{ if and only if } f(2) \; R' \; f(1),$$
but we also have $1 \; R \; 2$ and $2 \; R \; 1$, whence
$$f(1) \; R' \; f(2) \text{ and } f(2) \; R' \; f(1), \tag{1}$$
but this is impossible, for if $f(1) = 1$, then $f(2) = 2$, and thus from (1) $2 \; R' \; 1$, but we do not have $2 \; R' \; 1$. On the other hand, as the only other possible one-one function, if $f(1) = 2$ then $f(2) = 1$, and again we must have from (1) $2 \; R' \; 1$, contrary to the definition of R'.

3.3 Representation Theorems

In attempting to characterize the nature of the models of a theory, the notion of isomorphism enters in a central way. Perhaps the best and strongest characterization of the models of a theory is expressed in terms of a significant representation theorem. As outlined informally earlier, by a *representation theorem* for a theory the following is meant. A certain class of models of a theory, distinguished for some intuitively clear conceptual reason, is shown to exemplify within isomorphism every model of the theory. More precisely, let $\mathfrak{M}$ be the set of all models of a theory, and let $\mathfrak{B}$ be some distinguished subset of $\mathfrak{M}$. A representation theorem for $\mathfrak{M}$ with respect to $\mathfrak{B}$ would consist of the assertion that given any model M in $\mathfrak{M}$ there exists a model in $\mathfrak{B}$ isomorphic to M. In other words, from the standpoint of the theory every possible variation of model is exemplified within the restricted set $\mathfrak{B}$. It should be apparent that a trivial representation theorem can always be proved by taking $\mathfrak{B} = \mathfrak{M}$. A representation theorem is just as interesting as the intuitive significance of the class $\mathfrak{B}$ of models and no more so. An example of a simple and beautiful representation theorem is Cayley's theorem that every group is isomorphic to a group of transformations. One source of the concept of a group, as it arose in the nineteenth century, comes from consideration of the one-one functions which map a set onto itself. Such functions are usually called transformations. It is interesting and surprising that the elementary axioms for groups are sufficient to characterize transformations in this abstract sense, namely, in the sense that any model of the axioms, i.e., any group, can be shown to be isomorphic to a group of transformations. (For a discussion and proof of this theorem, see Suppes 1957/1999, Ch. 12.)

Certain cases of representation theorems are of special interest. When the set $\mathfrak{B}$ is a unit set, i.e., a set with exactly one element, then the theory is said to be categorical. Put another way, a theory is categorical when any two models are isomorphic. Thus, a categorical theory has within isomorphism really only one model. Examples of categorical theories are the elementary theory of numbers when a standard notion of set is used, and the elementary theory of real numbers with the same standard notion of set. It has sometimes been asserted that one of the main differences between nineteenth- and twentieth-century mathematics is that nineteenth-century mathematics was concerned with categorical mathematical theories, while the latter deals with noncategorical theories. It is doubtful that this distinction can be made historically, but there is certainly a rather sharp conceptual difference between working with categorical and noncategorical theories. There is a clear sense in which noncategorical theories are more abstract.

From a psychological standpoint, a good case can probably be made for the view that a theory is regarded as abstract when the class of models becomes so large that any simple image or picture of a typical model is not possible. The range of models is too diverse; the theory is very noncategorical. Another closely related sense of 'abstract' is that certain intuitive and perhaps often complex properties of the original model of the theory have been dropped, as in the case of groups, and we are now prepared to talk about models which satisfy a theory even though they have a much simpler internal structure than the original intuitive model.

Homomorphism of models. In many cases within pure mathematics a representation theorem in terms of isomorphism of models turns out to be less interesting than a representation theorem in terms of the weaker notion of homomorphism. A good example of this sort within the philosophy of science is provided by theories of measurement, and the generalization from isomorphism to homomorphism can be illustrated in this context. When we consider general practices of measurement it is evident that in terms of the structural notion of isomorphism we would, roughly speaking, think of the isomorphism as being established between an empirical model of the theory of measurement and a numerical model. By an *empirical model* we mean a model in which the basic set is a set of empirical objects and by a *numerical model* one in which the basic set is a set of numbers. However, a slightly more detailed examination of the question indicates that difficulties about isomorphism quickly arise. In all too many cases of measurement, distinct physical objects are assigned the same number, and thus the one-one relationship required for isomorphism of models does not hold. Fortunately, this is the only respect in which we must change the general notion, in order to obtain an adequate account for theories of measurement of the relation between empirical and numerical models. The general notion of homomorphism is designed to accommodate exactly this situation. To obtain the formal definition of homomorphism for two algebras or two simple relation structures as previously defined, we need only drop the requirement that the function establishing the isomorphism be one-one. When this function is many-one but not one-one, we have a homomorphism that is not an isomorphism.[5]

These remarks may be made more concrete by considering the theory of weak orderings as a theory of measurement. It is easy to give a simple example of two weak orderings such that the first is homomorphic to the second, but not isomorphic to it. Let

$$A = \{1, 2\}$$
$$R = \{(1,1)(2,2)(1,2)(2,1)\}$$
$$A' = \{1\}$$
$$R' = \{(1,1)\}$$

---

[5] A weaker notion of homomorphism is generally used in algebra. The condition that, e.g., structures (A, R) and (A', R') be homomorphic, with f being the mapping from A onto A', is that if xRy then $f(x)R'f(y)$, rather than if and only if. However, in the theory of measurement and in other applications in the philosophy of science, the definition used here is more satisfactory.

3.3 Representation Theorems

and
$$f(1) = 1$$
$$f(2) = 1.$$

From these definitions it is at once obvious that the weak ordering $\mathfrak{A} = (A, R)$ is homomorphic under the function f to the weak ordering $\mathfrak{A}' = (A', R')$. The point is that we have

$$1 \; R \; 2 \text{ if and only if } f(1) \; R' \; f(2),$$

as well as

$$2 \; R \; 1 \text{ if and only if } f(2) \; R' \; f(1),$$

and both these equivalences hold just because

$$f(1) = f(2) = 1.$$

On the other hand, it is also clear simply on the basis of cardinality considerations that $\mathfrak{A}$ is not isomorphic to $\mathfrak{A}'$, because the set A has two elements and the set A' has one element. It is also evident that $\mathfrak{A}'$ is not homomorphic to $\mathfrak{A}$. This also follows from cardinality considerations, for there is no function whose domain is the set A' and whose range is the set A. As this example illustrates, the relation of homomorphism between models of a theory is not an equivalence relation; it is reflexive and transitive, but not symmetric.

By a *numerical* weak ordering I shall mean a weak ordering $\mathfrak{A} = (A, \leq)$ where A is a set of numbers. The selection of the numerical relation $\leq$ to represent the relation R in a weak ordering is arbitrary, in the sense that the numerical relation $\geq$ could just as well have been chosen, but it is slightly more natural to use $\leq$ if there is a first element but not a last one, as is the case for the set of natural numbers. However, choice of one of the two relations $\leq$ or $\geq$ is the only intuitively sound possibility. The following theorem provides a homomorphic numerical representation theorem for finite weak orderings, and thus makes the theory of finite weak orderings a theory of measurement, because of its numerical representation.

THEOREM 1. *Every finite weak ordering is homomorphic to a numerical weak ordering.*[6]

---

[6]*Proof of Theorem 1.* Let $\mathfrak{A} = (A, \preceq)$ be a finite weak ordering. Probably the simplest approach is first to form equivalence classes of objects in A, with respect to the obvious equivalence relation $\approx$ defined in terms of $\preceq$:

$$a \approx b \text{ if and only if } a \preceq b \; \& \; b \preceq a.$$

Thus, using the standard notation '$[a]$' for equivalence classes, i.e.,

$$[a] = \{b \; : \; b \in A \; \& \; a \approx b\},$$

we first order the equivalence classes according to $\preceq$. Explicitly, we define

$$[a] \preceq^* [b] \text{ if and only if } a \preceq b.$$

It is straightforward to prove that $\preceq^*$ is reflexive, antisymmetrical, transitive, and connected in the set $A/\approx$ of equivalence classes, or, in other words, that it is a simple ordering of $A/\approx$. By hypothesis A is a finite set, and so necessarily $A/\approx$ is finite. Let $[a_1]$ be the first element of $A/\approx$ under the ordering $\preceq^*$, $[a_2]$ the second, ... and $[a_n]$ the last element. Consider now the numerical function g defined on $A/\approx$, defined as follows:

$$g([a_i]) = i \text{ for } i = 1, ..., n.$$

Theorem 1 was restricted to finite weak orderings for good reason; it is false if this restriction is removed. The classic counterexample is the lexicographical ordering of the plane. Let A be the set of all ordered pairs of real numbers, and let the relation $\preceq$ be defined by the equivalence $(x_1, x_2) \preceq (y_1, y_2)$ if and only if $x_1 < y_1$, or $x_1 = y_1$ and $x_2 \leq y_2$. Suppose that there exists a real-valued function f satisfying the equivalence:

$$f(x_1, x_2) \leq f(y_1, y_2) \text{ if and only if } (x_1, x_2) \preceq (y_1, y_2). \tag{1}$$

We fix x_2 and y_2 with $x_2 < y_2$ and define for each x_1:

$$f'(x_1) = f(x_1, x_2)$$

$$f''(x_1) = f(x_1, y_2).$$

In terms of these functions we define the following function g from real numbers to intervals:

$$g(x_1) = [f'(x_1), f''(x_1)].$$

On the assumption that the ordering is lexicographic, g must be one-one since two distinct numbers are mapped into two disjoint intervals. For instance, if $x_1 > x_1'$ then $f'(x_1) = f(x_1, x_2) > f(x_1', y_2) = f''(x_1')$. But it is well known that there can be no one-one correspondence between the uncountable set of real numbers and the countable set of nondegenerate disjoint intervals. Thus no such function g can exist, and a fortiori there can be no function f satisfying (1) for the lexicographic ordering.

The remarks earlier about models of arbitrarily large cardinality of first-order theories also provide a proof that the restriction to finite weak orderings in Theorem 1 is necessary. In the preceding section, one advantage of formulating theories within a set-theoretical framework was brought out in our consideration of the axioms that must be added to those for a weak ordering in order to guarantee that an infinite weak ordering is homomorphic to a numerical weak ordering. It was stated at that time that the necessary and sufficient conditions adequate to guarantee this result had no simple first-order statement. The reason for the difficulty of formulating the necessary conditions within a first-order framework is the need to quantify over infinite subsets of the basic set A of the weak ordering.

In order to state the desired theorem one preliminary notion is needed. Let $\mathfrak{A} = (A, \preceq)$ be a simple relation structure, and let B be a subset of A. Define the strict ordering relation $\prec$ in terms of $\preceq$ by the equivalence

$$a \prec b \text{ if and only if } a \preceq b \ \& \ \text{not} \ b \preceq a.$$

Then we say that B is $\preceq$-*order-dense* in A if and only if for every a and b in A and not in B such that $a \prec b$ there is a c in B such that $a \preceq c$ and $c \preceq b$. Note that the denumerable set of rational numbers is order-dense in the nondenumerable set of all real numbers with respect to the natural numerical ordering $\leq$. This relationship

---

Then g establishes an isomorphism between the ordering $\mathfrak{A}/\approx = (A/\approx, \preceq^*)$ and the numerical ordering $\mathfrak{N} = (N, \leq)$, where N is the set of first n positive integers. (The details of this part of the proof are tedious but obvious.) We then define the numerical function f on A, for every b in A, by:

$$f(b) = i \text{ if and only if } b \in [a_i],$$

i.e., if b is in the ith equivalence class under the ordering $\preceq^*$. The function f establishes a homomorphism between $\mathfrak{A}$ and $\mathfrak{N}$, as desired.

3.3 Representation Theorems

between the denumerable rational numbers and all real numbers is just the one that is necessary and sufficient for the existence of a homomorphism between an infinite and a numerical weak ordering. A minor complication arises in applying the denumerability condition, namely, the elements of the order-dense subset must not be equivalent. This additional condition is made precise in the statement of the theorem.

THEOREM 2. *Let $\mathfrak{A} = (A, \preceq)$ be an infinite weak ordering. Then a necessary and sufficient condition that $\mathfrak{A}$ be homomorphic to a numerical weak ordering is that there is a denumerable subset B of A such that (i) B is $\preceq$-order-dense in A and (ii) no two elements of B stand in the relation $\approx$.*[7]

---

[7] *Proof of Theorem 2.* The proof is related to the classical ordinal characterization of the continuum by Cantor (1895). We do not give all details here, but sketch the main outlines; for some additional details and related theorems, see Sierpenski (1958, Ch. 11).

To prove the sufficiency of the condition, let B be a denumerable subset with properties (i) and (ii). Moreover, if A has end points relative to the ordering we may without loss of generality include them in B. First, we know that there exists a real-valued function f on B, establishing that $\mathfrak{B} = (B, \preceq)$ is homomorphic to a numerical weak ordering, just because B is denumerable (for proof of this see, e.g., Suppes and Zinnes, 1963, Theorem 6). Now by the $\preceq$-order-dense condition on B, each element b of A that is not in B defines a *cut* in B, i.e., the partition of B into two sets, $X = \{a : a \in B \ \& \ a \preceq b\}$ and $Z = \{c : c \in B \ \& \ b \preceq c\}$. Let

$$r_1 = \underset{a \in X}{g.l.b.} \ f(a)$$

and

$$r_2 = \underset{c \in Z}{l.u.b.} \ f(c).$$

We then extend f to b by defining

$$f(b) = \frac{r_1 + r_2}{2}.$$

It is easy to show that the function f thus extended from B to A establishes the homomorphism of $\mathfrak{A}$ to a numerical ordering. For example, if a_1 and a_2 are in A but not in B, then if $a_1 \preceq a_2$, there is a c in B such that $a_1 \preceq c \preceq a_2$ and thus

$$f(a_1) \leq f(c) \leq f(a_2).$$

The remaining details of this part of the argument may be supplied by the reader.

To prove the necessity of (i) and (ii), we assume we have a function f establishing that $\mathfrak{A}$ is homomorphic to a numerical ordering. The set of nonempty intervals I_i of the real numbers with rational end points is a denumerable set, because of the denumerability of the rational numbers. We next construct corresponding intervals J_i of A by taking the inverse image under f of each interval I_i. From each J_i that is not empty we select an element a_i. Since the set of intervals is denumerable, the set X of elements a_i of A is denumerable. Secondly, let Re be the set of real numbers r such that for some b in A, $f(b) = r$ and

$$f(b) - \underset{a \in Y}{l.u.b.} f(a) \succ 0$$

where

$$Y = \{a : a \in A \ \& \ a \preceq b\}.$$

Because the set Re of real numbers defines a set of nonoverlapping intervals of real numbers, it is at most denumerable (of course, Re can be, and in some cases would be, empty). Let X' be the inverse image under f of Re. Then $X \cup X'$ is denumerable. To show that $X \cup X'$ is order-dense in A, let c_1 and c_2 be two elements of A but not in $X \cup X'$ such that $c_1 \prec c_2$. Now if there are no elements of A between c_1 and c_2 then c_1 is in X', contrary to hypothesis. On the other hand, if there are elements between c_1 and c_2 then at least one, say d, will be such that $f(d)$ lies in an interval with rational end points which is nested in the interval $[f(c_1), f(c_2)]$, and thus, $X \cup X'$ is order-dense in A. This completes the proof.

It has already been remarked that there is no direct way to formulate the existence of the denumerable set B, as required in Theorem 2, within first-order logic. Perhaps a still more obvious point, but one that has not been sufficiently recognized within the philosophy of science, is that the proof of Theorem 2 uses a great deal more mathematical apparatus than elementary logic. The restriction of the discussion of theoretical issues in the philosophy of science to the framework of first-order logic has too often meant a restriction of the extent to which the discussion could come into contact with other than highly simplified scientific theories especially constructed for the purpose at hand. An advantage of the set-theoretical formulation of theories is that all the necessary mathematical tools are immediately available for both the formulation of the theory and the deductive analysis of the structure of its models.

Embedding of models. We have seen that the notion of two models being homomorphic is a generalization of the notion of two models being isomorphic. A still more general and therefore weaker relation between models is that of one model being embedded in another. To prove an embedding theorem for a theory is to prove that there is an interesting class $\mathfrak{B}$ of models such that every model of the theory is isomorphic, or at least homomorphic, to a submodel belonging to $\mathfrak{B}$. The exact definition of submodel will vary slightly from one theory to another depending on the set-theoretical character of its models. For example, if $\mathfrak{A} = (A, \circ, e, ^{-1})$ is an algebra as defined above, then an algebra $\mathfrak{A}' = (A', \circ', e', ^{-1'})$ is a subalgebra of $\mathfrak{A}$ if A' is a subset of A, $\circ'$ is the operation $\circ$ restricted to A' (i.e., $\circ' = \circ \cap (A' \times A')$), $e' = e$, and $^{-1'}$ is the operation $^{-1}$ restricted to A'. In the case of simple relation structures the definition is still simpler. Let $\mathfrak{A} = (A, R)$ and $\mathfrak{A}' = (A', R')$ be two such structures. Then $\mathfrak{A}'$ is a submodel of $\mathfrak{A}$ if A' is a subset of A and R' is the relation R restricted to A', i.e., $R' = R \cap (A' \times A')$.

Theorem 1 could have been formulated as an embedding theorem along the following lines. Let Re be the set of real numbers. Then it is apparent at once that $(Re, \leq)$ is a numerical weak ordering as defined earlier, and every finite weak ordering can be homomorphically embedded in $(Re, \leq)$, i.e., is homomorphic to a submodel of $(Re, \leq)$.

At this point I close the general discussion of models and theories.[8] The next two sections of this chapter are devoted to examples of representation theorems, the first for elementary theories of measurement, and the next the important representation of any computable function by a program for an abstract universal computer. The section following these two, the final one of this chapter, returns to a more general topic, in

---

[8]I do add as a final general remark in this footnote that the language of a representation theorem often does not make explicit the nature of the isomorphism, but instead familiar idioms of a particular discipline are used. Here are two examples of what I mean.

(i) "The center of mass of a system of particles moves like a particle that has mass equal to the mass of the system ..." instead of "The center of mass of a system of particles is in its motion isomorphic to the motion of a one-particle system that has mass ..." (Theorem 4 of Section 7.1)

(ii) "The four observables satisfy the Bell inequalities if and only if they have a joint probability distribution", instead of "The given six pairwise distributions of the four observables can be isomorphically embedded in a common joint distribution of the four observables". (Theorem 6 of Section 7.2)

3.4 Representation of Elementary Measurement Structures

In this section I consider some of the simplest nontrivial examples of measurement structures. The basic sets of objects or stimuli will in all cases be finite, and the adequacy of the elementary axioms for various structures depends heavily on this finiteness. In addition to their finiteness, the distinguishing characteristic of the structures considered is that the objects are equally spaced in an appropriate sense along the continuum, so to speak, of the property being measured. The restrictions of finiteness and equal spacing enormously simplify the mathematics of measurement, but it is fortunately not the case that the simplification is accompanied by a total separation from realistic empirical applications. Finiteness and equal spacing are characteristic properties of many standard scales, for example, the ordinary ruler, the set of standard weights used with an equal-arm balance in the laboratory or shop, or almost any of the familiar gauges for measuring pressure, temperature, or volume.

Four kinds of such structures, and their representations, are dealt with. Each of them corresponds to a more general set of structures analyzed in the comprehensive treatise of Krantz, Luce, Suppes and Tversky (1971). The four kinds of structures are for extensive, difference, bisection, and conjoint measurement. The analysis given here is taken from Suppes (1972).

Extensive measurement. The distinction between extensive and intensive properties or magnitudes is a very old one in the history of science and philosophy. Extensive magnitudes are ones that can be added; e.g., mass and length are extensive magnitudes or quantities. Intensive magnitudes, in contrast, cannot be added, even though they can be measured. Two volumes of gas, e.g., with the same temperature, do not combine to form a gas with twice the temperature. It has been claimed by some theorists, e.g., Campbell (1920, 1928), that fundamental measurement of intensive magnitudes is not possible. The negative arguments of Campbell and others are not at all persuasive, and many examples of measurement structures provide a concrete refutation of Campbell's thesis, including those given later in this section.

I develop the axioms of extensive measurement in this section with three specific interpretations in mind. One is for the measurement of mass on an equal-arm balance, one is for the measurement of length of rigid rods, and one is for the measurement of subjective probabilities.[9] Other interpretations are certainly possible, but I shall restrict detailed remarks to these three.

From a formal standpoint the basic structures are triples $\langle \Omega, \mathcal{F}, \succeq \rangle$ where Ω is a nonempty set, $\mathcal{F}$ is a family of subsets of Ω and the relation $\succeq$ is a binary relation on $\mathcal{F}$. By using subsets of Ω as objects, we avoid the need for a separate primitive concept of concatenation. As a general structural condition, it shall be required that $\mathcal{F}$ be an

---

[9]Because of the extensive use of this probability interpretation in Chapter 5, I adopt below the standard notation 'Ω' for the set of possible outcomes of a probability space as the general notation for extensive measurement in this section.

algebra of sets on Ω, which is just to require that $\mathcal{F}$ be nonempty and be closed under union and complementation of sets, i.e., if A and B are in $\mathcal{F}$ then $A \cup B$ and $-A$ are also in $\mathcal{F}$.

The intended interpretation of the primitive concepts for the three cases mentioned is fairly obvious. In the case of mass, Ω is a set of physical objects, and for two subsets A and B, $A \succeq B$ if and only if the set A of objects is judged at least as heavy as the set B. It is probably worth emphasizing that several different uses of the equal-arm balance are appropriate for reaching a judgment of comparison. For example, if $A = \{a, b\}$ and $B = \{a, c\}$ it will not be possible literally to put A on one pan of the balance and simultaneously B on the other, because the object a is a member of both sets. But we can make the comparison in at least two different ways. One is just to compare the nonoverlapping parts of the two subsets, which in the present case just comes down to the comparison of $\{b\}$ and $\{c\}$. A rather different empirical procedure that even eliminates the need for the balance to be equal arm is to first just balance A with sand on the other pan (or possibly water; but in either case, sand or water in small containers), and then to compare B with this fixed amount of sand. Given the standard meaning of the set-theoretical operations of intersection, union, and complementation, no additional interpretation of these operations is required, even of union of sets, which serves as the operation of concatenation.

In the case of the rigid rods, the set Ω is just the collection of rods, and $A \succeq B$ if and only if the set A of rods, when laid end to end in a straight line, is judged longer than the set B of rods also so laid out. Variations on exactly how this qualitative comparison of length is to be made can easily be supplied.

In the case of subjective probabilities or objective propensities, the set Ω is the set of possible outcomes of the experiment or empirical situation being considered. The subsets of Ω in $\mathcal{F}$ are just events in the ordinary sense of probability concepts, and $A \succeq B$ if and only if A is judged at least as probable as B.

Axioms for extensive measurement, subject to the two restrictions of finitude and equal spacing, are given in the following definition. In the definition, and subsequently, we use the standard definitions for equivalence $\approx$ in terms of a weak ordering and also of a strict ordering. The definitions are just these: $A \approx B$ if and only if $A \succeq B$ and $B \succeq A$; $A \succ B$ if and only if $A \succeq B$, and not $B \succeq A$.

DEFINITION 1. *A structure* $\mathbf{\Omega} = (\Omega, \mathcal{F}, \succeq)$ *is a* finite, equally spaced extensive structure *if and only if Ω is a finite set, $\mathcal{F}$ is an algebra of sets on Ω, and the following axioms are satisfied for every A, B, and C in $\mathcal{F}$:*

1. *The relation $\succeq$ is a weak ordering of $\mathcal{F}$;*
2. *If $A \cap C = \emptyset$ and $B \cap C = \emptyset$, then $A \succeq B$ if and only if $A \cup C \succeq B \cup C$;*
3. *$A \succeq \emptyset$;*
4. *Not $\emptyset \succeq \Omega$;*
5. *If $A \succeq B$ then there is a C in $\mathcal{F}$ such that $A \approx B \cup C$.*

From the standpoint of the standard ideas about the measurement of mass or length, it would be natural to strengthen Axiom 3 to assert that if $A \neq \emptyset$, then $A \succ \emptyset$, but

3.4 ELEMENTARY MEASUREMENT STRUCTURES

because this is not required for the representation theorem and is unduly restrictive in the case of probabilities, the weaker axiom seems more appropriate.

In stating the representation theorem, we use the notion of an *additive measure* μ from $\mathcal{F}$ to the real numbers, i.e., a function μ such that, for any A and B in $\mathcal{F}$,

i. $\mu(\emptyset) = 0$,
ii. $\mu(A) \geq 0$,
iii. if $A \cap B = \emptyset$ then $\mu(A \cup B) = \mu(A) + \mu(B)$.

THEOREM 1. *Let $\mathbf{\Omega} = (\Omega, \mathcal{F}, \succeq)$ be a finite, equally spaced extensive structure. Then there exists an additive measure μ such that for every A and B in $\mathcal{F}$*

$$\mu(A) \geq \mu(B) \text{ if and only if } A \succeq B.$$

Moreover, all singleton sets fall into at most two equivalence classes in $\mathcal{F}$; if there are two, one of these equivalence classes contains the empty set.

The proof of this theorem and some of the preceding discussion is to be found in Suppes (1969a, pp. 4–8).

It is important to point out that, for all three of the empirical interpretations of the primitive concepts used in the theory of extensive measurement characterized in Definition 1, many important empirical procedures required for each interpretation are missing. Students in elementary engineering or physics laboratories are taught many details about measuring mass, or weight if you will, with equal-arm balances, or length with measuring rods or tapes. But even these details are inadequate as an account of fundamental measurement. For such measurement must be used at an earlier stage to validate the weights or rods used by the students. The empirical procedures are much more elaborate for creating a set of standard weights, for example. The axioms of Definition 1 come into play in creating 'equal' standard weights or checking that a given set of standard weights does seem satisfactory. I say 'seem' deliberately, for there is no natural end to the empirical investigation of standard measurement procedures or instrumentation in physics or other sciences. The art of creating good measuring instruments is a complicated and subtle one, certainly so at the frontier of many sciences. Such art, I have argued repeatedly, cannot be adequately described in a manual, but requires an apprenticeship of training, many components of which are nonverbal. Can you imagine, for example, becoming a good tennis player by just reading a book about it or listening to a few lectures? The same is true of any laboratory or manufacturing set of skills required to produce good measurement instruments.[10]

The elementary theory of measurement of different formal kinds set forth in this section is an important formal aspect of measurement, but in no sense provides a full foundation, which, as is evident from what has been said thus far, I believe impossible to formalize fully in an explicit way. But exactly where to stop is not formally clear. We can build and then program a smart robot to do the job of producing standard measuring instruments that satisfy to a good approximation the axioms of Definition 1. We can, of course, formalize the program the robot is given, even though I have reservations on

---

[10] Even Descartes recognized this, as can be seen in his description of how to cut lenses, i.e., make lenses, in the Tenth Discourse of his *Optics* (1967/2001, pp. 162–173).

this point if the robot is taught by the standard method of tracing out its trajectory for each step of construction. But, this objection aside, the design and construction of the robot will call for experience and learning going beyond any verbal manual of instruction. More detailed remarks on these matters are found in an old article of mine on models of data (Suppes 1962). See also the discussion of coordinating definitions in Section 1.2.

Difference measurement. Referring to the distinction between extensive and intensive properties discussed earlier, I could easily make a case for entitling this subsection *intensive measurement*, for it is characteristic of difference measurement that no operation corresponding to addition is present, and no meaningful combination of objects or stimuli is postulated for the difference structures.

As before, the basic set of objects or stimuli will be nonempty and finite, but in the case of difference structures the relation on the set will be a quaternary one. I denote the basic set of objects by A and the quaternary relation by $\succeq$. The idea behind the quaternary relation $\succeq$ is that $ab \succeq cd$ holds when and only when the qualitative (algebraic) difference between a and b is at least as great as that between c and d. In the case of similarity judgments, for example, the relation $\succeq$ would hold when the subject in a psychological experiment judged that the similarity between a and b was at least as great as the similarity between c and d, due account being taken of the algebraic sign of the difference. The inclusion of the algebraic difference requires some care in interpretation; for example, in many similarity experiments a natural algebraic sign is not attached to the similarity. Instances that satisfy the present requirement are judgments of utility or of pitch or of intensity of sound; in fact, any kind of judgments in which the subject will recognize and accept that the judgments naturally lie along a one-dimensional continuum.

We define for the quaternary relation $\succeq$ just as for a binary relation, $\succ$ and $\approx$:

$ab \succ cd$ if and only if not $cd \succeq ab$,

$ab \approx cd$ if and only if $ab \succeq cd$ and $cd \succeq ab$.

It is also convenient to have at hand certain elementary definitions of the binary relation of strict precedence or preference and the relation $\approx$ of indifference or indistinguishability. These definitions are the following.

DEFINITION 2. *$a \succ b$ if and only if $ab \succ aa$.*

DEFINITION 3. *$a \approx b$ if and only if $ab \approx ba$.*

In order to express the equal-spacing part of our assumptions, we need one additional definition, namely, the definition that requires that adjacent objects in the ordering be equally spaced. For this purpose we introduce the definition of the binary relation J. This binary relation is just that of immediate predecessor.

DEFINITION 4. *aJb if and only if $a \succ b$ and for all c in A if $a \succ c$, then either $b \approx c$ or $b \succ c$.*

I now turn to the definition of finite equal-difference structures. The axioms given follow those given by Suppes and Zinnes (1963).

3.4 ELEMENTARY MEASUREMENT STRUCTURES

DEFINITION 5. *A quaternary structure* $\mathfrak{A} = (A, \succeq)$ *is a finite, equally spaced difference structure if and only if A is a finite set and the following axioms are satisfied for every a, b, c, and d in A:*

1. *The relation $\succeq$ is a weak ordering of $A \times A$;*
2. *If $ab \succeq cd$, then $ac \succeq bd$;*
3. *If $ab \succeq cd$, then $dc \succeq ba$;*
4. *If aJb and cJd, then $ab \approx cd$.*

Keeping in mind the empirical interpretations mentioned already, it is easy to grasp the intuitive interpretation of each axiom. The first axiom just requires that the quaternary relation $\succeq$ be a weak ordering in terms of the qualitative difference between objects or stimuli. Axiom 2 is the most powerful and fundamental axiom in many ways. It expresses a simple necessary property of the intended interpretation of the relation $\succeq$. Axiom 3 just expresses a necessary algebraic fact about the differences. Notice that Axioms 1–3 are necessary axioms. Only Axiom 4 is sufficient but not necessary; it relates J to the quaternary relation $\approx$. The intuitive idea of this axiom is that if a stands in the relation J to b, and c stands in the relation J to d, then the difference between a and b is judged to be the same as the difference between c and d, due account being taken of algebraic sign.

From these four axioms we can prove the following representation theorem.

THEOREM 2. *Let $\mathfrak{A} = (A, \succeq)$ be a finite, equally spaced difference structure. Then there exists a real-valued function φ on A such that for every a, b, c, and d in A*

$$\varphi(a) - \varphi(b) \geq \varphi(c) - \varphi(d) \quad \text{if and only if} \quad ab \succeq cd.$$

The proof of this theorem is given at the end of the section. In addition, a number of elementary properties are organized in a series of elementary lemmas leading up to the proof of the theorem.

Upon casual inspection it might be supposed that the first three axioms of Definition 5 would characterize all finite-difference structures for which a numerical representation could be found. However, Scott and Suppes (1958) showed that the theory of all representable finite difference structures is not characterized by these three axioms and indeed cannot be characterized by any simple finite list of axioms.

On the other hand, it might be thought that with the addition of the non-necessary Axiom 4 it would be difficult to satisfy the axioms, because an arbitrary collection of stimuli or objects would not. However, if the stimuli being studied lie on a continuum, then it will be possible to select a standard sequence that will satisfy the axioms, just as is done in the case of selecting a standard set of weights for use on an equal-arm balance.

Bisection measurement. Relational structures closely related to the finite difference structures are bisection structures $\mathfrak{A} = (A, B)$ where B is a ternary relation on the finite set A with the interpretation that $B(a, b, c)$ if and only if b is the midpoint of the interval between a and c. The method of bisection has a long history in psychophysics, but it is important to emphasize that satisfaction of the axioms given below requires no assumptions of an underlying physical measurement. All we need is the intuitive idea of a qualitative continuum, and even that is not needed for formal purposes. After the fundamental psychological measurement in terms of the method of bisection has been made, it is desirable when possible to find a computationally simple psychophysical function relating physical measurements of the same magnitude to the psychological measurements.

The axioms given below for the method of bisection imply a number of checks that should be satisfied before it is asserted that a numerical representing function exists, but these checks have often been ignored in the experimental literature that reports use of the method of bisection. For the simplest set of axioms and definitions, we take both the bisection relation B and the ordering relation $\succeq$ as primitive, but it is easy to eliminate $\succeq$ by definition. We use the binary relation J as defined earlier (Definition 4).

DEFINITION 6. *A structure* $\mathfrak{A} = (A, \succeq, B)$ *is a finite, equally spaced* bisection structure *if and only if the set A is finite and the following axioms are satisfied for every a, a', b, c, and c' in A:*

1. *The relation $\succeq$ is a weak ordering of A;*
2. *If $B(abc)$ and $B(abc')$ then $c \approx c'$;*
3. *If $B(abc)$ and $B(a'bc)$ then $a \approx a'$;*
4. *If $B(abc)$ then $a \succ b$ and $b \succ c$;*
5. *If aJb and bJc then $B(abc)$;*
6. *If $B(abc)$ and $a'Ja$ and cJc' then $B(a'bc')$.*

The intuitive interpretation of the axioms is relatively transparent. The first axiom is already familiar. Axioms 2 and 3 require uniqueness of the endpoints up to equivalence, which clearly separates bisection from betweenness. Axiom 4 relates the ternary bisection relation and the binary ordering relation in a natural way, although it imposes a formal constraint on the bisection relation which would often be omitted. Inclusion of this order property as part of the relation B simplifies the axioms. Axiom 5 is a strong assumption of equal spacing, and Axiom 6 expresses an additional feature of this equal spacing. In view of the axioms given earlier for difference structures, it is somewhat surprising that Axiom 6 can be shown to be independent of Axiom 5, but it is easy to give a model of Axioms 1-5 to show that this is the case. For we can take a model with

$$B(abc) \text{ if and only if } aJb \text{ and } bJc$$

and satisfy all of the first five axioms.

The representation theorem assumes the following form.

THEOREM 3. *Let $\mathfrak{A} = (A, \succeq, B)$ be a finite equally spaced bisection structure. Then there exists a real-valued function φ defined on A such that for every $a, b,$ and c in A*

(i) $\varphi(a) \geq \varphi(b)$ *if and only if $a \succeq b$,*

(ii) $2\varphi(b) = \varphi(a) + \varphi(c)$ *and $\varphi(a) > \varphi(b) > \varphi(c)$ if and only if $B(a, b, c)$.*

The proof of this theorem is given at the end of the section.

3.4 ELEMENTARY MEASUREMENT STRUCTURES

Conjoint measurement. In many kinds of experimental or observational environments, it turns out to be the case that the measurement of a single magnitude or property is not feasible or theoretically interesting. What is of interest is the joint measurement of several properties simultaneously. In this case we consider axioms for additive *conjoint* measurement. The intended representation is that we use ordered pairs of objects or stimuli. The first members of the pairs are drawn from one set and consequently represent one kind of property or magnitude. The second members of the pairs are objects drawn from a second set representing a different magnitude or property. Given the ordered-pair structure, we shall only require judgments of whether or not one pair jointly has more of the 'conjoined' attribute than a second pair.

It is easy to give examples of interpretations for which this way of looking at ordered pairs is natural. Suppose we are asked to judge the capabilities of individuals to assume a position of leadership in an organization. What we are given about the individuals are evaluations of their technical knowledge on an ordinal scale and a charisma measure on an ordinal scale. Thus for each individual we can say how he compares on each scale with any other individual. The problem is to make judgments as between the individuals in terms of their overall capabilities. The axioms given below indicate the kind of further conjoint ordinal conditions that are sufficient to guarantee finite equally spaced conjoint measurement, where in this case the equal spacing is along each dimension.

As a second example, a pair (a, p) can represent a tone with intensity a and frequency p, and the problem is to judge which of two tones sounds louder. Thus the subject judges $(a,p) \succeq (b,q)$ if and only if tone (a,p) seems at least as loud as (b,q). Other examples from disciplines as widely separated as economics and physics are easily given, and are discussed in considerable detail in Krantz, Luce, Suppes and Tversky (1971, Ch. 6).

It is to be stressed that the additive representation sought in this section is a special case. Generalizations of additivity are discussed in the reference just cited. It is also to be noted that the restriction in this section to ordered pairs rather than ordered n-tuples is not essential.

Before turning to the axioms of (additive) conjoint measurement, we need a couple of elementary definitions that permit us to define ordering relations on the individual components. On the basis of the axioms on the ordering relation between pairs, we shall be able to prove that these ordering relations on the components are also weak orderings. In the following elementary definitions A_1 is the set of first components and A_2 the set of second components. Thus, when reference is made to an ordered pair (a, p), it is understood that a is in A_1 and p is in A_2.

DEFINITION 7. *$a \succeq b$ if and only if for all p in A_2, $(a, p) \succeq (b, p)$.*

In terms of this relation we define $a \succ b$ and $a \approx b$ in the usual fashion. Also, a similar definition is needed for the second component.

DEFINITION 8. *$p \succeq q$ if and only if for all a in A_1, $(a, p) \succeq (a, q)$.*

We also use the notation already introduced for the relation $\succeq$ on $A_1 \times A_2$, namely, $(a,q) \succ (b,q)$ if and only if not $(b,q) \succeq (a,p)$, and $(a,p) \approx (b,q)$ if and only if $(a,p) \succeq (b,q)$ and $(b,q) \succeq (a,p)$. The axioms for additive conjoint measurement in the finite, equally spaced case are embodied in the following definition.

DEFINITION 9. *A structure $(A_1, A_2, \succeq)$ is a finite, equally spaced additive conjoint structure if and only if the sets A_1 and A_2 are finite and the following axioms are satisfied for every a and b in A_1 and every p and q in A_2:*

1. *The relation $\succeq$ is a weak ordering on $A_1 \times A_2$;*
2. *If $(a, p) \succeq (b, p)$ then $(a, q) \succeq (b, q)$;*
3. *If $(a, p) \succeq (a, q)$ then $(b, p) \succeq (b, q)$;*
4. *If aJb and pJq then $(a, q) \approx (b, p)$.*

The intuitive content of the four axioms of Definition 9 is apparent, but requires some discussion. Axiom 1, of course, is the familiar requirement of a weak ordering. Axioms 2 and 3 express an independence condition of one component from the other. Thus Axiom 2 says that if the pair (a, p) is at least as great as the pair (b, p) then the same relationship holds when p is replaced by any other member q of A_2, and Axiom 3 says the same thing about the second component. Axiom 4 is sufficient but not necessary. It states the equal-spacing assumption, and corresponds closely to the corresponding axiom for finite, equally spaced difference structures.

It might be thought the monotonicity assumption that if $(a, p) \approx (b, q)$ and $a \succ b$, then $q \succ p$, also needs to be assumed as an axiom, but as we show in the proof of the representation theorem, this additional assumption is not necessary: it can be proved from the first four axioms alone.

The statement of the representation theorem, to which we now turn, assumes exactly the expected form. The only thing to note is that the two real-valued functions on each component are welded together by the same unit. This is reflected by the common change of unit α in the theorem, but a different origin is permitted, as is shown in Chapter 4 (Theorem 4.5.4).

THEOREM 4. *Let $(A_1, A_2, \succeq)$ be a finite, equally spaced additive conjoint structure. Then there exist real-valued functions φ_1 and φ_2 on A_1 and A_2 respectively such that for a and b in A_1 and p and q in A_2*

$$\varphi_1(a) + \varphi_2(q) \geq \varphi_1(b) + \varphi_2(p) \text{ if and only if } (a, q) \succeq (b, p).$$

Proofs of Theorems 2–4.[11]

Proof of Theorem 2. Although the following elementary lemmas are not necessary to give a proof of Theorem 2, they are needed in a completely explicit discussion, and their inclusion will perhaps be useful in organizing the reader's thinking about difference structures, which are not as familiar as extensive structures. Indications of the proofs of the elementary lemmas are given only in a few instances.

All of the lemmas refer to a fixed quaternary structure $\mathfrak{A} = (A, \succeq)$, and the binary relations $\succ$, $\approx$, and J defined earlier.

LEMMA 1. *The relation $\succ$ is asymmetric and transitive on A.*

---

[11] The proofs given, which occupy the remainder of this section, may be omitted without loss of continuity.

3.4 Elementary Measurement Structures

LEMMA 2. *The relation $\approx$ is reflexive, symmetric, and transitive on A.*

LEMMA 3. *Exactly one of the following holds for any a and b in A: $a \succ b$, $b \succ a$, $a \approx b$.*

LEMMA 4. *If $aJ^n b$, then $a \succ b$. (The proofs require use of induction on n in this and most of the following lemmas.)*[12]

LEMMA 5. *If $a \succ b$, then there is a (positive integer) n such that $aJ^n b$.*

LEMMA 6. *If $aJ^n b$ and $aJ^n c$, then $b \approx c$.*

LEMMA 7. *If $aJ^m b$ and $bJ^n c$, then $aJ^{m+n} c$.*

LEMMA 8. *If $aJ^m b$ and $aJ^{m+n} c$, then $bJ^n c$.*

LEMMA 9. *If $aJ^{m+n} b$, then there is a c in A such that $aJ^m c$.*

LEMMA 10. *If $aJ^n b$ and $cJ^n d$, then $ab \approx cd$.*

LEMMA 11. *If $ab \approx cd$ then either there is some n such that $aJ^n b$ and $cJ^n d$, or there is some n such that $bJ^n a$ and $dJ^n c$, or $a \approx b$ and $c \approx d$.*

We turn now to a sketch of the proof of Theorem 2. Let c^* be the first element of A with respect to the ordering $\succ$.[13] Define the numerical function φ on A as follows for every a in A:[14]

$$\varphi(a) = \begin{cases} 1 & \text{if } a \approx c^*, \\ -n+1 & \text{if } c^* J^n a. \end{cases}$$

Then using the elementary lemmas we may prove:

(i) $\varphi(a) > \varphi(b)$ if and only if $a \succ b$;

(ii) $\varphi(a) - \varphi(b) \geq \varphi(d) - \varphi(e)$ if and only if $ab \succeq de$.

Proof of Theorem 3. We begin with the proof of two lemmas. The first corresponds to Lemma 10 in the proof of Theorem 2 and the second to Lemma 11. It should be noted that the lemmas of Theorem 2 which are just about the relations $\succ$ and J also apply here.

LEMMA 12. *If $aJ^n b$ and $bJ^n c$ then $B(abc)$.*

Proof. We proceed by induction. For $n = 1$, we have Axiom 5. Suppose now that our inductive hypothesis holds and we have

(1) $\quad aJ^{n+1} b$ and $bJ^{n+1} c$.

---

[12] In this and subsequent lemmas, as well as in the proof of Theorem 2 and later theorems, the concept of the nth power of the binary relation J is repeatedly used. This concept is defined recursively: $aJ^1 b$ if and only if aJb; $aJ^n b$ if and only if there is a c such that $aJ^{n-1} c$ and cJb.

[13] As the definition of $\varphi(a)$ makes clear, c^* is not unique as a first element. So it is better to say 'let c^* be a first element of A.' So all members of A that are equivalent to c^* in the ordering are first elements. In more technical language they are members of the equivalence class $[c^*] = \{a : a \approx c^* \,\&\, a \in A\}$.

[14] As the definition of $\varphi(a)$ shows, it is actually more natural to use $\preceq$ rather than $\succeq$ throughout this section, so that c^* is a minimum rather than a maximal element of the ordering and then when $c^* J^n a$, $\varphi(a) = n+1$. I have used $\succeq$ here, however, because we used it in the treatise on measurement (Krantz et al. 1971).

Then we know at once from properties of J that there are elements a' and c' in A such that

(2) aJa' and $a'J^n b$,

(3) $bJ^n c'$ and $c'Jc$.

Whence by inductive hypothesis from (2) and (3)

(4) $B(a'bc')$,

and then from (2) and (3) again, as well as (4) and Axiom 6, we infer

$$B(abc)$$

as desired.

LEMMA 13. *If $B(abc)$ then there is an n such that $aJ^n b$ and $bJ^n c$.*

Proof. From the hypothesis of the theorem and Axiom 4 we have

$$a \succ b \text{ and } b \succ c,$$

whence from familiar properties of J, there are m and n such that

$$aJ^m b \text{ and } bJ^n c.$$

Suppose now $m \neq n$; for definiteness and without loss of generality we may suppose that $m < n$. Then there is d such that

$$bJ^m d,$$

whence by Lemma 1

$$B(abd),$$

but by hypothesis $B(abc)$, whence by Axiom 2

$$c \approx d.$$

But then we have

$$bJ^m c \text{ and } bJ^n c,$$

which is impossible, and so we conclude $m = n$, as desired.

Given Lemmas 1 and 2, the proof of the existence of a function φ such that

(i) $\varphi(a) > \varphi(b)$ if and only if $a \succ b$

and

(ii) $\varphi(b) = \frac{1}{2}(\varphi(a) + \varphi(c))$ and $\varphi(a) > \varphi(b) > \varphi(c)$ if and only if $B(a,b,c)$

is similar to the proof of the corresponding part of Theorem 2 and need not be developed in detail.

Proof of Theorem 4. First of all, on the basis of Axioms 1–3 of Definition 9 the following elementary lemmas about the ordering induced on the two components A_1 and A_2 are easily proved.

LEMMA 14. *The relation $\approx$ on A_i, for $i = 1, 2$, is an equivalence relation, i.e., it is reflexive, symmetric, and transitive.*

LEMMA 15. *The relation $\succ$ on A_i, for $i = 1, 2$, is a symmetric and transitive.*

3.4 ELEMENTARY MEASUREMENT STRUCTURES

LEMMA 16. *For a and b in A_1 exactly one of the following is true: $a \approx b$, $a \succ b$, $b \succ a$. For p and q in A_2, exactly one of the following is true: $p \approx q$, $p \succ q$, $q \succ p$.*

We next prove the two lemmas mentioned earlier in the discussion of the axioms of Definition 9.

LEMMA 17. *If $(a,p) \approx (b,q)$ and $a \succ b$ then $q \succ p$.*

Proof. Suppose it is not the case that $q \succ p$. Then by Lemma 3 either $p \approx q$ or $p \succ q$. If $p \approx q$, then $(a,p) \approx (a,q)$, whence by transitivity and the hypothesis of the lemma, $(b,q) \approx (a,q)$, and thus $b \approx a$, which contradicts Lemma 3 and the hypothesis that $a \succ b$. On the other hand, a contradiction also follows from the supposition of the other alternative, i.e., $p \succ q$. For we have $(a,p) \succ (a,q)$, whence by familiar properties of weak orderings and the hypothesis of the lemma, $(b,q) \succ (a,q)$ and thus $b \succ a$, which again contradicts Lemma 3 and the hypothesis that $a \succ b$. Thus, we conclude that from the hypothesis of the lemma it follows that $q \succ p$, as desired.

LEMMA 18. *If $(a,p) \approx (b,q)$ and $p \succ q$ then $b \succ a$.*

Proof. Identical in structure to that for Lemma 4.

We turn next to the proof of Theorem 4. The proof closely resembles that of Theorem 2. Let a_1 be a first element of A_1 with respect to the ordering $\succeq$ on A_1, and let p_1 be a first element of A_2 with respect to the ordering $\succeq$ on A_2. Define, then, the numerical functions φ_1 and φ_2 on A_1 and A_2 as follows (for a in A_1 and p in A_2):

$$\varphi_1(a) = \begin{cases} 1 \text{ if } a \approx a_1, \\ -n+1 \text{ if } a_1 J^n a, \end{cases}$$

$$\varphi_2(p) = \begin{cases} 1 \text{ if } p \approx p_1, \\ -n+1 \text{ if } p_1 J^n p. \end{cases}$$

As in the case of the proof of Theorem 2, it is easy to show:

$$\varphi_1(a) > \varphi_1(b) \text{ if and only if } a \succ b,$$

$$\varphi_2(p) > \varphi_2(q) \text{ if and only if } p \succ q.$$

Moreover, Lemmas 1–9 proved in preparation for the proof of Theorem 2 also hold in the present setting, for they just depend on the binary relations on the components. Of course, for each of these lemmas, there is, strictly speaking, now a pair of lemmas, one for the ordering on each component.

Corresponding to Lemma 10 of this earlier list, we can prove by the same inductive argument, using Axiom 4 of Definition 9 that if $aJ^n b$ and $pJ^n q$ then $(a,q) \approx (b,p)$. Second, we can prove the elementary fact that if $(a,q) \approx (b,p)$ then either (i) there is some n such that $aJ^n b$ and $pJ^n q$, or (ii) there is some n such that $bJ^n a$ and $qJ^n p$, or (iii) $a \approx b$ and $p \approx q$. From these results we prove then the fundamental result that $\varphi_1(a) + \varphi_2(q) = \varphi_1(b) + \varphi_2(p)$ if and only if $(a,q) \approx (b,p)$, which completes the proof of Theorem 4.

3.5 Machine Representation of Partial Recursive Functions[15]

The representation of most mathematical objects as sets with some particular structure is rightly regarded as a great philosophical achievement in the foundations of mathematics at the beginning of the twentieth century. In many ways an even deeper accomplishment several decades later was the proof that seemingly different definitions of computable functions all lead to the same class of functions. Here I restrict the formal developments to showing that any partial recursive function (*partial recursive* being one important formulation of computable) can be represented by a program of an unlimited register machine. Many other equivalent representations are known in the literature.

It is necessary to recognize that two computer programs that compute the same mathematical function can differ in many observable ways: the space or time needed for the computation, the number of lines or characters in the programs, etc. As in a variety of other representations considered in this and other chapters, structural isomorphism need not imply, and usually does not imply, observational or computational equivalence in all respects.

Before giving a formal definition of effective calculability or partial recursiveness, let us consider some examples of functions which are effectively calculable.[16] We may begin with addition. The recursive definition of addition in terms of the successor function S is given by the following pair of equations for the function f of addition. The recursion is on n.

$$f(0, n) = n$$

$$f(S(n), m) = S(f(n, m)).$$

For example, in the usual notation for numerals,

$$S(0) = 1,$$

$$S(1) = 2,$$

etc., where S is the successor function. Then in a finite number of steps we can find the answer to a sum such as $f(3, 2)$, or $3 + 2$.

$$\begin{aligned} f(3,2) &= S(f(2,2)) \\ &= SS(f(1,2)) \\ &= SSS(f(0,2)) \\ &= SSS(2) \\ &= SS(3) \\ &= S(4) \\ &= 5. \end{aligned}$$

---

[15]This rather technical section may be omitted without substanial loss of continuity, except for Sections 8.2 and 8.3. The general reader may find it useful to look at the first two paragraphs, for the clarification of the concept of being computable is one of the great twentieth-century achievements in the foundations of mathematics.

[16]The class of partial recursive functions was first characterized, but in different ways, by Church (1936), Kleene (1936), Post (1936), Turing (1936), Markov (1951) and others in the same period.

3.5 MACHINE REPRESENTATION OF PARTIAL RECURSIVE FUNCTIONS

The second function in the sequence of primitive recursive functions would most naturally be multiplication, μ.

$$\mu(n, 0) = 0$$
$$\mu(m, S(n)) = f(m, \mu(m, n)).$$

The primitive recursive functions will not be formally defined here, for the formal definition is rather complicated, but they are the most important subset of partial recursive functions, being total functions where recursion is elementary. Several other examples of primitive recursive functions follow. The first is exponentiation.

$$p(n, 0) = 1$$
$$p(m, S(n)) = \mu(p(m, n), m).$$

The factorial function $n!$ is just:

$$0! = 1$$
$$(n+1)! = (n+1)n!.$$

Arithmetic subtraction of 1 is designated: $n \dot{-} 1$. It is recursively defined by:

$$0 \dot{-} 1 = 0$$
$$S(n) \dot{-} 1 = n$$

and, in general,

$$m \dot{-} 0 = m$$
$$m \dot{-} S(n) = (m \dot{-} n) \dot{-} 1.$$

The iteration, or generalized exponential, function, due to Ackermann, provides an example of a function which is not primitive recursive but partial recursive.

$$f(0, 0, n) = n$$
$$f(0, m+1, n) = f(0, m, n) + 1$$
$$f(1, 0, n) = 0$$
$$f(p+2, 0, n) = 1$$
$$f(p+1, m+1, n) = f(p, f(p+1, m, n), n).$$

Note that

$$f(0, m, n) = n + m$$
$$f(1, m, n) = n \cdot m$$
$$f(2, m, n) = n^m.$$

A second function, which is not primitive recursive, but is partial recursive, is defined

by the following three equations:
$$U(0, n) = n + 1$$
$$U(m + 1, 0) = U(m, 1)$$
$$U(m + 1, n + 1) = U(m, U(m + 1, n)).$$

But we shall not build up the apparatus to prove this fact.

DEFINITION 1. *A function whose arguments and values are natural numbers is* partial recursive *iff it can be obtained from the functions of I, II and III by a finite number of applications of IV, V and VI:*

I. $S(x_1) = x_1 + 1$ (*Successor Function*)
II. $0^n(x_1, \ldots, x_n) = 0$ (*Constant Function*)
III. $U_i^n(x_1, \ldots, x_n) = x_i$ (*Projection Functions*)
IV. *If* $h, g_1, \ldots, g_m$ *are partial recursive, so is the function* f *such that*
$$f(x_1, \ldots, x_n) = h(g_1(x_1, \ldots, x_n), \ldots, g_m(x_1, \ldots, x_n)) \quad (Composition)$$
V. *If* g, h *are partial recursive, so is the function* f *such that*
$$f(0, x_2, \ldots, x_n) = g(x_2, \ldots, x_n)$$
$$f(z + 1, x_2, \ldots, x_n) = h(z, f(z, x_2, \ldots, x_n), x_2, \ldots, x_n)) \quad (Primitive\ Recursion)$$
VI. *Let* μy *mean 'the least* y *such that'. If* g *is partial recursive so is the function* f *such that*
$$f(x_1, \ldots, x_n) = \mu y[g(x_1, \ldots, x_n, y) = 0],$$
where if no such y *exists,* f *is undefined for the given arguments.*

The concept of *computable* function of natural numbers is now identified with that of partial recursive function. As a result of work of Church, Kleene, Markov, Post and Turing, mentioned earlier, and others, many equivalences are known.

Rather than use universal Turing machines (UTMs) directly for the representation of partial recursive functions we shall first use the more easily understandable and intuitive unlimited register machines (URMs) introduced by Shepherdson and Sturgis (1963). I follow their development.

Unlimited register machines (URM). I begin with some notation: $<n>$ is the content of register n before carrying out an instruction; $<n'>$ is the content after carrying out an instruction.

DEFINITION 2. *A URM has (i) a denumerable sequence of registers numbered* $1, 2, 3, \ldots$, *each of which can store any natural number, but each program uses only a finite number of registers and (ii) six basic instructions:*

(a) $P(n):$ *add* 1, $<n'> \ = \ <n> +1,$
(b) $D(n):$ *subtract* 1, $<n'> \ = \ <n> -1$ *if* $<n> \neq 0,$
(c) $0(n):$ *clear register*, $<n'> \ = \ 0,$ *i.e., only the number* 0 *is stored in the register.*
(d) $C(m, n):$ *copy from register* m *to* n, $<n'> \ = \ <m>,$
(e) $J[E1]:$ *jump to exit* 1,

3.5 MACHINE REPRESENTATION OF PARTIAL RECURSIVE FUNCTIONS

(f) $J(m)[E1]$: *jump to exit* 1 *if* $<m> = 0$.[17]

In programs we use (e) and (f) in form J[3]: jump to line (3). If jump is to a nonexistent line, machine stops. Thus E1 is a variable for line number to jump to.

As an example, the following program puts 'half' of a number stored in register 1 in register 2, and the remainder in register 3.

1. $J(1)[8]$ Go to line 2 if $<1> \neq 0$. Else to line 8 (i.e., stop)
2. $D(1)$ Subtract 1 from number in register 1
3. $P(2)$ Add 1 to number in register 2
4. $J(1)[8]$
5. $D(1)$
6. $P(3)$
7. $J[1]$ Jump to line 1

We now define *lines* and *programs* for a URM.

DEFINITION 3. *A* line (*of a program of a URM*) *is either an ordered couple consisting of a natural number* $n \geq 1$ (*the line number*) *and one of the instructions* $(a) - (d)$, *or an ordered triple consisting of a natural number* $n \geq 1$, *one of the instructions* (e) *or* (f), *and a natural number* $m \geq 1$.

DEFINITION 4. *A* program (*of a URM*) *is a finite sequence of* l *lines such that*

(i) *the first member of the* ith *line is* i,

(ii) *the numbers* m *that are third members of lines are such that* $1 \leq m \leq l+1$.

We also refer to programs as *routines*. How a URM *follows* a program is intuitively obvious and will not be formally defined.

Subroutines are defined like programs, except

(i) subroutines may have several exits,

(ii) third members of triples may range over $E1, \ldots, Ek$, with these variables being assigned values in a given program.

Example of subroutine $\bar{0}(n)[E1]$: Clear register n and proceed to exit E1.

1. $J(n)[E1]$
2. $D(n)$
3. $J[1]$

The dispensability of subroutines is easily proved.

DEFINITION 5. *Let* f *be a number-theoretic function of* n *arguments. Then* f *is computable by a URM iff for every set of natural numbers* $\{x_1, \ldots, x_n, y, N\}$ *with* $y \neq x_i$ *for* $i = 1, \ldots, n$, *and* $x_1, \ldots, x_n, y \leq N$ *there exists a routine* $R_N(y = f(x_1, \ldots, x_n))$ *such that if* $<x_1>, \ldots, <x_n>$ *are the initial contents of registers* $x_1 \ldots, x_n$ *then*

(i) *if* $f(<x_1>, \ldots, <x_n>)$ *is undefined the machine will not stop;*

---

[17]Note that it is convenient to have 'empty register' mean that only 0 is stored in it, because if no number at all is stored, special definitions must be given for $<n> -1$ and other expressions.

(ii) if $f(<x_1>,\ldots,<x_n>)$ is defined, the machine will stop with $<y>$, the final content of register y, equal to $f(<x_1>,\ldots,<x_n>)$, and with the final contents of all registers $1, 2, \ldots, N$, except y, the same as initially.

Because the proof of the main theorem consists of writing subroutines for each of the six schemata used to define the set of partial recursive functions, some more remarks about subroutines may be useful.

From a logical standpoint, subroutines are similar to definitions in a theory. The mathematical development of set theory just in terms of the primitive concept of membership would be not only intolerable but practically impossible. The same remark applies to programming. To write really complex programs in bare machine language is intolerable. From a formal standpoint the theory of definitions is thoroughly developed, and it is easy to give rules of definition that guarantee an absence of problems arising from idiosyncrasies or unsuspected peculiarities in the definitions introduced. Good definitions should satisfy the criteria of noncreativity and eliminability. The criterion of noncreativity requires that it not be possible to prove any theorem formulated in the original notation only because of the introduction of the definition. Here is a simple example to illustrate the need for this criterion. We take as the single axiom of our theory just the axiom of associativity:

$$(1) \qquad x \circ (y \circ z) = (x \circ y) \circ z.$$

We now introduce the 'definition' of the constant e:

$$(2) \qquad xoe = x.$$

From this definition, but not from the axioms above, we can prove

$$(\exists y)(\forall x)(x \circ y = x),$$

and thus the proposed definition is unsatisfactory because it violates the criterion of noncreativity. Intuitively it is clear that (2) is unsatisfactory as a definition in conjunction with (1) because it adds conceptual content to the theory.

In similar fashion a subroutine must be noncreative. Any function that is computable by a program using the subroutine must be computable by a program that does not use it.

The criterion of eliminability demands that in all contexts a defined symbol be eliminable in terms of previously defined or primitive symbols, and, of course, that an equivalence hold between the statement using the symbol and the corresponding statement from which the symbol is eliminated. (For a more detailed discussion of these two criteria of definition, one reference is Suppes (1957/1999, Ch. 8.))

Here is an example of a pair of subroutines that do not satisfy the criterion of eliminability.

$$
\begin{array}{ll}
0_1(n): & 0_2(n): \\
\quad 1. \ 0_2(n) & \quad 1. \ 0_1(n) \\
\quad 2. \ P(n) & \quad 2. \ D(n)
\end{array}
$$

In other words, subroutine $0_2(n)$ is defined in terms of $0_1(n)$, and $0_1(n)$ is defined in terms of $0_2(n)$, so that we have a vicious circle with no way of defining or writing a program for either $0_1(n)$ or $0_2(n)$ in terms of the six basic instructions. Circular subroutines, just like circular definitions, are prohibited by requiring that the subroutines be introduced in a fixed linear order, and the nth subroutine can use in its program only basic instructions and the preceding $n-1$ subroutines. (As is apparent, this linear requirement can be weakened.)

We turn now to the representation theorem.

THEOREM 1. (Machine Representation). *A number-theoretic function is partial recursive iff it is computable by a URM.*

Proof. The proof that functions computable by a URM are partial recursive is not difficult: check that each line of a program causes a change in registers given by a partial recursive function and then show that the sequence of changes is partial recursive because of closure properties of the set of partial recursive functions.

The argument the other way is the interesting part, because of the great simplicity of instructions of a URM. The proof, due to Shepherdson and Sturgis (1963), is simpler than the corresponding proof for Turing machines. It just consists of writing URM subroutines for Schemata I-VI of Definition 1 that define partial recursive functions.

I. Subroutine $R_N(y = S(x))$
   1. $C(x, y)$
   2. $P(y)$

II. Subroutine $R_N(y = 0^n(x_1, \ldots, x_n))$
   1. $0(y)$

III. Subroutine $R_N(y = U_i^n(x_1, \ldots, x_n))$
   1. $C(x_i, y)$

IV. Composition $R_N(y = f(x_1, \ldots, x_n))$ where we use subroutines for g_i, h to define $f : f(x_1, \ldots, x_n) = h(g_1(x_1, \ldots, x_m), \ldots, g_m(x_1, \ldots, x_n))$
   1. $R_{N+1}(N+1 = g_1(x_1, \ldots, x_n))$
   $\vdots$
   m. $R_{N+m}(N+m = g_m(x_1, \ldots, x_n))$
   m + 1. $R_{N+m}(y = h(N+1, \ldots, N+m))$.

VI. Subroutine for $R_N(y = f(x_1, \ldots, x_n))$ where we use subroutine for g to define f by VI: $f(x_1, \ldots, x_n) = \mu y[g(x_1, \ldots, x_n, y) = 0]$
   1. $0(y)$
   2. $R_{N+1}(N+1 = g(x_1, \ldots, x_n, y))$
   3. $J(N+1)[6]$
   4. $P(y)$
   5. $J[2]$

Note that after instruction $1, <y> = 0$. If $<N+1> = 0$ (step 3), then we are done and go to nonexistent line 6. If not, we augment $<y>$ by 1 and go back to line 2 to try again for the least y.

Schema V is somewhat more complicated. There are two matters of notation to deal with. First, we now only number necessary lines–subroutines or jumps. Second, we use a notation for iteration of instructions. If I is a single-exit instruction or subroutine which does not affect register n, then $I^{<n>}$ stands for the result of performing I $<n>$ times and reducing $<n>$, the number in register n, to zero.

V. Subroutine $R_N(y = f(x_1, \ldots, x_n))$ where we use subroutines for g and h to define f by V:
$$f(0, x_2, \ldots, x_n) = g(x_2, \ldots, x_n),$$
$$f(z+1, x_2, \ldots x_n) = h(z, f(z, x_2, \ldots, x_n), x_2, \ldots, x_n) :$$

1. $R_N(y = g(x_2, \ldots, x_n)), 0(N+1)$
2. $\{R_{N+2}(N+2 = h(N+1, y, x_2, \ldots, x_n)), C(N+2, y), P(N+1)\}^{<x_1>}$
3. $C(N+1, x_1)$.

Partial recursive functions over an arbitrary finite alphabet. For a more general setting and to have a formulation not using 'abstract' numbers as arguments, we may define partial recursive functions over a finite alphabet V. V is a finite set, and V^* is the set of all finite sequences of elements of V; in the present context, we shall call the elements of V^* *words*. In the following, variables a_i range over V, and variables $x, y, x_1, \ldots, x_n$, etc., over V^*. Also xa_i is the concatenation of x and a_i.

The Schemata I'-VI' are as follows:

I'. $S_{a_i}(x) = xa_i$

II'. $0^n(x_1, \ldots, x_n) = \emptyset$

III'. $U_i^n(x_1, \ldots, x_n) = x_i$

IV'. = IV (*Composition*).
$$f(x_1, \ldots, x_n) = h(g_1(x_1, \ldots, x_n), \ldots, g_m(x_1, \ldots, x_n))$$

V'. $f(\emptyset, x_2, \ldots, x_n) = g(x_2, \ldots, x_n)$
$f(za_i, x_2, \ldots, x_n) = h_i(z, f(z, x_2, \ldots, x_n), x_2, \ldots, x_n) \qquad i = 1, \ldots, s.$

VI'. $f(x_1, \ldots, x_n) = \mu_i y[g(x_1, \ldots, x_n, y) = \emptyset]$

where $\mu_i y$ means the shortest word composed just of a_i.
We define in the same fashion as before a URM over V.

The basic instructions can be reduced, and also in the previous case, to three:

a') $P_N^{(i)}(n)$: place a_i on the right-hand end of $<n>$,

b') $D_N(n)$: delete the left-most letter of $<n>$ if $<n> \neq \emptyset$,

c') $J_N^{(i)}(n)[E1]$: jump to exit 1 if $<n>$ begins with a_i.

(As before, the parameter N refers to the registers left unchanged except for above operations.) The same representation theorem holds as before for numerical partial functions.

Next, we move toward Turing machines (TMs) by using just one single register (SRM) and storing information or data in the word of V^* being processed.[18] Proof of the adequacy of the reduction is omitted. For punctuation we add , to V. The three instructions of a SRM over $V \cup \{,\}$ are:

---

[18] A more detailed treatment of Turing machines is given in Chapter 8 beginning with Definition 8.2.9. There is also additional use of register machines in Section 8.3.

a") $P^{(i)} : x \to xa_i$ (add a_i to right-hand end of x)

b") $D : a_i x \to x$ (delete first letter)

c") $J^{(i)}[E1]$: jump to exit 1 if x begins with a_i.

Let $V = \{1\}$ and take , as 0. Then the results for a SRM show that any partial recursive function of natural numbers is computable by a machine with a single one-way tape (on which x is printed), two tape symbols 0 and 1, and two heads, a reading head at the left and writing head at the right.

Many other detailed results about register machines are given in Shepherdson and Sturgis (1963). For comparable detailed results on Turing machines there are many sources, e.g., Davis (1958), Rogers (1967) and at a more advanced level Soare (1987).

3.6 Philosophical Views of Mental Representations

Without attempting anything like a detailed and accurate account of the long and complicated history of the concept of mental representation in philosophy, it can be enlightening and relevant to review some of the standard conceptions and issues from Aristotle onward. The main point will not be to assess the correctness or to criticize the adequacy of the analysis given, but to reflect on whether or not there is a notion of isomorphism in the background, as reflected in such concepts as likeness or resemblance. Comment will also be made as to how such notions are tied to representations.

Aristotle. That the defining feature of sense perception is receiving the form of a physical object is, in general terms, the view of perception for which Aristotle is most famous. As he works out the details it is not a bad view at all, even though it is quite obvious from a modern viewpoint that it cannot be entirely correct.

The background of Aristotle's discussion of perception and implicitly, therefore, of mental representation, is his distinction between form and matter, which applies to objects and phenomena of all kinds. For example, the form of an axe is different from the matter that receives that form. This distinction is also relative in the sense that, for example, the matter of a house can be made up of bricks, which, in turn, have their own form and matter, of which they are made up. I defend the continued viability of this general concept of matter in Suppes (1974c).

Aristotle makes use of the distinction between form and matter in his theory of perception. It is the role of the sense organ to have the potential to receive the form but not the matter of a physical object, as described in this passage from the second book of the *De Anima*:

> We must understand as true generally of every sense (1) that sense is that which is receptive of the form of sensible objects without the matter, just as the wax receives the impression of the signet-ring without the iron or the gold, and receives the impression of the gold or bronze, but not as gold or bronze; so in every case sense is affected by that which has colour, or flavour, or sound, but by it, not *qua* having a particular identity, but *qua* having a certain quality, and in virtue of its formula; (2) the sense organ in its primary meaning is that in which this potentiality lies.   (*De Anima*, 424a17–424a25)

So, if in perceiving the candle we receive exactly the form of the candle, but not the matter, then we have a representation of the physical candle that is isomorphic to it in the mind, just because of the sameness of form. The relevant point here is that the notion of the sameness of form corresponds very closely to the concept of isomorphism that I have been arguing for as a central concept of representation. The concept of form catches nicely that of isomorphism in the sense that the form does not thereby refer to all the properties of the candle, but only to the properties perceived by the various senses of sight, touch, etc. This kind of restriction to the properties considered is, as already noted in earlier discussions, characteristic of any workable notion of isomorphism.

Upon reading the above passage and my comments, someone might reflect that this is not a very rich theory of the mental, but only of immediate sense perception. Aristotle, however, goes on to extend the same ideas about forms to the intellect. It is not possible here to discuss all the subtleties involved in his worked-out theory, but the following passage makes clear how the part of the soul that thinks and judges operates in the same way with forms and, thus, the characteristic notion of isomorphism is again applicable:

> Concerning that part of the soul (whether it is separable in extended space, or only in thought) with which the soul knows and thinks, we have to consider what is its distinguishing characteristic, and how thinking comes about. If it is analogous to perceiving, it must be either a process in which the soul is acted upon by what is thinkable, or something else of a similar kind. This part, then, must (although impassive) be receptive of the form of an object, *i.e.*, must be potentially the same as its object, although not identical with it: as the sensitive is to the sensible, so must mind be to the thinkable. ...Hence the mind, too, can have no characteristic except its capacity to receive. That part of the soul, then, which we call mind (by mind I mean that part by which the soul thinks and forms judgements) has no actual existence until it thinks. So it is unreasonable to suppose that it is mixed with the body; for in that case it would become somehow qualitative, *e.g.*, hot or cold, or would even have some organ, as the sensitive faculty has; but in fact it has none. It has been well said that the soul is the place of forms, except that this does not apply to the soul as a whole, but only in its thinking capacity, and the forms occupy it not actually but only potentially.   (*De Anima*, 429a10-18, a23-30)

It is a definite part of Aristotelian thought that the forms are the focus of the contemplation of the intellect but do not exist separate from individual bodies. There is no separate universe of Platonic forms. A very clear statement on this Aristotelian view is made by St. Thomas Aquinas in the following passage.

> I say this, because some held that the form alone belongs to the species, while matter is part of the individual, and not of the species. This cannot be true, for to the nature of the species belongs what the definition signifies, and in natural things the definition does not signify the form only, but the form and the matter. Hence, in natural things the matter is part of the species; not, indeed, signate matter, which is the principle of individuation, but common matter. For just as it belongs to the nature of this particular man to be composed of this soul, of this flesh, and of these bones, so it belongs to the nature of man to be composed of soul, flesh, and bones; for whatever belongs in common to the substance of all the individuals contained under a given species must belong also to the substance of the species.   (*Summa Theologica*, I, Q.75. Art. 4)[19]

---

[19]Translation taken from *Basic Writings of Saint Thomas Aquinas* (1944).

3.6 PHILOSOPHICAL VIEWS OF MENTAL REPRESENTATIONS

Descartes. Without holding to all of the psychological machinery introduced by Aristotle and developed further by Aquinas, Descartes still holds to the scholastic view that the visual perception of an object reflects the form of the object. This comes from the physical fact that 'the light reflected from its body depicts two images of it, one in each of our eyes.' So he is here using natural facts about physiological optics to end up with an account of images or forms in the soul. Here is the best passage I know of in the *Passions of the Soul*.

> Thus, for example, if we see some animal approach us, the light reflected from its body depicts two images of it, one in each of our eyes, and these two images form two others, by means of the optic nerves, in the interior surface of the brain which faces its cavities; then from there, by means of the animal spirits with which its cavities are filled, these images so radiate towards the little gland which is surrounded by these spirits, that the movement which forms each point of one of the images tends towards the same point of the gland towards which tends the movement which forms the point of the other image, which represents the same part of this animal. By this means the two images which are in the brain form but one upon the gland, which, acting immediately upon the soul, causes it to see the form of this animal. ... (Article XXXV)

Descartes, in general, certainly rejects much scholastic doctrine, but in his psychology of vision, very much stays with the natural idea of the soul or the mind having images that reflect just the form of the object seen. This is the core of the Aristotelian doctrine of perception and remains constant in Descartes in spite of many other changes. It is radically changed by Hume.

Hume. On various grounds, in the *Treatise of Human Nature* (1739/1951), Hume argues vigorously against any direct concept of mental representation between objects in the external world and what is in the mind. He has many things to say about these matters. I only will quote and comment upon some key passages.

The first point to emphasize is that the mechanism of the mind for Hume is association, and he proudly, and in many ways, rightly, thinks of association as playing the role in the theory of the mind that gravitation plays in Newton's system of the world. In other words, the causal principle, beyond which it is not possible to seek for explanation, is, in the case of the problem of falling bodies on Earth and of motion of the planets in the heavens, gravitational attraction between bodies. In the mind the mechanism is that of association, both between impressions, which in many cases can be identified with perceptions, and ideas, which are always derived from impressions. There are three characteristics of impressions and ideas which give rise to associations. They are resemblance, contiguity in time or place, and cause and effect. It is important to emphasize that for Hume, the only characteristic giving rise to associations of impression is that of resemblance. So that all inferences, for example, about cause and effect, must be between association of ideas. After naming these three properties, Hume defends resemblance:

> The qualities, from which this association arises, and by which the mind is after this manner convey'd from one idea to another, are three, *viz.* RESEMBLANCE, CONTIGUITY in time or place, and CAUSE and EFFECT. ...

> ...I believe it will not be very necessary to prove, that these qualities produce an association among ideas, and upon the appearance of one idea naturally introduce another. 'Tis plain, that in the course of our thinking, and in the constant revolution of our ideas, our imagination runs easily from one idea to any other that *resembles* it, and that this quality alone is to the fancy a sufficient bond and association. (*Treatise*, p. 11)

The rest of the paragraph makes a similar defense of contiguity and signals the more extensive discussion of cause and effect later.

The important point relevant to the discussion here is the pride of place that Hume gives to resemblance, since only it is a quality, as he puts it, that holds between impressions. And, of course, resemblance is exactly Hume's notion close to that of the modern notion of isomorphism as a basis of representation.

I said earlier that Hume denies a direct resemblance between fragmentary perceptions and the real objects posited to generate those perceptions, but he does defend a realistic notion of real objects and their continued identity in time. His long and complicated argument about these matters is one of the most interesting parts of Book I of his *Treatise*. I want to point out various ways in which he uses the concept of resemblance or isomorphism. The following passage is particularly striking, because he argues for the constancy of our perception of something like the sun or other very stable objects by using the fact that our successive perceptions, though interrupted, resemble each other and justify, therefore, as he argues in detail, the inference to identity across time of the object. Here is the key passage about this use of resemblance.

> When we have been accustom'd to observe a constancy in certain impressions, and have found, that the perception of the sun or ocean, for instance, returns upon us after an absence or annihilation with like parts and in a like order, as at its first appearance, we are not apt to regard these interrupted perceptions as different, (which they really are) but on the contrary consider them as individually the same, upon account of their resemblance. But as this interruption of their existence is contrary to their perfect identity, and makes us regard the first impression as annihilated, and the second as newly created, we find ourselves somewhat at a loss, and are involv'd in a kind of contradiction. In order to free ourselves from this difficulty, we disguise, as much as possible, the interruption, or rather remove it entirely, by supposing that these interrupted perceptions are connected by a real existence, of which we are insensible. This supposition, or idea of continu'd existence, acquires a force and vivacity from the memory of these broken impressions, and from that propensity, which they give us, to suppose them the same; and according to the precedent reasoning, the very essence of belief consists in the force and vivacity of the conception. (*Treatise*, p. 199)

Hume does not stop here. Over several pages he explains four different aspects of this set-up in the inference from interrupted perceptions to identity of objects through time. It is one of the most sustained arguments in the *Treatise* and a wonderful piece of philosophical analysis. He first explains the basis for the principle of the individuation of objects, and he introduces here something that is very much in the spirit of isomorphism and a relevant notion of invariance. This is the final short paragraph on discussing the principle.

3.6 PHILOSOPHICAL VIEWS OF MENTAL REPRESENTATIONS

> Thus the principle of individuation is nothing but the *invariableness* and *uninterruptedness* of any object, thro' a suppos'd variation of time, by which the mind can trace it in the different periods of its existence, without any break of the view, and without being oblig'd to form the idea of multiplicity or number. (*Treatise*, p. 201)

The principle of invariance is a principle of invariance through time, justifying the inference to constancy of the object. The second explanation is the inference of constancy to the concept of perfect numerical identity. Here is the final paragraph of this long argument. Hume summarizes very well why we come to believe in the continued existence of objects.

> Our memory presents us with a vast number of instances of perceptions perfectly resembling each other, that return at different distances of time, and after considerable interruptions. This resemblance gives us a propension to consider these interrupted perceptions as the same; and also a propension to connect them by a continu'd existence, in order to justify this identity, and avoid the contradiction, in which the interrupted appearance of these perceptions seems necessarily to involve us. Here then we have a propensity to feign the continu'd existence of all sensible objects; and as this propensity arises from some lively impressions of the memory, it bestows a vivacity on that fiction; or in other words, makes us believe the continu'd existence of body. If sometimes we ascribe a continu'd existence to objects, which are perfectly new to us, and of whose constancy and coherence we have no experience, 'tis because the manner, in which they present themselves to our senses, resembles that of constant and coherent objects; and this resemblance is a source of reasoning and analogy, and leads us to attribute the same qualities to the similar objects. (*Treatise*, pp. 208–209)

In this closing passage, the key role of resemblance as Hume's concept of isomorphism is evident throughout.

The conclusion of the argument at this point is not the end of the matter for Hume. His detailed and complicated attack on the problem of the existence of external objects, including at a later stage the problem of personal identity, continues for another sixty some pages to the end of Book I of the *Treatise*. It would be diversionary to expose the details, but there are one or two points directly relevant to the consideration of his use of isomorphism in the concept of resemblance. In the remaining part of Book I, resemblance, contiguity and cause and effect keep their central position as the important relations between ideas or impressions. A good example of this extended use is in the characterization of memory in terms of resemblance.

> For what is the memory but a faculty, by which we raise up the images of past perceptions? And as an image necessarily resembles its object, must not the frequent placing of these resembling perceptions in the chain of thought, convey the imagination more easily from one link to another, and make the whole seem like the continuance of one object? In this particular, then, the memory not only discovers the identity, but also contributes to its production, by producing the relation of resemblance among the perceptions.
> (*Treatise*, pp. 260–261)

In this respect, Hume's view, as he so often remarks, resembles the vulgar opinion, namely, the important feature of memory is just that of producing images resembling the perceptions of the past. In my own view, it is hard to have a general view of memory radically different from this. The details, of course, are vastly complicated and now much studied. It is characteristic of some of Hume's basic mechanisms that they both contribute to and also distort accurate memories. I have in mind the extensive studies of the way in which associations contribute to accurate recall and also to misleading intrusions and inaccuracies. What is surprising is how little the theory of memory has advanced in describing in modern terms specific neural mechanisms of memory. A good example would be a detailed physical or chemical description of the mechanisms by which items are retrieved from memory—for example, words when reading or listening to speech. Of all mechanisms of memory, this is one of those most constantly in demand in every kind of situation. It is important not to be misunderstood on this point. The research on characteristics of memory is enormous in scope. For an excellent recent review see Tulving and Craik (2000). What is still missing is much deeper insight into the physical and chemical mechanisms supporting the cognitive activities of memory. (Something is said about these matters in Chapter 8 in the closing section on brain-wave representations of language.)

To finish with Hume's complicated further arguments, I mention only one general line of thought that is important in discussions of representation and, even more, of concepts of isomorphism. Hume's central attack on overly simple ideas of external or personal identity consists of pointing out how we use the concept of sameness in ordinary language and experience. Thus we will say that a ship is the same even though many parts have been repaired or replaced. We recognize that plants, animals and children change enormously over short periods of time, yet we refer to them as the same tree, same horse or same child. In some cases, there can be quite drastic changes. He mentions a church being completely rebuilt in a different style. We may still talk about it being the same parish church. The point is that the notion of sameness, which we think of as a surrogate for identity, really is relative to many aspects of perception, use, intention and custom. It is different with resemblance. Resemblance is something fundamental for Hume and it remains so in any sound philosophy. The real point is to be skeptical of claims of identity and to recognize what is critical is the notion of approximate isomorphism or resemblance between experiences at different times and places, as the basis on which we build our more elaborate theories of persons, places and things.

Kant. Kant was the first major philospher to break the longstanding link between isomorphism and representation. He has many things to say about representation, but offers no version of the concept of isomorphism. It is remarkable that it has not been more noted that, in the long history of thought on these matters reaching from Aristotle to Hume, Kant is really the first major figure to move away from any working concept of isomorphism as central to his concept of representation. The vitality of this link, broken by Kant, has not been completely restored in subsequent philosophy, full as it is of skepticism about the Aristotelian theory of perception resting on the concept of sensible and intelligible forms, and equally so of the empirical tradition culminating in Hume.

My reading of the history is that both Hume and Kant took as their model of science the natural philosophy of Newton, and what it had achieved, as worked out in such detail in Newton's *Principia Mathematica* (1687). But they took very different turns. Hume had as his goal the development of an empirical theory of human nature as a science comparable to that of Newton, with Hume's concept of association playing the role that attraction did in the Newtonian system. For Hume, as for Newton, there was no explanation at present possible beyond that of association in the case of human nature or gravitation (and like forces) in the case of nature. They were both, in Hume's terms, original qualities. Kant had a different agenda. He was concerned that the Humean approach to science led only to a contingent relation of cause and effect and, more generally, merely contingent knowledge of the world. For him the virtue of Newton's work was its establishment in clear and definitive form of necessary laws of nature. To give knowledge in its best sense such a necessary foundation was the goal of Kant's enterprise. In working out the details of his elaborate system, he had much to say about representation and little about isomorphism. Hume wanted to give what he thought was a genuinely empirical account of how the mind works. Kant was concerned, rather, to establish by transcendental argument how the mind is able to establish necessary truths about the world, but also how these truths must be limited and fenced off from metaphysical speculation.

In what is surely meant to be a reference to Hume, Kant has this to say in the division of the *Critique of Pure Reason* (1781/1997) entitled "Transcendental Analytic":

> It is, to be sure, a merely empirical law in accordance with which representations that have often followed or accompanied one another are finally associated with each other and thereby placed in a connection in accordance with which, even without the presence of the object, one of these representations brings about a transition of the mind to the other in accordance with a constant rule. This law of reproduction, however, presupposes that the appearances themselves are actually subject to such a rule, and that in the manifold of their representations an accompaniment or succession takes place according to certain rules; for without that our empirical imagination would never get to do anything suitable to its capacity, and would thus remain hidden in the interior of the mind, like a dead and to us unknown faculty.   (*Critique*, A100-A101)[20]

Two pages later Kant puts himself in direct opposition to Hume's deep use of resemblance between successive perceptions to permit the 'fancy of permanent objects' to be made. He gives a transcendental argument that there must be something more given a priori in the mind beyond the mere succession of resembling perceptions. (A102)[21]

There is an important and significant aspect of Kant's working out of his system that I do not want to neglect or undervalue. He did try, in a magnificent way, to provide an a priori synthetic foundation for Newtonian science in his *Metaphysical Foundations of Natural Science* (1786/1970). He claims to derive, in a priori synthetic fashion, Proposition I of the chapter "Metaphysical Foundations of Dynamics" which asserts that matter fills the space, not by its mere existence, but by a special moving force, and then in the explanation of this proposition he introduces the two possibilities of attractive and

---
[20] p. 229 in the recent translation of Guyer and Wood (1997).
[21] p. 230.

repulsive forces. He does not attempt to claim he can derive, in this a priori synthetic framework, the actual inverse square law of gravitational attraction.

A final comment on Kant. I close these remarks with a revealing quotation from the beginning of "The Analytic of Concepts" in the first *Critique*. It shows how Kant is not concerned to use associations to understand how concepts are empirically constructed. It is the unity of apperception, to use the technical Kantian phrase, that pulls everything together and permits relating one concept to another, but the unity of apperception is not an empirical feature of the way the mind works. It is an a priori concept made necessary by the need for understanding or knowledge. Notice that in the last sentence of this quotation we come to representations of representations and judgments, but there is no detailed claim about any kind of isomorphism in the relation between these or other representations.

> The understanding has been explained above only negatively, as a nonsensible faculty of cognition. Now we cannot partake of intuition independently of sensibility. The understanding is therefore not a faculty of intuition. But besides intuition there is no kind of cognition than through concepts. Thus the cognition of every, at least human, understanding is a cognition through concepts, not intuitive but discursive. All intuitions, as sensible, rest on affections, concepts therefore on functions. By a function, however, I understand the unity of the action of ordering different representations under a common one. Concepts are therefore grounded on the spontaneity of thinking, as sensible intuitions are grounded on the receptivity of impressions. Now the understanding can make no other use of these concepts than that of judging by means of them. Since no representation pertains to the object immediately except intuition alone, a concept is thus never immediately related to an object, but is always related to some other representation of it (whether that be an intuition or itself already a concept). Judgment is therefore the mediate cognition of an object, hence the representation of a representation of it. (*Critique*, A68-B93)[22]

James. Writing a century after Kant and a century and a half after Hume, James has much to say about the nature of laws of association. I confine myself here to Chapter XIV of his *Principles of Psychology* (1890/1950), although there are many pertinent discussions and analyses in other chapters, especially Chapter IX, The Stream of Thought, and Chapter X, The Consciousness of Self. The one central aspect of the thought of Hume and of British philosophers following Hume, like James Mill, of which James is skeptical, is that complex ideas are built up from simple ideas by association or by any other mechanism. He totally rejects the composition of complex thoughts from simple ones. Apart from that, however, he very much agrees on the importance of association as the fundamental causal account of how the mind or, as he often puts it, the brain works. Here is what he says in the last page of the chapter on association:

> In the last chapter we already invoked association to account for the effects of use in improving discrimination. In later chapters we shall see abundant proof of the immense part which it plays in other processes, and shall then readily admit that few principles of analysis, in any science, have proved more fertile than this one, however vaguely formulated it often may have been. Our own attempt to formulate it more definitely,

---

[22]pp. 204-205.

3.6 PHILOSOPHICAL VIEWS OF MENTAL REPRESENTATIONS

and to escape the usual confusion between causal agencies and relations merely known, must not blind us to the immense services of those by whom the confusion was unfelt. From this practical point of view it would be a true *ignoratio elenchi* to flatter one's self that one has dealt a heavy blow at the psychology of association, when one has exploded the theory of atomistic ideas, or shown that contiguity and similarity between ideas can only be there after association is done. The whole body of the associationist psychology remains standing after you have translated 'ideas' into 'objects', on the one hand, and 'brain-processes' on the other; and the analysis of faculties and operations is as conclusive in these terms as in those traditionally used.   (*Principles*, p. 604)

James's discussion of the laws of association and related matters is too extensive and varied to try to summarize in a serious way here. I will mention only salient points related to my theme of isomorphism being at the heart of one central notion of representation. To begin with, James refers, not only in this chapter but in many other places, to images and to imagery. It is quite evident from the context of these many different passages that almost always he has in mind something like an isomorphic relationship between the object or phenomena generating the image and the image itself. This continual use of some notion of isomorphism seems almost inevitable in any detailed study of the psychology of visual images. It is also supported by James' adoption in the opening pages of the *Principles* (p. vi of the Preface) of an attitude of scientific realism toward the external world. At the same time, it is important to emphasize that these references to images in the mind, casual or systematic, are almost never accompanied by any serious formal or quantitative attempt to characterize exactly the relation between the image and that which generates it, or, as in Hume's case, the relation between successive perceptions or images. As we know already from projective geometry and the theory of perspective, the exact characterization of certain necessary constraints in the representation of physical objects, for example by binocular vision, is intricate and complicated. It is a specialized subject to investigate this topic, and it is not ordinarily a part of the general discussion of images or of mental representations. Formal developments of related, but not exactly corresponding, problems of perspective are discussed in some mathematical detail in Sections 4–7 of Chapter 6, the sections on visual space.

I have six summary points concerning James's analysis and remarks on the psychological phenomena of association. For each, I indicate relevant page numbers in the standard edition already cited.

First, James makes the point in several places (p. 554) that association is not between ideas but between things thought of. Although he is not clear about this, he perhaps accepts that the associative relation is between images of these things. He does consider it part of his general denial of there being simple ideas as the atoms of psychological thought.

Second, for James there are many kinds of association and it is important to recognize this, whatever may be said about general laws. He formulates this in terms of connections 'thought of.'

> The jungle of connections *thought of* can never be formulated simply. Every conceivable connection may be thought of—of coexistence, succession, resemblance, contrast, contradiction, cause and effect, means and end, genus and species, part and whole, sub-

stance and property, early and late, large and small, landlord and tenant, master and servant,—Heaven knows what, for the list is literally inexhaustible. (*Principles*, p. 551)

Third, the one important elementary law of association is that of contiguity, especially in time, either simultaneous or in close succession (pp. 563-566). Moreover, fourth, for this one fundamental law of association, that of contiguity, he does give a qualitative explanation in terms of the law of habit in the nervous system. Here is a summary of a long discussion:

> *The psychological law of association* of objects thought of through their previous contiguity in thought or experience *would thus be an effect, within the mind, of the physical fact that nerve-currents propagate themselves easiest through those tracts of conduction which have been already most in use.* (*Principles*, p. 563)

What James says in this quotation and in the discussion accompanying it is modern in spirit and follows from his thorough treatment of brain activity, as known in his time, in Chapter II. On the other hand, the discussion is so qualitative that it adds little in terms of actually offering a scientific explanation of association, something which, of course, Hume deemed not possible, at least within the conceptual framework of his time.

Fifth, and importantly for our discussion here, James denies that there is a fundamental or elementary law of association by similarity, which I take to be similar to Hume's concept of resemblance (pp. 578ff). He says, in terms of compound thoughts or ideas, some positive things about association. Towards the end of the chapter, he has the following very definite statement about similarity not being of any true importance.

> The similarity between the objects, or between the thoughts (if similarity there be between these latter), has no causal agency in carrying us from one to the other. It is but a result—the effect of the usual causal agent when this happens to work in a certain particular and assignable way. But ordinary writers talk as if the similarity of the objects were itself an agent, co-ordinate with habit, and independent of it, and like it able to push objects before the mind. This is quite unintelligible. The similarity of two things does not exist till both things are there—it is meaningless to talk of it as an *agent of production* of anything, whether in the physical or the psychical realms.
>
> (*Principles*, p. 591)

In contrast to this passage, however, he does have positive things to say about a concept of congruity earlier in the chapter:

> *Habit, recency, vividness, and emotional congruity* are, then, all reasons why one representation rather than another should be awakened by the interesting portion of a departing thought. We may say with truth that *in the majority of cases the coming representation will have been either habitual, recent, or vivid, and will be congruous.* ..... In spite of the fact, however, that the succession of representations is thus redeemed from perfect indeterminism and limited to a few classes whose characteristic quality is fixed by the nature of our past experience, it must still be confessed that an immense number of terms in the linked chain of our representations fall outside of all assignable rule.
>
> (*Principles*, p. 577)

3.6 PHILOSOPHICAL VIEWS OF MENTAL REPRESENTATIONS

In characterising what he means by emotional congruity, he gives vivid examples in terms of the relation between the images we bring to mind and the emotional mood we are in. So, for example, if we are joyous we do not think of depressing images. On the other hand, if we are in a melancholic or sad state of mind, then we will have images of a negative and dark sort.

James's thought on these matters is tangled and complex. As the above passage shows, he does not think that similarity can be the basis of association. Yet he quotes with approval in the last section of the chapter, a long passage from Alexander Bain, who gives this beautifully detailed example. The discoverers of new sources of energy or power, as exemplifed in water mills and, later, steam engines, must have associated in terms of use, and not in terms of physical properties, because the physical properties of human methods of generating energy, of using objects to move energy, for example, is so very different from either a water mill or a steam engine, and a steam engine is itself so physically different from a water mill. In quoting this kind of example, he is also saying something similar to what he said about emotional congruity. And as in the earlier passage on the many kinds of association, James, with his very pluralistic attitude, seems willing to accept that whenever there is a quality in two phenomena or two things that is in ordinary terms, similar, then that form of isomorphism, as I would put it, can form the basis of an association. The property does not have to be a defined physical property, or even a defined physical relation of the object or phenomena in question, but can be something arising out of intentions and uses. When we put the matter this way, we can even go further back and recognize that the principle of contiguity is itself a special form of isomorphism, because of the shared space and time. But as James points out, it is not that we associate anything to anything else equally, but we have great selectivity even in what we associate by contiguity.

James' unsystematic and wide-ranging discussion of many examples brings various concepts of resemblance or similarity back into play after their neglect by Kant. But he does not do this in a systematic and formal way. There is no doubt that in some sense, resemblance, similarity or isomorphism, call it what you will, has lost ground in the move from Hume to James, more or less independent of what happened in the case of Kant.

A final point about James. He makes, towards the end of the chapter, the important distinction between spontaneous association and what he calls voluntary association, what we might currently call intentional association (pp. 583–588). Spontaneous or free association is very often the focus of these discussions, but, as James points out, focused or intentional association is very much the important mechanism in recall, especially when the recall does not come easily without some intentional effort, as in the case of forgetting someone's name. The other important intentional use of association is in problem solving where new associations are required in a focused way to find the solution. James mentions rightly in this context the importance of association in mathematical and scientific thinking, as a subject as yet too little investigated in a systematic way. I won't develop the theme here. It is touched upon by James, but it is clear that such intentional associations make use of coherent or congruous associations in succession to put together the conceptual pieces needed for a given piece of problem solving. Here again we have moved the concept of isomorphism away from the visual or physical properties of objects or phenomena to association between the way they are used.

Special case of images. The modern literature, in the more than 100 years since James published *Principles of Psychology*, is enormous and varied, both in philosophy and psychology. I will not attempt any kind of survey of most aspects of this literature, but only concentrate here on mental imagery, with a particular effort to contrast the things that in recent years philosophers have said, as opposed to the kinds of questions upon which psychologists have focused their experiments and theoretical analyses.

First of all, perhaps the most striking conclusion is how little discussion there is of imagery in spite of its long philosophical history as a topic, in the writings of many philosophers concerned with mental activity. A good example would be Peacocke (1983). This serious and complicated work, focused on sense and content, with special reference to experience, does not even have a single index entry under imagery. This absence of explicit references does not mean Peacocke is not discussing what is ordinarily thought of as perception or what, in even more ordinary parlance, would be talked about as the mental image of a physical object at which one was looking. In many ways he wants to replace remarks about imagery by talk about content. And talk about content, as he would like to put it, is the kind of thing that is said as finishing the phrase 'it visually appears to the subject that ...' (Peacocke 1983, p. 8). That he wants not to associate content with an image in any direct sense can be seen from his characterization of what he means by representational content, which must have two features.

> The first is that the representational content concerns the world external to the experiencer, and as such is assessable as true or false. The second feature is that this content is something intrinsic to the experience itself—any experience which does not represent to the subject the world as being the way that this content specifies is phenomenologically different, an experience of a different type. It is quite consistent with these two features that the presence of experiences with a given representational content has been caused by past experience and learning. (Peacock 1983, pp. 9–10)

Peacocke has many interesting things to say about distinctions, some traditional, some new, between sensation, perception and judgment, to use old-fashioned terminology that he mainly avoids. His concept of representational content is most certainly to be distinguished from sensation, as he points out in many different ways and many different places. In spite of the fact that he concentrates on visual representation and has many familiar examples, he avoids not simply using the term 'image' but, more importantly, avoids making any claims about any kind of isomorphism between representational content and physical properties of objects in the external world. Again, so as not to be misunderstood in what I am claiming about Peacocke's analysis, this is not to say he avoids mentioning physical properties, but, as can be seen from the quotation above, he mainly wants to refer to these properties simply as being properties of the object, and not to distinguish between a systematic visual image in the mind and the physical object being looked at. It is exactly the point of much of the psychological literature, which I review below, to concentrate on just the salient properties of the visual images in the mind, and how these relate systematically by various kinds of isomorphisms to physical objects themselves.

An outsider, casually familiar with the traditional literature from Hume and earlier, might think, as he approached a modern careful study like that of Peacocke, that the general notion of Hume of resemblance, or the earlier Aristotelian notion of form, would now be spelled out in sophisticated formal language of projective transformations and their invariance in a way that would bring a needed deeper, systematic element to the earlier philosophical thought. In spite of the carefulness and seriousness of many aspects of Peacocke's analysis, nothing could be further from the endeavor. His concerns are elsewhere, in terms of epistemic, linguistic and semantic distinctions. Perhaps the most striking evidence that Peacocke avoids any complex problems of spatial imagery, and of the relation of those images to the geometrical aspects of physical objects, is in his third chapter, entitled, 'Spatial Contents and Constraints', which is remarkable for the absence of all but the most trivial kinds of geometrical distinctions. It is characteristic of Peacocke's viewpoint that he makes contact only with the older, more elementary, psychological literature and not the modern one. It would be interesting to know how he would approach the analysis of one of Roger Shepard's (1960) most intriguing examples of the use of imagery. In this experiment, a person is asked to tell how many windows there are in the house in which he lives. Ordinarily, this number is not immediately known to a person and ordinarily, most subjects given this task will report that what they do is imagine looking at each room while counting up the total number of windows. Now, in order for this imagining to work, there must be back of it appropriate notions of isomorphism between imagined room and the real room itself, for each of the rooms in the house. And, of course, most of us can do a pretty good job of answering Shepard's question. In this framework the theoretical point of great importance is the study of what isomorphic relations we do and do not have, in terms of how our mental imagery of familiar but complex physical objects works.

The imagery debate has gone on for some time in psychology, with lots of experiments and analysis brought to bear. Recently, a number of philosophers have waded into this debate as well. A good example that analyzes both sides of the argument is Rollins (1989). He presents in a reasonable way the argument for some kind of isomorphism, what he calls 'nomic similarity', much supported by those believing in images. But he also summarizes the arguments against such isomorphism by descriptionists, who maintain that cognitive content is not to be found in images of objects, but in descriptions of the objects, represented best of all by sentences describing the objects. Support for this view is to be found in a well-known article by Fodor and Pylyshyn (1988). These and other philosophical aspects of the debate are also well reviewed in the book by Michael Tye (1991). On the other hand, this literature, fair as it often is to both sides, is marked by an absence of detailed consideration of what a real theory of isomorphism, similarity or resemblance would look like. In other words, no serious details are developed from a geometrical or structural standpoint.

Psychological views of imagery. So, having emphasized this point about the absence of details in the philosophical literature—absent in a way they are not absent in

Hume or James—I turn finally to a good review of the psychological literature by Finke (1989), who surveys a vast number of psychological experiments on imagery and also takes a fair-minded attitude toward the many intricacies, arguments and counterarguments that have been raised in the debates about imagery. Finke organizes his analysis of the psychological literature on mental imagery around five principles, for each of which he offers both supporting experiments and analysis as well as corresponding criticism.

The first concerns information retrieval using mental imagery. Here experiments on memory take center stage. The second is on visual characteristics of mental images. He states, as the principle to be defended or criticized, a principle of perceptual equivalence:

> Imagery is functionally equivalent to perception to the extent that similar mechanisms in the visual system are activated when objects or events are imagined as when the same objects or events are actually perceived. (*Principles of Mental Imagery*, p. 41)

It is easy enough to name some philosophers who would have apoplexy at the suggestion that this principle is really true of human mental imagery.

The third principle concerns spatial characteristics of mental images. Labelling it the principle of spatial equivalence, Finke gives it the following formulation:

> The spatial arrangement of the elements of a mental image corresponds to the way objects or their parts are arranged on actual physical surfaces or in an actual physical space. (*Principles of Mental Imagery*, p. 61)

It is easy enough to think of objections to this principle, but it also has a substantive appeal that can be defended both by ordinary experience and a great variety of psychological experiments. Probably the psychologist most identified with this principle is Stephen Kosslyn, who has defended it vigorously on many different occasions (1973, 1975, 1976, 1978, 1980).

The fourth principle concerns transformations of mental images and is especially associated with the experiments of Roger Shepard and his associates on mental rotation of images, e.g., Shepard and Metzler (1971). Shepard and Cooper (1982) contains a full description of earlier experiments and much other detail. Finke gives the principle of transformational equivalence the following formulation:

> Imagined transformations and physical transformations exhibit corresponding dynamic characteristics and are governed by the same laws of motion.
>
> (*Principles of Mental Imagery*, p. 93)

Although Shepard's experiments certainly demonstrate something like a commonsense notion of mental rotation takes place, the stark form and strong isomorphic supposition underlying Finke's statement of the principle is, to experimental psychologists who do not really believe strongly in images, rather like throwing raw meat to lions—they can't wait to get at it and show by experiments what is wrong with it. Any one who has dabbled in the imagery debate, either philosophically or psychologically, will know where opponents strike at principles such as those formulated by Finke. They will bring in the many ways in which interpretation and context affect the correspondences found in different situations. So, at the very least, strong *ceteris paribus* conditions would need to be formulated at the beginning to accept Finke's formulations as they are stated without reservation. On the other hand, those who defend the principles have much on

3.6 PHILOSOPHICAL VIEWS OF MENTAL REPRESENTATIONS

their side. The difficulty of providing a reasonable theory without any such approximate isomorphisms is much harder than many of the critics recognize.

Finally, and perhaps most controversally, Finke states as the final of his five principles that of structural equivalence. "The structure of mental images corresponds to that of actual perceived objects" (p. 120).

The idea of this principle is that structural relations, relations among the parts of complex geometrical patterns, can often be found to be preserved in mental images. As Finke points out, one of the common misconceptions about mental imagery is that images are generated in flashes or all at once, rather than being constructed in parts over time. (I must say that construction in parts is very much more in the general spirit of how memories are constructed. Much research shows that memories are not simply data accessed, as in a digital computer, but are constructed step by step by processes of association.)

An experiment that brings out how structure plays a role is one by Kosslyn, Reiser, Farah and Fliegel (1983). Subjects were shown drawings of patterns that were themselves constructed out of simple geometric forms, for example, squares, rectangles and triangles. Subjects were given different descriptions which varied in the number of parts used to describe the pattern. For example, two rectangles, one vertical and one horizontal, centered about their midpoint can be described as five squares or as two overlapping rectangles. Subjects were given both kinds of descriptions, both minimal in number for the parts and maximal. Subjects were then instructed to generate an image of the given pattern according to the description. Their generation times were recorded. There was, to a remarkable degree, a very nice linear relation between the mental time needed to generate the image and the number of parts used in the description of the pattern. Much of this kind of thing, of course, is reminiscent of the Gestalt psychological experiments of more than half a century ago. I would find it astounding if all aspects of use of imagery in such experiments could be boldly cast in doubt by alternative explanations of descriptionists or those believing in purely proposition-like accounts of mental activity. But this is not the place to carry that debate further.

What I have tried to defend in this final section is the importance, even if controversial, of some notion of sameness of structure or isomorphism, not just in straightforward mathematical examples, as in the theory of measurement or in geometry, but also in something as complex and rich as mental imagery. In this case the complexity is very much reflected in the complexity of the as yet unending debate, both in philosophy and in psychology, over the status of mental images. I have made my own sympathies clear, but I acknowledge that the critics of imagery have done much to help deepen both experiments and the thinking about them.[23]

---

[23]Further complexities abound. When visual diagrams or sketches are used in mathematical or scientific research or teaching, psychological features that I have not tried to consider here, even at the most superficial level, play a key role. I mention here only the eye movements that can be observed during problem-solving activity. They tell us much about the continual and rapid interaction between scanning the diagrams and storing images or descriptions in necessarily limited working memory. For extensive details on these matters, see Epelboim and Suppes (2001). The often quicksilver nature of animal and human problem solving has not as yet been adequately reflected in the debate on mental imagery.

4

Invariance

Intricately associated with the idea of representation is that of invariance of the representation. In this chapter I begin by exploring in the first section the close relation between invariance and symmetry, as reflected in visual designs. Already, even for the most simple designs, we can add to our understanding by computing their elementary symmetry groups, as I hope to make clear with appropriate examples and definitions. It is easy to move from these intuitive ideas to more formal ideas of invariance, symmetry and meaning.

The second section examines questions of invariance of qualitative visual perceptions as expressed in ordinary language, especially prepositions in English. These first two sections are meant to be relatively informal and intuitive in spirit. In the third section I return to more formal but elementary matters, the problem of invariance for the theories of equal-interval measurement analyzed from the standpoint of the representation theorems in Chapter 3. In the fourth section the lack of invariance of fundamental physical equations is examined, and in the fifth section, the final one, the elegant results on entropy as a complete invariant for many ergodic processes are outlined.

4.1 Invariance, Symmetry and Meaning

The ordinary or common meaning of invariance gives a very nice sense, qualitatively, of its technical meaning. Something is invariant if it is unalterable, unchanging, or constant. Of course, the first question that then arises, is this: "What is unalterable or unchanging?" The intuitive answer, which again matches well, in a general way, the technical answers I shall examine later, is that a property of an object, collection of objects or more generally, some phenomenon, is constant. Familiar examples are the shape of a cup as we move the cup around, rotate it, turn it upside down, etc. The shape of the cup does not change, nor does its weight. An important physical example is that the speed of light in a vacuum is invariant with respect to any inertial frame of reference with which we measure it, contrary to the simple addition of velocities familiar in classical physics.

It is the idea of invariance expressing constancy that is especially important. Thus, we say, psychologically, someone's attitudes are invariant or unchanging, just because they are constant over the years. Of course, all of the meanings I've stated are approximate synonyms for each other. Something that is constant is unalterable. Something that is constant is unchanging and so forth.

Closely associated with invariance is symmetry, as has been stressed by many commentators on invariance. The example closest to us is the approximate bilateral symmetry of the human body. It is sometimes plausibly argued, although I shall not do so here, that the psychological appeal of bilateral symmetry in designs, and in architectural structures, is a consequence of our own bilateral symmetry. Whether this psychological claim is true or not is not easy to determine, but the appeal of bilateral symmetry in structures and in design is very evident. I show just two figures here. One is a Chinese design (Jones, 1867, Plate XIX). The vertical line bisecting the figure is the axis of bilateral symmetry.

FIGURE 1   Chinese design from a china dish, probably nineteenth century.

The second figure is of a plan and elevation of a villa by Palladio (1570/1965). The plan and elevation drawings, even more than the structure itself, make evident the effort to have as much bilateral symmetry as possible.

We can generalize from the particular case of bilateral symmetry to the geometric viewpoint derived from Felix Klein. It can be stated this way. "For a given group of transformations of a space, find its invariant properties, quantities or relations." For a given geometry, which, from the standpoint formulated here, means a set-theoretical structure consisting of the basic set, constituting the points of the space, and the rela-

4.1 INVARIANCE, SYMMETRY AND MEANING

tions and functions on this set, a transformation that carries one geometric structure into another isomorphic to it, on the same space, is an *automorphism* of the space. Klein's viewpoint was that the group of automorphisms of a geometry provides another way of defining the geometry. Or, put in a more purely Kleinian way, any group of automorphisms of a space defines a geometry and there will be relations and functions of this geometry that are invariant under the group of automorphisms.

Here is a famous quotation from Klein from his 1872 Erlangen address (1893). I have made occasional minor changes in the English translation.

> For geometric properties are, from their very idea, independent of the position occupied in space by the configuration in question, of its absolute magnitude, and finally of the sense in which its parts are arranged. The properties of a configuration remain therefore unchanged by any motions of space, by transformation into similar configurations, by transformation into symmetrical configurations with regard to a plane (reflection), as well as by any combination of these transformations. The totality of all these transformations we designate as the *principal group* of space-transformations: *geometric properties are not changed by the transformations of the principal group*. And, conversely, *geometric properties are characterized by their remaining invariant under the transformations of the principal group*. For, if we regard space for the moment as immovable, etc., as a rigid

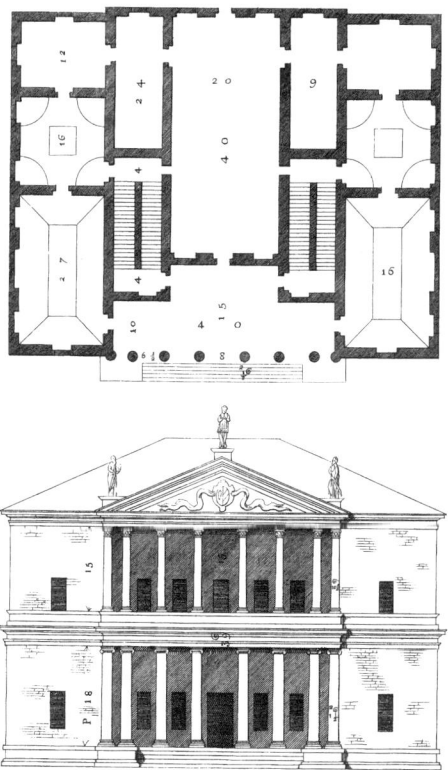

FIGURE 2   Plan and elevation of a Palladio villa.

manifold, then every figure has an individual character; of all the properties possessed by it as an individual, only the properly geometric ones are preserved in the transformations of the principal group.

(Klein, p. 218)

Within this framework, we say that a geometric figure is *symmetric* under a given group if every automorphism of the group maps the figure into itself. For example, a bilaterally symmetric figure is so mapped by a mirror reflection along what we familiarly call a line of symmetry, in the case of plane figures. Figures symmetric under various rotations are also familiar examples of symmetry, the circle above all, but, of course, the square is also symmetric under any 90-degree rotation about its center. Notice the difference. The square has a very small finite group of rotations under which it is invariant, whereas the circle is invariant under an infinite group of rotations.

In similar fashion three-dimensional figures like spheres and cubes are also invariant under various groups of motions or transformations. Idealization is important here. Most real physical objects look, or are different, in some observable, even if minute, way, under rotations or reflections.

Familiar examples of geometric invariants are the relation of betweenness and the relation of congruence of Euclidean geometry, which can be axiomatized just in terms of these two concepts.

It is worth remarking why symmetry and groups naturally go together. We may think of it this way. A geometric figure F in a space A is *symmetric* under a single transformation φ mapping A onto A if φ maps F onto itself. Now if this holds, so should the inverse mapping φ^{-1}, since $\varphi^{-1} \circ \varphi$ is just the identity map of A. In addition, it is natural to expect closure of this symmetry property. If φ_1 and φ_2 both map F onto itself then so should their composition $\varphi_1 \circ \varphi_2$. These two properties of a set G of transformations:

(i) If $\varphi \in G$ then $\varphi^{-1} \in G$,

(ii) If $\varphi_1, \varphi_2 \in G$ then $\varphi_1 \circ \varphi_2 \in G$,

are sufficient to guarantee G is a group of transformations or automorphisms of the given space A, since composition of transformations is necessarily associative, and $\varphi \circ \varphi^{-1}$ is the identity. This same argument also shows why invariance and groups so naturally go together.[1]

When artists or architects draw a symmetric figure or design, it is undoubtedly usually the case that they do not explicitly think in terms of the automorphisms of the space in which the figure or design is implicitly embedded. Even sophisticated mathematicians of ancient Greek times did not have an explicit group concept, but once uncovered, it seems a natural formalization of the intuitive idea of symmetry.

Various kinds of symmetries in design have been recognized for thousands of years. The square with its two diagonals shown in Figure 3 looks the same, i.e., is invariant in

---

[1] The general axioms for groups, not just groups of transformations, were given in Section 2.3, but as was mentioned at the beginning of Section 3.3, every group is isomorphic to a group of transformations, which is one reason for thinking of groups of transformations as the most intuitive models of groups. For a detailed history of the development of the abstract concept of group, see Wussing (1984).

4.1 INVARIANCE, SYMMETRY AND MEANING

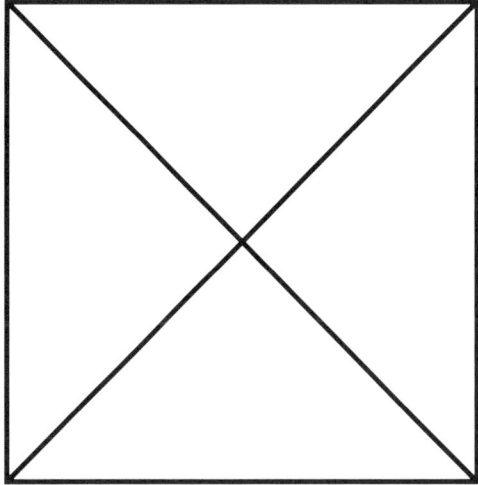

FIGURE 3   A square with two diagonals.

appearance under counterclockwise rotations in the plane about its center of 90°, 180°, 270°, and, of course, 360°. (Which way to rotate is, of course, a matter of convention. I have followed what is more or less standard.)

On the other hand, the square with one diagonal shown in Figure 4 is invariant in appearance only under counterclockwise rotations of 180° or 360° about its center.

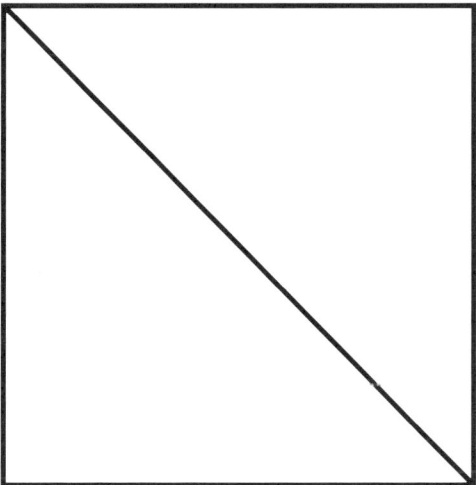

FIGURE 4   A square with one diagonal.

With only a moderate increase in formalism we can define the two groups of symmetries exhibited by the two figures under counterclockwise rotation. Here *groups* are meant in the technical mathematical sense defined above and in Chapter 2. Because it is simpler let us consider Figure 4 first. As a matter of notation, let $r(n)$ be a counterclockwise rotation of n degrees. Note first that $r(0) = r(360)$, and that $r(0)$ is the identity element of the group. The full set is obviously just $\{r(0), r(180)\}$ and the binary operation $\circ$ of the group is that of following one counterclockwise rotation $r(m)$ with another $r(n)$, which we write as $r(n) \circ r(m)$. The reader can easily verify that the group axioms of Section 2.3 are satisfied, once it is noted that $r(180)$ is the inverse of itself, as, of course, is $r(0)$.

In the case of Figure 3, the symmetry group of counterclockwise rotations is larger, being the set $\{r(0), r(90), r(180), r(270)\}$, and here the inverse of each "sums" to 360, except for $r(0)$. Thus, for example, the inverse of $r(270)$ is $r(90)$, since

$$r(90) \circ r(270) = r(0).$$

A thorough development of symmetry in regular figures, especially two-dimensional ones, is given in Toth (1964). This is a subject with a rich history in both mathematics and art, but I turn aside from it to consider concepts and results more directly relevant to the structures that characterize scientific theories.

General logical result on invariance. A classical result expresses very well the important idea that that which is invariant with respect to any one-one mapping of the world, or universe of any model under consideration, is just the logical relations. This is the fullest possible generalization of Klein's transformational and group-theoretic approach to geometry. Let me give two formulations, one that goes back to the early work of Lindenbaum and Tarski (1934-35/1983, p. 385):

> Roughly speaking, Th. 1 states that every relation between objects (individuals, classes, relations, etc.) which can be expressed by purely logical means is invariant with respect to one-one mapping of the 'world' (i.e., the class of all individuals) onto itself and this invariance is logically provable. The theorem is certainly plausible and had already been used as a premiss in certain intuitive considerations. Nevertheless it had never before been precisely formulated and exactly proved.

A closely related but different formulation is given much later by Tarski and Givant (1987, p. 57):

> (i) Given a basic universe U, a member M of any derivative universe $U^\approx$, [e.g., the Cartesian product $U \times U$] is said to be logical, or a logical object, if it is invariant under every permutation P of U. On the basis of (i) one can show, for example, that for every (nonempty) U there are only four logical binary relations between elements of U: the universal relation $U \times U$, the empty relation $\emptyset$, the identity relation ..., and the diversity relation ....

What this latter formulation brings out is that there are only four logical binary relations. No other binary relation is invariant, that is, constant, under arbitrary permutations of the universe.

Meaning. It is easy to pass from symmetry to meaning. Here are some familiar examples. Orientation, such as Aristotle's up and down for space, is not meaningful in

Euclidean geometry.[2] Why? It is not invariant under the Euclidean group of motions, i.e., rotations, translations and reflections.

In Euclidean geometry it is meaningful to compare the lengths of two line segments, regardless of whether they are parallel or not. In affine geometry for such a comparison to be meaningful, however, the line segments must be parallel, which includes lying on the same line. It is a distinguishing feature of affine geometry, in fact, that it is not a metric space. There is no common measure along lines as we change direction from one line to another. It is only for parallel segments that there is a meaningful concept of congruence, which is what we mean when we speak of affine congruence.

In projective geometry, even the affine concept of betweenness is not preserved under projections from one line to another from a given point of perspective. In this case, we must go to a four-place relation, the relation of separation: $ab\,S\,cd$ if and only if a and b separate c and d, and conversely. Figure 5 shows how separation is preserved under projection, but betweenness is not. For example, we can easily see in the figure that b lies between a and d, but the projection b' does not lie between a' and d'. On the other hand, the pair ab separates the pair cd, and conversely; moreover, under projection the pair $a'b'$ separates the pair $c'd'$, and conversely.

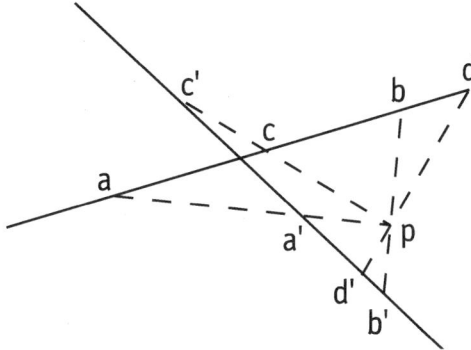

FIGURE 5   Drawing showing the noninvariance of betweenness and the invariance of separation under projection from one line to another from a given point of perspective p.

Objective meaning in physics. In physics, there is a tendency to go beyond simply talking about meaningfulness and to talk about *objective* meaning. The important concept is that of invariant relations, or invariance more generally, for observers moving relative to each other with constant velocity. So, for example, in classical physics, two inertial observers moving relative to each other should confirm the same measurements of that which is genuinely invariant. This doesn't mean that they don't make correct

---

[2] Here is a well-known passage from Aristotle in *On the Heavens*.

There are certain things whose nature it is always to move away from the centre, and others always towards the centre. The first I speak of as moving upwards, the second downwards. Some deny that there is an *up* or *down* in the world, but this is unreasonable.

(Aristotle, 308a14–17, Guthrie translation, 1960)

measurements in their own frames of reference, using their individual measurement procedures.[3] But, as we know, directions, for example, will in general be different for the two observers, as will other properties. On the other hand, what we expect to hold is that the numerical distance between two simultaneous spatial points is the same for all observers using the same calibration of their measurement instruments. The same is true of time. Temporal intervals are the same for all inertial classical observers using clocks with the same calibration. As is emphasized in foundational discussions of classical physics, spatial and temporal intervals are observer-independent, but the concepts of the center of the universe, the beginning of time, or being in a state of absolute rest are not invariant. They have no physical meaning, in spite of earlier views, such as those of Aristotelian physics, to the contrary. Of course, the use of these concepts by Aristotle was in no sense ridiculous. Phenomenological experience certainly gives us a natural concept of absolute rest, i.e., of being at rest with respect to the Earth, and the same for the center of the universe. On the other hand, as a matter much farther from familiar experience, Aristotle famously argued that the world is eternal, i.e., time has no beginning, creating problems for later Christian theologians, such as John Philoponus of the sixth century.

Here is a second example from special relativity. Invariance for two inertial observers of the 'proper' time of a moving particle is sufficient to derive that their observations with respect to their own frames of reference are related by a Lorentz transformation. The proper time interval τ_{12} of two points on the trajectory of an inertial particle satisfies the following equation:

$$\tau_{12}^2 = (t_1 - t_2)^2 - \frac{1}{c^2}\left[(x_1 - x_2)^2 + (y_1 - y_2)^2 + (z_1 - z_2)^2\right].$$

The important concept here is that proper time has objective meaning independent of the observer's particular frame of reference. Moreover, proper time has the important feature of being a complete invariant, since its invariance determines the group of Lorentz transformations (Suppes 1959). More details about these two examples are given in Chapter 6.

A third example is the classic theorem of Noether (1918). Roughly speaking, it says that to every conservation law of mechanics there corresponds an invariant quantity, which is a first integral of the equations of motion.

Example A. Conservation of momentum of a system of particles implies that the center of mass of the system moves with constant, i.e., invariant, velocity.

Example B. A system admits rotations around a given line. Then the angular momentum with respect to this line as axis is invariant, i.e., constant, in time.

These various geometric and physical examples make clear that the concept of invariance is relative not only to some group of transformations, but more fundamentally and

---

[3]Observers moving relatively to each other on trajectories that are not parallel record very different velocities for the same observed body, but this does not mean their observations are subjective in the usual epistemological sense. Many observers at rest with respect to each other could have full intersubjective agreement on their measurements. The concept of objective meaning introduced here is common in physics, but not in general philosophical discussions of meaning. Within the philosophy of science it is a good way of defining *meaningful with respect to a theory*, namely, that which is invariant with respect to the theory.

essentially, relative to some theory. Invariant properties of models of a theory are the focus of investigations of invariance. So, for example, Aristotle's theory of the heavens led naturally to their being eternal, since they were unchanging, and consequently there could be no beginning of time.

An invariance holding across many theories is the implicitly or explicitly assumed flatness of space in Ptolemaic, Copernican, Galilean, Newtonian and Einsteinian special-relativity theories of the physical universe. A technical geometric way of formulating the matter is the fundamental assumption that space-time is a four-dimensional affine space, which has many invariant consequences, as is explained in Chapter 6.

4.2 Invariance of Qualitative Visual Perceptions

In this section, various ordinary spatial expressions are introduced with their ordinary meaning, and the question asked is this. What is the natural geometrical theory with respect to which each expression is invariant? In Chapter 6, similar questions are asked about the results of various experiments on visual space, which challenge the thesis that visual space is Euclidean. As already hinted at, I shall follow the practice in geometry and physics of relating invariance under a given group of transformations to a principle of symmetry. Invariance implies symmetry and symmetry implies invariance.

The concepts, results and problems that fall under the general heading of this section have not been studied very intensely in the past but have received more attention in the last twenty years or so. I mention in this connection especially Bowerman (1989), Crangle and Suppes (1989), and Levelt (1982, 1984). The references in these publications provide good leads back into the older literature. A typical problem that I have in mind here is the sort of thing that Levelt has investigated thoroughly, for example, the way language is used to give spatial directions, and the limitations of our ability to give such directions or describe visual scenes with accuracy. In the case of my own earlier work with Crangle, we were concerned especially to analyze the geometry underlying the use of various common prepositions in English.

The results of that analysis can be easily summarized in Table 1 which is reproduced with some modifications from Crangle and Suppes (1989). The kinds of geometry referred to are standard, with the exception perhaps of the geometry of oriented physical space. For example, in the case of the sentence *The pencil is in the box* (where it is assumed the box is closed), it is clear that only a purely topological notion of invariance is needed. On the other hand, in the sentence *Mary is sitting between Jose and Maria*, it is easy to see that the geometry needs to be affine. And once the idea of a metric is introduced, as in the sentence *The pencil is near the box*, we must go from affine geometry to some underlying notion of congruence as reflected in Euclidean geometry. Although we may be more refined in the analysis, note that in this sentence it is quite satisfactory to use absolute geometry which is Euclidean geometry minus the standard Euclidean axiom. This axiom asserts that given a point a and a line L on which the point does not lie, then there exists at most one line through a in the plane formed by aL which does not meet the line. We get hyperbolic rather than Euclidean geometry by adding the negation of this axiom to absolute geometry. It seems to me that the notion of nearness used in ordinary talk is satisfied well enough by the congruence relation of

TABLE 1   Kinds of geometry and examples of prepositional use

Topology	*The pencil is in the box. (box closed)* *One piece of rope goes over and under the other.*
Affine geometry	*The pencil is in the box. (box open)* *Mary is sitting between Jose and Maria.*
The geometry of oriented physical space	*The book is on the table.* *Adjust the lamp over the table.*
Projective geometry	*The post office is over the hill.* *The cup is to the left of the plate.*
Geometries that include figures and shapes with orienting axes	*The dog is in front of the house.* *The pencil is behind the chair.*
Geometry of classical space-time	*She peeled apples in the kitchen.*

absolute geometry. In fact, relatively technical geometrical results are required to move us from absolute to Euclidean geometry, the sort of technical facts required, for example, in architecture and construction. Another way of noting the adequacy of absolute geometry to express many of the elementary results of Euclidean geometry is that the first 26 propositions of Book I of Euclid's *Elements* are provable in absolute geometry. On the other hand, in the case of the preposition *on*, it is clear that a notion of vertical orientation is required, a notion completely absent from Euclidean geometry, and in fact not definable within Euclidean geometry. A different kind of orientation is required in the case of objects that have a natural intrinsic orientation. Consider for instance, the sentence given in Table 1, *The dog is in front of the house.* Finally, in the case of many processes it is not sufficient to talk about static spatial geometry but for a full discussion one needs the assumption of space-time. An example is given at the end of Table 1.

What I now want to look at are the kinds of axioms needed to deal with the cases of geometry that are not standard. It would not be appropriate simply to repeat standard axioms for topological, projective, affine, absolute, and Euclidean geometry. A rather thorough recent presentation of these geometries is to be found in Suppes et al. (1989, Ch. 13). What I shall do is make reference to the primitive notions on which these various axioms are based.

Oriented physical space. Undoubtedly, the aspect of absolute or Euclidean geometry which most obviously does not satisfy ordinary talk about spatial relations is that there is no concept of vertical orientation. Moreover, on the basis of well-known results of Lindenbaum and Tarski (1934-35/1983) concerning the fact that no nontrivial binary relations can be defined in Euclidean geometry, the concept is not even definable in Euclidean geometry. For definiteness I have in mind as the primitive concepts of Euclidean geometry the affine ternary relation of betweenness for points on a line and the concept of congruence for line segments. In many ways I should mention, however, it is more convenient to use the notions of parallelism and perpendicularity, and in any case I shall assume these latter two notions are defined. There are many different ways, of course, of axiomatizing as an extension of Euclidean geometry the concept of verticality. One simple approach is to add as a primitive the set $\mathcal{V}$ of vertical lines, and then to add axioms of the following sort to three-dimensional Euclidean geometry. *Given any point a there is a vertical line through a. If K is a vertical line and L is parallel to K, then L is a vertical line.*

There are however, two different kinds of difficulties with this approach. The first and conceptually the most fundamental difficulty is that our natural notion of physical space, as we move around in our ordinary environment, is that its orientation, both vertically and horizontally, is fixed uniquely. We do not have arbitrary directional transformations of verticality nor of horizontal orientation. We naturally have a notion of north, east, south, and west, with corresponding degrees. Secondly, and closely related to this, very early in the discussion of the nature of physical space, it was recognized that we have difficulties with treating the surface of the earth as a plane with horizontal lines being in this plane and vertical lines being perpendicular to them. The natural geometry of oriented physical space in terms of ordinary experience—a point to be expanded upon in a moment—is in terms of spherical geometry, with the center of the earth being an important concept, as it was for Aristotle and Ptolemy in ancient times.

Aristotle directly uses perceptual evidence as part of his argument for the conclusion that the earth is spherical (*On the Heavens*, Book II, Ch. 14, 297b): "If the earth were not spherical, eclipses of the moon would not exhibit the shapes they do, and observation of the stars would not show the variation they do as we move to the north or south."

Ptolemy's argument in the *Almagest* is even better. Because it will not be familiar to many readers, I quote it in full, for the arguments are perceptual throughout. This passage occurs in Book I, Section 4 of the *Almagest*, written more than four hundred years after Aristotle's work.

> That the earth, too, taken as a whole, is sensibly spherical can best be grasped from the following considerations. We can see, again, that the sun, moon and other stars do not rise and set simultaneously for everyone on earth, but do so earlier for those more towards the east, later for those towards the west. For we find that the phenomena at eclipses, especially lunar eclipses, which take place at the same time [for all observers], are nevertheless not recorded as occurring at the same hour (that is at an equal distance from noon) by all observers. Rather, the hour recorded by the more easterly observers is always later than that recorded by the more westerly. We find that the differences in the hour are proportional to the distances between the places [of observation]. Hence one can reasonably conclude that the earth's surface is spherical, because its evenly curving surface (for so it is when considered as a whole) cuts off [the heavenly bodies] for each set of observers in turn in a regular fashion.
>
> If the earth's shape were any other, this would not happen, as one can see from the following arguments. If it were concave, the stars would be seen rising first by those

more towards the west; if it were a plane, they would rise and set simultaneously for everyone on earth; if it were triangular or square or any other polygonal shape, by a similar argument, they would rise and set simultaneously for all those living on the same plane surface. Yet it is apparent that nothing like this takes place. Nor could it be cylindrical, with the curved surface in the east-west direction, and the flat sides towards the poles of the universe, which some might suppose more plausible. This is clear from the following: for those living on the curved surface none of the stars would be ever-visible, but either all stars would rise and set for all observers, or the same stars, for an equal [celestial] distance from each of the poles, would always be invisible for all observers. In fact, the further we travel toward the north, the more of the southern stars disappear and the more of the northern stars appear. Hence it is clear that here too the curvature of the earth cuts off [the heavenly bodies] in a regular fashion in a north-south direction, and proves the sphericity [of the earth] in all directions.

There is the further consideration that if we sail towards mountains or elevated places from and to any direction whatever, they are observed to increase gradually in size as if rising up from the sea itself in which they had previously been submerged: this is due to the curvature of the surface of the water. (*Almagest*, pp. 40–41)

Aristotle's concept of natural motion, which means that heavy bodies fall toward the center of the earth, is well argued for again by Ptolemy in Section 7 of Book I.

...the direction and path of the motion (I mean the proper, [natural] motion) of all bodies possessing weight is always and everywhere at right angles to the rigid plane drawn tangent to the point of impact. It is clear from this fact that, if [these falling objects] were not arrested by the surface of the earth, they would certainly reach the center of the earth itself, since the straight line to the center is also always at right angles to the plane tangent to the sphere at the point of intersection [of that radius] and the tangent. (*Almagest*, pp. 43-44)

What is important to notice about this argument, and similar but less clear arguments of Aristotle, is that the notion of vertical or up is along radii extended beyond the surface of the earth, and not in terms of lines perpendicular to one given horizontal plane. Thus the perceptual notion of verticality is in terms of a line segment that passes through the center of the earth. The notion of horizontal is that of a plane perpendicular to a vertical line at the surface of the earth. The strongest argument for this viewpoint is the perceptual evidence of the nature of the natural falling motion of heavy bodies.

Ptolemy also uses observations of the motion of the stars and the planets to fix the direction of east and west, and also the poles of north and south. We use in ordinary experience such arguments as the rising and setting of the sun to fix the direction of east and west.

These perceptual arguments about physical space—including of course perceptual arguments about gravity—reduce the group of Euclidean motions to the trivial automorphism of identity. This means that from a global standpoint the concept of invariance is not of any real significance in considering the perceptual aspects of oriented physical space. On the other hand the strong sense of the concept of global that is being used here must be emphasized. From the standpoint of the way the term *global* is used in general, there remain many symmetries of a less sweeping sort that are important and that are

4.2 QUALITATIVE VISUAL PERCEPTIONS

continually used in perceptual or physical analysis of actual objects in space. Indeed this is a very intuitive outcome from the standpoint of ordinary perception. Combining both our visual sense of space and our sense of space arising from gravitational effects, it is wholly unnatural to think of anything like the group of Euclidean motions being the fundamental group for physical space in the sense of direct experience or perception. It is in fact a remarkable abstraction of the Greeks that orientation was not made a part of the original axioms of Euclid.

On the other hand, the notion of invariance in perception arises continually in a less global fashion in considering the symmetry of perceived figures or perceived phenomena arising from not just visual but also auditory or haptic data.

The classification of geometries for the prepositions used in Table 1 may be regarded as an *external* view from the standpoint of space as a whole. Perceptually, however, this is not the way we deal with the matter. In ordinary experience, we begin with the framework of a fixed oriented physical space as described above. Within this framework we can develop an *internal* view of spatial relations as expressed in ordinary language.

As an example, I examine the invariance of the preposition *in*. In this and subsequent analysis the relation expressed by *in* shall be restricted to pairs of rigid bodies. Even though it is ultimately important to characterize invariance for other situations, e.g., the sugar being in the water, a good sense of the fundamental approach to invariance being advocated can be conveyed with consideration only of familiar situations with rigid bodies.

Let a and b be rigid bodies with a in b. As in Table 1 there are two cases of body b to consider. The first is when b has a closed hollow interior and a is in this interior. For simplicity of notation and analysis, I introduce the restrictive assumption that a is a spherical ball, so that the orientation of a can be ignored. Define $I(a,b)$ = the set of points in the closed hollow interior of b that the center of a can occupy. Let $\Phi(a,b)$ be the set of all transformations of $I(a,b)$ onto itself that represent possible changes in the position of a relative to b. For example, if the position of the center of a is on one side of the hollow interior of b it could be transformed, i.e., moved, to the other side of the interior without affecting the truth of the assertion that a is in b.

The set $\Phi(a,b)$ is the symmetry group (under the operation of function composition) of the position of a inside b, but because the hollow interior of b can be quite irregular in shape $\Phi(a,b)$ may not have standard properties of Euclidean isometries. It does have the expected invariance properties, namely, if $\varphi \in \Phi(a,b)$, then a with center c at point p is in b if and only if a with center c at point $\varphi(p)$ is in b.

Body b is itself subject to rigid motion in the fixed physical space. If b is closed, rotations around a horizontal axis are permitted, but if b is open, possible rigid motions of b must be restricted to those that preserve vertical orientation. And, of course, when b is subject to a rigid motion ψ, body a must be subject to the same transformation ψ, in order to preserve the invariance of the relation of being inside. Obviously, a may also be transformed by $\varphi \in \Phi(a,b)$, so that its full transformation could be the composition $\varphi \circ \psi$ with invariance of *in* preserved.

Similar analyses can be given of the internal invariance properties of the other prepositions listed in Table 1. They all have different symmetry groups, but the exact nature of each group is determined by the particular shape and size of the relevant bodies, which may vary drastically from one context to another. It is not surprising that no computationally simple theory of invariance works for ordinary spatial relations as expressed in natural language, for the robust range of applicability of such relations is far too complex to have a simple uniform geometry.

Further applications. In addition to providing an analysis of spatial terms, the kind of approach outlined above can be used in a comparison of different languages. A natural question is whether there is a universal semantics of spatial perception or different languages have intrinsically different semantics. Certainly there are subtle differences in the range of prepositions. Bowerman (1989) points out, for example, that whereas we can say in English both *The cup is on the table* and *The fly is on the window*, in German we use the preposition *auf* for being on the table, but *an* for being on the vertical window.

Such differences are to be expected, although it is a matter of considerable interest to study how the prepositions of different languages overlap in their semantic coverage. It is a well-known fact that learning the correct use of prepositions is one of the most difficult aspects of learning a second language within the family of Indo-European languages.

One hope might be that there is a kind of universal spatial perception, so we might search for geometrical invariance across languages. But if we fit the geometries closely to individual languages, the primitives but not the theorems may be different in the sense of having a different geometrical meaning. A related set of questions can be asked about the order of developmental use of spatial prepositions by children. This seems to present a particularly good opportunity for careful examination of what happens in different languages, because of the relatively concrete and definite rules of usage that govern spatial prepositions in each language in their literal use. Bowerman (1989) has an excellent discussion of many details, but does not focus on systematic geometrical aspects.

4.3 Invariance in Theories of Measurement

In connection with any measured property of an object, or set of objects, it may be asked how unique is the number assigned to measure the property. For example, the mass of a pebble may be measured in grams or pounds. The number assigned to measure mass is unique once a unit has been chosen.

The measurement of temperature in °C or °F has different characteristics. Here an origin as well as a unit is arbitrarily chosen. Other formally different kinds of measurement are exemplified by (1) the measurement of probability, which is absolutely unique, and (2) the ordinal measurement of such physical properties as hardness of minerals, or such psychological properties as intelligence and racial prejudice.

Use of these different kinds of transformations is basic to the main idea of this chapter. An empirical hypothesis, or any statement in fact, which uses numerical quantities is empirically meaningful only if its truth value is invariant under the appropriate transformations of the numerical quantities involved. As an example, suppose a psychologist has an ordinal measure of I.Q., and he thinks that scores $S(a)$ on a certain new test T

have ordinal significance in ranking the intellectual ability of people. Suppose further that he is able to obtain the ages $A(a)$ of his subjects. The question then is: Should he regard the following hypothesis as empirically meaningful?

HYPOTHESIS 1. *For any subjects a and b, if $S(a)/A(a) < S(b)/A(b)$, then $I.Q.(a) < I.Q.(b)$.*

From the standpoint of the invariance characterization of empirical meaning, the answer is negative. To see this, let I.Q.$(a) \geq$ I.Q.(b), let $A(a) = 7$, $A(b) = 12$, $S(a) = 3$, $S(b) = 7$. Make no transformations on the I.Q. data, and make no transformations on the age data. But let ϕ be an increasing transformation which carries 3 into 6 and 7 into itself. Then we have
$$\frac{3}{7} < \frac{7}{12},$$
but
$$\frac{6}{7} \geq \frac{7}{12},$$
and the truth value of Hypothesis 1 is not invariant under ϕ.

The empirically significant thing about the transformation characteristic of a quantity is that it expresses in precise form how unique the structural isomorphism is between the empirical operations used to obtain a given measurement and the corresponding arithmetical operations or relations. If, for example, the empirical operation is simply that of ordering a set of objects according to some characteristic, then the corresponding arithmetical relation is that of less than (or greater than), and any two functions which map the objects into numbers in a manner preserving the empirical ordering are adequate.

It is then easy to show that, if f_1 and f_2 are adequate in this sense, then they are related by a monotone-increasing transformation. Only those arithmetical operations and relations which are invariant under monotone-increasing transformations have any empirical significance in this situation.

When we turn from the examination of numerical quantities to models of more complex theories, we obtain results of a similar character. For example, in examining classical mechanics we get a representation that is unique, when units of measurement are fixed, up to a Galilean transformation, that is, a transformation to some other inertial system. In the case of relativistic structures of particle mechanics, the uniqueness is up to Lorentz transformations. These two kinds of transformations are analyzed in Chapter 6.

To give an elementary example, we can state the uniqueness result corresponding to the representation theorem (Theorem 3.3.1) for finite weak orders. *Let $\mathfrak{A} = (A, \preceq)$ be a finite weak order. Then any two numerical weak orderings to which it is homomorphic are related by a strictly increasing numerical function.* Put in other language, the numerical representation of finite weak orders is unique up to an ordinal transformation. Invariance up to ordinal transformations is not a very strong property of a measurement, and it is for this reason that Hypothesis 1 turned out not to be meaningful.

Moving back to the general scheme of things, a representation theorem should ordinarily be accompanied by a matching invariance theorem stating the degree to which a representation of a structure is unique. In the mathematically simple and direct cases it is easy to identify the group as some well-known group of transformations. For more complicated structures, for example, structures that satisfy the axioms of a scientific theory, it may be necessary to introduce more complicated apparatus, but the objective is the same, to wit, to characterize meaningful concepts in terms of invariance.

One note to avoid confusion: it is when the concepts are given in terms of the representation, for example, a numerical representation in the case of measurement, or representation in terms of Cartesian coordinates in the case of geometry, that the test for invariance is needed. When purely qualitative relations are given, which are defined in terms of the qualitative primitives of a theory, for example, those of Euclidean geometry, then it follows at once that the defined relations are invariant and therefore meaningful. On the other hand, the great importance of the representations is the reduction in computations and notation they achieve, as well as understanding of structure. This makes it imperative that we have a clear conception of invariance and meaningfulness for representations which may be, in appearance, rather far removed from the qualitative structures that constitute models of the theory.

In the case of physics, the primitive notions themselves of a theory are not necessarily invariant. For example, if we axiomatize mechanics in a given frame of reference, then the notion of position for a particle is not invariant but is subject to a transformation itself. A more complicated analysis of invariance and meaningfulness is then required in such cases (see Chapter 6 for such an analysis). The general point is clear, however: the study of representation is incomplete without an accompanying study of invariance of representation. We amplify this point with a detailed analysis of measurement theories.

Second fundamental problem of measurement: invariance theorem. Solution of the representation problem for a theory of measurement does not completely lay bare the structure of the theory, for it is a banal fact of methodology that there is often a formal difference between the kind of assignment of numbers arising from different procedures of measurement. As an illustration, consider the following five statements:

(1) The number of people now in this room is 7.

(2) Stendhal weighed 150 on September 2, 1839.

(3) The ratio of Stendhal's weight to Jane Austen's on July 3, 1814 was 1.42.

(4) The ratio of the maximum temperature today to the maximum temperature yesterday is 1.1.

(5) The ratio of the difference between today's and yesterday's maximum temperature to the difference between today's and tomorrow's maximum temperature will be .95.

The empirical meaning of statements (1), (3), and (5) is clear, provided we make the natural assumptions, namely, for (3) that the same scale of weight, whether it be avoirdupois or metric, was being used, and for (5) that the same temperature scale is being used, whether it be Fahrenheit or Celsius. In contrast, (2) and (4) do not have a clear empirical meaning unless the *particular* scale used for the measurement is specified. On the basis of these five statements we may formally distinguish three kinds of measurement. Counting is an example of an *absolute* scale. The number of members of a given collection of objects is determined uniquely. There is no arbitrary choice of a unit or zero available. In contrast, the measurement of mass or weight is an example of a *ratio* scale. Any empirical procedure for measuring mass does not determine the unit of mass. The choice of a unit is an empirically arbitrary decision made by an individual

4.3 THEORIES OF MEASUREMENT

or group of individuals. Of course, once a unit of measurement has been chosen, such as the gram or pound, the numerical mass of every other object in the universe is uniquely determined. Another way of stating this is to say that the measurement of mass is unique up to multiplication by a positive constant. (The technical use of 'up to' will become clear later.) The measurement of distance is a second example of measurement of this sort. The ratio of the distance between Palo Alto and San Francisco to the distance between Washington and New York is the same whether the measurement is made in miles or kilometers.

To avoid certain common misunderstandings, amplification is needed of the claim that no empirical procedure for measuring mass determines the unit of mass. A chemist, measuring a sample of a certain ferric salt on an equal-arm balance with a standard series of metric weights, might find this claim surprising, for he might suppose that the selection of a standard series of metric weights had fixed as part of his empirical procedure one gram as the unit of measurement. There are at least two lines of argument that should prove effective in convincing the chemist that his supposition was incorrect. In the first place, it would be pertinent to point out that exactly the same information could be expressed by converting this final measurement into a measurement in pounds or ounces. To express the measurement in this latter form, no further empirical operations with the equal-arm balance would have to be performed. But, it must be admitted, the chemist could well reply that, although no further empirical operations had to be performed with the present balance, the use of the conversion factor from grams to pounds entails appealing to empirical operations previously performed by some Bureau of Standards in determining the conversion factor. An analysis of his retort takes us to the second line of argument, which goes deeper and is more fundamental. To begin with, his appeal to previous empirical operations on other balances may be turned back on him, for to justify labeling his measurement with his given standard series as a measurement in grams, appeal must also be made to previously performed empirical operations, namely, those which constituted the calibration of his series of weights as a standard metric series. The important point is that the empirical operations performed by the chemist himself establish no more or less than the ratio of the mass of the ferric salt sample to a subset of weights in his standard series. And the same kind of ratio statement may be made about the empirical operations that led to the calibration of his standard series by the technicians of the firm which produced the series. In other words, the statement by the chemist:

(6) This sample of ferric salt weighs 1.679 grams

may be replaced by the statement:

(7) The ratio of the mass of this sample of ferric salt to the gram weight of my standard series is 1.679, and the manufacturer of my series has certified that the ratio of my gram weight to the standard kilogram mass of platinum-iridium alloy at the International Bureau of Weights and Measures, near Paris, is .0010000.[4]

The measurement of temperature is an example of the third formally distinct kind

---

[4] As is evident, I am ignoring problems of the accuracy of measurement, which are of great importance in many areas of science. The subject, when carefully discussed, uses extensively the probability concepts developed in Chapter 5.

of measurement mentioned earlier. An empirical procedure for measuring temperature by use of a thermometer determines neither a unit nor an origin.[5] In this sort of measurement the ratio of any two intervals is independent of the unit and zero point of measurement. For obvious reasons, measurements of this kind are called *interval* scales. Examples other than measurement of temperature are measurements of temporal dates, linear position, or cardinal utility.[6]

In terms of the notion of absolute, ratio, and interval scales we may formulate the second fundamental problem for any exact analysis of a procedure of measurement: *determine the scale type of the measurements resulting from applying the procedure.* As the heading of this section we have termed this problem the *invariance* problem for a theory of measurement. From a mathematical standpoint the determination of the scale type of measurement arising from a given algebra of empirical operations and relations is the determination of the way in which any two numerical models of the algebra isomorphic to a given empirical model are related. For example, in the case of mass, we may make four equivalent statements:

(8) Mass is a ratio scale.

(9) The measurement of mass is unique up to multiplication by a positive number (the number corresponding to a change of unit).

(10) The measurement of mass is unique up to a similarity transformation (such a transformation is just multiplication by a positive number).

(11) Given any model of an appropriate algebra of operations and relations for measuring mass, then any two numerical models (of this algebra) that are isomorphic to the given model are related by a similarity transformation.

The precise meaning of 'isomorphic' in (11) was given in Chapter 3.[7] At this point the only aim is to give a general sense of what is meant by the invariance problem. The 'unique up to' terminology of (9) and (10) is often used in purely mathematical contexts. Another formulation of the uniqueness problem is one which uses the mathematical notion of invariance: *determine the set of numerical transformations under which a given procedure of measurement is invariant.*[8] In detailed work in later chapters the notion of invariance will often not be used in a technical way, but it will be obvious what definition would be appropriate in a given context and how the uniqueness results obtained could be sharply formulated in terms of invariance.

Classification of scales of measurement. The general notion of a scale is not something we need to define in an exact manner for subsequent developments. For the purpose of systematizing some of the discussion of the uniqueness problem, we may define a scale as a class of measurement procedures having the same transformation

---

[5] Naturally we are excluding from consideration here the measurement of absolute temperature whose zero point is not arbitrary.

[6] *Cardinal utility* is economists' term for an interval-scale measurement of utility, as opposed to merely ordinal measurement of utility.

[7] Sometimes, when two different physical objects have the same weight measure, the mapping to a numerical algebra is homomorphic, rather than isomorphic, as is also explained in Chapter 3.

[8] I did not say *group* of transformations because there are procedures of measurement for which the invariant set of transformations is not a group.

properties. Examples of three different scales have already been mentioned, namely, counting as an absolute scale, measurement of mass as a ratio scale, and measurement of temperature as an interval scale.[9]

Still another type of scale is one which is *arbitrary except for order*. A number is assigned to give a rough-and-ready ordinal classification. Moh's hardness scale, according to which minerals are ranked in regard to hardness as determined by a scratch test, and the Beaufort wind scale, whereby the strength of a wind is classified as calm, light air, light breeze, etc., are examples of this fourth, very weak kind of scale. The social sciences abound with ordinal scales, the significance of which is discussed below.

Numbers are also sometimes used for pure *classification*. For example, in some states the first number on an automobile license indicates the county in which the owner lives. The assignment of numbers in accordance with such a scale is arbitrary except for the assignment of the same number to people in the same county, and distinct numbers to people in distinct counties.

The weakest scale is one where numbers are used simply to name an object or person. The assignment is completely arbitrary. Draft numbers and the numbers of football players are examples of this sort of measurement. Such scales are usually called *nominal* scales.

We have distinguished, in an intuitive fashion, six types of scales. We will in the following pages primarily confine our discussion to an analysis of the first four. They are not, however, the only scales of measurement, and we shall upon occasion mention others.

We now want to characterize each of these six scales in terms of their transformation properties. For future reference each of the six transformations mentioned is formally defined. To begin with, an absolute scale is unique up to the *identity transformation*.

DEFINITION 1. *The* identity transformation *on the set of real numbers is the function f such that for every real number x, $f(x) = x$.*

So far, counting has been our only example of an absolute scale. For the moment we shall also include probability, but later we shall discuss the naturalness of the usual normalizing assumption, that is, that probabilities add up to one.

A ratio scale is unique up to a *similarity transformation*.

DEFINITION 2. *Let f be a real-valued function whose domain is the set of real numbers. Then f is a* similarity transformation *if and only if there exists a positive real number α such that for every real number x*

$$f(x) = \alpha x.$$

Measurement of mass and of distance have been our two examples of ratio scales. For the case of mass, if we wish to convert from pounds to ounces, the appropriate similarity transformation is to take $\alpha = 16$. If in the measurement of distance we wish to change feet to yards we set $\alpha = \frac{1}{3}$. The identity transformation is the similarity transformation with $\alpha = 1$.

---

[9]Counting, etc., is not *the* absolute scale, but technically a member of the absolute scale under our class definition. The usage here conforms to ordinary usage, which tends to treat classes as properties.

Third, an interval scale is unique up to a *linear transformation*.

DEFINITION 3. *Let f be a real-valued function whose domain is the set of real numbers. Then f is a* linear transformation *if and only if there is a positive number α and a number β such that for every number x*

$$f(x) = \alpha x + \beta.$$

If in the measurement of temperature we wish to convert x in degrees Fahrenheit to Celsius we use the linear transformation defined by $\alpha = \frac{5}{9}$ and $\beta = -\frac{160}{9}$. That is,

$$y = \frac{5}{9}(x - 32) = \frac{5}{9}x - \frac{160}{9}.$$

Obviously every similarity transformation is a linear transformation with $\beta = 0$.

An ordinal scale is unique up to a *monotone transformation*. Rather than define monotone transformation directly it will be convenient first to define monotone-increasing and monotone-decreasing transformations.

DEFINITION 4. *Let f be a real-valued function whose domain is some set of real numbers. Then f is a* monotone-increasing transformation *if and only if for every x and y in the domain of f if $x < y$ then $f(x) < f(y)$.*

Obviously every linear transformation is a monotone-increasing transformation on the set of all real numbers. The squaring function, that is, the function f such that

$$f(x) = x^2, \tag{1}$$

is not a linear transformation but it is monotone increasing on the set of non-negative real numbers. Notice that it does not have this property on the set of all real numbers for $-5 < 4$ but

$$f(-5) = 25 > 16 = f(4).$$

It is important to realize that a monotone-increasing transformation need not be definable by some simple equation like (1). For example, consider the set

$$A = \{1, 3, 5, 7\}$$

and let f be the function defined on A such that

$$\begin{aligned} f(1) &= -5 \\ f(3) &= 5 \\ f(5) &= 289 \\ f(7) &= 993. \end{aligned}$$

Clearly f is monotone increasing on A, but it does not satisfy any simple equation.

DEFINITION 5. *Let f be a real-valued function whose domain is some set of real numbers. Then f is a* monotone-decreasing transformation *if and only if for every x and y in the domain of f if $x < y$ then $f(x) > f(y)$.*

Two examples of monotone-decreasing transformations on the set of all real numbers are:

$$f(x) = -x$$

and

$$f(x) = -x^3 + 2.$$

4.3 THEORIES OF MEASUREMENT

As another instance, consider the set A again, and let f be defined on A such that

$$f(1) = 6$$
$$f(3) = 4$$
$$f(5) = 2$$
$$f(7) = -10.$$

Obviously f is monotone decreasing on A.

DEFINITION 6. *Let f be a real-valued function whose domain is some set of real numbers. Then f is a* monotone transformation *if and only if f is a monotone-increasing transformation or a monotone-decreasing transformation.*

Obviously monotone transformations characterize ordinal scales. The obvious invariance theorem for finite weak orders, as important examples of ordinal scales, is the following.

THEOREM 1. *Let $\mathfrak{A} = (A, \preceq)$ be a finite weak order. Then any two numerical representations of $\mathfrak{A}$ are related by a monotone transformation.*

In practice it is often convenient to consider only monotone-increasing or monotone-decreasing transformations, but this restriction is mainly motivated by hallowed customs and practices rather than by considerations of empirical fact.

Classificatory and nominal scales are unique up to a 1-1 *transformation* which is simply a one-to-one function.

In addition to these six classical scale types, which are essentially those proposed by Stevens (1947), there are others which will be discussed later. An example is the class of procedures which we shall call *hyperordinal* scales. They are characterized by transformations which preserve first differences and which we label *hypermonotone*. Here we may roughly characterize them as empirical structures which just lack being interval scales. Various methods of measuring sensation intensities or utility provide examples.

DEFINITION 7. *Let f be a real-valued function whose domain is some set of real numbers. Then f is a* hypermonotone transformation *if and only if f is a monotone transformation and for every x, y, u and v in the domain of f, if*

$$|x - y| < |u - v|$$

then

$$|f(x) - f(y)| < |f(u) - f(v)|.$$

Naturally every linear transformation is a hypermonotone-increasing transformation, but the converse is not true. Consider, for example,

$$A = \{1, 2, 3, 8\}$$

and the function f such that

$$f(1) = 1$$
$$f(2) = 2$$
$$f(4) = 5$$
$$f(8) = 15.$$

Clearly f is hypermonotone increasing but not linear on A.

There is, strictly speaking, a nondenumerable infinity of scales which are characterized by various groups of numerical transformations, but most of them are not of any real empirical significance. Also, it is possible to extend the notion of measurement to order structures like lattices and partial orderings which cannot be represented numerically in any natural way. My own view is that this extension is undesirable, for in the language of Scott and Suppes (1958) it identifies the class of theories of measurement with the class of finite relational structures, and the general theory of finite relational structures extends far beyond any intuitive conceptions of measurement.[10]

The classificatory analysis of this section is summarized in Table 2.

TABLE 2   Classification of Scales and Measurement

Scale	Transformation Properties	Examples
Absolute	Identity	Counting, relative frequency
Ratio	Similarity (Multiplication by a positive number)	Mass, distance, electric current, voltage
Interval	Linear (multiplication by a positive number and addition of an arbitrary number)	Temperature, potential energy, linear position, cardinal utility
Hyperordinal	Hypermonotone (preservation of order of first differences)	Pitch, loudness, utility
Ordinal	Monotone or monotone increasing	Moh's scale, Beaufort wind scale, qualitative preferences
Nominal	1-1	Phone numbers, Social Security numbers

I now turn to the invariance theorems corresponding to the four measurement representation theorems of Section 3.4

THEOREM 2. *Let $\mathfrak{A} = (A, \mathcal{F}, \succeq)$ be a finite, equally spaced extensive structure. Then a numerical representation of $\mathfrak{A}$ is invariant under the group of similarity transformations of Table 2.*

The proof of this elementary theorem is left to the reader.

THEOREM 3. *Let $\mathfrak{A} = (A, \succeq)$ be a finite, equally spaced difference structure. Then a numerical representation of $\mathfrak{A}$ is invariant under the group of linear transformations*

---

[10] As is probably obvious, a *finite* relational structure is a relational structure whose domain is a finite set. Given this finiteness, some numerical representation can be constructed in a more or less arbitrary manner, which will consequently be of little if any scientific interest. Only those reflecting some actual procedures of measurement, perhaps somewhat idealized, will be of any interest.

4.3 THEORIES OF MEASUREMENT

of Table 2.[11]

THEOREM 4. *Let $\mathfrak{A} = (A, \succeq, B)$ be a finite bisection structure. Then a numerical representation of $\mathfrak{A}$ is invariant under the group of linear transformations of Table 2.*[12]

It is worth noting that the next invariance theorem (Theorem 5) has a natural geometrical interpretation. If we think of the functions φ_1 and φ_2 mapping pairs into the Cartesian plane, then the uniqueness theorem says that in the standard geometrical sense, any change of scale must be uniform in every direction, but the origin can be translated by a different distance along the different axes.

THEOREM 5. *Let $(A_1, A_2, \succeq)$ be a nontrivial finite equally spaced additive conjoint*

---

[11] *Proof of Theorem 3.* To prove that the numerical function φ of Theorem 3 is unique up to a linear transformation, we define for every a in A two functions h_1 and h_2:

$$h_1(a) = \frac{\varphi_1(a) - \varphi_1(c^*)}{\varphi_1(c^*) - \varphi_1(c^{**})}$$

$$h_2(a) = \frac{\varphi_2(a) - \varphi_2(c^*)}{\varphi_2(c^*) - \varphi_2(c^{**})}$$

where φ_1 and φ_2 are two functions satisfying the representation construction and c^* is the first element of A under the ordering $\succeq$ and c^{**} the second element. We can easily show that h_1 is a linear transformation of φ_1 and h_2 is a linear transformation of φ_2 and also that h_1 is identical to h_2. It is then easy to prove that φ_1 is a linear transformation of φ_2, that is, there are numbers α, β with $\alpha > 0$ such that for every a in A

$$\varphi_1(a) = \alpha \varphi_2(a) + \beta.$$

Strictly speaking, c^* is not necessarily *the* first element and c^{**} *the* second element under the ordering $\succeq$, for there could be other members of the domain A equivalent in the ordering to c^* and still others equivalent to c^{**}. The simplest technical way to make this precise is to introduce equivalence classes with respect to the ordering, as was done in Chapter 3 in a footnote to the proof of Theorem 3.

[12] *Proof of Theorem 4.* For the proof of the uniqueness of the numerical function φ of Theorem 4 up to a linear transformation, as in the case of the proof of Theorem 3, we assume we have two functions φ_1 and φ_2 both satisfying (i) and (ii). We then define h_1 and h_2, just as in that proof. By the very form of the definition it is clear that h_1 is a linear transformation of φ_1, and h_2 a linear transformation of φ_2. We complete the proof by an inductive argument to show that $h_1 = h_2$ (whence φ_2 is a linear transformation of φ_1).

The induction is with respect to the elements of A ordered by $\succ$, with a_1 the first element.

Now by definition

$$h_1(a_1) = h_2(a_1) = 0.$$

Suppose now that for a_m, with $m \leq n$,

$$h_1(a_m) = h_2(a_m).$$

We prove that

$$h_1(a_{n+1}) = h_2(a_{n+1}).$$

Now we know at once that $a_{n-1} J a_n$ and $a_n J a_{n+1}$, whence by virtue of Axiom 5

$$B(a_{n-1}, a_n, a_{n+1}),$$

and therefore by hypothesis

$$2\varphi_i(a_n) = \varphi_i(a_{n-1}) + \varphi_i(a_{n+1}),$$

whence

$$\varphi_i(a_{n+1}) = 2\varphi_i(a_n) - \varphi_i(a_{n-1}),$$

for $i = 1, 2$. Now since h_i is a linear transformation of φ_i, it follows that we also have

$$h_i(a_{n+1}) = 2h_i(a_n) - h_i(a_{n-1}),$$

but by the inductive hypothesis the right-hand side of this last equation is the same for h_1 and h_2, and so we conclude that $h_1(a_{n+1}) = h_2(a_{n+1})$.

structure. Then a numerical representation (φ_1, φ_2) as in Theorem 3.4.4 is invariant up to the group of pairs (f_1, f_2) of linear transformations of (φ_1, φ_2), but with the restriction that f_1 and f_2 have a common multiplicative constant, i.e., there are numbers α, β and γ such that

$$f_1 \circ \varphi_1 = \alpha \varphi_1 + \beta \text{ and } f_2 \circ \varphi_2 = \alpha \varphi_2 + \gamma.^{13}$$

4.4 Why the Fundamental Equations of Physical Theories Are Not Invariant

Given the importance, even the deep physical significance, attached to the concept of invariance in physics, and often also in mathematics, it is natural to ask why theories are seldom written in an invariant form.

My answer will concentrate on physics, and somewhat more generally, on theories of measurement. One obvious and practically important point in physics is that it is simpler and more convenient to make and record measurements relative to the fixed framework of the laboratory, rather than to record them in a classical or relativistic invariant fashion.

Moreover, this point is even more obvious if we examine the use of units of measurements. Analysis of measurements of length and mass will suffice. In both these cases,

---

[13] *Proof of Theorem 5.* To prove the invariance results on φ_1 and φ_2, we may proceed as in the case of Theorem 3. We define four functions:

$$g(a) = \frac{\varphi_1(a) - \varphi_1(c^*)}{\varphi_1(c^*) - \varphi_1(c^{**})} \qquad g'(a) = \frac{\varphi_1'(a) - \varphi_1'(c^*)}{\varphi_1'(c^*) - \varphi_1'(c^{**})}$$

$$h(p) = \frac{\varphi_2(p) - \varphi_2(r^*)}{\varphi_2(r^*) - \varphi_2(r^{**})} \qquad h'(p) = \frac{\varphi_2'(p) - \varphi_2'(r^*)}{\varphi_2'(r^*) - \varphi_2'(r^{**})},$$

where c^* is the first element of A, under the ordering $\succeq$ on A_1, c^{**} is the second element, r^* is the first element of A_2, and r^{**} the second. It is, as before, obvious that g is a linear transformation of φ_1, g' a linear transformation of φ_1', h a linear transformation of φ_2, and h' a linear transformation of φ_2'. Secondly, we can show that $g = g'$ and $h = h'$ by an inductive argument similar to that used in the proof of Theorem 4. So we obtain that there are numbers α, α', β and γ with $\alpha, \alpha' > 0$ such that for every a in A_1 and every p in A_2

(iii) $\varphi_1'(a) = \alpha \varphi_1(a) + \beta$ and $\varphi_2'(p) = \alpha' \varphi_2(p) + \gamma$.

It remains to show that $\alpha = \alpha'$ when A_1 and A_2 each have at least two elements not equivalent in order. Without loss of generality we may take $a \succ b$ and $p \succ q$. Then we have, from $(a, q) \approx (b, p)$,

$$\varphi_1'(a) - \varphi_1'(b) = \varphi_2'(p) - \varphi_2'(q),$$

and thus by (iii)

$$\frac{\alpha \varphi_1(a) - \alpha \varphi_1(b)}{\alpha' \varphi_2(p) - \alpha' \varphi_2(q)} = 1,$$

and so

$$\frac{\alpha}{\alpha'} \left(\frac{\varphi_1(a) - \varphi_1(b)}{\varphi_2(p) - \varphi_2(q)} \right) = 1;$$

but by hypothesis

$$\varphi_1(a) - \varphi_1(b) = \varphi_2(p) - \varphi_2(q),$$

whence

$$\frac{\alpha}{\alpha'} = 1;$$

i.e.,

$$\alpha = \alpha',$$

which completes the proof.

4.4 FUNDAMENTAL EQUATIONS OF PHYSICAL THEORIES

selection of the unit of measurement is arbitrary, i.e., not invariant. What is invariant is the ratio of distances or masses. The awkward locutions of continual use of ratios rather than the shorthand of conventional standards of units can be seen in ancient Greek mathematics. In fact, elementary computations show the advantage of units. Suppose we need to compare about 1000 distances between cities with airports. If only invariant ratios were recorded, then about half a million ratios as pure numbers would be required, as opposed to 1000 physical quantities, i.e., 1000 numbers with a common unit, such as meters or kilometers.

The passion for writing ratios in the Greek style lasted a long time. Newton's *Principia* (1687/1946, p. 13) is a fine example. Here is Newton's classic geometrical formulation of his second law of motion: "The change of motion is proportional to the motive force impressed; and is made in the direction of the right line in which that force is impressed." It is often claimed, and quite sensibly, that moving away from the Greek geometrical style of expression to algebraic and differential equations was of great importance in introducing more understandable and efficient methods of formulation and computation in physics.

Apart from the rather special case of ratios, the same distinction is evident in synthetic and analytic formulations of standard geometries. A synthetic and invariant formulation of Euclidean geometry is easily formulated, just in terms of the qualitative relations of betweenness and congruence. The analytic representations, in contrast, are relative to an arbitrary choice of coordinate system and are therefore in and of themselves not invariant. An invariance theorem is ordinarily proved to show how the various appropriate analytic representations are related, i.e., by which transformations. These transformations, in fact, constitute the analytic form of the group of Euclidean motions already mentioned.

The advantages of the analytic representations, particularly in terms of vector spaces, is universally recognized in physics. It would be considered nothing less than highly idiosyncratic and eccentric for any physicist to recommend a return to the synthetic geometric formulations used before the eighteenth century.

Put in this context, it can be seen immediately that there is an inevitable conflict between invariance, as in synthetic formulations, and efficient computations, as in analytic formulations. In fact, selection of the 'right' coordinate system to facilitate, and particularly simplify, computation is recognized as something of significance to be taught and learned in physics.[14]

---

[14] A beautiful kinematical example in classical mechanics, originally due to Siacci (1878), is the resolution of the motion of a particle restricted to a plane using the radius vector, at each time t, from a fixed origin in the plane to the particle's position. This vector is represented by two components, not the usual (x, y) coordinates, but one along the tangent and the other along the normal of the vector, i.e., the perpendicular to the tangent. Now consider a particle whose normal and tangential components of acceleration are constant in time. So we have constant acceleration along the two moving axes and it is easy in this representation to solve the equation of motion to show that its position s at each instant is given by
$$s = ae^{b\varphi},$$
where φ is the angle made by the tangent to a fixed line. The details of this example can be found in Whittaker (1904/1937, pp. 21–22).

Beyond symmetry. From what I have said, it might be misleadingly inferred that in choosing convenient laboratory coordinate systems physicists neglect considerations of invariance. In practice, physicists hold on to invariance by introducing and using the concept of covariants. Before introducing this concept, it will be useful to survey the natural generalizations beyond symmetry in familiar geometries.

The most general automorphisms of Euclidean space are transformations that are compositions of rotations, reflections and translations. We have already considered rotations and reflections in Section 4.1. Familiar figures such as circles, squares and equilateral triangles are invariant under appropriate groups of rotations and reflections—an infinite group for circles, and finite groups for squares and equilateral triangles.

On the other hand, none of these familiar figures are invariant under automorphisms that include translations. In fact, no point is, so that the only invariant is that of the entire space being carried onto itself. But translations are important not only in pure geometry but in many applications, especially in physics, where the Galilean transformations of classical physics and the Lorentz transformations of special relativity play an important role, and both include translations.

In the last paragraph an unnoted shift was made from automorphisms of a qualitative form of Euclidean geometry, based on such primitive concepts as betweenness and congruence, to the numerical coordinate frames of reference of the Galilean and Lorentz groups of transformations. Such frames are needed to give an explicit account of the physicists' concept of covariants of a theory.

Typical examples of such covariants are velocity and acceleration, both of which are not invariant from one coordinate frame to another under either Galilean or Lorentzian transformations, because, among other things, the direction of the velocity or acceleration vector of a particle will in general change from one frame to another. (The scalar magnitude of acceleration is invariant.)

Covariants. The laws of physics are written in terms of such covariants. Without aiming at the most general formulation, the fundamental idea is conveyed by the following. Let $Q_1, \ldots, Q_n$ be quantities that are functions of the space-time coordinates, with some Q_i's being derivatives of others, for example. Then in general, as we go from one coordinate system to another, $Q'_1, \ldots, Q'_n$ will be covariant, rather than invariant, and so their mathematical form is different in the new coordinate system. But any physical law involving them, say,

$$F(Q_1, \ldots, Q_n) = 0 \tag{1}$$

must have the same form

$$F(Q'_1, \ldots, Q'_n) = 0 \tag{2}$$

in the new coordinate frame. This requirement of same form is the important invariant requirement. Equations (1) and (2) are, in the usual language of physicists, also called covariant. So, the term *covariant* applies both to quantities and equations. I omit an explicit formal statement of these familiar ideas.

Here is a simple example from classical mechanics. Consider the conservation of momentum of two particles before and after a collision, with v_i the velocity before, w_i the velocity afterward, and m_i the mass, $i = 1, 2$, of each particle. The law, in the form of (1), looks like this:

$$v_1 m_1 + v_2 m_2 - (w_1 m_1 + w_2 m_2) = 0,$$

and the transformed form will be, of course,

$$v'_1 m_1 + v'_2 m_2 - (w'_1 m_1 + w'_2 m_2) = 0,$$

but the velocities v_i and w_i will be, in general, covariant rather than invariant. The masses m_1 and m_2 are, of course, invariant.

Finally, I mention that in complicated problems, proving that a physical law is covariant, i.e., the invariance of its form, can be, at the very least, quite tedious.

4.5 Entropy as a Complete Invariant in Ergodic Theory[15]

I now turn to an extended example that is one of the most beautiful cases of invariance, with consequences both in mathematics and in physics. We will need to build up some apparatus for this discussion, which anticipates the detailed development in the first section of Chapter 5.

Let us first begin with a standard probability space $(\Omega, \Im, P)$, where it is understood that $\Im$ is a σ-algebra of subsets of Ω and P is a σ-additive probability measure on $\Im$. (See Definition 2 of Section 5.1.) We now consider a mapping T from Ω to Ω. We say that T is *measurable* if and only if whenever $A \in \Im$ then $T^{-1}A = \{\omega : T\omega \in A\} \in \Im$, and even more important, T is *measure preserving* if and only if $P(T^{-1}A) = P(A)$. The mapping T is *invertible* if the following three conditions hold: (i) T is $1-1$, (ii) $T\Omega = \Omega$, and (iii) if $A \in \Im$ then $TA = \{T\omega : \omega \in A\} \in \Im$. In the applications we are interested in, each ω in Ω is a doubly infinite sequence and T is the *right-shift* such that if for all n, $\omega_n = \omega'_{n+1}$ then $T(\omega) = \omega'$. Intuitively this property corresponds to stationarity of the process—a time shift does not affect the probability laws of the process, and we can then use T to describe orbits or sample paths in Ω.

I now introduce the central concept of entropy for ergodic theory. To keep the mathematical concepts and notations simple, I shall restrict myself to discrete-time processes and, in fact, to processes that have a finite number of states as well. So, for such processes we have a simple definition of entropy.

First, the entropy of a random variable $\mathbf{X}$ with a (discrete) probability density $p(x)$ is defined by

$$H(\mathbf{X}) = -\Sigma p_i \log p_i.$$

In a similar vein, for a stochastic process $\mathcal{X} = \{\mathbf{X}_n : 0 < n < \infty\}$

$$H(\mathcal{X}) = \lim_{n \to \infty} \frac{1}{n} H(\mathbf{X}_1, \ldots, \mathbf{X}_n). \qquad (1)$$

Notice that what we have done is define the entropy of the process as the limit of the entropy of the joint distributions. To illustrate ideas here, without lingering too long, we can write down the explicit expression for the joint entropy $H(\mathbf{X}, \mathbf{Y})$ of a pair of

---

[15] This section may be omitted without loss of continuity. A good detailed reference for the results outlined here and a more general survey of invariance in ergodic theory are to be found in Ornstein and Weiss (1991). The use of probability concepts in this section goes beyond the concepts introduced in Chapter 5, which concentrates on foundations rather than applications of probability.

discrete random variables with a joint discrete density distribution $p(\mathbf{X}, \mathbf{Y})$.

$$H(\mathbf{X}, \mathbf{Y}) = -\sum_x \sum_y p(x,y) \log p(x,y),$$

which can also be expressed as

$$H(\mathbf{X}, \mathbf{Y}) = -E \log p(\mathbf{X}, \mathbf{Y})$$

where E stands for expectation. Notice, of course, there is a requirement for (1) that I have been a little bit casual about. It is important to require that the limit exist. So the entropy of the stochastic process will not be defined if the limit of the entropies of the finite joint distributions does not exist as n goes to infinity.

For a Bernoulli process, that is, a process that has on each trial identically and independently distributed random variables (i.i.d.), we have a particularly simple expression for the entropy. It is just the entropy of a single one of the random variables, as the following equations show:

$$\begin{aligned} H(\mathcal{X}) &= \lim_{n \to \infty} \frac{1}{n} H(\mathbf{X}_1, \mathbf{X}_2, \ldots, \mathbf{X}_n) \\ &= \frac{n H(\mathbf{X}_1)}{n} = H(\mathbf{X}_1) = -\Sigma p_i \log p_i. \end{aligned}$$

We require only a slightly more complex definition for the case of a Markov chain, again, of course, with the provision that the limit exists.

$$\begin{aligned} H(\mathcal{X}) &= \lim_{n \to \infty} H(\mathbf{X}_n \mid \mathbf{X}_{n-1}, \ldots, \mathbf{X}_1) \\ &= H(\mathbf{X}_2 \mid \mathbf{X}_1) \\ &= -\Sigma_i p_i \Sigma_j p_{ij} \log p_{ij}. \end{aligned}$$

Now we give, without detailed technical explanations, because the examples will be simple, the definition of ergodicity. Let $(\Omega, \mathcal{F}, P, T)$ be a probability space with measure-preserving transformation T. Then T is *stationary* if for any measurable set A, $\mu(TA) = \mu(A)$, and transformation T is called *ergodic* if for every set A such that $TA = A$, the measure of A is either 0 or 1. We obtain a stochastic process by defining the random variable $\mathbf{X}_n$ in terms of T^n, namely, for every ω in Ω, $\mathbf{X}_n(\omega) = \mathbf{X}(T^n \omega)$. It is easy to see that Bernoulli processes as defined above are stationary and also ergodic. A more intuitive characterization of stationarity of a stochastic process is that every joint distribution of a finite subset of random variables of the process is invariant under time translation. So, for every k,

$$p(x_{t_1+k}, x_{t_2+k}, \ldots, x_{t_m+k}) = p(x_{t_1}, x_{t_2}, \ldots, x_{t_m}).$$

Perhaps the simplest example of a nonergodic process is the Markov chain in heads and tails:

	h	t
h	1	0
t	0	1

Given that the initial probability $p_1(h) = p_1(t) = \frac{1}{2}$, the event A consisting of the sample path having heads on every trial has probability $\frac{1}{2}$, i.e., $p(A) = \frac{1}{2}$, but $T(A) = A$ and thus the process is not ergodic. It is intuitively clear, already from this example, what

4.5 ENTROPY AS A COMPLETE INVARIANT

we expect from ergodic processes: a mixing that eliminates all dependence on the initial state.

On the other hand, we have an ergodic process, but the entropy is 0, when the matrix is slightly different and the probability of heads on the first trial is now 1.

$$\begin{array}{c|cc} & h & t \\ \hline h & 0 & 1 \\ t & 1 & 0 \end{array} \qquad p(h_1) = 1.$$

It is clear enough from the one possible sequence $hththt\ldots$ that all measurements or outcomes are predictable and so the entropy must be 0. Note what this process is. It is not a Bernoulli process, but a very special deterministic Markov chain with two states. In connection with this example, I mention that the standard definition in the literature for ergodic Markov chains does not quite require stationarity. The definition ordinarily used is that a Markov chain is ergodic if it has a unique asymptotic probability distribution of states independent of the initial distribution. So it is the independence of initial distribution rather than stationarity that is key. Of course, what is obvious here is that the process must be asymptotically stationary to be ergodic.

There are some further important distinctions we can make when a stochastic process has positive entropy; not all measurements are predictable, but some may be. We can determine, for example, a factor of the process, which is a restriction of the process to a subalgebra of events. A factor is, in the concepts used here, always just such a subalgebra of events. The concept of a K process (K for Kolmogorov), is that of a stochastic process which has no factor with 0 entropy, that is, has no factor that is deterministic. Bernoulli processes are K processes, but there are many others as well.

One important fact about ergodic processes is the fundamental ergodic theorem. To formulate the theorem we need the concept of the indicator function of a set A. For any event A, let $\mathbf{I}_A$ be the indicator function of A. That is, for $\omega \in A$

$$\mathbf{I}_A(\omega) = \begin{cases} 1 & \text{if } \omega \in A, \\ 0 & \text{if } \omega \notin A. \end{cases}$$

The meaning of the fundamental ergodic theorem is that the time averages equal the space or ensemble averages, that is, instead of examining a temporal cross-section for all ω's having a property expressed by an event A, we get exactly the same thing by considering the shifts or transformations of a single sample path ω. The theorem holds for all ω's except for sets of measure 0.

THEOREM 1. *For almost any ω in Ω*

$$\lim_{n \to \infty} \frac{1}{n} \sum_{k=0}^{n-1} \mathbf{I}_A(T^k \omega) = P(A).$$

Isomorphism of ergodic processes. We define two stationary stochastic processes as being *isomorphic (in the measure-theoretic sense)* if we can map events of one to events of the other, while preserving probability. The intuitive meaning of this is that their structural uncertainty is the same.

More technically, we define isomorphism in the following way. Given two probability spaces $(\Omega, \Im, P)$ and $(\Omega', \Im', P')$ and two measure-preserving transformations T and T' on Ω and Ω' respectively, then we say that $(\Omega, \Im, P, T)$ is *isomorphic in the measure-theoretic sense* to $(\Omega', \Im', P', T')$ if and only if there exists a function $\varphi : \Omega_0 \to \Omega'_0$, where $\Omega_0 \in \Im, \Omega'_0 \in \Im', P(\Omega_0) = P(\Omega'_0) = 1$, that satisfies the following conditions:

(i) φ is $1-1$,
(ii) If $A \subset \Omega_0$ and $A' = \varphi A$ then $A \in \Im$ iff $A' \in \Im'$, and if $A \in \Im$
$$P(A) = P'(A'),$$
(iii) $T\Omega_0 \subseteq \Omega_0$ and $T'\Omega'_0 \subseteq \Omega'_0$,
(iv) For any ω in Ω_0
$$\varphi(T\omega) = T'\varphi(\omega).$$

The mapping of any event A to an event A' of the same probability, i.e, condition (ii), is conceptually crucial, as is the commutativity expressed in (iv).

A problem that has been of great importance has been to understand how stochastic processes are related in terms of their isomorphism. As late as the middle 1950s, it was an open problem whether two Bernoulli processes, one with two states, each with a probability of a half, $B(\frac{1}{2}, \frac{1}{2})$, and one with three states, each having a probability of one third, $B(\frac{1}{3}, \frac{1}{3}, \frac{1}{3})$, are isomorphic, in accordance with the definition given above. The problem was easy to state, but difficult to solve. Finally, the following theorem was proved by Kolmogorov and Sinai.

THEOREM 2. (Kolmogorov 1958, 1959; Sinai 1959). *If two Bernoulli processes are isomorphic then their entropies are the same.*

Since $B(\frac{1}{2}, \frac{1}{2})$ and $B(\frac{1}{3}, \frac{1}{3}, \frac{1}{3})$ have different entropies, by contraposition of Theorem 2, they are not isomorphic. It was not until twelve years later that Ornstein proved the converse.

THEOREM 3. (Ornstein 1970). *If two Bernoulli processes have the same entropy they are isomorphic.*

With these two theorems, extension to other processes, in particular, Markov processes, was relatively straightforward. And so, the following theorem was easily proved.

THEOREM 4. *Any two irreducible, stationary, finite-state discrete Markov processes are isomorphic if and only if they have the same periodicity and the same entropy.*

Out of this came the surprising and important invariant result. Entropy is a complete invariant for the measure-theoretic isomorphism of aperiodic, ergodic Markov processes. Being a complete invariant also means that two Markov processes with the same periodicity are isomorphic if and only if their entropy is the same. What is surprising and important about this result is that the probabilistic structure of a Markov process is large and complex. The entropy, on the other hand, is a single real number. It is hard to think of an invariance result for significant mathematical structures in which the disparity between the complexity of the structures and the easy statement of the complete invariant is so large. Certainly, within probability theory, it is hard to match as an important invariance result. It is also hard to think of one in physics of comparable simplicity in terms of the nature of the complete invariant.

On the other hand, it is important to note that even though a first-order ergodic Markov process and a Bernoulli process may have the same entropy rate and, therefore, be isomorphic in the measure-theoretic sense, they are in no sense the same sort of process. We can, for example, show by a very direct statistical test whether a given sample path of sufficient length comes from the Bernoulli process or the first-order Markov process. There is, for example, a simple chi-square test for distinguishing between the two. It is a test for first-order versus zero-order dependency in the process.

But this situation is not at all unusual in the theory of invariance. We can certainly distinguish, as inertial observers, between a particle that is at rest in our inertial frame of reference, and one that is moving with a constant positive speed. What is observed can be phenomenologically very different for two different inertial observers, Galilean or Lorentzian. This way of formulating matters may seem different from the approach used in the geometrical or ergodic cases, but this is so only in a superficial way. At a deeper and more general level, all the cases of invariance surveyed here have the common theme of identifying some special properties that are constant or unchanging.

5

Representations of Probability

No major contemporary school of thought on the foundations of probability disputes any of the axioms, definitions, or theorems considered in Section 1, with some minor exceptions to be noted. From the standpoint of the concepts introduced in Section 1, the long-standing controversies about the foundations of probability can be formulated as disputes about which models of the theory of probability as characterized in Definition 2 are to be regarded as genuine probability spaces. Put another way, the controversies center around how the axioms of Definition 2 are to be extended to include only the "genuine" cases of probability spaces. In this and the following sections, we shall divide the many variant positions on the nature of probability into the six most important classes of opinion. Section 2 is devoted to the classical definition, which was given a relatively definitive formulation by Laplace at the beginning of the nineteenth century. The next section is devoted to the relative-frequency view of probability, first formulated in a reasonably clear fashion by John Venn at the end of the nineteenth century, and later investigated and extended by Richard von Mises and others. Section 4 is focused on the more recent theories of randomness in finite sequences and measures of their complexity as defined by Kolmogorov. Section 5 is devoted to logical theories of probability, with particular emphasis on Jeffreys and on confirmation theory as developed by Carnap and others. Section 6 considers propensity theories of probability, especially in terms of concrete physical or psychological examples, which lead to qualitative invariant relations that are often not noticed. Section 7 concerns subjective theories, with which de Finetti, I. J. Good, Lindley and L. J. Savage are particularly associated. Section 8, the final section, surveys pragmatic views about the nature of probability, which do not wholeheartedly endorse any one of the six main positions analyzed in the preceding sections.

In the statistical, more than the philosophical, literature on foundations of probability there is a close association with the foundations of statistical inference, or, what some philosophers would call, inductive inference. Because the literature on these two related foundational topics is now so large, I have limited myself in the chapter to the foundations of probability, with only some superficial remarks about statistical inference, where the nature of probability is deeply intertwined with a specific form of statistical inference. The present chapter, already very long, is still far from adequate in conveying many important aspects of the various representations or interpretations of probability.

A chapter of similar length would be required to give a comparable analysis of the main alternative approaches to the foundations of statistics.

5.1 The Formal Theory

Our axiomatization of probability proceeds by defining the set-theoretic predicate 'is a probability space'. The intellectual source of the viewpoint presented is Kolmogorov's classic work (1933/1950). The main purpose of this section is to set forth the standard axioms and concepts of probability theory as background for the study of the foundations of probability in the rest of this chapter. Readers familiar with the standard formal development, as taught in modern probability courses, should omit this section. Almost all of the philosophical discussion of probability is to be found later in the chapter. Also, some much less standard formal theory is to be found in Section 7.2 on hidden variables in quantum mechanics, in Section 7.3 on concepts of reversibility for stochastic processes and in Sections 8.3 and 8.4 on stochastic learning models.

Primitive notions. The axioms are based on three primitive notions: a nonempty set Ω of possible outcomes, a family $\Im$ of subsets of Ω representing possible events,[1] and a real-valued function P on $\Im$, i.e., a function whose range is a set of real numbers; for $E \in \Im, P(E)$ is interpreted as the probability of E. These three notions may be illustrated by a simple example. Let Ω be the set of all possible outcomes of three tosses of a coin.

$$\Omega = \{hhh, hht, hth, htt, thh, tht, tth, ttt\},$$

where hhh is the event of getting three heads in a row, etc. Let $\Im$ be the family of all subsets of Ω. The *event* of getting at least two heads is the set

$$A = \{hhh, hht, hth, thh\}.$$

The event of getting exactly two heads is the set

$$B = \{hht, hth, thh\},$$

and so on. The important point is this. Any event that could occur as a result of tossing the coin three times may be represented by a subset of Ω. This subset has as elements just those possible outcomes, the occurrence of any one of which would imply the occurrence of the event. Thus, the event of getting at least two heads occurs if any one of four possible outcomes occurs.

This representation of events as sets is one of the more important aspects of the application of probability theory to empirical data. To decide what sets are to represent the events is to make the major decision about the canonical form of the data from an experiment. As we can see in the simple case of tossing a coin three times, we represent the outcomes of the experiment in such a way that most of the physical detail of the actual tossing of the coin is ignored. We do not, for example, record the time at which the coin came to rest, nor the time between tosses. We do not describe the mechanism used to make the tosses. We are interested only in a very simplified, and, in fact, very simpleminded abstraction of what actually happened.

---

[1] In what follows, possible events will be referred to simply as 'events'.

5.1 THE FORMAL THEORY

The language of events and the representation of events as subsets of the set of all possible outcomes are now standard in textbooks on probability theory. Perhaps the only major exception is to be found in the literature on logical theories of probability by philosophers and logicians; but it is easy to give a set-theoretical version of this theory that in no way affects its substantive content, as we shall see later. However, this language of events has not always been universally accepted. The earlier difficulties with this language stem from the lack of a clearly defined formal basis for it, as is given by the standard set-theoretical interpretation now used. Prior to the characterization of events as subsets of the set of all possible outcomes, or, as we shall often put it, as subsets of the sample space or probability space, the logical structure of an event was not well defined, and it was not always clear what was to count as an event.[2] Objection to the vagueness of the notion of the occurrence of an event is found in the following passage from Keynes' treatise on probability (1921, p. 5).

> With the term "event", which has taken hitherto so important a place in the phraseology of the subject, I shall dispense altogether. Writers on probability have generally dealt with what they term the "happening" of "events". In the problems which they first studied this did not involve much departure from common usage. But these expressions are now used in a way which is vague and ambiguous; and it will be more than a verbal improvement to discuss the truth and the probability of propositions instead of the occurrence and the probability of *events*.

It is to be noted that Keynes' objections appeared more than ten years before the concept of event was clarified in Kolmogorov's classical treatise. Many of the difficulties in Keynes' own views of probability arise from the vagueness of the notion of proposition. The essentially intensional character of this notion makes it difficult to use as a fundamental concept in probability theory if standard mathematical methods are subsequently to be applied. I shall consider this point again in discussing logical theories of probability in Section 5.

Let us return to the coin-tossing example. If the coin is fair, then for each ω in Ω,

$$P\{\omega\} = \frac{1}{8}.$$

The probability of any other event may be obtained simply by adding up the number of elements of Ω in the subset. Thus the probability of getting at least two heads in three tosses of the coin is $P(A)$ and

$$\begin{aligned} P(A) &= \{hhh, hht, hth, thh\} \\ &= P\{hhh\} + P\{hht\} + P\{hth\} + P\{thh\} \\ &= \tfrac{1}{8} + \tfrac{1}{8} + \tfrac{1}{8} + \tfrac{1}{8} = \tfrac{1}{2}. \end{aligned}$$

The additive property just exemplified is one of the fundamental properties of probability postulated as an axiom: the probability of either one of two mutually exclusive

---

[2] I make no formal distinction between 'sample space' and 'probability space', but it is natural to refer to the sample space of an experiment, which is usually finite, and the probability space of a theory, which is often infinite. The set-theoretical definition that follows (Definition 2) mentions only 'probability space'.

events is equal to the sum of their individual probabilities, that is, if $A \cap B = \emptyset$ then
$$P(A \cup B) = P(A) + P(B).$$

Language of events. Since throughout the remainder of this book we shall apply probability language to sets which are interpreted as events, it will be useful to have a table relating set-theoretical notation to probability terms. To begin with, the assertion that $A \in \Im$ corresponds to asserting that A is an event. At first glance the notion of the family $\Im$ of subsets may seem superfluous. It may seem reasonable to treat as events all possible subsets of the sample space Ω. When Ω is a finite set, this is a reasonable and natural approach; but it is not an acceptable treatment for very large spaces of the sort that arise in many applications of probability theory. An example is given below to show why it is unreasonable to assume that all subsets of Ω may be considered as events. To give another example from the table, the assertion that $A \cap B = \emptyset$ corresponds to the assertion that events A and B are incompatible. In this table and subsequently, we use a compact notation for the intersection or union of a finite or infinite number of sets:

$$\bigcap_{i=1}^{n} A_i = A_i \cap A_2 \cap \cdots \cap A_n$$

$$\bigcup_{i=1}^{n} A_i = A_1 \cup A_2 \cup \cdots \cup A_n$$

$$\bigcap_{i=1}^{\infty} A_i = A_1 \cap A_2 \cap \cdots$$

$$\bigcup_{i=1}^{\infty} A_i = A_1 \cup A_2 \cup \cdots .$$

We use the same sort of notation for sums:

$$\sum_{i=1}^{n} P(A_i) = P(A_1) + P(A_2) + \cdots + P(A_n).$$

Also, if Ω is a finite set, for example, and f is a numerical function defined on Ω, then if $\Omega = \{a, b, c\}$,

$$\sum_{\omega \in \Omega} f(\omega) = f(a) + f(b) + f(c).$$

SET-THEORETICAL NOTATION AND PROBABILITY TERMS

Set Theory

(a) $A \in \Im$
(b) $A \cap B = \emptyset$
(c) $A_i \cap A_j = \emptyset$ for $i \neq j$
(d) $\bigcap_{i=1}^{n} A_i = B$
(e) $\bigcup_{i=1}^{n} A_i = B$
(f) $B = -A$
(g) $A = \emptyset$
(h) $A = \Omega$
(i) $B \subseteq A$

Probability Theory

(a) A is an event.
(b) Events A and B are incompatible.
(c) Events $A_1, A_2, \ldots, A_n$ are pairwise incompatible.
(d) B is the event which occurs when events $A_1, A_2, \ldots, A_n$ occur together.
(e) B is the event which occurs when at least one of the events $A_1, A_2, \ldots, A_n$ occurs.
(f) B is the event which occurs when A does not.
(g) Event A is impossible.
(h) Event A is inevitable or certain.
(i) If event B occurs, then A must occur.

In (f), complementation is, of course, with respect to Ω. For instance, if A is the event of getting at least two heads as in our example above, then $-A$ is the event of not getting at least two heads and

$$-A = \Omega - A = \{tht, tth, htt, ttt\}.$$

Algebras of events. In order that the family of events has the appropriate set-theoretical structure, we must postulate certain closure properties for the family. By postulating that the union of any two events is also an event and that the complement of any event is also an event, the family of sets representing events becomes in the usual set-theoretical language an algebra of sets. By postulating in addition that the union of any infinite sequence of elements of $\Im$ is also in $\Im$, the algebra becomes a σ-algebra. Without these closure properties awkward situations arise in talking about events that naturally should be available. For example, if the event A is a head on the first toss of a coin and B is the event of a tail on the second toss, it would be strange not to be able to consider the event $A \cap B$, which is the event of the joint occurrence of A and B. The countable unions required for σ-algebras are not intuitively too obvious, but this property is often required for more advanced work. We shall include it explicitly here in order to discuss later the kind of examples that do or do not have this property, and the extent to which in more complicated contexts it is an intuitively desirable property to have. These notions are formalized in the following definition.

DEFINITION 1. *$\Im$ is an* algebra of sets *on Ω if and only if $\Im$ is a nonempty family of subsets of Ω and for every A and B in $\Im$:*

1. $-A \in \Im$;
2. $A \cup B \in \Im$.

Moreover, if $\Im$ is closed under countable unions, that is, if for $A_1, A_2, \ldots, A_n, \ldots \in \Im$, $\cup_{i=1}^{\infty} A_i \in \Im$, then $\Im$ is a σ-algebra on Ω.

The proof of the following theorem about elementary properties of algebras of sets is left as an exercise for the reader.

THEOREM 1. *If $\Im$ is an algebra of sets on Ω then*

(i) $\Omega \in \Im$,

(ii) $\emptyset \in \Im$,

(iii) *If $A \in \Im$ and $B \in \Im$, then*
$$A \cap B \in \Im,$$

(iv) *If $A \in \Im$ and $B \in \Im$, then*
$$A - B \in \Im,$$

(v) *If $\Im$ is a σ-algebra as well, and $A_1, A_2, \cdots \in \Im$, then*
$$\bigcap_{i=1}^{\infty} A_i \in \Im.$$

Theorem 1 tells us a good deal about the structure of algebras of sets. It should be apparent that an algebra of sets can be very trivial in character. For example, if Ω is the sample space on which we plan to define probability, we can choose as an algebra of sets the family $\Im$ consisting just of Ω and the empty set. It is obvious, of course, that there is no interest whatsoever in this particular algebra. It is also not difficult to select some intuitively natural families of sets that are not algebras of sets. In the case of our coin-tossing example, the family of sets consisting just of the unit sets of the outcomes, which are sometimes called the atomic events, is not an algebra of sets, because the union of two of these outcomes is not itself in the family of sets.[3] Again, considering this example, the family of all subsets of Ω containing at least one h is also not an algebra of sets, because the complement of the largest such subset is not itself in the family.

An example of an algebra which is not a σ-algebra is easily constructed from the natural numbers. Let $\Im$ be all finite subsets of Ω together with the complements of such finite subsets. Then the set of even natural numbers is not in the algebra, though it is in the σ-algebra, because it is the countable union of finite sets in the algebra.

Axioms of probability. We are now ready to define probability spaces. In the definition we assume the set-theoretical structure of Ω, $\Im$, and P already mentioned: Ω is a nonempty set, $\Im$ is a nonempty family of subsets of Ω, and P is a real-valued function on $\Im$. Thus the definition applies to ordered triples $(\Omega, \Im, P)$, which we shall call *set function structures*, since P is a function on a family of sets.

DEFINITION 2. *A set function structure $\boldsymbol{\Omega} = (\Omega, \Im, P)$ is a finitely additive probability space if and only if for every A and B in $\Im$:*

P1. $\Im$ is an algebra of sets on Ω;
P2. $P(A) \geq 0$;
P3. $P(\Omega) = 1$;
P4. If $A \cap B = \emptyset$, then $P(A \cup B) = P(A) + P(B)$.

---

[3] A *unit set* is a set having just a single element as its only member.

5.1 THE FORMAL THEORY

Moreover, Ω is a probability space (without restriction to finite additivity) if the following two axioms are also satisfied:

P5. $\Im$ is a σ-algebra of sets on Ω;

P6. If $A_1, A_2, \ldots$ is a sequence of pairwise incompatible events in $\Im$, i.e., $A_i \cap A_j = \emptyset$ for $i \neq j$, then

$$P(\bigcup_{i=1}^{\infty} A_i) = \sum_{i=1}^{\infty} P(A_i).$$

The properties expressed in P5 and P6 can be separated. For example, a probability measure P satisfying P6 is said to be countably additive, and we can concern ourselves with countably additive probability measures on algebras that are not σ-algebras. On the other hand, for full freedom from a mathematical standpoint it is desirable to have both P5 and P6 satisfied, and for this reason we have assumed that probability spaces satisfy both P5 and P6.[4] In particular, this full freedom is essential for the standard development of the asymptotic theory as sample size or number of trials approaches infinity. It should be apparent that if the set Ω is finite then every finitely additive probability space is also a probability space. It may be mentioned that an additional postulate is sometimes required of probability spaces. The postulate asserts that if $A \subseteq B, B \in \Im$, and $P(B) = 0$, then $A \in \Im$. When a probability space satisfies this additional postulate it is called *complete*.

We now turn to a number of elementary concepts and theorems. Most of them hold for finitely additive probability spaces, and, thus, do not require any use of Axioms P5 and P6.

THEOREM 2.

(i) $P(\emptyset) = 0$.

(ii) For every event $A, P(A) + P(-A) = 1$.

(iii) If the events A_i are pairwise incompatible for $1 \leq i \leq n$, then

$$P(\bigcup_{i=1}^{n} A_i) = \sum_{i=1}^{n} P(A_i).$$

Proof. We prove only (iii), which requires use of mathematical induction. For $n = 1$, clearly

$$P(A_1) = P(A_1).$$

Our inductive hypothesis for n is that

$$P(\bigcup_{i=1}^{n} A_i) = \sum_{i=1}^{n} P(A_i).$$

Now consider

$$P(\bigcup_{i=1}^{n+1} A_i).$$

---

[4] A simple example, given after Definition 14 of this section, is the computation of the mean of the Poisson distribution. The computation is for an infinite sequence of probabilities and requires Axioms P5 and P6.

Since
$$\bigcup_{i=1}^{n+1} A_i = (\bigcup_{i=1}^{n} A_i) \cup A_{n+1},$$

$$\begin{aligned}
P(\bigcup_{i=1}^{n+1} A_i) &= P((\bigcup_{i=1}^{n} A_i) \cup A_{n+1}) \\
&= P(\bigcup_{i=1}^{n} A_i) + P(A_{n+1}) \quad &&\text{By P4 and hypothesis of the theorem} \\
&= \sum_{i=1}^{n} P(A_i) + P(A_{n+1}) \quad &&\text{By inductive hypothesis} \\
&= \sum_{i=1}^{n+1} P(A_i)
\end{aligned}$$

Q.E.D.

Discrete probability densities. We next define discrete probability densities on a set Ω. Such a discrete probability density on a set Ω is a point function analogous to the set function P. For many kinds of applications the use of a discrete or continuous probability density function is essential, or at least highly desirable, in carrying out detailed computations.

DEFINITION 3. *A discrete probability density on a nonempty set Ω is a non-negative real-valued function p defined on Ω such that for all $\omega \in \Omega$, $p(\omega) = 0$ except for an at most countable number of elements of Ω and $\sum_{\omega \in \Omega} p(\omega) = 1$.*

It should be apparent that a discrete probability density can always be used to generate a unique probability measure on a given algebra of sets, as expressed in the following theorem, whose proof is omitted.

THEOREM 3. *Let Ω be a nonempty set, let p be a discrete probability density on Ω, let $\Im$ be a σ-algebra of sets on Ω, and let P be a real-valued function defined on $\Im$ such that if $A \in \Im$ then $P(A) = \sum_{\omega \in A} p(\omega)$. Then $\mathbf{\Omega} = (\Omega, \Im, P)$ is a probability space.*

For the present, we shall defer consideration of continuous densities on infinite sets. Here are four of the more important discrete densities. First, there is the distribution of a simple toss of a coin, usually called a *Bernoulli* density, because there are only two possible values, i.e., only two elements in the probability space Ω. The sample space $\Omega = \{0, 1\}$, and $p(1) = p$, $p(0) = q = 1 - p$, where $0 \leq p \leq 1$.[5] For a fair coin, $p = \frac{1}{2}$. Second, there is the closely related *binomial* density, which arises from n independent Bernoulli trials. The probability space $\Omega = \{0, 1, ..., n\}$, i.e., the probability space is the

---

[5] A grammatical ambiguity is tolerated continually in the notation used here, and in many other places in this book. In the sentence "$\mathbf{\Omega} = (\Omega, \mathcal{F}, P)$ is a probability space" the expression '$\mathbf{\Omega} = (\Omega, \mathcal{F}, P)$' is a noun phrase, but in the sentence "The sample space $\Omega = \{0, 1\}$" the main verb is '='. Without use of this ambiguity, resolved easily in each particular context, we need to use too many lengthy explicit formulations.

set of non-negative integers $0 \leq k \leq n$, and
$$p(k) = \binom{n}{k}p^k(1-p)^{n-k},$$
where $\binom{n}{k}$ is the binomial coefficient:
$$\binom{n}{k} = \frac{n!}{k!(n-k)!}.$$
The classic case of the binomial density is tossing a coin n times with p the probability of a head on each trial. The probability $p(k)$ is the probability of k heads. It is common to refer to the binomial *distribution* as frequently as the binomial *density*, when Ω is a set of numbers, so we need to define this important associated concept of *distribution*. Let f be a discrete density on a nonempty numerical set Ω. Define on Ω
$$F(x) = \sum_{y \leq x} f(y).$$
Then F is the (cumulative) probability distribution defined in terms of f. Notice that when Ω is an arbitrary finite set, the concept of distribution is not well defined. To generalize beyond numerical sets, a meaningful linear ordering of the elements of Ω is needed.

Third, there is the *Poisson* density with parameter λ, where $\lambda \geq 0$.[6] The probability space $\Omega = \{0, 1, 2, ...\}$ and
$$p(k) = \frac{e^{-\lambda}\lambda^k}{k!}.$$
The applications of the Poisson density or distribution complement those of the binomial distribution. In applying the binomial distribution, we count the occurrences of some phenomena such as heads on given trials or presence of a disease on a given date. So, for example, a head may occur on the first trial but not the second. In contrast, there are many other phenomena for which our interest is in the number of occurrences in a given time interval, for instance an hour, a day, a year. Other examples would be the number of phone calls going through a given exchange in 24 hours, the number of airplanes landing at an airport in one week, and the number of discharges of a Geiger counter in two minutes. The Poisson distribution can also be applied to spatial phenomena; for instance, the number of bacteria of a given kind found in one-liter samples of water from a city's water supply. Implicit in these examples of either a spatial or temporal sort is that successive samples will be taken. For instance, we might look at telephone exchange data for 200 successive days. Although none of the phenomena described would ever have even a large finite number of occurrences, say 10^{100}, in a realistic temporal interval or spatial extent, it is mathematically convenient to define the density for every non-negative integer. A reason, of course, is that there is no natural stopping point, but this is certainly not the main reason. It is always easy to pick a finite domain more than adequate to deal with actual observations, but when a large finite domain is used it is harder to determine the mathematical or theoretical properties of the distribution.

---

[6]Poisson (1781–1840) was one of the great French mathematical physicists of the nineteenth century. His derivation of the Poisson density was published in 1837, but not really recognized as a significant contribution until the end of the century.

Fourth, there is the *geometric* density with parameter p, with $0 \leq p \leq 1$. The probability space $\Omega = \{1, 2, 3, ...\}$ and

$$p(k) = p(1-p)^{k-1}.$$

A theoretically important example is the density of the number k of trials to get the first head in successive Bernoulli trials with probability p of success. For instance, the probability of getting the first head on the fifth toss of a biased coin with $p = .25$ is 0.079 (approximately). The geometric density is the natural one to use in studying radioactive decay when discrete trials are used, as we shall see in the section on the propensity interpretation of probability.

Conditional probability. We next consider the important concept of conditional probability. Its central role comes from the fact that dependencies, especially temporal ones, are expressed in conditional terms.

DEFINITION 4. *If A and B are in $\Im$ and $P(B) > 0$, then*

$$P(A|B) = \frac{P(A \cap B)}{P(B)}.$$

We call $P(A|B)$ the *conditional probability* of event A, given the occurrence of event B. As an example, consider the situation in which a coin is tossed twice. $\Omega = \{hh, ht, th, tt\}$, $\Im$ is the set of all subsets of Ω, and $p(\omega) = \frac{1}{4}$ for all ω in Ω. There are two events A and B such that A is the event of exactly one head in two tosses and B is the event of at least one head in two tosses. Clearly, $P(A) = \frac{1}{2}$ whereas $P(B) = \frac{3}{4}$. Since $A \cap B = A$, $P(A \cap B) = P(A) = \frac{1}{2}$. Therefore, $P(A|B) = \frac{P(A)}{P(B)} = \frac{(\frac{1}{2})}{(\frac{3}{4})} = \frac{2}{3}$.

For reasons that are apparent, the following theorem about conditional probabilities is usually labeled the multiplication theorem.

THEOREM 4. (Multiplication Theorem). *If $A_1, A_2, A_3, \ldots, A_n$ are in $\Im$ and $P(\cap_{i=1}^{n} A_i) > 0$, then*

$$P(\bigcap_{i=1}^{n} A_i) = P(A_1) \cdot P(A_2|A_1) \cdot P(A_3|\bigcap_{i=1}^{2} A_i) \cdots P(A_n|\bigcap_{i=1}^{n-1} A_i).$$

Proof. We use induction on n. For $n = 1$, clearly $P(A_1) = P(A_1)$. Our inductive hypothesis for n is that

$$P(\bigcap_{i=1}^{n} A_i) = P(A_1) \cdots P(A_n|\bigcap_{i=1}^{n-1} A_i). \tag{1}$$

But by the definition of conditional probability

$$P(\bigcap_{i=1}^{n+1} A_i) = P(\bigcap_{i=1}^{n} A_i) \cdot P(A_{n+1}|\bigcap_{i=1}^{n} A_i).$$

Therefore, using (1)

$$P(\bigcap_{i=1}^{n+1} A_i) = P(A_1) \cdot P(A_2|A_1) \cdots P(A_n|\bigcap_{i=1}^{n-1} A_i) \cdot P(A_{n+1}|\bigcap_{i=1}^{n} A_i). \qquad \text{Q.E.D.}$$

5.1 THE FORMAL THEORY

Applications of the multiplication theorem usually involve other simplifying assumptions. Markov chains provide good examples. Intuitively, a sequence of trials in which the probability of what happens on a trial depends on what occurred on the immediately preceding trial, but not on earlier trials, is a (first-order) Markov chain. (A formal definition will not be given at this point.[7]) A slightly artificial example will provide a simple illustration. Suppose that remembering a given algorithm, let us say, depends only on whether it was remembered on the last trial of learning, and no earlier one. Let r be remembering and f forgetting. Then the transition matrix is:

$$\begin{array}{c c} & \begin{array}{cc} r & f \end{array} \\ \begin{array}{c} r \\ f \end{array} & \begin{pmatrix} .75 & .25 \\ .10 & .90 \end{pmatrix} \end{array}$$

where the two rows show the state on trial $n-1$ and the two columns the possible states on trial n. Thus if on trial $n-1$, the algorithm was remembered, the probability is .75 it will be remembered on trial n. Assume we have the sequence of observations $ffffrrfrrr$, and suppose further that on trial 1, the probability of forgetting is 1.0—on the first exposure. Then we can use these specialized assumptions and the multiplication theorem to compute the probability of our observed sequence.

$$\begin{aligned} p(ffffrrfrrr) &= p(f)p(f|f)^3 p(r|f)p(r|r)p(f|r)p(r|f)p(r|r)^2 \\ &= (1)(.9)^3(.10)(.75)(.25)(.10)(.75)^2 \\ &= .0007689 \qquad \text{(approximately)}. \end{aligned}$$

That conditional probabilities behave like ordinary probabilities is shown by the next two theorems.

THEOREM 5. *If A, B, and C are in $\mathfrak{S}$ and if $P(A) > 0$, then:*

(i) $P(B|A) \geq 0$,

(ii) $P(\Omega|A) = 1$,

(iii) $P(A|A) = 1$,

(iv) *If $B \cap C = \emptyset$, then $P(B \cup C|A) = P(B|A) + P(C|A)$,*

(v) *If $P(B) > 0$, then $P(A|B) = P(A) \cdot \frac{P(B|A)}{P(B)}$,*

(vi) *If $A \cap B = \emptyset$, then $P(B|A) = 0$,*

(vii) *If $A \subseteq B$, then $P(B|A) = 1$.*

THEOREM 6. *If $A \in \mathfrak{S}$ and $P(A) > 0$, then $(\Omega, \mathfrak{S}, P_A)$ is a finitely additive probability space such that for every B in $\mathfrak{S}, P_A(B) = P(B|A)$. Moreover, if $(\Omega, \mathfrak{S}, P)$ is countably additive, so is $(\Omega, \mathfrak{S}, P_A)$.*

Proof. It is given that $\mathfrak{S}$ is an algebra on Ω. From Theorem 5(i) we have $P_A(B) \geq 0$, which satisfies P2 of Definition 2. From Theorem 5(ii) $P_A(\Omega) = 1$, which satisfies P3. From Theorem 5(iv), if $B \cap C = \emptyset$, then $P_A(B \cup C) = P_A(B) + P_A(C)$, which satisfies P4, and similarly for countable additivity. Q.E.D.

---

[7] A detailed formal treatment of Markov chains is given in the last section of Chapter 7, which is on reversibility of causal processes. A less thorough development, but one emphasizing ergodic properties, is to be found in Section 4.5.

The interest of Theorem 6 is that it shows how we may use knowledge of an event A to generate a new probability space.

THEOREM 7. (Theorem on Total Probability). *If $\Omega = \cup_{i=1}^{n} A_i$ and if $A_i \in \Im$, $P(A_i) > 0$ and $A_i \cap A_j = \emptyset$ for $i \neq j$ and for $1 \leq i,j \leq n$, then for every $B \in \Im$,*

$$P(B) = \sum_{i=1}^{n} P(B|A_i)P(A_i).$$

Proof. Using the hypothesis of the theorem and distributive laws of logic we have:

$$B = B \cap \Omega = B \cap (\bigcup_{i=1}^{n} A_i) = \bigcup_{i=1}^{n} (B \cap A_i). \tag{2}$$

Hence

$$P(B) = P(\bigcup_{i=1}^{n} (B \cap A_i)). \tag{3}$$

Since by hypothesis $A_i \cap A_j = \emptyset$ for $i \neq j$, we may apply the theorem on finite additivity (Theorem 2(iii)) to (3) and obtain

$$P(B) = \sum_{i=1}^{n} P(B \cap A_i). \tag{4}$$

But by Definition 4

$$P(B \cap A_i) = P(B|A_i)P(A_i), \tag{5}$$

and our theorem follows immediately from (4) and (5). Q.E.D.

We use Theorem 7 in an essential way to prove the next theorem, which, in spite of its simple proof, occupies an important place in probability theory. In the statement of the theorem, the H_i's are thought of as hypotheses, but this is not, obviously, a formal requirement.

THEOREM 8. (Bayes' Theorem).[8] *If $\cup_{i=1}^{n} H_i = \Omega, P(H_i) > 0, H_i \cap H_j = \emptyset$, for $i \neq j$ and for $1 \leq i,j \leq n$, if $B \in \Im$ and $P(B) > 0$, then*

$$P(H_i|B) = \frac{P(B|H_i)P(H_i)}{\sum_{j=1}^{n} P(B|H_j)P(H_j)}.$$

Proof. By Theorem 5(v)

$$P(H_i|B) = \frac{P(B|H_i)P(H_i)}{P(B)}. \tag{6}$$

But by Theorem 7,

$$P(B) = \sum_{j=1}^{n} P(B|H_j)P(H_j). \tag{7}$$

Substituting (7) into (6) we obtain the desired result at once. Q.E.D.

As we shall see in the section on subjective probability, this mathematically trivial theorem has far-reaching applications. The terminology of Bayes' Theorem deserves some comment. The H_i's are ordinarily a set of mutually exclusive and exhaustive hypotheses, one of which offers the relatively best explanation of the event B. The

---

[8] The Reverend Thomas Bayes (1763) wrote an important early memoir on this theorem and its use.

5.1 THE FORMAL THEORY

probability $P(H_i)$ is called the *prior* probability of the hypothesis H_i. These prior probabilities are the center of much of the controversy concerning applications of probability theory. The conditional probability $P(B|H_i)$ is called the *likelihood* of H_i on B, or the likelihood of H_i on the evidence B. The conditional probability $P(H_i|B)$ is called the *posterior* probability of the hypothesis H_i given the observed event B. It should be understood that there are no controversies over the mathematical status of Bayes' Theorem. Indisputably the theorem is a logical consequence of the axioms of Definition 2.

Bayes' postulate. This postulate is the classical procedure of taking all prior possibilities, $P(H_i)$, as being equal. It is important to notice that this postulate is concerned only with the *applications* of probability theory; it is not relevant to the purely mathematical theory of probability spaces. It is to be emphasized that Bayes' Theorem is logically independent of Bayes' Postulate. Also independent of Bayes' Postulate, as given above, is the *Bayesian Rule of Behavior:* accept the hypothesis with the largest posterior probability, $P(H_i|B)$.[9] Of course, in applying this rule we sometimes use Bayes' Postulate to calculate the prior probabilities, $P(H_i)$, which are needed to compute the posterior probabilities, $P(H_i|B)$.

An example applying Bayes' postulate. Suppose there is an urn containing four balls about which we know that exactly one of the following is true:

H_1: All four balls are white;
H_2: Two balls are white and two balls are black.

Suppose further that one white ball is withdrawn from the urn. The problem of interest is that of establishing a probability space suitable for the computation of the posterior probabilities of H_1 and H_2.

There is no general rule for constructing the set Ω of possible outcomes in the theory of probability, although the theory of probability does indicate how Ω is to be manipulated once it has been specified. Some criteria for generating the probability space will be considered later. For the moment, however, Ω can be constructed in such a manner that four consecutive selections of balls from the urn without replacement will necessarily and uniquely determine which hypothesis, H_1 or H_2, is true. Thus,

$$\Omega = \{wwww, wwbb, bbww, wbbw, bwwb, wbwb, bwbw\}.$$

It will be observed that H_1 and H_2 are subsets of Ω.

$$H_1 = \{wwww\}$$
$$H_2 = \{wwbb, bbww, wbbw, bwwb, wbwb, bwbw\} = -H_1.$$

The event of selecting one white ball on the first draw is designated by W.

$$W = \{wwww, wbwb, wwbb, wbbw\}.$$

---

[9] It is fair to say, however, the practice is more to report the posterior probabilities than to recommend action, which may well depend on other factors such as utility or practice.

Bayes' Postulate tells us that the probabilities of the hypotheses are, a priori, equal. That is,

$$P(H_1) = \tfrac{1}{2}$$
$$P(H_2) = \tfrac{1}{2}.$$

The likelihood of H_i on event W can be calculated by a consideration of the ratio of white to total balls on each hypothesis;

$$P(W|H_1) = 1$$
$$P(W|H_2) = \tfrac{1}{2}.$$

This calculation of likelihood is based on what we shall call the Postulate of Simple Randomness of Sampling. The intuitive idea of the postulate is that if H is the true hypothesis and if $\omega_1, \omega_2 \in H$ (i.e., ω_1 and ω_2 are elements of the sample space Ω, and are also elements of the subset H), then

$$P(\{\omega_1\}) = P(\{\omega_2\}).$$

(Since $\{\omega_1\}$ or $\{\omega_2\}$ might not be in $\Im$, this is not an exact statement of the Postulate.) Put another way, we use a procedure of sampling such that if H is the true hypothesis, any possible outcome of sampling compatible with H is as probable as any other such outcome. It is important to note that in many cases our experimental procedure permits us adequate flexibility to guarantee satisfaction of this postulate. On the other hand, no such control can usually be exercised over the prior probabilities, since they are a function of our information prior to sampling. Hence, the controversy about Bayes' Postulate.

The posterior probability of each hypothesis can be calculated by Theorem 8, which, for this example, can be stated as follows:

$$P(H_1|W) = \frac{1 \cdot \tfrac{1}{2}}{1 \cdot \tfrac{1}{2} + \tfrac{1}{2} \cdot \tfrac{1}{2}} = \frac{2}{3}$$

$$P(H_2|W) = \frac{\tfrac{1}{2} \cdot \tfrac{1}{2}}{1 \cdot \tfrac{1}{2} + \tfrac{1}{2} \cdot \tfrac{1}{2}} = \frac{1}{3}.$$

Substitution into the proper formulas is made in an obvious way.

We may construct another space that will yield the same results as Ω in the previous example. The difficulty that arises in the definition of Ω is not intrinsic to probability theory but is illustrative of the ambiguity in getting an actual characterization of a probability space. In the preceding example, it is clear that we could decide which of the intuitive hypotheses is correct simply on the basis of at most three draws without replacement. Thus, we might use:

$$\Omega' = \{www, wwb, wbw, bww, bbw, bwb, wbb\}.$$

Observe that the number of elements in Ω' is equal to that in Ω. That is, their cardinality is the same. Similarly, H_1 and H_2 are designated as

$$H_1 = \{www\}$$
$$H_2 = \{wwb, wbw, bww, bbw, bwb, wbb\} = -H_1.$$

5.1 THE FORMAL THEORY

A moment's reflection is sufficient to see that Ω and Ω' will be the same in all remaining aspects and that H_1 will again have the higher posterior probability.

If sets instead of ordered n-tuples were used to represent elements of Ω or Ω', a difficulty would obviously arise. We would then have:

$$\Omega'' = \{\{w\}, \{w, b\}\}$$

and H_1 and H_2 would become

$$H_1 = \{\{w\}\}$$
$$H_2 = \{\{w, b\}\}.$$

The algebra appropriate to this example is

$$\Im'' = \{\emptyset, \{\{w\}\}, \{\{w, b\}\}, \Omega''\}.$$

The problem then is: what subset of $\Im$ is W, the event of getting one white ball? Clearly W cannot be appropriately characterized when Ω'' is the basic space.

However, Ω and Ω' do not represent the only legitimate possibilities. For several reasons, many statisticians would prefer the following space:

$$\Omega''' = \{(4, w), (2, w), (2, b)\},$$

and thus

$$H_1 = \{(4, w)\}$$
$$H_3 = \{(2, w), (2, b)\}.$$

The number which is the first member of each ordered couple indicates the total number of white balls in the urn on the given hypothesis (H_1 or H_2). The second member indicates a possible outcome of drawing one ball. Note that there are only three elements in Ω''', which is an advantage over Ω and Ω'. A second reason for preferring Ω''' is that the possible outcome of just one drawing is indicated, where elements of Ω indicate the outcome of four samples without replacement, and Ω' of three samples without replacement. Thus the redundancy in the number of elements in Ω and Ω'. The space Ω''' is tailored to the problem in a way in which Ω and Ω' are not.

Whatever sample space is used, the critical point lies in accepting Bayes' Postulate. Many experts are critical of the reasonableness of this postulate, particularly experts who advocate an objective interpretation of probability. Unfortunately, without Bayes' Postulate or some analogue which yields prior probabilities, it is impossible directly to apply the Bayesian rule of behavior previously stated.

One suggestion for meeting this difficulty is the following principle, the *principle of maximum likelihood*: accept that hypothesis whose likelihood for the observed event is maximum. H_j is accepted if and only if

$$P(E|H_j) = \max_i P(E|H_i).$$

We have as an obvious result:[10]

---

[10] For generalization of this theorem to infinite sets, see Theorem 3 of Section 5, formulated within the framework of infinitesimal, i.e., nonstandard analysis.

THEOREM 9. *If the number of hypotheses H_i is finite, then the Bayesian rule of behavior together with Bayes' Postulate leads to acceptance of the same hypothesis as the principle of maximum likelihood.*

Concept of an experiment. We next turn to the concept of an experiment. We define an *experiment*, relative to the sample space Ω, as any *partition* of Ω, i.e., any collection of nonempty subsets of Ω, such that (i) the collection is a subset of the algebra $\Im$ of sets, (ii) the intersection of any two distinct sets in the collection is empty, and (iii) the union of the collection is Ω itself. This definition of an experiment originates, as far as I know, with Kolmogorov (1933/1950). It is to be noticed that it is a technical concept within probability theory and is not meant to capture many of the nuances of meaning implied in the general scientific use of the term. Nonetheless, it is easy to give examples to show that the technical notion defined here corresponds to many of our intuitive ideas about experimental data. Consider the complicated, continuous-time record from a satellite recording many sorts of data. The full sample space Ω has as its elements all possible observable records for the time period and the variables under study. A meteorologist interested only in sunspot data defines an experiment with respect to Ω by partitioning the possible observable records according and only according to sunspot data. Data on other variables of any sort are ignored in this data reduction. Such reduction is usually necessary in any detailed application of a quantitative theory to the results of an experiment.

Independence. We now define the important concept of mutual independence of experiments.

DEFINITION 5. *The n experiments $M^{(1)}, \ldots, M^{(n)}$ are (mutually) independent if and only if for every $A_i \in M^{(i)}$ where $1 \leq i \leq n$,*

$$P(\bigcap_{i=1}^{n} A_i) = \prod_{i=1}^{n} P(A_i).$$

Notice that A_1 is any set in $M^{(1)}$, and so forth, and $\prod_{i=1}^{n}$ is the arithmetical product of the n probabilities.

The n events $A_1, A_2, \ldots, A_n$ are *(mutually) independent* if and only if the n experiments $M^{(1)}, M^{(2)}, \ldots, M^{(n)}$ are mutually independent, where

$$M^{(i)} = \{A_i, -A_i\}.$$

We consider two examples to illustrate the concept of independence.

First example. Consider three consecutive tosses of a coin, for which

$$\Omega = \{hhh, hht, hth, htt, thh, tht, tth, ttt\}.$$

Let $\Im$ be the set of all subsets of Ω, and for all $\omega \in \Omega$, let $p(\omega) = \frac{1}{8}$. This latter condition guarantees that the coin is fair. Experiment $M^{(1)}$ is defined as follows: for $i = 0, 1, 2, 3, A_i \in M^{(1)}$ if and only if A_i is the event of getting exactly i heads. We can easily verify that $M^{(1)}$ is an experiment since $M^{(1)}$ is obviously a partition of Ω.

5.1 THE FORMAL THEORY

Similarly, experiment $M^{(2)}$ is defined for $i = 0, 1, 2$ such that $B_i \in M^{(2)}$ if and only if B_i is the event of getting exactly i alternations of heads and tails.

The question is then to determine if $M^{(1)}$ and $M^{(2)}$ are mutually independent experiments. There are 12 cases to consider, since $M^{(1)}$ has four elements and $M^{(2)}$ has three elements. But take, for example, the case of $i = 0$. In this case, $A_0 = \{ttt\}$ for which $P(A_0) = \frac{1}{8}$ and $B_0 = \{hhh, ttt\}$ for which $P(B_0) = \frac{1}{4}$. It is clear that

$$P(A_0 \cap B_0) = P(A_0) = \frac{1}{8},$$

but

$$P(A_0) \cdot P(B_0) = \frac{1}{32}.$$

Thus experiments $M^{(1)}$ and $M^{(2)}$ are not independent.

Second example. Consider the simultaneous toss of a pair of coins twice. One coin is red and one black. For the red coin a head is designated as 'H', whereas for the black coin a head is designated 'h', etc. For this example, the probability space is

$$\Omega = \{HhHh, HhHt, HhTh, ..., TtTt\};$$

as before, let $\Im$ be the set of all subsets of Ω and for all $\omega \in \Omega, p(\omega) = \frac{1}{16}$. An experiment $M^{(1)}$ is described for $i = 0,1,2$ in which $A_i \in M^{(1)}$ if and only if A_i is the event of exactly i heads on the red coin. Experiment $M^{(2)}$ for $i = 0, 1$ is defined so that $B_i \in M^{(2)}$ if and only if B_i is the event of exactly i alternations of heads and tails for the black coin. There are, of course, six separate cases to consider. Take, for example, A_0 and B_1. Then

$$A_0 = \{ThTh, ThTt, TtTh, TtTt\},$$

for which $P(A_0) = \frac{1}{4}$ and

$$B_1 = \{HhHt, HtHh, HtTh, HhTt, ThHt, TtHh, TtTh, ThTt\},$$

for which $P(B_1) = \frac{1}{2}$. Furthermore,

$$\begin{aligned} A_0 \cap B_1 &= \{TtTh, ThTt\} \\ P(A_0 \cap B_1) &= \tfrac{1}{8} \\ P(A_0) \cdot P(B_1) &= \tfrac{1}{4} \cdot \tfrac{1}{2} = \tfrac{1}{8}. \end{aligned}$$

Similar computations for the other five cases establish that experiments $M^{(1)}$ and $M^{(2)}$ are mutually independent.

The following theorem is obvious, and its converse is false.

THEOREM 10. *If n experiments are mutually independent, then any m of them are mutually independent, for $m < n$.*

It is clear that pairwise independence of events $A_1, A_2, \ldots, A_n$ does not guarantee mutual independence of these events. Consider the following example (due to S. Bernstein). Let $\Omega = \{a_1, a_2, a_3, a_4\}$ and $A_1 = \{a_1, a_2\}$, $A_2 = \{a_1, a_3\}$, and $A_3 = \{a_1, a_4\}$. For every ω in $\Omega, p(\omega) = \frac{1}{4}$. It is easy to compute that

$$P(A_1) = P(A_2) = P(A_3) = \frac{1}{2}$$

$$P(A_1 \cap A_2) = P(A_2 \cap A_3) = P(A_1 \cap A_3) = \frac{1}{4},$$

which establishes pairwise independence. But in contrast,

$$P(A_1 \cap A_2 \cap A_3) = \tfrac{1}{4} \text{ and not } \tfrac{1}{8},$$

as would be expected. The three events are pairwise independent, but they are not mutually independent.[11]

In the important case of two events A and B, it is easily shown from the preceding definitions that they are independent if and only if

$$P(A \cap B) = P(A)P(B).$$

The concept of independence is one of the most profound and fundamental ideas, not only of probability theory but of science in general. The exactly defined probability notions of independence express very well the central intuitive idea back of the general concept. The two most basic ideas are concerned with the independence of what is happening here and now from the remote past or far distant places. Any claim that there is immediate causation at a spatial or temporal distance is met with skepticism and usually by an effort to explain away such action at a distance by intermediate proximate causes.[12] The probabilistic concept of independence provides a central tool for analyzing these ideas about causation in a quite general way (Suppes 1970a).

Random variables. We now turn to the important concept of random variable. In many applications of probability theory in the sciences, the underlying sample space has a very complicated set-theoretical structure, and yet, at any given moment, we are interested only in certain aspects of that space. To this end, we define certain real-valued functions on the sample space and study the probabilistic behavior of these functions.

To conform to a notation common in statistics we shall henceforth use boldface '**X**', '**Y**', and so forth, to denote random variables.

DEFINITION 6. **X** *is a* random variable *if and only if* **X** *is a real-valued function whose domain is the probability space* Ω, *and for each real number* x, *the set* $\{\omega : \mathbf{X}(\omega) < x\}$ *is in* $\Im$.

The requirement that $\{\omega : \mathbf{X}(\omega) < x\}$ be in $\Im$ is no restriction at all on finite probability spaces for which $\Im$ is the set of all subsets. How a random variable may represent an experiment is made clear in the next definition.

DEFINITION 7. *Let M be an experiment and let* **X** *be a real-valued function defined on* Ω. *Then* **X** *is* a random variable with respect to the experiment M *if and only if for each A in M there is a real number k_A such that for every ω in A,* $\mathbf{X}(\omega) = k_A$.

In discussing a non-numerical discrete sample space such as

$$\Omega = \{hh, ht, th, tt\}$$

---

[11] For another example, one physically motivated that comes directly from the theory of quantum entanglement, see footnote 15.

[12] It was not always so. In the eighteenth century gravitation was the great example of action at a distance, and there were respectable action-at-a-distance theories of electricity and magnetism in the nineteenth century.

5.1 THE FORMAL THEORY

it is natural to suggest replacing it by the numerical sample space

$$\Omega_1 = \{0, 1, 2\}$$

where each integer represents a possible number of heads in two tosses of a coin. An alternative suggestion, as might be expected, is to define a random variable on Ω rather than to replace Ω by Ω_1. Thus we may define the random variable $\mathbf{X}$ on Ω such that $\mathbf{X}(hh) = 2$, $\mathbf{X}(ht) = 1$, $\mathbf{X}(th) = 1$, and $\mathbf{X}(tt) = 0$, and the random variable $\mathbf{X}$ now serves the purpose originally suggested for Ω_1. As already remarked, the use of boldface capital letters '$\mathbf{X}$', '$\mathbf{Y}$', '$\mathbf{Z}$' for random variables is customary in statistics. On the other hand, it is not customary to write down the arguments of a random variable, i.e., in the statistical literature it would be surprising and unusual to encounter an expression like '$\mathbf{X}(hh)$'. To conform to the usual shorthand notation of statistics we shall introduce some abbreviating notation.

DEFINITION 8. *Let $\mathbf{X}$ be a random variable on Ω. Then*

$$P(\mathbf{X} \leq x) = P(\{\omega : \mathbf{X}(\omega) \leq x\})$$

and

$$P(a \leq \mathbf{X} \leq b) = P(\{\omega : a \leq \mathbf{X}(\omega) \leq b\}).$$

Corresponding definitions may be given for the notation: $P(\mathbf{X} = x)$, $P(\mathbf{X} < x)$, $P(\mathbf{X} \geq a)$, and $P(\mathbf{X} > x)$. (Note that we use the letters 'a', 'b', 'x' as numerical variables.)

For the case of two tosses of a coin, assuming the coin is fair, i.e., for $\omega \in \Omega, p(\omega) = \frac{1}{4}$, we have:

$$\begin{aligned} P(\mathbf{X} \leq 1) &= P(\{\omega : \mathbf{X}(\omega) \leq 1\}) \\ &= P(\{HT, TH, TT\}) \\ &= P\{HT\} + P\{TH\} + P\{TT\} \\ &= \tfrac{1}{4} + \tfrac{1}{4} + \tfrac{1}{4} \\ &= \tfrac{3}{4}. \end{aligned}$$

Similarly, we easily derive that

$$\begin{aligned} P(\mathbf{X} > 0) &= \tfrac{3}{4} \\ P(\mathbf{X} < 1) &= \tfrac{1}{4}. \end{aligned}$$

It is easy to use this notation to define the *distribution* F of a random variable $\mathbf{X}$ as

$$F(x) = P(\mathbf{X} \leq x).$$

The distribution F of a random variable is also sometimes called a *cumulative* distribution. Note that random variables always have distributions, contrary to an arbitrary finite set Ω which may have a density defined on it, but not a distribution, because there is no relevant or natural ordering of the elements of Ω, a point made earlier. Since the range of a random variable is a set of numbers, the problem of an ordering to define the distribution does not arise.

Although the classification is not exhaustive, it is useful to distinguish between discrete and continuous random variables. The definition does not allow us to infer from the fact that a random variable is discrete that the underlying probability space Ω is also discrete.

DEFINITION 9. *A random variable* **X** *is* discrete *iff the set of values of* **X** *is countable.*

In other words, a discrete random variable is a function whose range is a countable set. The discrete density of a discrete random variable is easily defined.

DEFINITION 10. *Let* **X** *be a discrete random variable. Then the* discrete density *of* **X** *is the function p such that for any value x of* **X**

$$p(x) = P(\mathbf{X} = x).$$

In the case of the experiment consisting of tossing a fair coin twice in which the values of the numerical random variable **X** represent the number of heads obtained in two tosses, the discrete density p of **X** is:

$$p(0) = \tfrac{1}{4}$$
$$p(1) = \tfrac{1}{2}$$
$$p(2) = \tfrac{1}{4}.$$

Note how the phrase 'density of a random variable **X**' is to be interpreted. The density is actually on the set of possible values of the random variable, and the density f or distribution F may always be defined in terms of the random variable **X** and the probability measure P on the basic sample space.

Piecewise continuous densities and distributions. In the next definition we use the notion of a *piecewise* continuous function. This a function that is continuous except at a finite number of points and at each point of discontinuity the right-hand and left-hand limits exist.

DEFINITION 11. *A random variable* **X** *is* piecewise continuous *iff there is a piecewise continuous density f such that for every two real numbers a and b with $a \leq b$*

$$P(a \leq \mathbf{X} \leq b) = \int_a^b f(x)dx.$$

Ordinarily, we shall always say what the density of a piecewise continuous random variable is. Generalizations of Definition 11 are important in the mathematical literature but will not be considered here.

Note, once again, the obvious relation between the density f of a random variable **X** and its distribution F.

$$F(x) = \sum_{y \leq x} f(y) \qquad \text{(discrete case)}$$

$$F(x) = \int_{-\infty}^{x} f(x)dx \qquad \text{(piecewise continuous case)}.$$

5.1 THE FORMAL THEORY

From these equations it is obvious why a distribution F is also called a cumulative distribution.

An example of a continuous random variable on a non-numerical sample space is the following. (When the random variable has a continuous, not just piecewise continuous density, we shall refer to it simply as continuous.) Consider an experiment which is designed to test the accuracy of a new type of rocket. The rocket is aimed at a target three thousand miles away, and the experiment consists of determining where the rocket actually landed. Our basic sample space is the set of points of the earth's surface. One numerical random variable of interest here is the one whose value, for any outcome of the experiment, is the distance between the target and the actual location of the rocket.

Expectation of a random variable. We now turn to the definitions of several fundamental properties of random variables. The definitions are given for both discrete and piecewise continuous random variables. We begin with the expected value or mean of a random variable.

DEFINITION 12. *Let* **X** *be a discrete or piecewise continuous random variable with density f. Then the* expected value *or mean $E(\mathbf{X})$ of* **X** *is given by:*[13]

$$E(\mathbf{X}) = \sum x f(x) \qquad (discrete\ case)$$

$$E(\mathbf{X}) = \int_{-\infty}^{\infty} x f(x)\,dx \qquad (piecewise\ continuous\ case)$$

provided the sum or integral exists.

It should be apparent that the expectation of a random variable does not necessarily exist. For example, the expectation of the Cauchy density

$$f(x) = \frac{1}{\pi(1+x^2)}$$

does not exist.

It is also desirable to extend the definition of expected value to any continuous numerical function of a random variable.

DEFINITION 13. *Let* **X** *be a discrete or piecewise continuous random variable with density f, and let h be a continuous function in the latter case. Then the* expected value *or mean $E(h(\mathbf{X}))$ of $h(\mathbf{X})$ is given by:*

$$E(h(\mathbf{X})) = \sum h(x)f(x) \qquad (discrete\ case)$$

$$E(h(\mathbf{X})) = \int_{-\infty}^{\infty} h(x)f(x)\,dx \qquad (piecewise\ continuous\ case)$$

provided the sum or integral exists.

We may use Definition 14 to define moments about the origin and about the mean of a random variable.

DEFINITION 14. *Let* **X** *be a discrete or piecewise continuous random variable with density f. Then:*

---
[13] The sum is taken over the set of possible numerical values of the random variable.

(i) $\mu = \mu_{\mathbf{X}} = \mu_1' = E(\mathbf{X}) = $ mean of $\mathbf{X}$,

(ii) $\mu_r' = \mu_r(\mathbf{X}) = r^{th}$ moment about the origin of $\mathbf{X} = E(\mathbf{X}^r)$,

(iii) $m_r = m_r(\mathbf{X}) = r^{th}$ moment about the mean of $\mathbf{X} = E[(\mathbf{X} - \mu_{\mathbf{X}})^r]$,

provided the expectations exist.

It is customary to let $\sigma_{\mathbf{X}}^2$ denote the *variance* of the random variable $\mathbf{X}$, which is the second moment $m_2(\mathbf{X})$ about the mean. A simple but useful equation is:

$$Var(\mathbf{X}) = \sigma_{\mathbf{X}}^2 = E[(\mathbf{X} - E(\mathbf{X}))^2] = E(\mathbf{X}^2) - (E(\mathbf{X}))^2.$$

An example will be useful in clarifying several minor points. Let $\mathbf{X}$ be a piecewise continuous random variable with density f defined by:

$$f(x) = \begin{cases} \frac{1}{2} & \text{if } x \in [1,3], \\ 0 & \text{otherwise}. \end{cases} \tag{8}$$

Note that the density f defined by (8) could also be characterized as the uniform density on the sample space [1,3]. As it is, f is discontinuous at two points, $x = 1$ and $x = 3$. Following the standard practice in statistics, we henceforth usually define the density or distribution on the set of all real numbers, that is, the interval $(-\infty,\infty)$. In the case of f defined by (8) and in other such cases where the function is identically zero outside some given interval, we know the area under the curve representing the function is zero outside the interval, and, thus, we need only consider the integral over the given interval when finding moments, etc. Thus, for any function h of the random variable $\mathbf{X}$ with density f given by (8):

$$E(h(\mathbf{X})) = \int_{-\infty}^{\infty} h(x)f(x)dx$$

$$= \int_1^3 \frac{h(x)}{2} dx.$$

Thus, for the expected value of $\mathbf{X}$:

$$E(\mathbf{X}) = \int_1^3 \frac{x}{2} dx = \left. \frac{x^2}{4} \right|_1^3 = \frac{1}{4}(9 - 1) = 2,$$

and for the variance of $\mathbf{X}$:

$$\sigma_{\mathbf{X}}^2 = E[(\mathbf{X} - 2)^2] = \int_1^3 \frac{(x-2)^2}{2} dx$$

$$= \left. \frac{(x-2)^3}{6} \right|_1^3 = \frac{1}{6} - (-\frac{1}{6}) = \frac{1}{3}.\,^{14}$$

Of considerably more interest are the means and variances of random variables that have any one of the four classic discrete densities considered earlier. The mean of the Bernoulli density is simply p, and the variance $p(1-p)$. The binomial density has mean np and variance $np(1-p)$. The Poisson density is unusual in that its mean and variance are the same, namely, λ. The geometric density has mean $\frac{1}{p}$ and variance $\frac{(1-p)}{p^2}$.

To see how to derive these results, let us first consider a random variable $\mathbf{X}$ having the geometric density. From our earlier discussion of this distribution and the definition

---

[14] Even these simple calculations show why we use densities much more than distributions in applications. The regimes of computation naturally use densities. Of course, both densities and distributions for given random variables are definable in terms of the probability measure P of the sample space.

5.1 THE FORMAL THEORY

of the mean of a random variable, we have at once

$$\begin{aligned} E(\mathbf{X}) &= \sum_{k=1}^{\infty} kp(1-p)^{k-1} \\ &= p(1 + 2q + 3q^2 + \ldots), \text{ where } q = 1-p \\ &= \frac{p}{(1-q)^2} \\ &= \frac{1}{p}, \end{aligned}$$

as desired. We have used here a familiar and elementary fact about summing the series $1 + 2q + 3q^2 + \ldots$. For the variance of the geometric density, we first compute the second moment about the origin:

$$\begin{aligned} \mu_2'(\mathbf{X}) &= \sum_{k=1}^{\infty} k^2 pq^{k-1} \\ &= qp\Sigma k(k-1)q^{k-2} + p\Sigma kq^{k-1} \\ &= qp\frac{d^2}{dq^2}\Sigma q^k + p\frac{d}{dq}\Sigma q^k \\ &= qp\frac{d^2}{dq^2}(1-q)^{-1} + p\frac{d}{dq}(1-q)^{-1} \\ &= \frac{2qp}{p^3} + \frac{p}{p^2} \\ &= \frac{2q}{p^2} + \frac{1}{p}. \end{aligned}$$

(Note the "trick" of finding an infinite series that can be easily summed and whose first or second derivative is the given series.) Subtracting off $E(\mathbf{X})^2$, we then obtain

$$\begin{aligned} \sigma_{\mathbf{X}}^2 &= \frac{2q}{p^2} + \frac{1}{p} - \frac{1}{p^2} \\ &= \frac{q}{p^2} \end{aligned}$$

as desired.

The computations for the Poisson density are even simpler:

$$\begin{aligned} E(\mathbf{X}) &= \sum_{k=0}^{\infty} k p(k) \\ &= \frac{\Sigma k e^{-\lambda} \lambda^k}{k!} \\ &= \frac{\lambda \Sigma e^{-\lambda} \lambda^{k-1}}{(k-1)!} \\ &= \lambda \Sigma p(k-1) \\ &= \lambda. \end{aligned}$$

For the variance we have:

$$\begin{aligned} \sigma_{\mathbf{X}}^2 &= E(\mathbf{X}^2) - \lambda^2 \\ &= \sum_{k=0}^{\infty} k^2 p(k) - \lambda^2 \\ &= \sum_{k=1}^{\infty} k \frac{e^{-\lambda} \lambda^k}{(k-1)!} - \lambda^2 \\ &= \lambda \Sigma (k-1) p(k-1) + \lambda \Sigma p(k-1) - \lambda^2 \\ &= \lambda^2 + \lambda - \lambda^2 \\ &= \lambda, \end{aligned}$$

as desired.

To illustrate these ideas, we also take a look at three of the most common and important continuous or piecewise continuous distributions.

First, the *uniform* distribution on an interval $(a, b), a < x < b$, has density

$$f(x) = \begin{cases} \frac{1}{b-a} & \text{for } a < x < b, \\ 0 & \text{otherwise.} \end{cases}$$

The mean of a random variable $\mathbf{X}$ with such a density is easy to compute:

$$\begin{aligned} E(\mathbf{X}) &= \int_a^b \frac{x}{b-a} dx \\ &= \frac{1}{b-a} [\frac{x^2}{2}]_a^b \\ &= \frac{a+b}{2}, \end{aligned}$$

5.1 THE FORMAL THEORY

and by similar methods we easily show:

$$\sigma_{\mathbf{X}}^2 = \frac{(b-a)^2}{12}.$$

Second, the *normal* or *Gaussian* distribution on $-\infty < x < \infty$ has density

$$f(x) = \frac{1}{\sigma\sqrt{2\pi}} e^{-\frac{1}{2}(\frac{x-\mu}{\sigma})^2},$$

with mean μ and variance σ^2. (Statisticians call this the normal distribution, and physicists call it the Gaussian distribution–two distinct names for the same object.)

Third, the *exponential* distribution for $-\infty < x < \infty$ has the density

$$f(x) = \begin{cases} \lambda e^{-\lambda x} & \text{for } x > 0, \\ 0 & \text{otherwise.} \end{cases}$$

The mean is $\frac{1}{\lambda}$ and the variance $\frac{1}{\lambda^2}$.

Joint distributions. Extension of these ideas to several random variables is important for the analysis of many phenomena. For simplicity of notation I shall restrict myself to continuous or piecewise continuous variables, but the concepts defined all have immediate generalizations to arbitrary random variables.

DEFINITION 15. *Let* $\mathbf{X}$ *and* $\mathbf{Y}$ *be two random variables.*

(*i*) *Their* joint distribution *is the function* F *such that for all real numbers* x *and* y:

$$F(x, y) = P(\mathbf{X} \leq x, \mathbf{Y} \leq y);$$

(*ii*) *Their* joint density *function* f *is defined by:*

$$f(x, y) = \frac{\partial^2 F(x, y)}{\partial x \partial y};$$

(*iii*) *Their* marginal densities $f_1(x)$ *and* $f_2(x)$ *are defined by:*

$$f_1(x) = \int_{-\infty}^{\infty} f(x, y) dy$$

$$f_2(y) = \int_{-\infty}^{\infty} f(x, y) dx,$$

provided that the derivatives and integrals exist.

We may show that the two random variables are independent if for all x and y

$$f(x, y) = f_1(x) f_2(y).$$

In many ways the most theoretically important measure of dependence of two random variables is their covariance, which we define together with their correlation.

DEFINITION 16. *Let* $\mathbf{X}$ *and* $\mathbf{Y}$ *be two random variables, whose means and variances exist. The* covariance *of* $\mathbf{X}$ *and* $\mathbf{Y}$ *is defined as:*

$$\text{cov}(\mathbf{X}, \mathbf{Y}) = \int_{-\infty}^{\infty} \int_{-\infty}^{\infty} (x - E(\mathbf{X}))(y - E(\mathbf{Y})) f(x, y) dx dy,$$

provided that the double integral exists, and the correlation *of* **X** *and* **Y** *is defined as:*

$$\rho(\mathbf{X}, \mathbf{Y}) = \frac{cov(\mathbf{X}, \mathbf{Y})}{\sigma(\mathbf{X})\sigma(\mathbf{Y})},$$

provided that $\sigma(\mathbf{X}) \neq 0$ *and* $\sigma(\mathbf{Y}) \neq 0$.

It is important to note that if **X** and **Y** are independent, then the covariance and correlation equal zero, but the converse is not always true.[15]

These ideas generalize directly to n random variables $\mathbf{X_1}, \ldots, \mathbf{X_n}$. First, the *joint distribution* is the function F such that for all real numbers $x_1, \ldots, x_n$

$$F(x_1, \ldots, x_n) = P(\mathbf{X}_1 \leq x_1, \ldots, \mathbf{X}_n \leq x_n).$$

Second, the *joint density* function f is defined by:

$$f(x_1, \ldots, x_n) = \frac{\partial^n}{\partial x_1 \ldots \partial x_n} F(x_1, \ldots, x_n)$$

given that the derivatives exist. Given the joint density of the n random variables, they are *independent* iff

$$f(x_i, \ldots, x_n) = \prod_{i=1}^{n} f_i(x).$$

Modal aspects of probability. Questions of modality are often raised about statements of probability, and I examine briefly some of the issues. To illustrate modal ideas, it will be sufficient to consider once again the simple example of tossing a coin three times. The sample space consists of eight outcomes corresponding to the eight possible sequences of heads and tails that may be observed in three trials, as discussed earlier in detail. We can make a number of modal statements about the experiment represented by the sample space. In particular, the following statements, one about possibility and the other about necessity, are certainly true:

(1) It is possible that the event of getting heads on all three trials will occur.

(2) It is necessary that either at least one head or one tail occur on the three trials.

Other statements of possibility and necessity are easily constructed, but such statements are ordinarily of little interest in a probability context, because of their obviousness or triviality. The probability statements in which we are ordinarily interested, however, also have a clear modal character. Suppose that in our simple experiment the trials are Bernoulli with $p = \frac{1}{2}$. We would then accept the following statements as also true:

(3) The probability of getting a head on the third trial is independent of the outcome of the second trial.

---

[15]A physically interesting, but simple example of absence of a converse arises in quantum mechanics. The physical setting of this example is the kind of quantum entanglement analyzed for GHZ-type experiments at the end of Section 7.2. But the example can be completely understood on its own terms without any reference to quantum phenomena. Let $\mathbf{X}, \mathbf{Y}, \mathbf{Z}$ be three random variables whose values are ± 1, let $E(\mathbf{X}) = E(\mathbf{Y}) = E(\mathbf{Z}) = 0$, and let the third moment $E(\mathbf{XYZ}) = 1$. Then it is easy to prove that the covariances equal zero, i.e., $E(\mathbf{XY}) = E(\mathbf{YZ}) = E(\mathbf{XZ}) = 0$. Since the third moment is a direct statement of deterministic dependence, that the covariances are zero obviously does not imply independence of $\mathbf{X}, \mathbf{Y}$ and $\mathbf{Z}$.

5.1 THE FORMAL THEORY

(4) The probability of getting exactly two heads, given that at least two heads occur, is greater than $\frac{1}{2}$.

From a modal standpoint the semantics of this setup is relatively straightforward because the set of possible worlds is just the set of possible outcomes in the sample space. On the other hand, in almost all cases of application the set of possible outcomes as reflected in the sample space is much more restricted than the set of possible worlds ordinarily discussed in modal logic. Many set-theoretical and other mathematical relations are accepted as fixed across the set of possible outcomes in the sample space, and are not subject to variation as they often are in accounts of possible worlds.

A more important divergence from modal logic is apparent in the mathematical and scientific practice of using probability concepts. The probability measure and statements about probability are all handled in a purely extensional fashion, and the extensional status of probability concepts is no different from that of other scientific concepts. This is even true of subjectivistic theories of probability where strong intensional notions corresponding to assertions about belief would seem to play a natural role. (Occasionally some subjectivists have taken account of this problem but it is certainly not standard.)

There is a deeper reason, however, why the modality of probability is left implicit. This is the fact that in most complicated applications the explicit sample space is not set forth. In fact, the problem is so formulated that there is not a unique probability space but a plethora of them. A problem, either theoretical or applied, is stated in terms of a family of random variables. The consistency requirement is that the family of random variables have a joint distribution compatible with the joint distributions of all finite subsets of the random variables, but there is no uniqueness to the definition of sample space. What I have said, taken literally, would seem to run against the earlier definition of random variables as requiring them to be functions defined on the sample space. In practice, however, it is clear that the specific sample space plays no real role and what is important is the distribution of the random variable, that is, the distribution of the values of the random variable. From a modal standpoint this means that the set of possible worlds is left implicit. A sustained effort to avoid having a sample space anywhere in sight is to be found in de Finetti's two-volume treatise on probability (1974), where to avoid confusion of terminology he talks about *random quantities* rather than random variables. All the same, it is to be emphasized that de Finetti's terminology and conceptual analysis are not in agreement with that generally used by almost all contemporary statisticians, including many that are quite sympathetic to de Finetti's subjectivistic view of probability.

Probabilistic invariance. In Section 4.1, I described an important logical result on invariance. As stated by Tarski and Givant (1987, p. 57), there are just four binary relations between elements of a given universe invariant under every permutation of the elements of the universe. These four relations are the universal relation, the empty relation, the identity relation and the diversity relation.

No such simple and general result holds for probability theory, but many particular invariance results for particular probabilistic theories are known. In Section 4.5 an important instance was described, entropy as a complete invariant for Bernoulli and ergodic Markov processes that are isomorphic in the sense that there is a mapping from one process to another preserving probability of events.

There are many important, but more elementary, invariant results in probability, several of which we will consider as the final topic of this section. There are usually several different ways to formulate the same invariant result. Here, for simplicity's sake, I restrict consideration to finite sample spaces, i.e., one for which the set Ω of possible outcomes is finite. The second restriction, in the spirit of Tarski's results for logic, and, more generally, the geometric invariance Erlangen program of Felix Klein, as discussed in Section 4.1, is to formulate the results in terms of one-one transformations of Ω onto itself, i.e., in terms of bijections of Ω. I give simple informal proofs for special cases only.

Independence. Let $\mathbf{X}_1, \ldots, \mathbf{X}_n$ be n random variables representing n tosses of a coin. So Ω has 2^n possible outcomes. Then probability is *invariant*, i.e., constant, under any bijection of Ω onto itself if and only if the random variables $\mathbf{X}_i$ are independent and fair, i.e., $P(\mathbf{X}_i = 1,$ for heads$) = \frac{1}{2}$. Here is the proof for $n = 2$. The random variables $\mathbf{X}_1$ and $\mathbf{X}_2$ are independent if and only if their joint density $p(x, y) = p(x)p(y)$, where we write 'p' instead of 'f' for the discrete density. Then using obvious notation for the possible outcomes, we first prove

$$p(h_1 h_2) = p(h_1)p(h_2). \tag{9}$$

Since all four possible outcomes are equal in probability by our invariance assumption,

$$p(h_1 h_2) = \frac{1}{4},$$

but $p(h_1) = p(h_1 h_2) + p(h_1 t_2) = \frac{1}{2}$ and by a similar argument $p(h_2) = \frac{1}{2}$. Since $\frac{1}{4} = \frac{1}{2} \cdot \frac{1}{2}$, (9) is established, and likewise by the same kind of argument, the other three, and we have already shown that the coin is fair, since $p(h_1) = p(h_2) = \frac{1}{2}$. This proof is trivial, but it exhibits the germ of the general method.

Notice that this case is close to the logical result and the reason is obvious. On probability grounds, when we have both independence and fairness there is no probabilistic way of distinguishing between two possible outcomes. They are probabilistically equivalent.

Exchangeability. Again, consider coin tossing, but this time, we do not assume fairness. The coin may be biased. But we may, under the usual assumptions about coin tossing, assume a weaker kind of invariance. We assume that if the number of heads in two different possible outcomes is the same, then a bijection of Ω onto itself will preserve probability, i.e., be invariant, for all outcomes with the same number of heads. In other words, as seems intuitively clear, the order in which heads appear in a sequence of n tosses does not affect the probability. Put in a way more congenial to bijections of Ω, let an outcome of n trials be given. Then any permutation of this outcome must have the same probability under our invariance hypothesis. For example, consider three tosses. Then, using $\approx$ for equivalence in probability:

$$hht \approx hth \approx thh,$$
$$htt \approx tht \approx tth,$$

but there are no such equivalences for *hhh* and *ttt*. We also have a simple representation possible for this exchangeable setup. Namely, a representation of the invariance is caught by a single counting random variable **Y**, whose values, for the example, are simply 0, 1, 2 and 3. So $\mathbf{Y}(hht) = 2$, the number of heads. There is much more to be said about exchangeability, which is touched on in Section 7, but here we see the dramatic reduction from the 2^n possible outcomes to just a single random variable **Y** with $n+1$ possible values. When the random variables $\mathbf{X}_1, \ldots, \mathbf{X}_n$ are exchangeable, all that can be said probabilistically can be said just with the random variable **Y**.

Exponential decay. A third example, given in detail in Section 6 on propensity theories of probability, introduces a different kind of invariance, but one quite familiar in many applications of probability theory. I have in mind the memoryless property of the exponential distribution, or its discrete analogue, the geometric distribution. If an event anticipated has not occurred up to time t, then its probability of occurring Δt later is no different than it was at some earlier time t'. Another way of putting the matter, there is no aging of the process, no becoming more likely to occur sooner as the present moves forward. Here is the notation used to develop this example in detail in Section 6. Let $p(a_1)$ be the probability of a radium atom decaying at time t_1, and let $p(a_{n+1}|A_n)$ be the conditional probability that the atom will decay on trial t_{n+1}, given the event A_n of not decaying by time t_n. The invariance condition is then

$$p(a_{n+1}|A_n) = p(a_1).$$

Imagine how different the life of organisms would be if the expected age of dying obeyed such an invariant exponential law. There are, of course, many physical processes that are of this exponential kind.

Note that in the last two examples of invariance, exchangeability and exponential decay, the invariant properties of the process do not uniquely determine the probability distribution, but rather place important constraints on it. It is characteristic of the general invariant properties of most physical or psychological processes that they determine the form of the density or distribution up to a small number of parameters that must be estimated empirically from data. This point is taken up in more detail in Section 6 on propensity theories of probability.

5.2 Classical Definition of Probability

Considering the ubiquitous character of probability, and the widespread interest in gambling in all segments of the population, both now and throughout a large part of recorded history, it is surprising to find that systematic discussions of probability constitute a very late intellectual development. Apart from a few minor things in the sixteenth century, there is no serious work in probability theory before the seventeenth century. In spite of the deep developments in Greek mathematics and astronomy, questions about probability apparently did not receive any systematic treatment in Greek thought. It appears that nowhere in Greek science or mathematics was the problem of handling errors of measurement discussed as a probabilistic phenomenon, subject to quantitative analysis.

Traditionally, the main origin of systematic probability theory is attributed to the work of Pascal and Fermat in the seventeenth century, and there does not seem to be much modern evidence to challenge this attribution. The traditional account is that the Chevalier de Méré proposed certain questions to Pascal involving the computation of odds in games of chance. Pascal corresponded with Fermat on these matters, and out of this correspondence was formed the mathematical theory of probability. It is interesting to note that the problem discussed in the first letter extant in the correspondence between Pascal and Fermat, namely, the letter dated July 29, 1654, is already of a fairly difficult sort. It is called the *problem of points*. Two players in a game each need a given number of points to win, and on each move have equal chances of winning a single point. Suppose the players separate without finishing the game. The problem is to determine how the stakes should be divided on the basis of their current scores. This is the same as asking for the probability of winning the game, which each player has at any given stage. Pascal did not give the complete solution of this problem, nor shall we discuss it here. He did exhibit all the cases that can occur in a game of six points, and he also covered the case when one player lacks only a single point of winning and the second player has no points, as well as the case when one player has one point and the second player has no points.

In the same letter to Fermat, Pascal also discusses a paradoxical problem that arises in computing the odds of throwing a 6 with one die in four throws, which are 671 to 625, and the odds of throwing two 6's with two dice. The odds of throwing two 6's are not in favor of doing it in twenty-four throws, in spite of the fact that 24 is to 36 as 4 is to 6, where 36 is the number of cases with two dice and 6, of course, is the number of cases with one die. This problem is a simple one and Pascal treats it lightly, even though de Méré claimed he found in this apparent paradox a source of contradiction within arithmetic. I mention the example here, because I want to emphasize later that already in the very early discussions of probability questions, the classical definition that the probability of an event is the number of favorable cases divided by the number of possible cases was being used, easily and naturally, probably because of the obvious symmetric shape of a die.

Readers interested in the history of the theory of probability during the seventeenth and eighteenth centuries are referred to the standard work of Todhunter (1865, reprinted in 1949), and also to the more recent, less mathematical and more philosophical account of Hacking (1975). A discussion of the early history of gambling, including some interesting speculations on why the theory of probability did not develop earlier, in spite of the great interest in gambling during ancient times, is to be found in David (1962). Also, an article by Sambursky (1956) contains a detailed discussion of the concepts of the possible and the probable in ancient Greek thought.

In the period between the middle of the seventeenth century, when Pascal and Fermat were writing, and the end of the eighteenth century, when Laplace was the dominant figure, the most important works on probability were probably: (i) an early treatise by Huygens, *De Ratiociniis in Ludo Aleae*, first published in 1657, (ii) Jacob Bernoulli's *Ars Conjectandi,* first published in 1713, and (iii) A. de Moivre's *The Doctrine of Chances,* first published in 1718. Important mathematical results were obtained during this period

5.2 CLASSICAL DEFINITION OF PROBABILITY

by Lagrange, Euler, and others. Also, the famous memoir of the Reverend Thomas Bayes was published posthumously in the Philosophical Transactions of the Royal Society in the volume for 1763.

Laplace. The work, however, that towers over all others in this period is that of Laplace, beginning with two long memoirs of 1774(a,b) and culminating in the treatise by Laplace, first published in 1812, *Théorie Analytique des Probabilités*. This large treatise is of great conceptual and mathematical depth and maturity. From the standpoint of our interest here, the important part is the conceptual analysis of the foundations of probability in the introduction. This introduction was published separately early in the nineteenth century and is available in an English translation under the title, *A Philosophical Essay on Probabilities* (1951). Because this essay of Laplace's is the standard source of the classical definition of probability, we shall want to analyze the main points of Laplace's views.

In the first place, it is important to emphasize that for Laplace the universe is deterministic, and probability arises simply from our ignorance of the exact causes of events. He puts the matter very succinctly in the opening lines of Chapter II:

> All events, even those which on account of their insignificance do not seem to follow the great laws of nature, are a result of it just as necessarily as the revolutions of the sun. In ignorance of the ties which unite such events to the entire system of the universe, they have been made to depend upon final causes or upon hazard, according as they occur and are repeated with regularity, or appear without regard to order; but these imaginary causes have gradually receded with the widening bounds of knowledge and disappear entirely before sound philosophy, which sees in them only the expression of our ignorance of the true causes.
>
> Present events are connected with preceding ones by a tie based upon the evident principle that a thing cannot occur without a cause which produces it. This axiom, known by the name of the principle of sufficient reason, extends even to actions which are considered indifferent ...
>
> (Laplace, *A Philosophical Essay on Probabilities*, p. 3)

Laplace follows this passage with his famous statement on the deterministic character of the universe. From a knowledge of the present state of the universe, an "intelligence", or as perhaps we might say in modern terminology 'a computer adequate to submit the data to computation', would be able to determine the entire past and future of the universe.

> We ought then to regard the present state of the universe as the effect of its anterior state and as the cause of the one which is to follow. Given for one instant an intelligence which could comprehend all the forces by which nature is animated and the respective situation of the beings who compose it—an intelligence sufficiently vast to submit these data to analysis—it would embrace in the same formula the movements of the greatest bodies of the universe and those of the lightest atom; for it, nothing would be uncertain and the future, as the past, would be present to its eyes. The human mind offers, in the perfection which it has been able to give to astronomy, a feeble idea of this intelligence. Its discoveries in mechanics and geometry added to that of universal gravity, have enabled

it to comprehend in the same analytical expressions the past and the future states of the system of the world. Applying the same method to some other objects of its knowledge, it has succeeded in referring to general laws observed phenomena and in foreseeing those which given circumstances ought to produce. All these efforts in the search for truth tend to lead it back continually to the vast intelligence which we have just mentioned, but from which it will always remain infinitely removed. This tendency, peculiar to the human race, is that which renders it superior to animals; and their progress in this respect distinguishes nations and ages and constitutes their true glory.[16]

(Laplace, *A Philosophical Essay on Probabilities*, pp. 4-5)

As we turn to the formal statement of the definition of probability itself, it is essential to realize that for Laplace probability has this subjective connection with ignorance. In this sense the classical definition of probability is very much in agreement with subjective views of probability, as opposed to the objective relative-frequency sort of view discussed in the next section.

In Chapter III Laplace turns to the general principles of the calculus of probability. He begins immediately with the statement of the first two principles, of which the first principle is just the classical definition of probability.

First Principle.—The first of these principles is the definition itself of probability, which, as has been seen, is the ratio of the number of favorable cases to that of all the cases possible.

Second Principle.—But that supposes the various cases equally possible. If they are not so, we will determine first their respective possibilities, whose exact appreciation is one of the most delicate points of the theory of chance. Then the probability will be the sum of the possibilities of each favorable case.

(Laplace, *A Philosophical Essay on Probabilities*, p. 11)

Laplace goes on to state a number of additional principles, which I shall discuss subsequently; but the first two just quoted constitute the heart of the classical definition of probability, and, thus, a detailed examination of their import is warranted at this point. In turning to an analysis of Laplace's definition, I should emphasize at the beginning that I shall no longer be concerned with the history of probability theory, and in particular of the criticisms of the classical theory as they have developed historically. Instead, I want to concentrate on the logical analysis of the definition and its defects.

From a formal standpoint, it is quite easy to see how Definition 2 of Section 1 can be extended so that the formal definition of probability agrees with the classical definition as formulated in Laplace's first principle. For simplicity, let us restrict ourselves to the case of finite sets of possible outcomes. Because the inclusion of Laplace's first principle renders redundant most of the axioms of Definition 2, it will perhaps be instructive to begin anew by first stating a definition and then asserting a simple theorem relating models of this definition to models of Definition 2. In the statement of this new definition we use Cantor's double-bar notation for the cardinality of a set; for instance, $\bar{\bar{A}}$ is the cardinality of the set A.

---

[16]Roger Hahn (1967) has made the important historical point that Laplace actually stated his views on determinism in one of his early memoirs (1773). He was obviously influenced, as Hahn shows, by an earlier statement of a similar sort by Condorcet in a published letter to D'Alembert (1768).

5.2 CLASSICAL DEFINITION OF PROBABILITY

DEFINITION 1. *A structure* $\boldsymbol{\Omega} = (\Omega, \Im, P)$ *is a finite Laplacean probability space if and only if:*

1. *Ω is a finite set;*
2. *$\Im$ is an algebra of sets on Ω;*
3. *For A in $\Im$,*

$$P(A) = \frac{\overline{\overline{A}}}{\overline{\overline{\Omega}}}.$$

It seems evident that Definition 1 expresses in a well-defined formal fashion the intent of Laplace's first principle. It is also clear that the following theorem is a trivial consequence of the definition.

THEOREM 1. *Any finite Laplacean probability space $(\Omega, \Im, P)$ is a finitely additive probability space in the sense of Definition 2 of Section 1.*

For the discussion of games of chance, certainly the most popular probability subject in the seventeenth and eighteenth centuries, Laplacean probability spaces seem natural and appropriate, because the symmetries imposed on dice or randomly shuffled decks of cards lead very easily to satisfaction of Axiom 3 of Definition 1.

Before we turn to specific reasons for rejecting Laplacean probability spaces as being an adequate device for formulating all probability questions, we should consider an important philosophical issue involved in the arguments we shall use. The nature of this issue may be made clear by drawing a parallel to the set-theoretical construction of the natural numbers. In set-theoretical developments of mathematics, it is now customary to identify the natural numbers with either the finite cardinals or the finite ordinals. In fact, if we follow the von Neumann construction of the ordinal numbers, we may have a three-way identification of natural numbers, finite cardinals, and finite ordinals. It is not appropriate to give technical details here of how this construction is made within the framework provided by general axioms about sets, except to emphasize that one of the touchstones of any such construction is satisfaction of Peano's axioms for the natural numbers. Now it is a familiar fact of the history of mathematics that difficult and subtle theorems about the natural numbers were proved before these axioms were formulated, or any subsequent set-theoretical construction of the natural numbers was carried out. Axiom systems like Peano's, or set-theoretical constructions like von Neumann's, which do not yield these theorems would be rejected, because quite independently of either the axioms or the construction, we have a clear notion of what is true or false of the natural numbers, or, more conservatively, of what we are willing to accept as true or false of the natural numbers. In my own view, the appropriate account of the reasons for this clarity is to be given in terms of the empirical facts at the basis of arithmetic, but reasons for holding that arithmetic is primarily an empirical science cannot be gone into in this work. What does have to be faced in a discussion of successive positions on the foundations of probability is that on the basis of ill-defined intuitions we shall produce arguments for accepting or rejecting various probability spaces. In most cases, I shall take it that the intuitive ground for the adequacy of the example or counterexample will be definite enough not to require any general methodological argument in support of it. We would reject any axiom system for the natural numbers which gave an account of even and odd numbers that did not square with the familiar truths accepted since Pythagoras; similarly, we reject any account of how probability spaces should be constructed if it does not give an intuitively acceptable account of experiments with biased dice or unfair coins. The reasons why such constructions are not acceptable I shall take to be obvious, once the explicit counterexample is produced.

As we turn to the detailed analysis of the classical definition of probability, it is worth noting initially that it is not too easy to square Laplace's first and second principles with one another. The first principle gives an unequivocal and well-defined, but highly restricted, definition of probability, just that which is embodied in Axiom 3 of Definition 1, earlier in this section. The second principle, on the other hand, gives a rather vague statement of what is to be done when the cases are not equally possible. For the moment, we shall ignore the second principle and emphasize only the first one. When we try to stick to the definition embodied in the first principle, difficulty arises at once concerning the representation of experiments with biased mechanisms. For example, within the classical definition, how are we to represent the possible outcomes of tossing a biased coin twice? For definiteness, suppose we are dealing with a coin that on the basis of considerable experience we have found to have a probability of .55 that in a standard toss a head will come up. It is obvious that there is no way of defining a Laplacean probability space in the sense of Definition 1 that will represent in a natural way the result of tossing such a coin twice; for the event structure determines in a definite fashion what the set Ω should be, and once this event structure is determined, Axiom 3 adds little. This observation about the event structure's determining the probability, according to Axiom 3 of Definition 1, is a conceptually significant, even though mathematically trivial, conclusion about Laplacean probability spaces. As should be apparent from Definition 1, the Laplacean concept of probability is definable in terms of the set Ω of possible outcomes and the algebra $\Im$ of events. This means that from a conceptual standpoint there are no substantive assumptions in the general formulation of Laplacean probability spaces beyond the requirement that $\Im$ be an algebra (or σ-algebra) of sets on Ω. Moreover, as is natural in the finite case, we simply take $\Im$ to be the set of all subsets of Ω, and we are left with no substantive axioms of any sort, except possibly finiteness of the set of possibilities.[17]

Now it might be thought that what should replace the obviously substantive axioms of Definition 2 of Section 1 are the principles of symmetry used in talking about games with dice or cards, but it is clear that from a formal standpoint these principles of symmetry are redundant. I mean by this that the sample space Ω is determined solely by the event structure. If we take the classical definition seriously, then the definition of the sample space Ω for the rolling of a die three times or the tossing of a coin twice, whatever the example may be, is solely determined by a characterization of the possible outcomes, and not at all by any considerations of symmetry. Principles of symmetry enter only in telling us intuitively whether or not a particular application of the classical definition is sensible. In order to incorporate any principles of symmetry into the formal definition

---

[17]Theorem 3 of Section 5 shows how from a formal standpoint Laplacean probability spaces can be generalized to provide a near equivalent to any probability space, once the apparatus of infinitesimals is introduced.

5.2 Classical Definition of Probability

of probability, it would be necessary to weaken the classical definition in the following way. We would need to make the application of Laplace's first principle conditional on the satisfaction of certain principles of geometrical and physical symmetry. From one standpoint this seems to be a reasonable procedure to follow, for it is clear in all cases of familiar application that some such principles are satisfied, and are, in fact, the key to the intuitively acceptable application of the classical definition. On the other hand, to bring such principles explicitly into the systematic definition of probability is to complicate enormously the formal task, for it is by no means clear how one can satisfactorily state the general principles of symmetry involved.[18]

Classical paradoxes. I now turn to the consideration of several classical paradoxes that arise when we attempt to apply Definition 1.

Bertrand's paradox. Consider the following simple problem. A bottle contains a mixture of water and wine. All that is known about the mixture is that the ratio of water to wine is not less than 1, and not more than 2. Extending Axiom 3 of Definition 1 in the obvious way to the continuous case, the principle of indifference or symmetry expressed in Axiom 3 says that we should assume that the probability of a mixture lying anywhere between a ratio of 1 and 2 of water to wine arises from a uniform distribution on this interval. Thus the probability of concentration of the mixture lying between 1 and 1.25 is .25. The probability of its lying between 1.6 and 2 is .4, etc. This all seems straightforward enough.

However, now examine the other way of looking at the problem. Instead of the ratio of water to wine, we consider the inverse ratio of wine to water. From the information given in the initial statement of the problem, we know that the ratio of wine to water lies between $\frac{1}{2}$ and 1. Once again we assume a uniform distribution on this interval. Thus the probability that the true ratio of wine to water lies in the interval from $\frac{1}{2}$ to $\frac{3}{4}$ is .50. Now the wine-to-water ratio $\frac{3}{4}$ corresponds to the water-to-wine ratio $\frac{4}{3}$, but on the assumption of a uniform distribution for both ways of putting the ratio, we have arrived at a contradiction. Let 'H' stand for water and 'W' for wine. For

$$P(\frac{4}{3} \leq \frac{W}{H} \leq 2) = \frac{2}{3}, \tag{1}$$

and

$$P(\frac{1}{2} \leq \frac{H}{W} \leq \frac{3}{4}) = \frac{1}{2}, \tag{2}$$

but

$$P(\frac{4}{3} \leq \frac{W}{H} \leq 2) = P(\frac{1}{2} \leq \frac{H}{W} \leq \frac{3}{4}), \tag{3}$$

because it is a truth of arithmetic that

$$\frac{4}{3} \leq \frac{W}{H} \leq 2 \text{ if and only if } \frac{1}{2} \leq \frac{H}{W} \leq \frac{3}{4}.$$

Clearly equations (1), (2), and (3) are mutually inconsistent. This paradox was given

---

[18]In contrast, particular principles of symmetry are easy to state and use in applications. See, for example, the discussion of invariance at the end of Section 1, or of qualitative discrete densities at the end of Section 6.

the name of 'Bertrand's Paradox' by Poincaré after a French mathematician who was the first to discuss this example (Bertrand 1888).

The source of the difficulty in Bertrand's Paradox is the introduction of a random variable whose range is an infinite set, in this case, a finite interval of real numbers. The uniform distribution is determined not by counting the number of possible values of the random variable and assigning the prior probability $\frac{1}{N}$ to each point, but by measuring the length of the interval and assigning equal prior measure to any two subintervals of equal length. Difficulty then arises because we may consider a transformation of the random variable that is not a linear transformation, in this case, the nonlinear transformation of passing from a number to its reciprocal. It might be thought that we could avoid the paradox by simply restricting ourselves to a finite number of points; but this is certainly not the case once we are in a situation like the present one where there is a natural metric or measure on the set of points. If, for example, we restricted ourselves in considering the ratio of water to wine to the finite set of points

$$1.1, 1.2, 1.3, 1.4, 1.45, 1.46, 1.47, 1.5, 1.8, 1.9,$$

most people would be startled indeed to find us assigning equal weight to all these points, particularly if we were proceeding on the general assumption of ignorance as to the nature of the mixture beyond the specification that it lies between 1 and 2. The reason for the surprise is apparent; namely, the points selected are not equally spaced in the natural metric we impose on the interval of possible values. If we take seriously the idea of such a natural metric then, of course, the paradox arises even for a finite set of points. For example, if we select the equally spaced ratios in the following set

$$1.0, 1.1, 1.2, 1.3, 1.4, 1.5, 1.6, 1.7, 1.8, 1.9, 2.0,$$

it is natural to assign an equal prior probability to each ratio, but when we consider the inverse problem and look at the ratio of wine to water, this same set then becomes the following

$$.50, .53, .56, .59, .63, .67, .71, .77, .83, .91, 1.00,$$

and it would seem very unnatural to assign an equal prior probability to members of this nonequally spaced set of numbers. A still older problem is the following.

Buffon's needle problem (1733, 1777). A number of parallel lines are drawn on the floor and a needle is dropped at random. What is the probability that the needle will cross one of the lines on the floor? We are concerned to predict the position of the needle in relation to the system of parallel lines on the floor. Now the position of the needle can be described by coordinates, and the contradiction arises from the fact that different kinds of coordinates related by nonlinear transformations may be used. Two familiar examples are ordinary Cartesian coordinates and polar coordinates. Certainly we will obtain different results if we assign a uniform distribution, first with respect to Cartesian coordinates, and then with respect to polar coordinates.

Bertrand's random chord paradox (1888). The problem is to find the probability that a "random chord" of a circle of unit radius has a length greater than $\sqrt{3}$ (where, of

5.2 CLASSICAL DEFINITION OF PROBABILITY

course, $\sqrt{3}$ is the length of the side of an inscribed equilateral triangle). As Bertrand remarks, there are three obvious and simple solutions of this problem.

1. Because any chord of the circle intersects the circle in two points, we may suppose these two points to be independently distributed on the circumference of the circle. Without loss of generality, we may suppose one of the two points to be at a vertex of an inscribed equilateral triangle. The three vertices of the equilateral triangle then divide the circumference into three equal parts, so there is just $\frac{1}{3}$ of the circumference in which the other point can lie in order that the resulting chord will be greater than $\sqrt{3}$. Thus, the probability of finding such a chord is $\frac{1}{3}$.

2. If we now consider the length of the chord directly, we see at once that its length depends on its distance from the center of the circle and not on its direction. To express this fact about its distance from the center, we may suppose that it has a fixed direction perpendicular to a given diameter of the circle, with distance along this diameter measuring its distance from the center. We may then suppose that its point of intersection with this diameter has a uniform distribution. For the chord to have a length greater than $\sqrt{3}$, the distance of the point of intersection from the center of the circle must be less than $\frac{1}{2}$, so that the probability of finding such a random chord is $\frac{1}{2}$.

3. Any chord is uniquely defined by the point of intersection of the chord and a radius perpendicular to the chord. One natural assumption is that this point of intersection is distributed uniformly over the circle. Then the probability of its lying in any region of area A is $\frac{A}{\pi}$, since the total area of the circle is π, and, hence, the probability of obtaining a random chord of length greater than $\sqrt{3}$ is $\frac{1}{4}$.

Without doubt other natural properties of chords could be used to generate distributions other than these three.[19] This example should make it clear that the application of the classical definition of probability is not determined in a conceptually definite and clear fashion. It is important to note that this problem does not arise after a particular probability space has been selected. The application of Definition 1 of this section, and in particular of Axiom 3 of that definition, is completely straightforward for a particular probability space and algebra of sets on that space. What is neither clear nor evident is what particular probability space should be selected to represent in the appropriate fashion a given problem. Once again we must go beyond Definition 1 to ask for the more fundamental principles of symmetry that should be used to construct the appropriate probability space. As far as I know, no extended discussion of these principles of symmetry has ever been undertaken, although, from one standpoint, the selection of a particular measure in confirmation theory, which is discussed in Section 5, hinges very much upon the application of some simple principles of symmetry. On the other hand, discussions of principles of symmetry in confirmation theory have not got beyond a very rudimentary stage, and are not adequate to decide what prior probability measure to use in a simple, but relatively sophisticated, problem like that of the random chord. Despair of finding any simple and natural principles of symmetry that would be accepted

---

[19]Note that without being explicit about it, we have moved to an implicit use of probability spaces with an infinite set of outcomes. Formal analysis of such cases was not made explicit before the twentieth century, and then scarcely before Kolmogorov's classic work (1933) provided a widely accepted approach, which is embodied in the definition of probability spaces given in Section 1 as Definition 2.

in all circumstances is perhaps one of the main lines of defense of the subjective theories of probability, which are discussed in Section 7.

Historical note on Laplace's principles 3–10. In connection with Laplace's first two principles of probability, which formulate the classical definition of probability, I mentioned that attention would also be given to the other principles stated by Laplace. It is, I think, a matter of some historical interest to review them briefly, and to see how they relate to the formal reconstruction of the classical definition given in this section. These additional principles are stated in Chapters III and IV of Laplace's *Philosophical Essay on Probabilities*, to which we have already referred.

The third principle simply states the product rule for independent events. The fourth principle states the product rule for events that are not independent, and, thus, is simply a form of the definition of conditional probability as given in Section 1. The form of the fourth principle is then
$$P(A \cap B) = P(A|B)P(B).$$
The fifth principle is just another formulation of the fourth principle, in this case in the form we used as a definition of conditional probability:
$$P(A|B) = \frac{P(A \cap B)}{P(B)}.$$
The sixth principle is a statement of Bayes' Theorem. It is perhaps important to note that in the formulation of the sixth principle Laplace takes explicit account of how the theorem is to be formulated in case the prior probabilities of the causes or hypotheses are unequally probable. Thus, his formulation just corresponds to the formulation of Bayes' Theorem given in Section 1 as Theorem 8. The seventh principle is a formulation of the theorem on total probability given as Theorem 7.

The eighth principle opens Chapter IV of his Introduction. With the consideration of the last three principles, we turn from questions purely of probability to questions of expectation or, as Laplace puts it, questions concerning hope. The eighth principle tells us how to calculate the mathematical expectation from the probability of occurrences of events that will cause a favorable outcome. Laplace's simple example will make the point clear:

> Let us suppose that at the play of heads and tails Paul receives two francs if he throws heads at the first throw and five francs if he throws it only at the second. Multiplying two francs by the probability $\frac{1}{2}$ of the first case, and five francs by the probability $\frac{1}{4}$ of the second case, the sum of the products, or two and a quarter francs, will be Paul's advantage. It is the sum which he ought to give in advance to that one who has given him this advantage; for, in order to maintain the equality of the play, the throw ought to be equal to the advantage which it produces.
>
> If Paul receives two francs by throwing heads at the first and five francs by throwing it at the second throw, whether he has thrown it or not at the first, the probability of throwing heads at the second throw being $\frac{1}{2}$, multiplying two francs and five francs by $\frac{1}{2}$ the sum of these products will give three and one half francs for Paul's advantage and consequently for his stake at the game.

(Laplace, *Philosophical Essay on Probabilities*, pp. 20-21)

5.3 INFINITE RANDOM SEQUENCES

The ninth principle extends the eighth principle to the case when losses as well as gains can accrue. Finally, the tenth principle is a not-too-clear verbal formulation of Daniel Bernoulli's principle of moral expectation (1738), which is discussed in Section 7 of this chapter under the name of expected utility.

It should be clear that Principles 3–8 are all fairly immediate consequences of the formal definition of probability spaces given in Section 1, and do not in fact depend on the classical definition of probability itself.

Finally, it is important to remark that even though the classical definition of probability is vague and difficult to apply, it continues to be important because it is, in effect, the definition of probability that is used in many important applications in which the guiding principles of symmetry are clear and generally well agreed upon. It is the classical definition of probability that is used to make standard calculations in games of chance.

5.3 Relative-frequency Theory for Infinite Random Sequences

This section is devoted to a discussion of the relative-frequency theory (see, e.g., Nagel 1939b). We begin with a theorem. This is followed by examples and criticisms which purport to examine the reasonableness of using the theorem for the application of probability theory. The theorem uses the notion of a finite sequence of length N. Such a sequence is just a function whose domain of definition is the first N positive integers. The formal statement of the theorem should not obscure its simple idea of just using relative frequencies to define probabilities.

THEOREM 1. *Let f be a finite sequence of length N, let $\Im$ be an algebra of sets on the range of f, that is, the set $R(f)$, and let the function P be defined for each A in $\Im$ as follows:*

$$P(A) = \frac{\overline{\{i : f_i \in A\}}}{N}$$

(that is, the number of i's such that f_i is in A, divided by N, is equal to the value $P(A)$). Then $(R(f), \Im, P)$ is a finitely additive probability space.

Proof. Clearly $P(A) \geq 0$. Since $\overline{\{i : f_i \in R(f)\}} = N$, then also $P(R(f)) = 1$. And if A and B are in $\Im$ and $A \cap B = \emptyset$, then

$$P(A \cup B) = \frac{\overline{\{i : f_i \in A \cup B\}}}{N}$$

$$= \frac{\overline{\{i : f_i \in A\}}}{N} + \frac{\overline{\{i : f_i \in B\}}}{N}$$

$$= P(A) + P(B).$$

As an example illustrating the above theorem, let f record 20 tosses of a coin. Thus f records an experiment. It is important to notice that f is not the set of all possible outcomes of the 20 tosses, but is some particular outcome. On a very simple relative-frequency theory, we would use this actual outcome to compute probabilities. Remember that we are now dealing with applications and want to know how to assign actual

probabilities. Given
$$f = thtttthhththhthhttht$$
Suppose A is the event of getting a head. Then,
$$P(A) = \frac{\overline{\overline{\{i : f_i \in A\}}}}{N} = \frac{9}{20}.$$

The major criticisms of the above method of empirically defining a probability—by the use of a finite sequence of trials—will be discussed. Alternative empirical methods will be presented subsequently.

The relative-frequency theory purports to be an objective theory. The probability of a given event is independent of our knowledge of the event. Put another way, for the relative-frequency theorist, the probability that a given coin will come up heads in one toss is a physical property of the coin and the experimental conditions under which the coin is tossed; it is not a function of our knowledge or lack of knowledge of the coin or the tossing (as a subjectivist would say).

From a formal standpoint the problem of applying a relative-frequency theory of probability is initially the problem of deciding which finitely additive probability spaces are admissible. The basic idea of the relative-frequency view is that probability spaces should be generated from sequences (finite or infinite) in the manner illustrated by Theorem 1. The difficult problem has been to find an adequate definition of the notion of *random* sequence. We shall consider several possibilities in this section, ending up with Church's definition, but first I want to mention the criticisms of using finite sequences as a basis.

The two major criticisms of finite sequences are: (i) instability of probabilities thus defined; (ii) difficulty of defining randomness for finite sequences. For example, in the case of the coin above, $P(A)$ would quite likely change if new trials were added to the 20 already given. Certainly the choice of exactly 20 trials is arbitrary. This criticism concerning instability can be met at least partially by insisting on a large number of trials. But even with a large number of trials it is difficult to give an adequate formal definition of randomness. Thus, from a mathematical standpoint, the natural thing is to "pass to the limit" and consider infinite sequences. Empirical criticisms of this introduction of infinite sequences will be mentioned later, and in the next section we return to the detailed consideration of finite random sequences.

Classically, infinite sequences have been the basis of the relative-frequency theory. Formal problems which arise in connection with them have attracted the attention of numerous mathematicians. After some mathematical preliminaries, three definitions of random sequences are examined, each definition being stronger than its predecessor.

An (*infinite*) *sequence* is a function whose domain of definition is the set of positive integers. If s is a sequence then s_n is the nth term of the sequence s. A sequence of *positive integers* is a sequence all of whose terms are positive integers.

Let s be a sequence of real numbers, then the $\lim_{n \to \infty} s_n = k$ if and only if for every $\epsilon > 0$, there is an integer N such that for all integers n if $n > N$ then
$$|s_n - k| < \epsilon.$$

DEFINITION 1. *If s and t are sequences of real numbers, then*

5.3 INFINITE RANDOM SEQUENCES

(i) $s + t$ is the sequence such that $(s + t)_n = s_n + t_n$,
(ii) $s \cdot t$ is the sequence such that $(s \cdot t)_n = s_n \cdot t_n$,
(iii) if $s_n \neq 0$ for every n, then $\frac{1}{s}$ is the sequence such that $(\frac{1}{s})_n = \frac{1}{s_n}$.

The proof of the following elementary theorem about limits is omitted.

THEOREM 2. *If s and t are sequences whose limits exist, then*

(i) $\lim (s + t) = \lim s + \lim t$,
(ii) $\lim (s \cdot t) = \lim s \cdot \lim t$,
(iii) *if $s_n \neq 0$ for every n, then if $\lim s \neq 0$ then* $\lim \frac{1}{s} = \frac{1}{\lim s}$.

We next introduce a frequency function.

DEFINITION 2. *Let s be a sequence and let $\Im$ be the family of all subsets of $R(s)$, the range of s. Let t be the function defined on $\Im \times I^+$ where I^+ is the set of all positive integers, such that for all A in $\Im$,*

$$t(A, n) = \overline{\{i : i \leq n \ \& \ s_i \in A\}}.$$

We call the number $\frac{t(A,n)}{n}$ the *relative frequency* of A in the first n terms of s. If the limit of the function $\frac{t(A,n)}{n}$ exists as $n \to \infty$, then this limit is the *limiting relative frequency* of A in s.

THEOREM 3. *Let s be a sequence and $\Im$ an algebra of sets on $R(s)$ such that if $A \in \Im$ then the limiting relative frequency of A in s exists. Let P be the function defined on $\Im$ such that for every A in $\Im$*

$$P(A) = \lim_{n \to \infty} \frac{t(A, n)}{n}.$$

Then $(R(s), \Im, P)$ is a finitely additive probability space.

Proof. Since for every n, $\frac{t(A,n)}{n} \geq 0$ then certainly

$$\lim_{n \to \infty} \frac{t(A, n)}{n} \geq 0.$$

Furthermore, for every n, $t(R(s), n) = n$ and hence

$$\lim_{n \to \infty} \frac{t(R(s), n)}{n} = 1.$$

Finally, if $A \cap B = \emptyset$ then

$$\lim \frac{t(A \cup B, n)}{n} = \lim \left[\frac{t(A, n)}{n} + \frac{t(B, n)}{n} \right]$$

$$= \lim \frac{t(A, n)}{n} + \lim \frac{t(B, n)}{n}.$$

And thus we see that $(R(s), \Im, P)$ is a finitely additive probability space.

An example of the above infinite sequence formulation is the following: Consider the coin sequence

$$(h, h, t, h, h, t, h, h, t, h, h, t, \ldots, h, h, t, \ldots).$$

Clearly, if A is the event consisting of the occurrence of heads, that is, if $A = \{h\}$, then

$$\lim \left[\frac{t(A, n)}{n} \right] = \frac{2}{3} = P(A).$$

If, on the other hand, $B = \{t\}$, then

$$\lim \left[\frac{t(B,n)}{n}\right] = \frac{1}{3} = P(B).$$

Although the intuitive concept of randomness is not entirely clear, it seems obvious that this sequence of tosses is not random. There is a repetition of the basic unit hht for which we would be tempted to search for some causal explanation. If probability is an objective characteristic of the system, then any simply selected subsequence should yield the same probability for an event as the initial infinite sequence. More precisely, since the probability of an outcome is an objective property of the sequence, we would expect to get the same numerical probability if we picked any infinite subsequence having the property that each term of the subsequence is chosen independently of the outcome of that term. But in our example, if we take the subsequence beginning with the third term and consisting of every third term thereafter, we get the infinite sequence $(t, t, t, t, t, \ldots, t, \ldots)$ and, in this sequence, the probability of tails is 1.

That limiting relative frequencies do not even always exist can be shown by the following example. Consider the sequence of 1's and 0's formed by the following pattern:

$$s = (1, 0, 0, 1, 1, 1, 1, 0, 0, 0, 0, 0, 0, 0, 0, 1, 1, \ldots)$$

where $s_n = 0$ if the k such that $2^{k-1} \le n < 2^k$ is even and $s_n = 1$ if the k such that $2^{k-1} \le n < 2^k$ is odd. If a is even, then the relative frequency of 1's in the first $2^a - 1$ terms is $\frac{1}{3}$. If a is odd, the relative frequency of 1's in the first $2^a - 1$ terms is greater than $\frac{2}{3}$, but approaches $\frac{2}{3}$ asymptotically as a increases.

Furthermore, it is possible to have a sequence s and a family of subsets of $R(s)$ such that the limiting relative frequency of each member of the family exists, but the family is not an algebra. In other words, $\lim(t(A,n)/n)$ and $\lim(t(B,n)/n)$ may exist in s but $\lim(t(A \cup B, n)/n)$ may not. The following example is due to Herman Rubin (oral communication).

Consider the elements 00, 11, 01, 10 which are alternated in pairs, e.g., 00, 11 and 01, 10, in a manner similar to that given for the sequence whose limiting relative frequency did not exist, i.e., in terms of powers of 2.

$$s = (00, 01, 10, 11, 00, 11, 00, 01, 10, \ldots).$$

Thus

$$s_n = \begin{cases} 00 \text{ if } n \text{ is odd and the } k \text{ such that } 2^{k-1} \le n < 2^k \text{ is odd,} \\ 11 \text{ if } n \text{ is even and the } k \text{ such that } 2^{k-1} \le n < 2^k \text{ is odd,} \\ 01 \text{ if } n \text{ is even and the } k \text{ such that } 2^{k-1} \le n < 2^k \text{ is even,} \\ 10 \text{ if } n \text{ is odd and the } k \text{ such that } 2^{k-1} \le n < 2^k \text{ is even.} \end{cases}$$

We define A as the event of getting a zero in the first member of a term and B as the event of getting a one in the second member of the term. Thus,

$$A = \{00, 01\}$$

$$B = \{01, 11\}$$

$$A \cap B = \{01\},$$

5.3 INFINITE RANDOM SEQUENCES

for which
$$P(A) = \frac{1}{2} \text{ and } P(B) = \frac{1}{2}.$$
For n even we have that for every $n > 1$
$$\frac{t[(A \cap B), 2^n - 1]}{(2^n - 1)} = \frac{1}{3},$$
whereas for n odd
$$\frac{t[(A \cap B), 2^n - 1]}{(2^n - 1)} \to \frac{1}{6} \text{ as } n \to \infty.$$
Thus $P(A \cap B)$ does not exist, and it easily follows that $P(A \cup B)$ does not exist.

Von Mises. Our next major problem is to characterize sequences which have the proper sort of randomness. The first systematic effort is arguably the important work of von Mises (1919).

Two intuitive principles are used by von Mises in his definition of a random sequence (collective). A sequence s is a random sequence if and only if (i) the limiting relative frequency of every real number r in R(s) exists, and (ii) if a gambler uses a system whereby he bets or doesn't bet (on a given $r \in R(s)$) on any arithmetic rule (such as betting only on odd numbered occasions), then the limiting relative frequency of r in the subsequence chosen by him is identical to the limiting relative frequency of r in the whole sequence.

The second principle is known as the principle excluding gambling systems, although its name is a bit misleading since it does not take care of all systems of gambling (e.g., doubling your bet each time). We shall see that it is far from easy to give an adequate formal reconstruction of this principle. We begin with several preliminary definitions.

A sequence s is a *monotone-increasing sequence of positive integers* if and only if its range $R(s)$ is a set of positive integers, and $s_m < s_n$ if $m < n$. A sequence s is a *subsequence* of t if, and only if, t is a sequence and there exists an r such that r is a monotone-increasing sequence of positive integers and $s = t \circ r$.

DEFINITION 3. *A* place-selection rule *is a monotone-increasing sequence of positive integers.*

Intuitively, a gambler always plays according to some place-selection rule (consciously or unconsciously).

DEFINITION 4. *Let s be a sequence and let $\Im$ be an algebra of sets on $R(s)$. Then s is a* random sequence in sense one *with respect to $\Im$ if and only if*

  (i) *for every A in $\Im$ the limiting relative frequency of A in s exists,*
  (ii) *for every A in $\Im$ and every place-selection rule g, the limiting relative frequency of A in $s \circ g$ exists and has the same value as the limiting relative frequency of A in s.*

This definition is, unfortunately, too strong. The following theorem shows that there are no nontrivial random sequences in the sense of Definition 4.

THEOREM 4. *Let s be a sequence and let $\Im$ be an algebra of sets on $R(s)$. If s*

is a random sequence in sense one with respect to $\Im$ then for every A in $\Im$

$$P(A) = 0 \text{ or } P(A) = 1.$$

(*That is, there are no nontrivial random sequences in sense one.*)

Proof. Suppose A is in $\Im$ and $P(A) \neq 0$ and $P(A) \neq 1$. Then we know that A occurs an infinite number of times in the sequence. So for the subsequence of terms where A occurs, there is a place selection rule g such that $s \circ g$ is that subsequence. But $P(A)$ in $s \circ g$ is 1, contrary to our definition of random sequences in sense one.

Definition 4 is formally precise, but its very general notion of a place-selection rule does not in any sense catch the intuitive idea of von Mises' principle excluding gambling systems, since there is no condition in the definition of the place-selection rule requiring that the nth term (choice) be a function only of n and the $n-1$ preceding outcomes. The next definition tries to express this idea.

Let I^+ be the set of positive integers, let S be the set of all sequences of 0's and 1's, i.e., the set of all sequences whose range is the set $\{0,1\}$, and for any sequence s in S, let $[s]_n$ be the set of all sequences in S that are identical to s in their first n terms. In other words,

$$[s]_n = \{t : t \text{ in } S \ \& \ (\forall \ m \leq n)(s_m = t_m)\}.$$

In order to try to catch the intuitive idea that a place-selection rule used on trial n should depend only on the trial numbers and $n-1$ preceding outcomes, we define

DEFINITION 5. *Let ϕ be a function that maps the set $S \times I^+$ onto the set $\{0,1\}$. Then ϕ is a* strict place-selection rule *if and only if whenever $[s]_{n-1} = [t]_{n-1}$ for any n and any two sequences s and t in S,*

$$\phi(s,n) = \phi(t,n).$$

We use this definition of a strict place-selection rule to construe a new sense of randomness. At first glance, Definition 5 may seem to be a definite improvement on the overly broad generality of Definition 3. The role of the function ϕ is to tell the gambler *when* to bet, just as in the case of a place-selection rule in the sense of Definition 3. In the intended interpretation the gambler should place a bet whenever the value of ϕ is 1, as is made clear in the following definition.

DEFINITION 6. *A sequence s of 0's and 1's is a* random sequence in sense two *if and only if*

(i) *the limiting relative frequency of 1's in s exists,*

(ii) *for every strict place-selection rule ϕ, the limiting relative frequency of 1's in the subsequence $s \circ \phi_s$ exists and has the same limiting value as in s itself.*

Unfortunately, the first-glance impression of randomness in sense two is mistaken. It does not represent any real improvement on sense one, because each monotonc increasing sequence of positive integers is just some strict place-selection rule ϕ_s. We thus have, corresponding to Theorem 4, the following:

THEOREM 5. *There are no nontrivial random sequences in sense two.*

There seems little hope of improving on randomness in sense two in any simple way. Our difficulty is that the uninhibited use of the obvious set-theoretical constructions of

5.3 INFINITE RANDOM SEQUENCES

subsequences or place-selection rules always ends up being the full, classical set of such rules, which in view of Theorems 4 and 5 is far too large. The depth of the difficulties for relative-frequency theory created by Theorems 4 and 5 has not always been appreciated by advocates of this theory, who have been reluctant to take their own ideas in a completely literal sense.

Church. We now turn to the best known approach to the difficulties encountered. The basic ideas are due to Church (1940). We shall end up restricting place-selection rules to rules that are recursive. Although several technical definitions are needed to make the development systematic and formal, we give at this point our final formal definition of randomness, and some readers who wish to skip the rather technical discussion following this definition may go at once to Theorem 6. We use the concept of partial recursive function introduced earlier in Section 5 of Chapter 3.

DEFINITION 7. *A sequence s of 0's and 1's is a random sequence in Church's sense if and only if*

(i) *the limiting relative frequency of 1's in s exists,*

(ii) *if ϕ is any partial recursive function mapping the set of positive integers onto $\{0, 1\}$, and if b is the sequence*

$$b_1 = 1,$$
$$b_{n+1} = 2b_n + s_n,$$

then the limiting relative frequency of 1's in the subsequence $s \circ \phi \circ b$ exists and has the same limiting value as in s itself.

As the following examples illustrate, the purpose of the sequence b is to represent a function of a variable number of variables as a function of one variable, for, as has already been remarked, it is the intent of von Mises's principle that the decision to play on the nth occasion should be a function only of the previous $n - 1$ outcomes and the number n. However, this means that the number of arguments of the place-selection function varies with each n, which is technically inconvenient in characterizing ϕ as partial recursive. Hence we introduce the sequence b. We shall usually write $\phi(b_n)$, rather than $(\phi \circ b)(n)$. As before, the role of the function ϕ in the definition is to tell the gambler *when* to bet, but not on what event or how much. The gambler should place a bet when the value of ϕ is 1. The following examples should make all this a good deal clearer.

First example of the ϕ function of Definition 7. This ϕ picks out the subsequence consisting of every odd-numbered term. Intuitively, we bet on the nth occasion if and only if $\phi(b_n) = 1$. We want to find a ϕ such that for every k, $\phi(b_{2k}) = 0$ and $\phi(b_{2k+1}) = 1$.

To begin, let us look at the b_n's for any possible sequence of 0's and 1's. We see that

$$b_1 = 1,$$

$$b_2 = 2b_1 + s_1 = \begin{cases} 2 \text{ for } s_1 = 0, \\ 3 \text{ for } s_1 = 1, \end{cases}$$

$$b_3 = 2b_2 + s_2 = \begin{cases} 4 \text{ for } s_1 = 0 \text{ and } s_2 = 0, \\ 5 \text{ for } s_1 = 0 \text{ and } s_2 = 1, \\ 6 \text{ for } s_1 = 1 \text{ and } s_2 = 0, \\ 7 \text{ for } s_1 = 1 \text{ and } s_2 = 1. \end{cases}$$

In general, the possible values of b_n are just the integers k such that $2^{n-1} \leq k < 2^n$. We know, for example,

$$2^{1-1} = 2^0 \leq b_1 < 2^1,$$

$$2^{2-1} = 2^1 \leq b_2 < 2^2,$$

$$2^{3-1} = 2^2 \leq b_3 < 2^3.$$

The intuitive idea of the b_n's is so to construct them that regardless of the actual outcomes, we always have $b_n \neq b_m$ if $n \neq m$.

For a sequence of all 0's the value of b_n is 2^{n-1}, whereas for all 1's it is $2^n - 1$. The ϕ that will select every odd term is the following:

$$\varphi(a) = \begin{cases} 0 \text{ if the unique } k \text{ such that } 2^{k-1} \leq a < 2^k \text{ is odd,} \\ 1 \text{ if the unique } k \text{ such that } 2^{k-1} \leq a < 2^k \text{ is even.} \end{cases}$$

For example,

$$\phi(b_1) = \phi(1) = 1 \text{ since } k = 1,$$

$$\phi(b_2) = \begin{cases} \phi(2) = 0 \text{ since } k = 2, \\ \phi(3) = 0 \text{ since } k = 2. \end{cases}$$

The above example depended only on the integer n.

Second example of the ϕ function of Definition 7. This example depends upon the outcome in a sequence of 1's and 0's. For a gambler who believes that the appearance of a zero will improve his luck, we want to find a ϕ such that $\phi(b_n) = 1$ if and only if $s_{n-1} = 0$. That is, he bets after each appearance of a zero. Here the appropriate ϕ is

$$\phi(a) = \begin{cases} 1 \text{ if } a \text{ is even,} \\ 0 \text{ if } a \text{ is odd.} \end{cases}$$

Third example of the ϕ function of Definition 7. Find the appropriate ϕ for the man who bets when and only when the previous two outcomes were zero.

$$\phi(a) = \begin{cases} 1 \text{ if } a \text{ is divisible by } 4, \\ 0 \text{ otherwise.} \end{cases}$$

General method. We can now derive a general method for finding the appropriate ϕ for the gambler who bets only after the occurrence of a given finite sequence $(t_1, t_2, \ldots, t_k)$ where the t_i's take on the values 0 and 1.[20]

---

[20]This derivation is due to Richard L. Jacob (oral communication).

5.3 INFINITE RANDOM SEQUENCES

More precisely, we want $\phi(b_n) = 1$ if and only if

$$s_{n-k} = t_1,$$
$$s_{n-k+1} = t_2,$$
$$\vdots$$
$$s_{n-1} = t_k.$$

By definition of the b_n's,

$$\begin{aligned}
b_{n-k+1} &= 2b_{n-k} + s_{n-k} = 2b_{n-k} + t_1, \\
b_{n-k+2} &= 2b_{n-k+1} + s_{n-k+1} = 2(2b_{n-k} + t_1) + t_2, \\
&= 2^2 b_{n-k} + 2t_1 + t_2, \\
b_{n-k+3} &= 2b_{n-k+2} + s_{n-k+2} = 2(2^2 b_{n-k} + 2t_1 + t_2) + t_3, \\
&= 2^3 b_{n-k} + 2^2 t_1 + 2t_2 + t_3, \\
&\vdots \\
b_n &= 2b_{n-1} + s_{n-1} = 2^k b_{n-k} + 2^{k-1} t_1 + \ldots + 2t_{k-1} + t_k \\
&= 2^k b_{n-k} + \sum_{i=1}^{k} 2^{k-i} t_i.
\end{aligned}$$

But by definition of the b_n's, b_{n-k} must be a positive integer. Hence the ϕ we want is:

$$\varphi(a) = \begin{cases} 1 \text{ if there exists a positive integer } m \text{ such that } a = 2^k m + \sum_{i=1}^{k} 2^{k-i} t_i, \\ 0 \text{ otherwise.} \end{cases}$$

Example: Find the ϕ for the gambler who bets only after the occurrence of the sequence $(0, 1, 1)$. Here $k = 3, s_{n-3} = t_1 = 0, s_{n-2} = t_2 = 1, s_{n-1} = t_3 = 1$.

$$2^k m + \sum_{i=1}^{k} 2^{k-i} t_i = 2^3 m + 2^2 t_1 + 2^1 t_2 + 2^0 t_3$$
$$= 2^3 m + 2^2 \cdot 0 + 2^1 \cdot 1 + 2^0 \cdot 1 = 8m + 3.$$

Hence the ϕ we want is

$$\varphi(a) = \begin{cases} 1 \text{ if there exists a positive integer } m \text{ such that } a = 8m + 3, \\ 0 \text{ otherwise.} \end{cases}$$

We can also use the expression we have derived for b_n in quite a different way. Consider the case where $k = n - 1$. Then $b_{n-k} = b_1 = 1$, and $t_i = s_{n-k-1+i} = s_i$. Therefore,

$$b_n = 2^k b_{n-k} + \sum_{i=1}^{k} 2^{k-i} t_i = 2^{n-1} + \sum_{i=1}^{n-1} 2^{n-1-i} s_i.$$

Now if we are given the value of b_n we can find the value of n and also the first $n-1$ terms of the sequence s simply by expressing b_n as a sum of powers of 2.

Example: We are given $b_n = 107 = 64 + 32 + 8 + 2 + 1 = 2^6 + 2^5 + 2^3 + 2^1 + 2^0$. Hence, $n - 1 = 6$ and so $n = 7$. Now if we write b_n as follows,

$$b_7 = 2^6 + 2^5 \cdot 1 + 2^4 \cdot 0 + 2^3 \cdot 1 + 2^2 \cdot 0 + 2^1 \cdot 1 + 2^0 \cdot 1$$

we can immediately write down the first six terms of the sequence since they are coefficients of the powers of 2:

$$s = (1, 0, 1, 0, 1, 1, \ldots).$$

It is worth noting that if we drop the requirement that ϕ be partial recursive, we can find a ϕ such that ϕ selects a subsequence all of whose terms are 1. Consider, for example, the following:

$$\phi(a) = s_{\mu(a)}$$

where $\mu(a)$ is equal to the least positive integer m such that 2^m is greater than a. Details are left to the reader.

Let us now turn to Church's proposal that ϕ be partial recursive. His solution hinges upon the requirement that ϕ be effectively calculable. Effective calculability of a function intuitively means that if we are given the arguments of the function we have an algorithm for computing the value of the function in a finite number of steps. It is obvious that the function ϕ last defined is not effectively calculable, for given a positive integer x, there is no method whatsoever of computing $\phi(x)$. We first must have given the sequence s.

Using results from analysis and probability theory, the following theorem may be proved.

THEOREM 6. *There exist nontrivial random sequences in Church's sense. Indeed, given any number a between 0 and 1, there exists a random sequence in Church's sense of 0's and 1's such that the limiting relative frequency of 1's is a.*

As Church (1940) has emphasized, the proof of Theorem 6 depends upon powerful nonconstructive methods of proof, which is one of the chief formal criticisms of the relative-frequency theory; namely, that elementary combinatorial problems in probability are made to depend on powerful, nonelementary methods of analysis. The probability, for example, of obtaining at least two heads in three tosses of a fair coin rests upon Theorem 6. It is true that dependence on the theorem could be avoided by using conditional assertions, but the central fact remains that it seems very unsatisfactory to require the concept of probability, ubiquitous as it is in ordinary experience, to depend upon such an elaborate framework of concepts that are not even familiar to many mathematical statisticians. For example, the treatment of these matters best known to statisticians is probably that of Wald (1936), which is far from satisfactory from a mathematical standpoint, because of the absence of an exact definition of what he means by requiring that a place-selection rule be definable within a fixed system of logic.[21]

---

[21] It should also be remarked that Copeland's concept of an admissible number (1928) and Reichenbach's equivalent concept of a normal sequence (1932) anticipate Church's analysis, but are not as satisfactory for technical reasons that will not be examined here.

5.3 INFINITE RANDOM SEQUENCES

From a formal standpoint probably the historically most important criticism of this characterization of random sequences has been that of Ville (1939, pp. 58–63). He shows that the von Mises-Church concept of a random sequence includes binary sequences of 1's and 0's in which all initial segments of the sequences contain at least as many 1's as 0's. As any casino owner or gambler would see at once, such sequences are unfair in the sense that betting on 1's is always a nonlosing proposition. Ville also used the important concept of a *martingale* to strengthen the concept of a fair gambling system.[22] A gambling system is a martingale if a gambler's expected fortune after the next play is the same as his actual fortune after the play just completed. More technically, but without being fully explicit about all the assumptions (see Feller 1966, p. 210 ff), a sequence of random variables $\{\mathbf{X}_n\}$ is a martingale if and only if

$$E(\mathbf{X}_{n+1}|\mathbf{X}_n, \mathbf{X}_{n-1}, \ldots, \mathbf{X}_1) = \mathbf{X}_n.$$

Still more demanding, a sequence of random variables $\{\mathbf{X}_n\}$ is *absolutely fair* if and only if

$$E(\mathbf{X}_1) = 0$$

and

$$E(\mathbf{X}_{n+1}|\mathbf{X}_n, \mathbf{X}_{n-1}, \ldots, \mathbf{X}_1) = 0.$$

(I have formulated these conditions in measure-theoretic terms, not in terms of collectives, because such measure-theoretic language is now the universal language of probability, regardless of the attitude toward foundations.)

For a spirited defense of von Mises against the criticisms of Ville, see van Lambalgen (1987a, pp. 43-54; 1996).

More recent work has strengthened the characterization of random sequences in a different direction. The most important viewpoint is that of the computational complexity of a sequence as a conceptually different way to judge its randomness. The detailed history is too tangled to enter into here. But the main points are the following. In 1964 Kolmogorov (1965) introduced the concept of the complexity of a finite sequence of symbols, e.g., 1's and 0's, drawn from a fixed finite alphabet. He defined the complexity of such a sequence as the length of a minimum program required to encode the sequence, with the minimum being relative to a given computer and programming language, as is discussed in more detail in the next section. The natural setting of Kolmogorov's ideas are finite rather than infinite sequences. Related work of Martin-Löf is also mentioned in the next section.

Indeed, the von Mises relative-frequency theory ultimately seems to fall between two stools. On the one hand, it is a very awkward and unnatural mathematical theory of probability as compared to the standard measure-theoretic approach outlined in Section 1 of this chapter. Yet, despite its technical complexity, it fails to provide an adequate theory of application for dealing with finite sequences. Casual philosophical writings on the foundations of probability have not emphasized sufficiently the mathematical difficulties involved in giving a complete and literal statement of this relative-frequency theory, adequate for the study of complicated probability phenomena. Certainly as far

---

[22] The term *martingale* was introduced by Ville, but the concept was anticipated in earlier work of P. Levy and possibly others.

as I know, there has not been a serious analysis of stochastic processes from a relative-frequency standpoint. Once dependencies in a sequence of events are admitted, as in stochastic processes, then, except for the special case of independent events, an appropriate definition of randomness becomes even more hair-raising in its complexity if the original ideas of von Mises are to be carried out. The mathematical problems besetting the theory of stochastic processes are more than sufficient in the much simpler measure-theoretic setting to justify remaining in this framework and avoiding the unnecessary complications of relative-frequency theory. With respect to questions of application, it is a patent fact that applications of probability theory deal with finite sequences and not with infinite sequences. What is needed for application is a more exact and satisfactory theory of randomness in finite sequences.

Finally, in explaining why the von Mises-Church relative-frequency theory is now not generally accepted as a theoretical basis for probability, we should reemphasize Church's important point already stated above. It seems clear that the subtle nonconstructive methods required to deal with random sequences as defined by Definition 7 are certainly not going to seem satisfactory for handling the simple intuitive problems of throwing dice or tossing coins, which receive such a natural treatment in the classical theory. Any theory that proposes to replace the classical theory must surely give a comparably simple treatment of games of chance in order to be judged ultimately satisfactory. This, the relative-frequency theory of probability, or at least current versions of it, cannot do in a fully formal way. Van Lambalgen (1987a) emphasizes the importance of the propensity interpretation as the most physically appealling alternative (see Section 6 of this chapter).

It is also important that when the complexity concepts introduced for finite sequences by Kolmogorov and others are extended to infinite sequences, then the definition of randomness is stricter than that of von Mises-Church, so that there are sequences random in the von Mises-Church sense but not in that of Kolmogorov.

The limited character of the analysis given here, in relation to a large and complicated literature, is to be emphasized. For an excellent overview, the reader is referred to van Lambalgen (1987a,b); unfortunately his (1987a) is not as accessible as it should be, but see also his (1990), (1992), and (1996). For more general references, see those given at the end of the next section.

5.4 Random Finite Sequences

A main argument for the move from infinite to finite sequences is that in scientific practice only finite sequences are encountered. A central idea of Kolmogorov's complexity definition (1963, 1965) is that a finite sequence is random if its description is of about the same length. Still a different intuitive idea is to give a procedural characterization of randomness. A finite sequence is random if it is produced by a random procedure. To a certain extent the contrast between actual finite sequences and procedures mirrors the contrast between collectives and the measure-theoretic approach, as reflected in a famous debate on these matters (Doob 1941; von Mises 1941).[23] A typical mathematical

---

[23]The term *measure-theoretic* just refers to the kind of formal probability measure introduced by Kolmogorov (1933) and used in Definition 2 of Section 1.

5.4 RANDOM FINITE SEQUENCES

move is to represent procedures by functions, and there is a literature on random functions (e.g., Goldreich, Goldwasser and Micali 1986). The intuitive idea of these authors is that a function is *poly-random* if no polynomial-time algorithm asking for the values of the function, for arguments chosen by the algorithm, can distinguish a computation that is valid for the function from a computation that just represents the outcome of independent coin tosses.

Although the details are not clear, from this definition it would seem to follow that Tarski's decision procedure for elementary algebra is poly-random, because it is an exponential-time algorithm. Obviously, the idea of poly-randomness is not in any sense an absolute notion of randomness, for processes that we clearly recognize as not intuitively random satisfy the definition. Another unsatisfactory feature of the definition of poly-random functions is that it reduces the notion to one that is dependent on the outcome of independent coin tosses, itself an intuitively random process. From a broad philosophical standpoint, such a reduction is not very helpful because one of our objectives is to characterize when a coin-tossing process, or a sequence of actual tosses, is to be regarded as random.

Another intuitive idea of random function is the idea back of the standard notion of random variable. We expect to have unexplainable variability in the values of a function. It is clear from how close the idea here is to standard statistical practice that something can be worked out. Martin-Löf's characterization of random (infinite) sequences proceeds very much along these lines and is part of the development of the preceding section. The core idea of Martin-Löf's (1966) is that a sequence is random with respect to a measure if it is not rejected at arbitrarily small levels of significance by any effective statistical test. There is much to be said about moving from this general characterization for infinite sequences to significance tests for finite samples. Although there are problems to be overcome, there would seem to me to be no essential obstacle to developing a theory of finite random sequences along the lines of significance tests for finite samples any more than there is an essential obstacle to taking Kolmogorov's route, if nonabsolute results are acceptable. Notice that there is a real conceptual difference between these two approaches. The latter approach, i.e., via significance tests, we might call the stochastic approach, and the route that Kolmogorov takes, following the literature, the complexity approach. There is no a priori reason that these two approaches should come out the same although we would expect some common asymptotic results, unless one of the approaches is clearly defective in catching the intuition of randomness we seem to have. For more on this point, see Theorem 4 below.

Kolmogorov complexity. One of Kolmogorov's earliest statements that some account of relative frequency as a key aspect of probability is needed is made in the context of proposing statistical tests for the randomness of finite sequences. Here is what he says.

> I have already expressed the view (see Kolmogorov 1933, Ch. 1) that the basis for the applicability of the results of the mathematical theory of probability to real 'random phenomena' must depend on some form of the *frequency concept of probability*, the unavoidable nature of which has been established by von Mises in a spirited manner. However,

for a long time I had the following views.

1. The frequency concept based on the notion of *limiting frequency* as the number of trials increases to infinity, does not contribute anything to substantiate the applicability of the results of probability theory to real practical problems where we have always to deal with a finite number of trials.
2. The frequency concept applied to a large but finite number of trials does not admit a rigorous formal exposition within the framework of pure mathematics.

Accordingly I have sometimes put forward the frequency concept which involves the conscious use of certain not rigorously formal ideas about 'practical reliability', 'approximate stability of the frequency in a long series of trials', without the precise definition of the series which are 'sufficiently large' etc. ...

...I still maintain the first of the two theses mentioned above. As regards the second, however, I have come to realise that the concept of random distribution of a property in a large finite population can have a strict formal mathematical exposition. In fact, we can show that in sufficiently large populations the distribution of the property may be such that the frequency of its occurrence will be almost the same for all sufficiently large sub-populations, when the *law of choosing these is sufficiently simple*. Such a conception in its full development requires the introduction of a measure of the complexity of the algorithm. I propose to discuss this question in another article. In the present article, however, I shall use the fact that there cannot be a *very large number of simple algorithms*. (Kolmogorov 1963, p. 369)

Kolmogorov never waivered from his conviction that finite, rather than infinite, sequences provide the proper scientific setting for the analysis of randomness.

I now turn to Kolmogorov's formal theory of complexity (1963, 1965). (Independently a similar concept was introduced by Solomonoff (1964) and also in related work by Chaitin (1966, 1969).)

We all recognize that some sequences are very simple. For example, considering now sequences of length n, a sequence of nothing but 1's, or a sequence of nothing but 0's, is clearly very simple in structure. We feel the same way about simple periodic sequences, for example, the sequence that simply alternates 1's and 0's. The problem is how to define complexity of finite sequences in a general way. The brilliant intuition of Kolmogorov is that a way of measuring the complexity of a sequence is by the length of the minimal program required to generate it. Before considering the formal definitions required to characterize Kolmogorov complexity, it is useful to issue a warning as to why the definitions are as complex as they are. The intuitive idea is quite natural. It would seem that we could pick some simple universal Turing machine, or some simple register machine, and use that as our fixed computer for computing effectively the complexity of any finite sequence. The matter, however, is very much more complicated, as any perusal of tricks of programming used to encode various complicated instructions will show. Because we insist on the minimal length program, we will be caught in a familiar problem for any computer of any power. It is not simple to deal with programs that recognize peculiar features of individual sequences. The upshot of all of this will be that the notion of Kolmogorov complexity is not an effectively computable function, that is, it is not a partial recursive function. If we think of the central feature of collectives it

5.4 RANDOM FINITE SEQUENCES

is perhaps not surprising that it is not effectively computable, although naive intuition might suggest that it should be.

With these complexities in mind we turn to the formal characterization. First of all, we shall take a suitable general class of universal computers such as unlimited register machines or universal Turing machines, which were characterized in Chapter 3, as having the capability of computing any partial recursive function.

Let x be a finite binary sequence of 0's and 1's, let U be a universal computer, let $U(p)$ be the output of U when program p is run by U, and let $l(p)$ be the length of program p.

DEFINITION 1. *The Kolmogorov complexity $K_U(x)$ of a string x with respect to U is defined as the minimum length taken over all programs that print x as output and halt, i.e., in symbols,*

$$K_U(x) = \min_{\{p:U(p)=x\}} l(p).$$

The most essential property of this measure of complexity is that it is universal, in the sense that any other universal computer V will give a measure of x's complexity that is within a constant that is independent of x.

THEOREM 1. (Universality). *Let V be any other universal computer. Then there is a constant $c(V)$ such that for any finite binary sequence x*

$$K_U(x) \leq K_V(x) + c(V).$$

The idea of the proof is that since U is a universal computer it can simulate V by a finite program of length $c(V)$, and so always satisfy the inequality.

In view of this theorem I drop subsequently references to U, even though for short programs there can be a big difference, because of the length of the simulation program. This should not be the case for the measure of long complex sequences.

Obviously, Definition 1 above is not really what we want because, roughly speaking, it is clear that complexity simply increases in general with the length of the sequence. What we want to be able to judge is, for sequences of fixed length, their relative complexity and to judge as random those of sufficient complexity. This means that we really need a conditional notion of complexity where some information is given about the sequence.

So, an important modification of Definition 1 is to conditionalize the measure of complexity given the length, $l(x)$, of the string x. In symbols

$$K_U(x|l(x)) = \min_{\{p:U(p,l(x))=x\}} l(p).$$

Here is a natural result relating the complexity of a sequence and its length.

THEOREM 2. *The conditional complexity of a finite binary sequence is less than its length plus a fixed constant independent of x.*

The proof rests on having the equivalent of a simple print command:

$$\text{PRINT } x_1, x_2 \ldots x_{l(x)}.$$

THEOREM 3. *The Kolmogorov complexity measure K is not a computable function.*

Proof. The proof depends upon the fact that as we examine programs P in search of the minimum program we are not able to effectively decide whether a given P is a candidate, because we are unable to decide in an effective way whether a given program P halts or not. If it does halt for input s, then it is a candidate; if it does not, it is not. Remember that to be effectively computable, we must be able to give an algorithm such that for any program and any sequence s we can compute whether the program will halt. This we cannot do, on the basis of the well-known undecidability of the halting problem. Obviously, for a particular sequence we can make a particular argument, but this is not what is meant in saying that the complexity measure is not computable. We have in mind giving an algorithm that works for all cases.

Universal probability. Let p be a program such that $U(p) = x$. If p is produced at random by a Bernoulli process of tossing a fair coin $l(p)$ times, then the probability of p is

$$2^{-l(p)}.$$

So, for the *universal probability* $P_U(x)$ of x we sum over the probability of producing at random, i.e., by coin tossing, at least one of the programs p such that $U(p) = x$:

$$P_U(x) = \sum_{\{p:U(p)=x\}} 2^{-l(p)}.$$

Although I have not been as formal as possible in this definition, it should be clear that the universal probability $P_U(x)$ is a measure-theoretic concept of probability in no way dependent on the Kolmogorov complexity of the finite sequence x.

This is the significance of the following theorem relating the two concepts, the universal probability of a finite sequence x and its Kolmogorov complexity.

THEOREM 4. *There exists a constant c such that for every finite binary sequence x*

$$2^{-K(x)} \leq P_U(x) \leq c2^{-K(x)}.$$

The full proof of this theorem is too intricate to give here. In fact, the whole development of this section has been less detailed than is desirable for a full grasp of the ideas. Fortunately, there are many good detailed expositions in the literature: Cover and Thomas (1991); Cover, Gacs and Gray (1989); Li and Vitanyi (1993); Martin-Löf (1969); Zvonkin and Levin (1970).

The final still unsettled question for this section is how we should define the concept of randomness for finite binary sequences. Kolmogorov (1968) proposes Theorem 2 as a basis for characterizing randomness of finite sequences, namely, a finite sequence x is random if its conditional complexity is close to the upper limit, i.e., close to the length of x plus a constant independent of x. Kolmogorov points out what is obvious about the definition. It is necessarily relative in nature. The same relative remark applies to statistical tests for randomness in a finite sequence. As Kolmogorov (1963) remarks, the number of tests and their simplicity must be considered, but again, there is no absolute definition. Too many tests and tests that are too algorithmically complex will reject a finite sequence as being random, no matter what.

Perhaps not surprising, it is generally recognized that an absolute notion of randomness can only be given for infinite sequences, but such an ideal result, without possibility of direct scientific applicability in either the design of experiments or the statistical test

5.4 RANDOM FINITE SEQUENCES

of experimental results relative to a hypothesis, model or theory, as the case may be, is typical of much scientific theorizing.

Relative frequencies as estimates of probability. In actual practice, pragmatic rules about the randomness of finite sequences work well when applied by experienced statisticians or scientists. Here is a simple example of the kind of pragmatism I mean. Suppose in a learning experiment I need to randomize 200 trials for rewarding an R_1 or an R_2 response. Independent of response, the reward schedule is Bernoulli with probability $\frac{1}{2}$ of rewarding R_1 on any trial, and therefore also $\frac{1}{2}$ for rewarding R_2 (but never both). Then the number of reward sequences, 2^{200}, is enormous. For computational purposes we want to preserve the Bernoulli measure of $p(R_1) = p(R_2) = \frac{1}{2}$, but without noticeably affecting this computation we can exclude from use the 2^{20} or so lowest complexity sequences. Now we would ordinarily not actually compute this set of low-complexity sequences, but would just reject one that seemed too regular. The story I am telling is too simple, but there is an obvious serious pragmatic point behind it. More careful elaboration can be supplied as needed or requested, in a given scientific context, but in practice the low probability of getting a sequence of low complexity is usually relied on, without specific complexity computations being made for a given finite sequence.

The more fundamental point, emphasized by numerous quotations in Section 8 from physicists, is that statisticians and scientists, even though holding widely varying views about the nature of probability, all use relative frequencies obtained from finite sequences of observations to estimate probabilities, the moments of random variables, parameters of theoretical models and so forth. There is really no serious school of thought about the nature of probability and its role in science which simply rejects any serious role for observed relative frequencies of phenomena. Nor is it common to say that probabilities are just relative frequencies obtained from observed and necessarily finite data.

All the same, there is a deep and serious controversy about the role of relative frequencies which goes back at least to the end of the nineteenth century. The controversy is at the heart of modern statistics since the middle of the twentieth century. It is now much too convoluted and elaborate to analyze in any detail in this book, but I hope a few paragraphs will show how one side in the controversy, *the frequentists*, as their name suggests, support a frequency representation or interpretation of probability. (For an excellent recent more detailed account of this controversy for the general scientific reader, see Rouanet et al. (1998).)

The classic nineteenth-century controversy was between epistemic probabilities expressing beliefs, especially about single events such as rain in Palo Alto tomorrow, and frequentist probabilities, meaning probabilities estimated from observed finite sequences of events. Thus *epistemic* represents a subjective viewpoint, which is analyzed in Section 7, and *frequentist* represents an objective view of probability. Bayes and Laplace are now viewed as the early proponents of the epistemic view, although other persons play an important role as well. The frequentist view, developing at the end of the nineteenth century and beginning of the twentieth century, was promoted by the British school of statisticians Karl Pearson (1857–1963), R. A. Fisher (1890–1962) and later by Jerzy Neyman and Egon Pearson, Karl Pearson's son. Apart from any technical details, the frequentist view rejected any subjective probabilities as having a systematic place in science. Broad acceptance of this objective viewpoint, endorsed by such influential figures as Kolmogorov, was dominant, and the Neyman-Pearson approach to statistics was, by 1950 or so, often labeled the orthodox view.[24] The Bayesian revival in the second half of the twentieth century is mainly a statistical story, but it is, not surprisingly, the source of the revival of subjective theories of probability, which are examined in Section 7.

5.5 Logical Theory of Probability

As opposed to the classical theory which I categorize as subjective, based as it is, in Laplace's phrase, on our ignorance of true causes, or the objective relative-frequency theory, there is also a substantial view, not so prominent now, of probability as an extension of logic. The logical relation of probability is the relation between propositions, which are to count as evidence, and conclusions, which are propositions on which the evidence bears. This logical relation, like rules of inference, is to be thought of in normative and not psychological terms. Advocates of this view can be found in the nineteenth century, but in this century the position was first set forth in the most sustained and persuasive manner by John Maynard Keynes (1921).

Keynes. He credits Leibniz for first conceiving the idea that probability is a branch of logic, but it is not my objective to trace the history prior to Keynes.[25] As was already mentioned in Section 1, Keynes strongly advocated that probability be thought of in terms of propositions rather than events. The only variation on this theme among those who have advocated a logical view of probability is whether to choose sentences, as syntactical objects, or propositions, as semantical objects, as the primary subject matter.

It will not be to the point to quote Keynes' axioms here. As opposed to his many interesting and controversial remarks about probability, the axioms are generally unexceptionable. They are mainly stated in conditional form and, as is so often characteristic of the formal side of various theories of probability, represent properties of probability that essentially all schools of thought agree on. From the standpoint of the objectives of this chapter, it is not possible to prove an interesting representation theorem for Keynes' viewpoint. However, there is one point about his axioms that is curious and worth noting. He says in the first axiom:

> Provided that a and h are propositions or conjunctions of propositions or disjunctions of propositions, and that h is not an inconsistent conjunction, there exists one and only

---

[24] For a spirited philosophical defense of the Neyman-Pearson approach to statistics, see Mayo (1996), and for an excellent retrospective account, Neyman (1971). Much more can be said about these matters, but they concern the foundations of statistics rather than the foundations of probability, so my remarks and references are mostly superficial, as far as statistics as such is concerned. This is regrettable, but a proper treatment would have required too much space.

[25] For an informative analysis of Leibniz's contribution to ideas about probability in the seventeenth century, see Hacking (1975, Ch. 10). The young Leibniz developed his ideas of probability in the context of legal reasoning, his own early education being the law rather than mathematics or philosophy.

5.5 LOGICAL THEORY OF PROBABILITY

one relation of probability P between a as conclusion and h as premise. Thus any conclusion a bears to any consistent premise h one and only one relation of probability.

(Keynes 1921, p. 135)

As ordinarily stated we would consider it a problem to prove that there exists such a relation of probability and, where possible, the stronger result that the relation of probability is unique. Keynes simply postulates this and thus eliminates any interesting analysis of what conditions are needed to guarantee existence and uniqueness.

A surprising aspect of Keynes' views is his holding that not all probabilities can be numerically expressed:[26]

> The recognition of the fact, that not all probabilities are numerical, limits the scope of the Principle of Indifference. It has always been agreed that a numerical measure can actually be obtained in those cases only in which a reduction to a set of exclusive and exhaustive *equiprobable* alternatives is practicable. ... But the recognition of this same fact makes it more necessary to discuss the principles which will justify comparisons of more and less between probabilities, where numerical measurement is theoretically, as well as practically, impossible.

(Keynes 1921, p. 65)

Jeffreys. A figure of greater historical importance than Keynes is Harold Jeffreys,[27] who is often cited as a subjective Bayesian, even though he clearly believed in objective probabilities.[28] He is often cited by Bayesians because of his many important applications in geophysics using prior probabilities. The depth and skill of his thinking about such applications is what marks him as historically important.[29] His defense of objective prior probabilities is much less well worked out. I mention some features. His theory of probability seems best characterized as a normative theory of reasonable beliefs and therefore, the natural inductive extension of deductive logic. Jeffreys says this in a very clear way.

> We must notice at the outset that induction is more general than deduction. The answers given by the latter are limited to a simple 'yes', 'no', or 'it doesn't follow'. Inductive logic must split up the last alternative, which is of no interest to deductive logic, into a number of others, and say which of them it is most reasonable to believe on the evidence available. Complete proof and disproof are merely the extreme cases. Any inductive inference involves in its very nature the possibility that the alternative chosen as the most likely may in fact be wrong. Exceptions are always possible, and if a theory does not provide for them it will be claiming to be deductive when it cannot be. On account of this extra

---

[26] This attitude of Keynes influenced the later and more detailed work on partially ordered qualitative probabilities by B. O. Koopman (1940a,b), although Koopman's views were more aligned to subjective rather than logical theories of probability.

[27] There have been four editions of Jeffreys' *Theory of Probability*: 1939, 1948, 1961 and 1983. Only the second edition of 1948 was seriously revised. My citations and quotations are from it.

[28] This subsection on Jeffreys has benefited from reading the recent article on Jeffreys of Galavotti (in press).

[29] Jeffreys (1948, 2nd edition) has this to say about Keynes' treatise. "This book is full of interesting historical data and contains many important critical remarks. It is not very successful on the constructive side, since an unwillingness to generalize the axioms has prevented Keynes from obtaining many important results."

generality, induction must involve postulates not included in deduction. Our problem is to state these postulates. It is important to notice that they cannot be proved by deductive logic. If they could, induction would be reduced to deduction, which is impossible. Equally they are not empirical generalizations; for induction would be needed to make them and the argument would be circular. We must in fact distinguish the general rules of the theory from the empirical content. The general rules are *a priori* propositions, accepted independently of experience, and making by themselves no statement about experience. Induction is the application of the rules to observational data.

(Jeffreys 1948, pp. 7–8)

The closeness to logic and mathematics is stated in a closely following paragraph:

The test of the general rules, then, is not any sort of proof. This is no objection because the primitive propositions of deductive logic cannot be proved either. All that can be done is to state a set of hypotheses, as plausible as possible, and see where they lead us. The fullest development of deductive logic and of the foundations of mathematics is that of *Principia Mathematica*, which starts with a number of primitive propositions taken as axioms; if the conclusions are accepted, that is because we are willing to accept the axioms, not because the latter are proved. The same applies, or used to apply, to Euclid. We must not hope to prove our primitive propositions when this is the position in pure mathematics itself. But we have rules to guide us in stating them, largely suggested by the procedure of logicians and pure mathematicians.

(Jeffreys 1948, p. 8)

Jeffreys' attitude toward judging reasonable degrees of belief is rather like judging the result of any procedure of measurement. Observers may differ, errors may occur, but just as in the case of physical measurement, this does not mean there is no impersonal standard of reasonableness. Here is what he says on this score.

Differences between individual assessments that do not agree with the results of the theory will be part of the subject-matter of psychology. Their existence can be admitted without reducing the importance of a unique standard of reference. It has been said that the theory of probability could be accepted only if there was experimental evidence to support it; that psychology should invent methods of measuring actual degrees of belief and compare them with the theory. I should reply that without an impersonal method of analysing observations and drawing inferences from them we should not be in a position to interpret these observations either. ... Nobody says that wrong answers invalidate arithmetic, and accordingly we need not say that the fact that some inferences do not agree with the theory of probability invalidates the theory. It is sufficiently clear that the theory does represent the main features of ordinary thought. The advantage of a formal statement is that it makes it easier to see in any particular case whether the ordinary rules are being followed.

This distinction shows that theoretically a probability should always be worked out completely. We have again an illustration from pure mathematics. What is the 1,000th figure in the expansion of e? Nobody knows; but that does not say that the probability that it is a 5 is 0.1. By following the rules of pure mathematics we could determine it definitely, and the statement is either entailed by the rules or contradicted; in probability language, on the data of pure mathematics it is either a certainty or an impossibility. Similarly, a guess is not a probability. Probability theory is more complicated than deductive logic,

5.5 LOGICAL THEORY OF PROBABILITY

and even in pure mathematics we must often be content with approximations.

(Jeffreys 1948, pp. 37–38)

Jeffreys' treatise is a wonderful source of carefully worked out case studies and is also excellent in its many critical remarks about other views of probability or of particular statistical methods. Here is a long passage criticizing Laplace's definition of probability and also relative-frequency definitions.

Most of current statistical theory, as it is stated, is made to appear to depend on one or other of various definitions of probability that claim to avoid the notion of degrees of reasonable belief. Their object is to reduce the number of postulates, a very laudable aim; if this notion could be avoided our first axiom would be unnecessary. My contention is that this axiom is necessary, and that in practice no statistician ever uses a frequency definition, but that all use the notion of degree of reasonable belief, usually without even noticing that they are using it and that by using it they are contradicting the principles they have laid down at the outset. I do not offer this as a criticism of their results. Their practice, when they come to specific applications, is mostly very good; the fault is in the precepts.

Three definitions have been attempted:

1. If there are n possible alternatives, for m of which p is true, then the probability of p is defined to be m/n.
2. If an event occurs a large number of times, then the probability of p is the limit of the ratio of the number of times when p will be true to the whole number of trials, when the number of trials tends to infinity.
3. An actually infinite number of possible trials is assumed. Then the probability of p is defined as the ratio of the number of cases where p is true to the whole number.

The first definition is sometimes called the 'classical' one, ... The second is the Venn limit, its chief modern exponent being R. Mises. The third is the 'hypothetical infinite population', and is usually associated with the name of Fisher, though it occurred earlier in statistical mechanics in the writings of Willard Gibbs, whose 'ensemble' still plays a ghostly part. The three definitions are sometimes assumed to be equivalent, but this is certainly untrue in the mathematical sense.

The first definition appears at the beginning of de Moivre's book. It often gives a definite value to a probability; the trouble is that the value is often one that its user immediately rejects. Thus suppose that we are considering two boxes, one containing one white and one black ball, and the other one white and two black. A box is to be selected at random and then a ball at random from that box. What is the probability that the ball will be white? There are five balls, two of which are white. Therefore, according to the definition, the probability is $\frac{2}{5}$. But most statistical writers, including, I think, most of those that professedly accept the definition, would give $\frac{1}{2} \cdot \frac{1}{2} + \frac{1}{2} \cdot \frac{1}{3} = \frac{5}{12}$. This follows at once on the present theory, the terms representing two applications of the product rule to give the probability of drawing each of the two white balls. These are then added by the addition rule. But the proposition cannot be expressed as the disjunction of 5 alternatives out of 12. ...

With regard to the second and third definitions, we must remember our general criteria with regard to a theory. Does it actually reduce the number of postulates, and can it be applied in practice? Now these definitions plainly do not satisfy the second criterion.

> No probability has ever been assessed in practice, or ever will be, by counting an infinite number of trials or finding the limit of a ratio in an infinite series. Unlike the first definition, which gave either an unacceptable assessment or numerous different assessments, these two give none at all. A definite value is got on them *only* by making a hypothesis about what the result would be. The proof even of the existence is impossible. On the limit definition, without some rule restricting the possible orders of occurrence, there might be no limit at all. The existence of the limit is taken as a postulate by Mises, whereas Venn hardly considered it as needing a postulate.
>
> (Jeffreys 1948, pp. 341–342, 345)

Jeffreys' criticisms are excellent, but violate the old adage that people in glass houses should not throw stones, in the sense that his own account of probability is quite lacking in clarity and rigor. Both in his treatise (2nd ed., 1948) and his small book *Scientific Inference* (2nd ed., 1957), he gives some clear qualitative axioms of probability, but never a set sufficient to prove a representation theorem, nor does he seem to recognize the need for such a theorem.

In a more general sense of representation, he has, as already indicated, many interesting things to say. For example, he proposes essentially two prior probability distributions to represent total ignorance, both familiar and used earlier by others, but his argument for them is worth noticing.[30]

> Our first problem is to find a way of saying that the magnitude of a parameter is unknown, when none of the possible values need special attention. Two rules appear to cover the commonest cases. If the parameter may have any value in a finite range, or from $-\infty$ to $+\infty$, its prior probability should be taken as uniformly distributed. If it arises in such a way that it may conceivably have any value from 0 to ∞, the prior probability of its logarithm should be taken as uniformly distributed. There are cases of estimation where a law can be equally well expressed in terms of several different sets of parameters, and it is desirable to have a rule that will lead to the same results whichever set we choose. Otherwise we shall again be in danger of using different rules arbitrarily to suit our taste. It is now known that a rule with this property of invariance exists, and is capable of very wide, though not universal, application.
>
> The essential function of these rules is to provide a formal way of expressing ignorance of the value of the parameter over the range permitted. They make no statement of how frequently that parameter, or other analogous parameters, occur within different ranges. Their function is simply to give formal rules, as impersonal as possible, that will enable the theory to begin. Starting with any distribution of prior probability and taking account of successive batches of data by the principle of inverse probability, we shall in any case be able to develop an account of the corresponding probability at any assigned state of knowledge. There is no logical problem about the intermediate steps that has not already been considered. But there is one at the beginning: how can we assign the prior probability when we know nothing about the value of the parameter, except the very vague knowledge just indicated? The answer is really clear enough when it is recognized that a probability is merely a number associated with a degree of reasonable confidence and has no purpose except to give it a formal expression. If we have no information

---

[30] These prior distributions of Jeffreys are well-known examples of *improper* priors, since they sum to ∞, not to 1. How they are used in practice can be seen from the example from Jeffreys given below.

5.5 LOGICAL THEORY OF PROBABILITY

> relevant to the actual value of a parameter, the probability must be chosen so as to express the fact that we have none. It must say nothing about the value of the parameter, except the bare fact that it may possibly, by its very nature, be restricted to lie within certain definite limits.
>
> (Jeffreys 1948, pp. 101–102)

Using these ideas, here is Jeffreys' solution to a problem that bothers relative-frequentists. How can we estimate a probability when only a single observation is given? On the assumption of complete ignorance apart from the one observation, it must be emphasized that Jeffreys fully understands his estimate is reasonable, but not necessarily highly accurate.

> The following problem was suggested to me several years ago by Professor M. H. A. Newman. A man travelling in a foreign country has to change trains at a junction, and goes into the town, of the existence of which he has only just heard. He has no idea of its size. The first thing that he sees is a tramcar numbered 100. What can he infer about the number of tramcars in the town? It may be assumed for the purpose that they are numbered consecutively from 1 upwards.
>
> The novelty of the problem is that the quantity to be estimated is a positive integer, with no apparent upper limit to its possible values. A uniform prior probability is therefore out of the question. For a continuous quantity with no upper limit the dv/v rule is the only satisfactory one, ...
>
> (Jeffreys 1948, pp. 213–214)

Here is Jeffreys' analysis, where I have changed the notation to fit this chapter. Let $\mathbf{X}$ be the random variable for the single numerical observation and let $\mathbf{Y}$ be the random variable for the total number of tramcars in the town. Then the prior for discrete values of $\mathbf{Y}$ is:

$$P(\mathbf{Y} = n) \propto \frac{1}{n},$$

i.e., is proportional to $\frac{1}{n}$. The likelihood is:

$$P(\mathbf{X} = m | \mathbf{Y} = n) = \frac{1}{n},$$

provided $m \leq n$. So, the posterior probability is:[31]

$$P(\mathbf{Y} = n | \mathbf{X} = m) \propto \frac{1}{n^2},$$

provided, of course, $m \leq n$. So we now compute:

$$P(\mathbf{Y} > 2m | \mathbf{Y} \geq m) = \sum_{n=2m+1}^{\infty} \frac{1}{n^2} \Big/ \sum_{n=m}^{\infty} \frac{1}{n^2} = \left(\frac{\pi^2}{6} - \sum_{n=1}^{2m} \frac{1}{n^2} \right) \Big/ \left(\frac{\pi^2}{6} - \sum_{n=1}^{m-1} \frac{1}{n^2} \right),$$

since $\sum_{n=1}^{\infty} \frac{1}{n^2} = \frac{\pi^2}{6}$. This probability is very close to $\frac{1}{2}$, and so the probability

$$P(\mathbf{Y} \leq 2m | \mathbf{Y} \geq m) \tag{1}$$

---

[31] Worth noting here is something many Bayesians accept. The prior is improper, but the posterior probability distribution, as the product of the prior and the likelihood, is proper, and this is what matters to Jeffreys.

is close to $\frac{1}{2}$. In particular if $m = 50$, $(1) = 0.5075$; if $m = 100$, $(1) = 0.5037$; if $m = 1000$, $(1) = 0.5004$; if $m = 10,000$, $(1) = 0.5000$.

It is Jeffreys' confidence in such objective priors that separates him from the leading subjectivists like de Finetti, discussed in Section 7, and, as I already remarked, why I have included notice of his work in the present section. Most subjectivists would insist that Jeffreys' prior is unrealistic, since without any information of a particular character about the problem, everyone knows that the number of tramcars in any city in the world cannot exceed 100,000 or so, under any conditions. But, of course, fixing the maximum value of n at 100,000 would not much change Jeffreys' posterior probability analysis. Other issues that arise about this kind of analysis cannot be gone into here.

The rest of this section will be devoted to two versions of the logical theory of probability that are more explicit. The first is Carnap's theory of confirmation, including the extension by Hintikka and later Kyburg, and the second is the model-theoretic approach of Gaifman, Krauss, and Scott, and also that of Chuaqui.

Carnap's confirmation theory. Three basic references to this subject are Carnap (1945, 1950, 1952). Excellent critiques of Carnap's ideas by Burks, Kemeny, Nagel and Putnam, and Carnap's replies, are to be found in Schlipp (1963). The fundamental idea of the theory is to develop the notion of degree of confirmation of a hypothesis with respect to a given body of evidence. For example, what degree of confirmation of the hypothesis that all sheep are black is given by the evidence consisting of the observation of six black cows and one black sheep? From a probability standpoint we expect the notion of degree of confirmation to have the formal properties of conditional probability.

Following the tradition of Carnap and others, we shall begin by working within an explicit linguistic framework. (However, this is not necessary, and, as we shall see, we can proceed purely by consideration of set-theoretical entities.) We consider languages which consist of the following:

(i) parentheses,
(ii) sentential connectives,
(iii) finite number of predicates,
(iv) individual names,
(v) individual variables which take as values an at most countable number of individuals,
(vi) quantifiers using the individual variables,
(vii) identity sign.

For the present, we shall restrict ourselves to one-place predicates. So, within this framework:

DEFINITION 1. L_N^m *is a language satisfying* (*i*)–(*vii*), *having* m *one-place predicates and* N *individual names.*

Since initially we shall be concerned with elementary notions of an arbitrary language L_N^m, we shall omit formal reference to L_N^m in definitions and theorems. We shall also require that a language L_N^m satisfy two requirements or conventions:

5.5 LOGICAL THEORY OF PROBABILITY

Convention 1. Distinct individual names designate distinct individuals.

Convention 2. All predicates (in a given language) are logically independent. For example, in the same language we would not permit the predicate 'is colored' and the predicate 'is green', since any green thing is colored.

We take the general notion of a well-formed formula to be already defined, as in Section 2.2, but some further special notions will be useful.

DEFINITION 2. *A sentence s is atomic if and only if s is a well-formed formula consisting of a predicate and an individual name.*

For purposes of illustration here and subsequently, let us consider a particular language, L_3^2, whose predicates are 'R' ('is red') and 'G' ('is glossy'). The three individual names are 'a', 'b', and 'c'. Examples of atomic sentences would include 'Ra' and 'Gc'.

DEFINITION 3. *A sentence s is a state description if and only if s is a well-formed formula consisting of a lexicographically ordered conjunction of atomic sentences and negations of atomic sentences such that in s each predicate followed by each individual name appears exactly once.*

Two examples of state descriptions in L_3^2 are the following.

$$s_1 = \text{'}Ga \ \& \ Gb \ \& \ -Gc \ \& \ Ra \ \& \ -Rb \ \& \ -Rc\text{'}$$
$$s_2 = \text{'}-Ga \ \& \ -Gb \ \& \ -Gc \ \& \ -Ra \ \& \ -Rb \ \& \ Rc\text{'}.$$

We understand that to make the conjunction of Definition 3 unique, we must fix on some lexicographic indexing of predicates and individual names. The number of state descriptions in L_3^2 is $((2)^2)^3$ or 64. And, in general, the number of state descriptions in any language L_N^m is $2^{m \cdot N}$. Intuitively, each state description corresponds to a possible world; distinct state descriptions correspond to distinct possible worlds, or, in the language of standard probability theory (see Section 1), to distinct possible outcomes. We now turn to the explicit definition of some standard logical notions.

DEFINITION 4. *The range of a sentence s is the set of all state descriptions in which the sentence holds. That is, if a state description in the range of s is true, s must be true.*

DEFINITION 5. *A sentence is L-true if its range is the set of all state descriptions.*

DEFINITION 6. *A sentence is L-false if its range is the empty set.*

For example, '$Ra \ \& \ -Ra$' is L-false.

DEFINITION 7. *Sentence s_1 L-implies sentence s_2 if and only if the range of s_2 contains the range of s_1, that is, $R(s_1) \subseteq R(s_2)$.*

DEFINITION 8. *X is the set of all sentences s.*

We next define the important concept of a regular measure (Carnap's terminology) on the set of sentences of a language L_N^m. Such a measure is just a discrete probability density on the set of state descriptions of the language, and the extension to other sentences of the language is by the obvious additivity condition.

DEFINITION 9. *A real-valued function m is a regular measure function if and only if*

(i) *The domain of m is X;*

(ii) *If s is a state description, $m(s) \geq 0$;*

(iii) $\sum_{s \in S} m(s) = 1$, *where S is the set of state descriptions;*

(iv) *If s is L-false then $m(s) = 0$;*

(v) *If s_1 is not L-false and $R(s_1)$ is the range of s_1 then $m(s_1) = \sum_{s \in R(s_1)} m(s)$.*

For a logical theory of probability, we need to develop, as an extension of the notion of logical deduction, a somewhat weaker notion relating evidence and hypothesis. Only rarely are evidence and hypothesis connected by a system of deductive inference. Thus, in practice, we need a device which will show a fixed relation between evidence and hypothesis. We could study a comparative notion of confirmation:

$$C(h, e; h', e').$$

The relation holds when the confirmation of h by e is greater than that of h' by e'. Instead, however, we start with a quantitative notion of a measure function. The central problem is what regular measure function to choose. Among the infinity of regular measure functions, which one best approximates our intuitive idea of a *rational* distribution prior to any information? We could make a qualitative approach to the class of feasible measures. For example, we could attempt to lay down qualitative postulates which would yield a regular measure function, but this approach will be developed in detail later for other concepts of probability. Following Carnap, we shall instead place a priori conditions directly on the possible measures.

First we note that with a measure function at hand, we may immediately define a confirmation function.

$$c(h, e) = \frac{m(h \ \& \ e)}{m(e)},$$

provided $m(e) > 0$. For example, suppose e L-implies h. Then, from our concept of measure,

$$m(h \ \& \ e) = m(e),$$

so the confirmation in this case is 1, i.e., c(h,e) = 1. On the other hand, if e L-implies $-h$, then

$$m(h \ \& \ e) = 0,$$

and the confirmation is 0, i.e., $c(h, e) = 0$. Because the confirmation function is immediately definable in terms of the measure function, the whole of our analysis will be concerned with what properties it is reasonable to require of a measure function. From Carnap's standpoint, this is not an empirical problem, but a normative or logical one.

We first introduce the notion of *state-symmetric* measure functions. (In Carnap's terminology, the measure is just *symmetric,* but I want to name two different kinds of symmetry.) Intuitively, the idea is that, a priori, we should not discriminate in any way between different individuals. For example, in L_3^2 a symmetrical measure function m would have the following property:

$$m(`Ra') = m(`Rb'), \tag{2}$$

5.5 Logical Theory of Probability

and:

$$m(`-Ra \ \& \ Gb') = m(`-Rc \ \& \ Ga'). \tag{3}$$

If we take a sentence and consider any permutation of the individual names, a symmetric measure of the sentence remains the same. In (3) above we use the permutation $a \to c$, $b \to a$, and $c \to b$. This corresponds intuitively to changing variables in deductive inferences. This concept is in itself not as strong as the indifference principle (Bayes' Postulate), because it does not determine a unique a priori probability distribution on the state descriptions.

DEFINITION 10. *Two state descriptions s_1 and s_2 are isomorphic if and only if s_1 may be obtained from s_2 by a permutation of the individual names.*

A permutation is, in this context, a 1-1 function mapping a set of names onto itself.

DEFINITION 11. *A regular measure function m on X is state-symmetric if and only if m assigns the same measure to isomorphic state descriptions.*

It has sometimes been suggested that we go beyond state symmetry and select as the a priori measure the unique regular state-symmetric measure function that assigns the same measure to all state descriptions. I call this measure m_w, because it is the one suggested somewhat obscurely by Wittgenstein in his treatise *Tractatus Logico-Philosophicus* (1922). (See *4.4, *4.26, *5.101, and *5.15; also Carnap 1945, pp. 80-81.)

An example can be given which intuitively establishes the inadequacy of the confirmation function which is based upon m_w. Consider the language L_{101}^1. The number of state descriptions is clearly 2^{101}. For any state description s the measure is given by:

$$m_w(s) = \left(\frac{1}{2}\right)^{101}.$$

Now let the evidence e be '$Pa_1 \ \& \ Pa_2 \ \& \ \ldots \ \& \ Pa_{100}$' and the hypothesis h be 'Pa_{101}'. Then $h \ \& \ e$ is a state description and so

$$m_w(h \ \& \ e) = \left(\frac{1}{2}\right)^{101}.$$

Of course, e holds in only two state descriptions; namely, $h \ \& \ e$, and $-h \ \& \ e$. Hence,

$$m_w(e) = \left(\frac{1}{2}\right)^{101}.$$

Thus,

$$c_w(h, e) = \frac{m_w(h \ \& \ e)}{m_w(e)} = \frac{1}{2}.$$

If e is replaced by e' which results from e by negating the m occurrences of 'P' in e, for $1 \leq m \leq 100$, we still get $c_w(h, e') = \frac{1}{2}$, which is highly unsatisfactory.

We now turn to Carnap's important concept of a structure description. His use of this concept is the most distinctive feature of his confirmation theory.

DEFINITION 12. *Let s be a state description. Then the structure description generated by s is the sentence that is the lexicographically ordered disjunction of all state descriptions isomorphic to s.*

Informally we shall often find it convenient to treat structure descriptions as *sets* of state descriptions.

For example, consider the language L_2^1. The state descriptions s_i and structure descriptions S_j are:

$$s_1 = \text{`}Ra \ \& \ Rb\text{'}$$
$$s_2 = \text{`}Ra \ \& \ -Rb\text{'}$$
$$s_3 = \text{`}-Ra \ \& \ Rb\text{'}$$
$$s_4 = \text{`}-Ra \ \& \ -Rb\text{'}$$
$$S_1 = \{s_1\}$$
$$S_2 = \{s_2, s_3\}$$
$$S_3 = \{s_4\}.$$

We now define the structure-symmetric measure.

DEFINITION 13. *The regular measure function m on X is* structure-symmetric *if and only if m assigns the same measure to all structure descriptions.*

Carnap did not formulate things exactly this way but it is an immediate consequence of the definitions we have given in this development that we have the following representation theorem whose proof is obvious.

THEOREM 1. (Carnap's Representation Theorem). *For any language L_N^m there is a unique regular measure that is both state-symmetric and structure-symmetric.*

This unique measure was labeled by Carnap m^*. Carnap did not use the language *structure-symmetric*, but it seems natural to have the two kinds of symmetry explicitly introduced in the terminology.

In the example of the language L_2^1 the measure m^* would be:

$$m^*(S_1) = m^*(S_2) = m^*(S_3) = \frac{1}{3},$$

from which

$$m^*(s_1) = m^*(s_4) = \tfrac{1}{3}$$
$$m^*(s_2) = m^*(s_3) = \tfrac{1}{6},$$

because m^* is also state-symmetric.

The following theorem applies to languages which have many-place predicates. Note that the theorem is independent of the exact number of predicates, the result depending only on the number of individual names. Later we shall use this theorem to establish an undesirable property of m^*.

THEOREM 2. *In a language that consists of a finite number of predicates (which need not be one-place) and N individual names, the number of state descriptions in a structure description is a divisor of $N!$.*

For example, in the case of $N = 5$ and a single one-place predicate, we have as possible numbers of state descriptions in any given structure description: 1, 2, 3, 4, 5, 6, 8, 10, 12, 15, 20, 24, 30, 40, 60, and 120.

5.5 Logical Theory of Probability

The proof of Theorem 2 involves some simple facts about finite groups which will not be explained here. The reader who has no knowledge of group theory is advised to omit the proof.

Proof. Let S be a structure description (of a language L_N satisfying our hypothesis) and let s be an arbitrary state description in S. Let $\mathcal{S}_N$ be the symmetric group of N letters, let π be a permutation in $\mathcal{S}_N$, and let $\pi^* s$ be the state description which results from s by applying the permutation π to the individual names (of L_N). It is easily verified that for every π in $\mathcal{S}_N$, $\pi^* s$ is in S. Also, for π, γ in $\mathcal{S}_N$

$$\pi^*(\gamma^* s) = \pi(\gamma)^* s.$$

We define the set $\mathcal{S}$ of permutations as follows:

$$\mathcal{S} = \{\pi : \pi \in \mathcal{S}_N \ \& \ \pi^* S = S\}.$$

If $\pi, \gamma \in \mathcal{S}$, then

$$s = \pi^* s = \gamma^* s = \pi^{-1} \gamma^* s.$$

Hence, it is clear that $\mathcal{S}$ is a subgroup of $\mathcal{S}_N$. Now if $\sigma \in \mathcal{S}_N$, then $\sigma G(s)$ is a left coset of G in $\mathcal{S}_N$ such that $\rho \in \sigma G(s)$ if and only if there is a π in G such that $\rho = \sigma \pi$.

We easily obtain that if ρ and σ are in $\mathcal{S}_N$, then

$$\rho^* s = \sigma^* s \text{ if and only if } \rho \in \sigma G(s),$$

for if $\rho^* s = \sigma^* s$ then $(\sigma^{-1} \rho)^* s = s$, $\sigma^{-1} \rho \in G(s)$, and $\rho \in \sigma G(s)$; and if $\rho \in \sigma G(s)$, then there is a π in G such that $\rho = \sigma \pi$ and hence

$$\rho^* s = \sigma \pi^* s = \sigma^* \pi^* s = \sigma^* s.$$

Therefore,

$$\rho^* s \neq \sigma^* s \text{ if and only if } \rho \notin \sigma G(s),$$

and the number of distinct state descriptions in S is simply the number of left cosets of $\mathcal{S}$ in $\mathcal{S}_N$, which is the index of G in $\mathcal{S}_N$. But the index of $G(s)$ is a divisor of $N!$, i.e., a divisor of the order of $\mathcal{S}_N$. Q.E.D.

We will conclude this analysis of confirmation theory with a few criticisms and remarks, moving in order from the particular to the general. First, we will consider the measure function m^*; next the state description concept, and then more general topics.

I. The predicates of the language must be complete in some strong sense, since $c^*(h, e)$ is *changed* by the addition to the language of a one-place predicate which occurs neither in h nor in e. This result is very counterintuitive.

II. In L_∞^m, the confirmation under c^* of any universal law as a result of a finite sample is zero. (Here L_∞ has an infinite sequence of distinct individual names.) For a variety of pertinent results, see Fagin (1976).

III. Let h be a sentence of L_N^m. Then m^* is *fitting* if for every N the measure of h in L_N^m is equal to the measure of h in L_{N+1}^m. Carnap proves that m^* is fitting when we restrict ourselves to one-place predicates. This is intuitively a quite desirable property, for surely the confirmation of h on evidence e should not vary with the introduction of a new individual name that occurs neither in h nor in e. The following counterexample shows that m^* is not fitting for a language having two-place predicates

(Rubin and Suppes 1955).

Let W be a binary predicate. Let 'a' and 'b' be the individual names of L_2^1 with 'W' the single predicate. Let i be the sentence

'Waa & Wab & Wba & Wbb'.

We easily verify that for L_2^1, $m^*(i) = 0.1$. Now let $s(N)$ be the number of structure descriptions in L_N^1. It follows immediately from Theorem 2 that the measure of any sentence i must equal

$$m_N^*(i) = \frac{x}{s(N) \cdot N!},$$

where x is a positive integer. In the case of m_3^*, we have

$$m_3^*(i) = \frac{x}{104 \cdot 3!}.$$

And it is clear that there is no positive integer x such that

$$x = (.1) \cdot (104) \cdot (3!).$$

In fact,

$$m_3^*(i) = \frac{23}{312}.$$

Using the results of R. L. Davis (1953), we have

$$s(4) = 3044.$$

Clearly there is no integer x such that

$$\frac{x}{(3044)4!} = \frac{23}{312},$$

and, consequently,

$$m_3^*(i) \neq m_4^*(i).$$

IV. It is unrealistic and impractical to require logical independence of one-place predicates. Even worse, it is undesirable to permit n-place predicates to have no formal properties. Thus, 'warmer than' would ordinarily be assumed to have the property of irreflexivity as a logical rather than a factual property. This problem has been extensively discussed in the literature.

V. All expressions must be definable in terms of basic predicates, but it is not clear how to give an adequate and complete set of basic predicates. A connected and equally fundamental problem is the introduction of numerical functions, as ordinarily used in the empirical sciences.

VI. I mentioned earlier that there was no real need to develop confirmation theory within the context of a formal language. Systematic metamathematical properties of the language are not used in any essential way in Carnap's development, and we may proceed just as well by talking about finite probability spaces of the sort defined in Section 1 and also in Section 2. Furthermore, in that context we can put in a somewhat different perspective the choice of a measure function and the relation of that choice to our earlier discussion of principles of symmetry in applying the classical definition of probability.

Let us take first the set-theoretical formulation of confirmation theory when dealing with a language with two one-place predicates and two individuals. The concept of a state description is simply replaced by the standard concept of a possible outcome, and the other definitions go through in terms of the standard relationships between logical concepts and set-theoretical concepts. For example, talk about one sentence L-implying another is replaced by talk about one event implying another, and this is interpreted to mean simply that the first event is a subset of the second.

Within this set-theoretical framework it is clear how we may express Carnap's particular choice of m^* as a preferred measure function. It is the measure function that arises from applying two principles of symmetry. In the case of possible outcomes defined in terms of a finite set of elementary properties, a structure description is now defined simply as the set of all possible outcomes that may be obtained from a given outcome by a permutation of the individuals—just the definition of state-symmetry given earlier. The other principle is that of structure-symmetry as already defined.

A simple example in which this a priori structuring of the set of possible outcomes does not seem natural is afforded by the tossing of a coin N times. The possible outcomes are just the finite sequences of length N reflecting every possible sequence of heads and tails. A structure description is the set of possible outcomes having the same number of heads. If we apply Carnap's measure m^* to this situation, we assign much more weight to the possible outcome all of whose tosses eventuate in heads than we do to one whose number of heads is, for example, approximately $\frac{N}{2}$. For this situation, in fact, we may prefer the Wittgenstein measure m_w to Carnap's measure m^*. In other words, Carnap's principles of symmetry do not seem nearly as natural and appropriate in this case.

Furthermore, in the design of many experiments, principles of symmetry are applied that force the choice of an a priori distribution in a fairly effective way, regardless of whether the set of possible outcomes is finite or infinite. For example, in an experiment dealing with a continuum of responses, the labeling of the parts of the continuum in terms of left-right and top-bottom is arbitrarily randomized across subjects. This is done in such a way that quite strong simple principles of symmetry are necessarily satisfied in the initial distribution of responses, regardless of initial biases of subjects. Of course, it is true that such an example is not the sort of thing envisaged by Carnap in his own discussion of a choice of a measure function; but in my own judgment, the relations between the two are closer than he might have been willing to admit.

VII. Carnap (1950, pp. 19–51) makes a point of emphasizing that in his view there are two distinct concepts of probability. One concept is that of empirical relative-frequency, and the other is the concept of probability expressed in confirmation theory. He cm phasizes most strongly that the statistical concept is empirical, while the confirmation-theory concept is analytical or logical in character. Although this distinction seems to have been accepted by a great many philosophers, it is not at all clear that a persuasive case has been made by Carnap for this dualistic approach.

I certainly would admit that in all theoretical science there are almost always two closely connected models of the theory, one model being given as a purely mathematical model, and the other as a model derived in some more-or-less direct fashion from empirical data. However, the existence of these two kinds of models does not justify the distinction between the two concepts of probability urged by Carnap. As Carnap suggests, there is a natural dichotomy between probabilities determined theoretically and probabilities determined empirically, but the thrust of this dichotomy is not at all the one he suggests. It is not a comparison between confirmation theory on the one hand and statistical theory on the other, but is simply the garden-variety comparison between a theoretical model and experimental data.

There is another way of looking at m^* that also runs against Carnap's. This is to look upon m^* as an objective prior distribution in the sense of Bayesians. Then the confirmation $c^*(h,e)$ of hypothesis h on evidence e is just the posterior probability in the Bayesian sense. On this view, there are definitely not two kinds of probability, logical and statistical, with some statistical probability to be used to supplement m^* and c^*, or the associated likelihood of e given h. As some Bayesians like to say, if an objective prior derived from principles of symmetry is used, then the statistics is purely mathematical or deductive in character. The evidence e is empirical, of course, but, it might be maintained, the likelihood of e given h is not. Looked at still another way, we can compute $c^*(h,e)$ for any possible evidence e. The computation is not empirical, and no statistical sense of probability enters, only the nonstatistical empirical question of what the *actual* evidence is. I emphasize that I have only touched on important and subtle issues here, and it is not possible within the framework of the present chapter to examine them in adequate detail. However, I hope I have said enough to generate skepticism about the validity of Carnap's dual conception of probability.[32]

Hintikka's two-parameter theory. Given the many difficulties in defending a unique choice such as m^*, Carnap retreated to a measure m_λ with λ a free parameter (Carnap 1952). This new measure, like m^*, has the property of state-symmetry. It seems obvious that unless particular arguments can be given for choosing a few special values of λ, and thereby selecting variant inductive logics, there is little reason to regard this new theory as a purely logical theory of probability. Inevitably, the selection of a particular value of λ to uniquely determine a measure seems in the grand tradition of subjective theories of probability.

But rather than explore Carnap's λ-theory of inductive logic, I think it is more useful to examine the later generalization of it to two parameters by Hintikka (1966), who also provides an excellent detailed comparison between his system and Carnap's.

Hintikka introduces two parameters α and λ, with λ being essentially the same as Carnap's λ. The main rationale for Hintikka's second parameter α is to provide a framework in which sentences that are not logically true may still receive a nonzero degree of confirmation, i.e., positive conditional probability, in an infinite universe of individuals, provided the finite evidence is positive enough. Roughly speaking, Carnap's and Hintikka's λ is an index of the weight given to the a priori probabilities determined by various kinds of symmetry. The larger λ the greater the weight of these a priori considerations.

Especially to deal with the asymptotic case of an infinite universe, it is essential

---

[32] For an excellent recent general review of confirmation theory, full of nice counterexamples and arguments against various proposals in the literature in the years after Carnap's long work on confirmation theory, see Fitelson (2001).

5.5 LOGICAL THEORY OF PROBABILITY

to move from state descriptions to a setup less dependent on particular individuals. Suppose the language is L_N^m, i.e., there are m one-place basic predicates $P_1 \ldots P_m$. Carnap introduces the term *Q-predicate* for a conjunction of the basic predicates or their negations. Thus there are 2^m Q-predicates when there are m basic predicates. In the example of L_3^2 considered after Definition 2, there are four Q-predicates:

$$Gx \,\&\, Rx,$$

$$Gx \,\&\, -Rx,$$

$$-Gx \,\&\, Rx,$$

$$-Gx \,\&\, -Rx.$$

A *constituent*, in Hintikka's terminology, is a sentence in L_N^m saying exactly which Q-predicates are instantiated. If the number of such is w, Hintikka calls (ambiguously but usefully) that constituent C_w. In the example of L_3^2 above, let us abbreviate the four Q-predicates in the order given, $Q_1(x), \ldots, Q_4(x)$. Then the sample '$Ga \,\&\, -Ra$', '$-Gb \,\&\, Rb$' and '$Gc \,\&\, -Rc$' would instantiate the constituent:

$$(\exists x)Q_1 x \,\&\, (\exists x)Q_2 x \,\&\, (\forall x)(Q_1 x \vee Q_2 x).$$

Hintikka focuses his theory on the a priori probability $P(C_w)$ and on the likelihood $P(e|C_w)$, with the a posteriori probability $P(C_w|e)$ then computed by Bayes' Theorem (Theorem 8 of Section 1). I restrict consideration here to the a priori probability $P(C_w)$.

To give a feeling for the asymptotic values of α and λ, which range from 0 to ∞, we have the following:

(i) We get Carnap's m^* when $\alpha = \infty$ and $\lambda = 2^m$, the number of Carnap's Q-predicates,

(ii) We get Wittgenstein's m_w when $\alpha = \infty$ and $\lambda = \infty$,

(iii) We get Carnap's λ-continuum when $\alpha = \infty$.

Hintikka starts with Carnap's λ-system. Given a fixed constituent C_w the a priori probability that α individuals are compatible with it, i.e., exemplify the Q-predicates of C_w, is:

$$\frac{b}{2} \cdot \frac{1+b}{1+\lambda} \cdot \frac{2+b}{2+\lambda} \cdots \frac{\alpha-1+b}{\alpha-1+\lambda}, \tag{4}$$

where $b = \frac{w\lambda}{2^m}$. In Hintikka's notation the numerator of (3) is $\pi(\alpha, b)$ and the denominator is $\pi(\alpha, \lambda)$. He sets the a priori probabilities of constituents proportional to (3). Thus

$$P(C_w) = \frac{\pi(\alpha, b)}{\sum_{i=0}^{K} \binom{K}{i} \pi(\alpha, \frac{i\lambda}{K})} \tag{5}$$

where $K = 2^m$. Obviously the symmetry of assigning the same a priori probability to two constituents that exemplify the same number of Q-predicates is part of the setup. It is also apparent that we cannot derive an equation like (4), dependent as it is on two real non-negative parameters α and λ, from general principles of symmetry. Consequently we cannot expect to have a representation theorem in the spirit of Theorem 1.

There are many good features of Hintikka's system, but it seems to me that it should be mainly regarded as a contribution to subjective theory. The choice of values of the two free parameters may be expected to vary from person to person and from one situation to another. This is exactly the sort of thing pure Bayesians, the strongest advocates of the subjective view, expect to be the case.

Kyburg. A richer and more realistic logical theory has been developed by Kyburg (1974). It is still in the tradition of an explicit logical language, emphasized so much in the earlier work of Carnap. Here is why Kyburg's formulation is both richer and more realistic than that begun by Carnap. First, he introduces a richer formal language, much more adequate to express in a formal way standard statistical concepts such as random sampling and testing hypotheses. Second, because of this richer language a much more realistic class of examples and problems from statistical practice—at least classroom practice in teaching statistics—can be analyzed. Third, the theory as developed is not formulated to require a single prior probability or distribution for a given problem or experiment, but will work well with a family of distributions.

Kyburg has many interesting remarks about the foundations of probability. His well worked-out treatise should have had more impact than it did over the past several decades. I am not certain why it has not, but I conjecture this is mainly because it makes heavy use of logical notation and formal logical concepts, such as defining randomness and probability as metalinguistic concepts. The great majority of statisticians and mathematicians have little taste for this heavily formal approach. Carnap had little influence on this majority of workers also, but Kyburg deserved a better hearing, for he made a detailed effort to deal with many of the standard puzzles about statistics and probability, as formulated independent of any explicit logically formal language.

Model-theoretic approach. Gaifman (1964), Scott and Krauss (1966), Krauss (1969), and also Gaifman and Snir (1982) have been concerned to develop the concept of a probabilistic measure for general first-order languages, and also to extend the developments to infinitary languages with the benefit of having a σ-additive theory. The spirit of this work is not to justify by presumably logical arguments the selection of a unique a priori probability measure. Rather, the intent is to generalize ordinary logic to *probability logic*. The central concept for Scott and Krauss is that of a sentence holding in a probability system Ω with probability α. This concept is the analogue of Tarski's concept of satisfaction for ordinary logic. From this and related concepts the authors construct an elegant formal theory.

However, this model-theoretic approach is not really to be taken as a contribution to the logical foundations of probability in Jeffreys' or Carnap's sense. In fact, most of the results of Gaifman, Scott, and Krauss have analogues in standard probability theory derived from Kolmogorov's axioms, which are essentially neutral toward foundations. In the sense of this section, their model-theoretic approach does not contribute a new representation theory to the foundations of probability.

Chuaqui. A significant representation theorem is proved by Rolando Chuaqui (1977, 1991) using a model-theoretic approach and a semantic definition of probability. But the representation theorem he proves does not really depend on his model-theoretic methods, but rather on their extension to nonstandard infinitesimal analysis. It is too much of a technical digression to introduce appropriate formal concepts and notation for infinitesimal analysis here, but I believe it is possible to give a clear intuitive description of Chuaqui's representation result. A *standard* stochastic process is just a family of random variables $\{\mathbf{X}_t : t \in T\}$ indexed by a set T, discrete or continuous. A *nonstandard* such process is indexed by infinitesimals or their reciprocals, hypereal numbers, as well. The point, made most clearly in Nelson (1987), is that by using nonstandard infinitesimal methods we can drastically simplify many parts of probability that seem burdened by scientifically unimportant, but mathematically real, technical complications. For example, it is easy to define a continuous-time, discrete-state stochastic process that can pass through an infinite number of states in finite time. Much worse are the endless measure-theoretic questions that burden continuous-time, continuous-state stochastic processes. Nelson provides important guidelines for a simpler theory of such matters, usable in all ordinary scientific contexts.

In Nelson's setup the usual continuous-time process is replaced by a hyperfinite one. For example, let the index set T be the set of natural numbers. It can be replaced by $\{0, 1, \ldots, \nu\}$, where ν is an infinite natural number. (A natural number ν is infinite if $\frac{1}{\nu}$ is an infinitesimal, say, ϵ, and ϵ has the property that for all standard finite real numbers $x, 0 < \epsilon < x$.) As a second example, let T be the closed interval $[0, 1]$. We define the discrete-time set $T' = \{0, dt, 2dt, \ldots, 1\}$, where $dt = \frac{1}{\nu}$ for some infinite natural number ν. So T' is hyperfinite. T' is said to be a near interval for $T = [0, 1]$. Nelson (1987) proves that any standard stochastic process is approximated within infinitesimal equivalence of all probabilistic quantities by a nearby process indexed by a near index set. This informally stated theorem is in itself a significant formal representation theorem for stochastic processes.

Chuaqui uses it to prove the following representation theorem (here I state the Corollary of his basic theorem, Chuaqui 1991, p. 351).

THEOREM 3. (Chuaqui 1991). *For any stochastic process indexed by a hyperfinite set T, there is a nearby infinitesimally equivalent stochastic process over a probability space whose elements are all equally probable.*

As Chuaqui notes, using the theorem of Nelson informally stated above, it is then easy to extend his theorem to what I call the Laplacean representation of any standard, stochastic process—Laplacean because the representing probability space satisfies the extended version of Laplace's first principle of probability, a uniform distribution over the set of possibilities.

This representation theorem is not as surprising as it may seem. It was emphasized already in the introduction of random variables in Section 1 that the probability space on which a random variable is defined is not uniquely determined by the distribution of the random variable, the object of major interest. So for finite probability spaces it is not hard to find an equivalent space with equiprobable atoms, for any random variable defined on the finite space. Nonstandard infinitesimal analysis permits a direct intuitive generalization of these ideas to any stochastic process with a hyperfinite index set.

5.6 Propensity Representations of Probability[33]

In recent years a propensity interpretation of probability, thought of primarily as an objective interpretation, has become popular with a number of philosophers, and there has developed a rather large philosophical literature on it. The concept of propensity itself is not a new one. The *Oxford English Dictionary* cites clear and simple uses of propensity and its general meaning already in the seventeenth century, for example, from 1660 'Why have those plants ... a propensity of sending forth roots?'[34] The idea of this interpretation of probability is to use the general idea of objects having various physical propensities, for example, a propensity to dissolve when placed in water, and to extend this idea to probability. As is also clear from these discussions, propensities can be looked upon as dispositions. (A rather detailed discussion of this point can be found in Chapter 4 of Mellor (1971).)

The most prominent advocate of the propensity interpretation of probability has been Popper, who set forth the main ideas in two influential articles (1957, 1959). Popper gives as one of his principal motivations for developing a propensity interpretation the need to give an objective interpretation of single-case probabilities in quantum mechanics, that is, an objective interpretation of the probabilities of individual events. Single-case probabilities, as they are called in the literature, are of course no problem for subjectivists, but they have been a tortuous problem for relative-frequency theorists. A good detailed discussion of how we are to think of propensities in relation both to single-case probabilities and to relative frequencies is to be found in Giere (1973). I agree with Giere that one of the real reasons to move away from the relative-frequency theory is the single-case problem and that therefore we should regard as fundamental or primary the propensity interpretation of singular events. Giere gives a number of textual quotations to show that Popper wavers on this point.

As I pointed out in Suppes (1974a), what is missing in these excellent intuitive discussions of the philosophical and scientific foundation of a propensity interpretation is any sense that there needs to be something proved about the propensity interpretation. Within the framework I propose, this would amount to proving a representation theorem. In Popper's (1974) response to my criticisms, he mentions his own ideas about conditional probability and how to handle the problem of evidence that has zero probability. There are some other fine points as well, but he misses the real point of my criticism by not giving a conceptually different analysis of propensities, so that something of genuine interest can be proved about the interpretation. The request for such a proof is not an idle or a merely formal request. In order for an interpretation of probability to be interesting, some clear concepts need to be added beyond those in the formal theory as axiomatized by Kolmogorov. The mere hortatory remark that we can interpret propensities as probabilities directly, which seems to be a strong propensity of some advocates, is to miss the point of giving a more thorough analysis.

Because I think there are many good things about the propensity interpretation, as I indicated in my 1974 article on Popper, I want to prove three different representation theorems, each of which is intended to give an analysis for propensity that goes beyond

---
[33] This section draws heavily on Suppes (1987).
[34] Moreover, there are many uses in Hume's *Treatise of Human Nature* (1739).

the formal theory of probability, and to state as well, as a fourth example, a surprising theorem from classical mechanics.

The first representation theorem is closest to probability itself and it applies to radioactive phenomena. The second is for psychological phenomena with propensity being realized as strength of response. The probabilities are then derived explicitly from response strengths. The third example, the most important historically, is the derivation of the behavior of coins, roulette wheels and similar devices, purely on the basis of considerations that belong to classical mechanics. The fourth example shows how random phenomena can be produced by purely deterministic systems.

These four different examples of propensity representation theorems do not in any direct sense force the issue between single-case and relative-frequency views of propensity. But I certainly see no general reason for not using them to compute single-case probabilities. It seems natural to do so whenever a relatively detailed account of the structure of a given propensity is given.

Propensity to decay. Before turning to technical developments, there are some general remarks to be made about the approach followed here, which are largely taken from Suppes (1973a).

The first remark concerns the fact that in the axioms that follow, propensities, as a means of expressing qualitative probabilities, are properties of events and not of objects. Thus, for example, the primitive notation is interpreted as asserting that the event A, given the occurrence of the event B, has a propensity to occur at least as great as the event C, given the occurrence of the event D. Moreover, the events B and D may not actually occur. What we are estimating is a tendency to occur or a propensity to occur, given the occurrence of some other event. If the use of the language of propensity seems awkward or unnatural in talking about the occurrence of events, it is easy enough simply to use qualitative probability language and to reserve the language of propensity for talking about the properties of objects, although I am opposed to this move myself. In any case, the issue is peripheral to the main problem being addressed.

The second remark concerns the clear distinction between the kind of representation theorem obtained here and the sort of theorem ordinarily proved for subjective theories. It is characteristic of subjective theories, as considered in Section 7, to introduce structural axioms strong enough to impose a unique probability measure on events. It is this uniqueness that is missing from the objective theory as formulated here, and in my own judgment, this lack of uniqueness is a strength, not a weakness. Take the case of radioactive decay. From the probabilistic axioms, without specific experimentation or identification of the physical substance that is decaying, we certainly do not anticipate being able to derive a priori the single parameter of the geometric distribution for decay. It is exactly such a parametric result, i.e., uniqueness up to a set of parameters, that is characteristic of objective theory, and, I would claim, characteristic of standard experimentation in broad areas of science ranging from physics to psychology. In other words, the structural axioms of probability, together with the necessary ones, fix the parametric form of the probability measure but do not determine it uniquely. Specific experiments and specific methods of estimation of parameters on the basis of empirical data are needed to determine the numerical values of the parameters. Subjectivists often end up with a working procedure similar to the present one by assuming a probability distribution over the space of parameters. Such procedures are close to what is being discussed here, and so they should be, because there is great agreement on how one proceeds in practice to estimate something like a parameter of a geometric or exponential distribution.

The important point I want to emphasize is that, in the fundamental underpinnings of the theory, subjectivists have ordinarily insisted upon a unique probability measure, and this commitment to an underlying unique probability measure seems to me to be an unrealistic premise for most scientific applications. It is not that there is any inconsistency in the subjectivistic approach; it is simply that the present objectivistic viewpoint is a more natural one from the standpoint of ordinary scientific practice. (As was already noted in Section 3, support for this view of the propensity representation is to be found in various remarks in van Lambalgen (1987a).)

I emphasize also that the intuitive meaning of the weaker structural axioms of objective theory is different from that of the stronger axioms of subjective theory. The objective structural axioms are used to express specific qualitative hypotheses or laws about empirical phenomena. Their form varies from one kind of application to another. A specific structural axiom provides a means of sharply focusing on what fundamental empirical idea is being applied in a given class of experiments. More is said about this point later for the particular case of radioactive decay.

I turn now to the formal developments. First of all, in stating the necessary axioms I shall use qualitative probability language of the sort that is standard in subjective theories of probability. The real place for propensity comes in the discussion of the rather particular structural axioms which reflect strong physical hypotheses about the phenomena of radioactive decay.

The axioms given in the following definition will not be discussed because the ones I give here represent only a minor modification of the first six axioms of Definition 8 of Krantz et al. (1971, p. 222), plus an alternative axiom that is Axiom 7 below.[35] Note that $\approx$ is equivalence, i.e.,

$$A|B \approx C|D \text{ iff } A|B \succeq C|D \text{ and } C|D \succeq A|B.$$

DEFINITION 1. *A structure* $\Omega = (\Omega, \Im, \succeq)$ *is a qualitative conditional probability structure if and only if* $\Im$ *is a σ-algebra of sets on* Ω *and for every* $A, B, C, D, E, F, G, A_i, B_i, i = 1, 2, \ldots$, *in* $\Im$ *with* $B, D, F, G \succ \emptyset$ *the following axioms hold:*

Axiom 1. If $A|B \succeq C|D$ *and* $C|D \succeq E|F$ *then* $A|B \succeq E|F$;

Axiom 2. $A|B \succeq C|D$ *or* $C|D \succeq A|B$;

Axiom 3. $\Omega \succ \emptyset$;

Axiom 4. $\Omega|B \succeq A|D$;

Axiom 5. $A \cap B|B \approx A|B$;

---

[35] Related but simpler necessary and sufficient axioms for qualitative conditional expectation are given in Definition 3 of Section 7 on subjective probability.

5.6 PROPENSITY REPRESENTATIONS OF PROBABILITY

Axiom 6. If $A_i \cap A_j = B_i \cap B_j = \emptyset$ for $i \neq j$ and $A_i|D \succeq B_i|F$ for $i = 1, 2, \ldots$, then $\cup A_i|D \succeq \cup B_i|F$; moreover, if for some i, $A_i|D \succ B_i|F$, then $\cup A_i|D \succ \cup B_i|F$;

Axiom 7. If $A \subseteq B \subseteq D, E \subseteq F \subseteq G$, $A|B \succeq E|F$ and $B|D \succeq F|G$ then $A|D \succeq E|G$; moreover, $A|D \succ E|G$ unless $A \approx \emptyset$ or both $A|B \approx E|F$ and $B|D \approx F|G$;

Axiom 8. If $A \subseteq B \subseteq D, E \subseteq F \subseteq G, B|D \succeq E|F$ and $A|B \succeq F|G$, then $A|D \succeq E|G$; moreover, if either hypothesis is $\succ$ then the conclusion is $\succ$.

To say that the axioms of Definition 1 are necessary means, of course, that they are a mathematical consequence of the assumption that a standard probability measure P is defined on $\Im$ such that

$$A|B \succeq C|D \text{ iff } P(A|B) \geq P(C|D). \tag{1}$$

Precisely which necessary axioms are needed in conjunction with the sufficient structural axioms to guarantee the existence of a probability measure satisfying (1) will vary from case to case. It is likely that most of the eight axioms of Definition 1 will ordinarily be needed. In many cases an Archimedean axiom will also be required; the formulation of this axiom in one of several forms is familiar in the literature. The following version is taken from Krantz et al. (1971, p. 223).

DEFINITION 2. *A qualitative conditional probability structure $\Omega = (\Omega, \Im, \succeq)$ is Archimedean if and only if every standard sequence is finite, where $(A_1, A_2, \ldots)$ is a standard sequence iff for all i, $A_i \succ \emptyset$, $A_i \subseteq A_{i+1}$ and $\Omega|\Omega \succ A_i|A_{i+1} \approx A_1|A_2$.*

I turn now to a theorem that seems to have been first stated in the literature of qualitative probability in Suppes (1973a). The reason it did not appear earlier seems to be that the solvability axioms ordinarily used in subjective theories of probability often do not hold in particular physical situations when the probabilistic considerations are restricted just to that situation. Consequently the familiar methods of proof of the existence of a representing probability measure must be changed. Without this change, the theorem is not needed.

The theorem is about standard sequences. Alone, the theorem does not yield a probability measure, but it guarantees the existence of a numerical function that can be extended to all events, and thereby it becomes a measure when the physical hypotheses expressing structural, nonnecessary constraints are sufficiently strong.

THEOREM 1. (Representation Theorem for Standard Sequences). *Let $(A_1, \ldots, A_n)$ be a finite standard sequence, i.e., $A_i \succ \emptyset, A_i \subseteq A_{i+1}$, and $\Omega|\Omega \succ A_i|A_{i+1} \approx A_1|A_2$. Then there is a numerical function Q such that*

(i) $A_i \subseteq A_j$ iff $Q(A_i) \leq Q(A_j)$,

(ii) if $A_i \subseteq A_j$ and $A_k \subseteq A_l$ then $A_i|A_j \approx A_k|A_l$ iff $\frac{Q(A_i)}{Q(A_j)} = \frac{Q(A_k)}{Q(A_l)}$.

Moreover, for any Q satisfying (i) and (ii) there is a q with $0 < q < 1$ and a $c > 0$ such that

$$Q(A_i) = cq^{n+1-i}.$$

Proof. Let $0 < q < 1$. Define $Q(A_i)$ as

$$Q(A_i) = q^{n+1-i}. \tag{2}$$

Then obviously (i) is satisfied, since the members of a standard sequence are distinct; otherwise there would be an i such that $A_i = A_{i+1}$ and thus $A_i|A_{i+1} \approx \Omega|\Omega$, contrary to hypothesis. So we may turn at once to (ii). First, note the following.

$$A_i|A_{i+m} \approx A_j|A_{j+m}. \tag{3}$$

The proof of (3) is by induction. For $m = 1$, it is just the hypothesis that for every i, $A_i|A_{i+1} \approx A_1|A_2$. Assume now it holds for $m-1$; we then have

$$A_i|A_{i+(m-1)} \approx A_j|A_{j+(m-1)},$$

and also for any standard sequence

$$A_{i+(m-1)}|A_{i+m} \approx A_{j+(m-1)}|A_{j+m},$$

whence by Axiom 7, $A_i|A_{i+m} \approx A_j|A_{j+m}$, as desired. Next, we show that if $A_i \subseteq A_j$, $A_k \subseteq A_l$ and $A_i|A_j \approx A_k|A_l$, then there is an $m \geq 0$ such that $j = i + m$ and $l = k + m$. Since $A_i \subseteq A_j$ and $A_k \subseteq A_l$, there must be non-negative integers m and m' such that $j = i + m$ and $l = k + m'$. Suppose $m \neq m'$, and without loss of generality suppose $m + h = m'$, with $h > 0$. Then obviously

$$A_i|A_{i+m} \approx A_k|A_{k+m}.$$

In addition,

$$A_{i+m}|A_{i+m} \approx \Omega|\Omega \succ A_{k+m}|A_{k+m+h},$$

and so again by Axiom 7

$$A_i|A_{i+m} \succ A_k|A_{k+m'},$$

contrary to our hypothesis, and so we must have $m = m'$.

With these results we can establish (ii). We have as a condition that $A_i \subseteq A_j$ and $A_k \subseteq A_l$. Assume first that $A_i|A_j \approx A_k|A_l$. Then we know that there is an m such that $j = i + m$ and $l = k + m$, whence

$$\begin{aligned}\frac{Q(A_i)}{Q(A_j)} &= \frac{q^{n+1-i}}{q^{n+1-i-m}} \\ &= \frac{q^{n+1-k}}{q^{n+1-k-m}} \\ &= \frac{Q(A_k)}{Q(A_l)}.\end{aligned}$$

Second, we assume that

$$\frac{Q(A_i)}{Q(A_j)} = \frac{Q(A_k)}{Q(A_l)}.$$

From the definition of Q it is a matter of simple algebra to see that there must be an m' such that $j = i + m'$ and $l = k + m'$, whence by our previous result, $A_i|A_j \approx A_k|A_l$.

Finally, we must prove the uniqueness of q as expressed in the theorem. Suppose there is a Q' satisfying (i) and (ii) such that there is no $c > 0$ and no q such that $0 < q < 1$ and for all i

$$Q'(A_i) = cq^{n+1-i}.$$

5.6 PROPENSITY REPRESENTATIONS OF PROBABILITY

Let
$$Q'(A_n) = q_1$$
$$Q'(A_{n-1}) = q_2.$$

Let
$$q = \frac{q_2}{q_1}$$
$$c = \frac{q_1^2}{q_2}.$$

Obviously,
$$Q'(A_n) = cq$$
$$Q'(A_{n-1}) = cq^2.$$

On our supposition about Q', let i be the largest integer (of course $i \leq n$) such that
$$Q'(A_i) \neq cq^{n+1-i}.$$

We have
$$A_i | A_{i+1} \approx A_{n-1} | A_n,$$
whence by (ii)
$$\frac{Q'(A_i)}{cq^{n-1}} = \frac{cq^2}{cq},$$
and so
$$Q'(A_i) = cq^{n+1-i},$$
contrary to hypothesis, and the theorem is proved.

I turn now to radioactive decay phenomena. One of the best-known physical examples of a probabilistic phenomenon, for which hardly anyone pretends to have a deeper underlying deterministic theory, is that of radioactive decay. Here I shall consider for simplicity a discrete-time version of the theory which leads to a geometric distribution of the probability of decay. Extension to continuous time is straightforward but will not be considered here. In the present context the important point is conceptual, and I want to minimize technical details. Of course, the axioms for decay have radically different interpretations in other domains of science, and some of these will be mentioned later.

In a particular application of probability theory, the first step is to characterize the sample space, i.e., the set of possible experimental outcomes, or as an alternative, the random variables that are numerical-valued functions describing the phenomena at hand. Here I shall use the sample-space approach, but what is said can easily be converted to a random-variable viewpoint.

From a formal standpoint, the sample space Ω can be taken to be the set of all infinite sequences of 0's and 1's containing exactly one 1. The single 1 in each sequence occurs as the nth term of the sequence representing the decay of a particle on the nth trial or during the nth time period, with its being understood that every trial or period is of the same duration as every other. Let E_n be, then, the event of decay on trial n. Let W_n be the event of no decay on the first n trials, so that
$$W_n = -\bigcup_{i=1}^{n} E_i.$$

The single structural axiom is embodied in the following definition. The axiom just asserts that the probability of decay on the nth trial, given that decay has not yet occurred, is equivalent to the probability of decay on the first trial. It thus expresses in a simple way a qualitative principle of constancy or invariance of propensity to decay through time.

DEFINITION 3. *Let Ω be the set of all sequences of 0's and 1's containing exactly one 1, and let $\Im$ be the smallest σ-algebra on Ω which contains the algebra of cylinder sets. A structure $\boldsymbol{\Omega} = (\Omega, \Im, \succeq)$ is a qualitative waiting-time structure with independence of the past iff $\boldsymbol{\Omega}$ is a qualitative conditional probability structure and in addition the following axiom is satisfied for every n, provided $W_{n-1} \succ \emptyset$:*

Waiting-time axiom. $E_n | W_{n-1} \approx E_1$.

The structures characterized by Definition 3 are called *waiting-time structures with independence of the past,* because this descriptive phrase characterizes their general property abstracted from particular applications like that of decay. A useful shorter description is *memoryless waiting-time structure.*

The simplicity of this single structural axiom may be contrasted with the rather involved axioms characteristic of subjective theories of probability (Section 7). In addition, the natural form of the representation theorem is different. The emphasis is on satisfying the structural axioms—in this case, the waiting-time axiom—and having a unique parametric form, rather than a unique distribution.

THEOREM 2. *(Representation Theorem for Decay). Let $\boldsymbol{\Omega} = (\Omega, \Im, \succeq)$ be a qualitative waiting-time structure with independence of the past. Then there exists a probability measure on $\Im$ such that the waiting-time axiom is satisfied, i.e.,*

$$(i) \qquad P(E_n | W_{n-1}) = P(E_1) \text{ iff } E_n | W_{n-1} \approx E_1,$$

and there is a number p with $0 < p \leq 1$ such that

$$(ii) \qquad P(E_n) = p(1-p)^{n-1}.$$

Moreover, any probability measure satisfying (i) is of the form (ii).

Proof. The events E_n uniquely determine an atom or possible experimental outcome ω of Ω, i.e., for each n, there is an ω in Ω such that

$$E_n = \{\omega\},$$

a situation which is quite unusual in sample spaces made up of infinite sequences, for usually the probability of any ω is strictly zero.

If $E_1 \approx \Omega$, then $P(E_1) = 1$, and the proof is trivial. On the other hand, if $\Omega \succ E_1$, then for each n, $(W_n, \ldots, W_1)$ is a standard sequence satisfying the hypotheses of the representation theorem for standard sequences. The numbering is inverted, i.e., $W_{i+1} \subseteq W_i$ and $W_{i+1} | W_i \approx W_2 | W_1$. (If $(W_1, \ldots, W_n)$ were a standard sequence, then so would be the infinite sequence $(W_1, \ldots, W_n, \ldots)$ in violation of the necessary Archimedean axiom.)[36] That $W_{i+1} \subseteq W_i$ is obvious from the definition of W_i. By virtue of the

---

[36] The Archimedean axiom is necessary, because the algebra of events is not finite.

5.6 PROPENSITY REPRESENTATIONS OF PROBABILITY

waiting-time axiom
$$W_{i+1}|W_i \approx W_1,$$
since $E_i = W_{i-1} - W_i$, and $W_1 = -E_1$, so
$$E_1 \approx E_{i+1}|W_i \approx W_i - W_{i+1}|W_i$$
$$\approx -W_{i+1}|W_i,$$
and then by elementary manipulations
$$W_1 \approx -E_1 \approx W_{i+1}|W_i.$$

Using the representation theorem for standard sequences, we know there is a numerical function P' and numbers c and q with $0 < q < 1$ and $c > 0$ such that
$$P'(W_i) = cq^i.$$
(Let i' be the numbering in the reverse order $(n, \ldots, 1)$; then $i' = n - (i-1)$, and the exponent $n + 1 - i$ in the representation theorem becomes i'.) Starting with a fixed standard sequence of length n, P' can be extended to every $i > n$ in an obvious fashion.

The next step is to extend P' as an additive set function to all atoms E_i by the equations
$$P'(E_i) = P'(W_{i-1} - W_i)$$
$$= P'(W_{i-1}) - P(W_i)$$
$$= c(q^{i-1} - q^i)$$
$$= cq^{i-1}(1-q).$$
The consistency and uniqueness of this extension is easy to check. Now
$$\sum_{i=1}^{\infty} P'(E_i) = c,$$
so we set $c = 1$ to obtain a measure P normed on 1 from P' and let $p = 1 - q$. We then have
$$P(E_i) = p(1-p)^{i-1}$$
$$P(W_i) = (1-p)^i.$$
The function P just defined is equivalent to a discrete density on Ω, since $E_i = \{\omega\}$ for some ω in Ω, and thus P may be uniquely extended to the σ-algebra of cylinder sets of Ω by well-known methods.

Finally, the proof of (ii) is immediate. If we suppose there is an n such that $P(E_n) \neq p(1-p)^{n-1}$, where $p = P(E_1)$, we may use the waiting-time axiom to obtain a contradiction, for by hypothesis $P(W_{n-1}) = (1-p)^{n-1}$ and $P(E_n|W_{n-1}) = p$, whence $P(E_n) = pP(W_{n-1}) = p(1-p)^{n-1}$, the standard density of the geometric distribution.

Q.E.D.

I quite accept that a criticism of this particular representation theorem is that the analysis of propensity is too close to the analysis of probability from a formal standpoint, a complaint I made earlier about some of the literature on propensity. In general, propensities are not probabilities, but provide the ingredients out of which probabilities are constructed. I think the favorable positive argument for what I have done has got to be put on a more subtle, and therefore more fragile, basis. The point has been made, but I will make it again to emphasize how propensities enter. The memoryless waiting-time axiom is a structural axiom that would never be encountered in the standard theory of subjective probability as a fundamental axiom. It is an axiom special to certain physical phenomena. It represents, therefore, a qualitative expression of a propensity. Second, the probabilities we obtain from the representation theorem are not unique but are only unique up to fixing the decay parameter. Again, this is not a subjective concept but very much an objective one. Identifying and locating the number of physical parameters to be determined is a way of emphasizing that propensities have entered and that a purely probabilistic theory with a unique measure has not been given.

Discrete qualitative densities. The qualitative point just made can be nicely amplified by restricting the qualitative axioms mainly to the discrete atoms, thereby simplifying the qualitative axioms considerably.

As is familiar in all sorts of elementary probability examples, when a distribution has a given form it is often much easier to characterize it by a density than by a probability measure on events. In the discrete case, the situation is formally quite simple. Each atom, i.e., each atomic event, has a qualitative probability and we need only judge relations between these qualitative probabilities. We require of a representing discrete density function p the following three properties:

(i) $p(a_i) \geq 0$.

(ii) $\sum_{i=1}^{n} p(a_i) = 1$.

(iii) $p(a_i) \geq (a_j)$ iff $a_i \succeq a_j$.

Note that the a_i are *not* objects or stimuli in an experiment, but qualitative atomic events, exhaustive and mutually exclusive.

We also need conditional discrete densities. For this purpose, we assume the underlying probability space Ω is finite or denumerable, with probability measure P on the given family $\mathcal{F}$ of events. The relation of the density p to the measure P is, for a_i an atom of Ω,

$$p(a_i) = P(\{a_i\}).$$

Then if A is any event such that $P(A) > 0$,

$$p(a_i|A) = P(\{a_i\}|A),$$

and, of course, $p(a_i|A)$ is now a discrete density itself, satisfying (i)–(iii). Here are some simple, but useful, examples of this approach.

Uniform density. The uniform density on Ω is characterized by all atoms being equivalent in the qualitative ordering $\succeq$, i.e.,

$$a_i \approx a_j.$$

5.6 PROPENSITY REPRESENTATIONS OF PROBABILITY

We may then easily show that the unique density satisfying this equivalence and (i)–(iii) is

$$p(a_i) = \frac{1}{n},$$

where n is the number of atoms in Ω.

Geometric density. We can simplify considerably the argument leading up to Theorem 2. The space Ω is that for Theorem 2. It is the discrete but denumerable set of atomic events $(a_1 \ldots, a_n \ldots)$ on each of which there is a positive qualitative probability of the change of state occurring. The numbering of the atoms intuitively corresponds to the trials of an experiment. The atoms are ordered in qualitative probability by the relation $\succeq$. We also introduce a restricted conditional probability. If $i > j$ then $a_i | A_j$ is the conditional atomic event that the change of state will occur on trial i given that it has *not* occurred on or before trial j. (Note that here A_j means, as before, no change of state from trial 1 through j.) The qualitative probability ordering relation is extended to include these special conditional events as well.

The two postulated properties, in addition to (i)–(iii) given above, are these:

(iv) Order property: $a_i \succeq a_j$ iff $j \geq i$;

(v) Memoryless property: $a_{i+1}|A_i \approx a_1$.

It is easy to prove that (iv) implies a weak ordering of $\succeq$. We can then prove that $p(a_n)$ has the form of the geometric density

$$p(a_n) = c(1-c)^{n-1} \quad (0 < c < 1),$$

but, of course, Axioms (i)–(v) do not determine the value of the parameter c.

Exchangeability. Sequences of outcomes that are exchangeable, i.e., invariant under a permutation of trial number, were mentioned at the end of Section 1. They also belong here, as an important physical invariant of many processes, but further analysis is delayed until the next section on subjective probability, where de Finetti's fundamental theorem on exchangeable sequences is analyzed.

Stationarity. Processes that are stationary in time are common in representing many fundamental physical processes, such as those of equilibrium statistical mechanics. Any process that is ergodic is necessarily required to be stationary, as was already made clear in Section 4.5. For processes with a discrete set of states and discrete time indices, stationarity has a simple qualitative formulation: for any finite sequence of trials $n_1 < n_2 < \cdots < n_m$, and any positive integer k,

$$a_{n_1} a_{n_2} \cdots a_{n_m} \approx a_{n_1+k} a_{n_2+k} \cdots a_{n_m+k},$$

where a_{n_1} is the discrete outcome on trial n_1, etc. The essence of stationarity is invariance under a time shift such as k.

Propensity to respond. There is a long history of various theoretical models being proposed in psychology to represent response strength, which in turn is the basis for choice probabilities. By a 'choice probability' I mean the probability that a given item a will be selected from a set A in some given experimental or naturalistic setting. The fact that the choice is to be represented by a probability is a reflection that the standard algebraic model of expected utility does not adequately represent much actual behavior. Whatever one may think individuals should do, it is a fact of life, documented extensively both experimentally and in other settings, that individuals, when presented with what appear to be repetitions of the same set of alternatives to choose from, do not repeatedly choose the same thing. The formal study of such situations has also been a matter of intensive work in the past several decades. One of the most simple and elegant models is the choice model proposed by Luce (1959). In Luce's own development he proceeds from his choice axiom, which is stated in terms of observable probabilities, to the existence of response strengths. To illustrate the kind of idea we want here we shall begin the other way, that is, by postulating a response strength, and then showing how from this postulate we easily derive his choice axiom. Second, an advantage of response strengths over response probabilities is to be seen in the formulation of Luce's alpha model of learning, where the transformations that represent learning from one trial to another are linear when formulated in terms of response strengths, but nonlinear when formulated in terms of choice probabilities.

We turn now to the formal development of these ideas. In the intended interpretation T is a presented set of alternatives to choose from, and the numerical function v is the measure of response (or stimulus) strength.

DEFINITION 4. *Let T be a nonempty set, and let v be a non-negative real-valued function defined on T such that for at least one x in T, $v(x) > 0$, and $\sum_{x \in T} v(x)$ is finite. Then $\mathcal{T} = (T, v)$ is a* response-strength model (of choice).

The general requirements put on a response-strength model are obviously very weak, but we can already prove a representation theorem that is more special than earlier ones, in the sense that in addition to satisfying the axioms of finitely additive probability spaces, Luce's Choice Axiom is satisfied as well. Moreover, to get various learning models, we impose further conditions.

THEOREM 3. (Representation Theorem). *Let $\mathcal{T} = (T, v)$ be a response-strength model, and define for U in $\mathcal{P}(T)$, the power set of T,*

$$P_T(U) = \frac{\sum_{x \in U} v(x)}{\sum_{x \in T} v(x)}.$$

Then $(T, \mathcal{P}(T), P_T)$ is a finitely additive probability space. Moreover, the probability measure P_T satisfies Luce's choice axiom, i.e., for V in $\mathcal{P}(T)$, with $\sum_{x \in V} v(x) \neq 0$, and with v' being v restricted to V, $\mathcal{V} = (V, v')$ is a response-strength model such that for $U \subseteq V$

$$P_V(U) = P_T(U|V).$$

Proof. The general representation part of the theorem is obvious. To prove Luce's

5.6 PROPENSITY REPRESENTATIONS OF PROBABILITY

axiom, we note that because $U \subseteq V$

$$\begin{aligned}
P_T(U|V) &= \frac{P_T(U \cap V)}{P_T(V)} \\
&= \frac{\sum_{x \in U} v(x)}{\sum_{x \in T} v(x)} \Big/ \frac{\sum_{x \in V} v(x)}{\sum_{x \in T} v(x)} \\
&= \sum_{x \in U} v(x) \Big/ \sum_{x \in V} v(x) \\
&= P_V(U).
\end{aligned}$$

Q.E.D.

The notation used in the theorem is slightly subtle and can be initially confusing. Note that the set referred to by the subscript represents the full physical set of choice alternatives. The set conditioned on, V in the case of $P_T(U|V)$, is information about the choice actually occurring from among the elements of a subset of T. It is not at all tautological that Luce's axiom should hold. In fact, there is a sizable statistical literature on how to best estimate response strengths for observed choice probabilities (Bradley and Terry 1952; Bradley 1954a,b, 1955; Abelson and Bradley 1954; Ford 1957).

To illustrate how the formulation of theory in terms of a propensity can simplify some analyses, we sketch the situation for Luce's alpha model of learning. Let f be the learning function mapping the finite vector of response strengths $v = (v_1, \ldots, v_r)$ from one trial to the next. Here we assume T is finite—in particular has cardinality r. We assume that response strengths are *unbounded*, i.e., for any real number α there is an n such that $|f^n(v)| > \alpha$, where f^n represents n iterations of the learning function. Second, *superposition* of learning holds, i.e., for any v, $v^* > 0$

$$f(v + v^*) = f(v) + f(v^*).$$

Third, *independence of scale or units* holds, i.e., for $v > 0$ and any real number $k > 0$

$$f(kv) = kf(v).$$

But it is a well-known result in linear algebra that the assumed conditions imply that f is a linear operator on the given r-dimensional vector space. In contrast, under these assumptions, but no stronger ones, the behavior from trial-to-trial of the response probabilities $P_T(U)$ is complicated; in particular, they are not related by a linear transformation of the probabilities.

Although the proof of Theorem 3 is very simple and in the development thus far little structure has been imposed on the response-strength function, the intended interpretation fits in very nicely with the propensity concept of probability—at least as I envisage the development of representation theorems for various propensities. In fact, a pluralistic aspect of propensities that I like is that there is no single natural representation theorem. Many different physical and psychological propensities should produce unique representation theorems. On the other hand, an obvious deficiency of Theorem 3, and other similarly "direct" representations of probability in terms of some propensity, is that no guarantee of randomness is provided. This is also a deficiency of the radioactive decay example as well. In both cases, adding axioms to deal with randomness is difficult. In the decay case, what is naturally a real-time phenomenon has to be dealt with, rather than the typical multinomial case. In the response-strength models, one would immediately expect learning from repetition and thus the obvious sequences of trials would not represent stationary processes. In principle the standard machinery for defining finite random sequences could be used, as analyzed in Section 4, but the details will not be repeated here.

Propensity for heads. There is a tradition that begins at least with Poincaré (1912) of analyzing physical systems that we ordinarily consider chance devices as classical mechanical systems. More detailed applications to chance devices were given by Smoluchowski (1918) and, in particular, by Hopf (1934). The history of these ideas has been nicely chronicled by von Plato (1983). The simple approach developed here, which requires only Riemann integration, is mainly due to Keller (1986).

We shall analyze the mechanics of coin tossing, but, as might be expected, under several simplifying assumptions that would be only partly satisfied in practice. First, we consider a circular coin of radius a whose thickness is negligible. Second, we assume perfect uniformity in the distribution of the metal in the coin, so that its center of gravity is also its geometrical center. The difference in the marking for a head and a tail is thus assumed to be negligible. Third, we neglect any friction arising from its spinning or falling. Fourth, we carry the analysis only to the first point of impact with the surface on which it lands. We assume it does not change the face up from this point on. We thereby ignore any problems of elasticity that might lead to bouncing off the surface, spinning and landing again before coming to rest. As Hopf points out, real systems are dissipative rather than conservative, but the mathematical analysis that replaces all of the above assumptions with the most realistic ones we can formulate is still not available in complete form. On the other hand, the idealizations we make are not totally unrealistic; there are good physical reasons for believing that more realistic assumptions about dissipation due to friction, etc., would not affect the conceptual character of the analysis at all, but only the quantitative details, which are not critical for our purposes.

It is also useful to ask whether the analysis to be given fits into the standard theory of particle mechanics. The answer is 'almost but not quite'. To take account of spin we treat the coin as a rigid body, not as a particle, although we could imitate the spin properties exactly by a finite collection of particles whose mutual distances remain constant.

Now to the formal details. We use a Cartesian coordinate system with x and z in the horizontal plane and with y being the measure of height, so that $y(t)$ is the height of the center of gravity of the coin at time t. The only vertical force is the force of gravity, so the Newtonian equation of motion is

$$\frac{d^2y(t)}{dt^2} = -g, \tag{4}$$

where g is the constant acceleration of gravity. As initial conditions at time $t = 0$, we suppose the height is a and the toss gives the coin an upward velocity u, i.e.,

$$y(0) = a, \quad \dot{y}(0) = u. \tag{5}$$

5.6 PROPENSITY REPRESENTATIONS OF PROBABILITY

Equations (4) and (5) uniquely determine $y(t)$ up to the point of impact. As is easily shown

$$y(t) = -\frac{gt^2}{2} + ut + a. \qquad (6)$$

As for spin, we assume the coin is rotating about a horizontal axis which lies along a diameter of the coin; we fix the z-coordinate axis to be parallel to this rotation axis, and we measure angular position as follows. The angle $\theta(t)$ is the angle between the positive y-axis and a line perpendicular to the head-side of the coin—both these lines lie in the x, y plane as can be seen from Figure 1, taken from Keller (1986). We assume

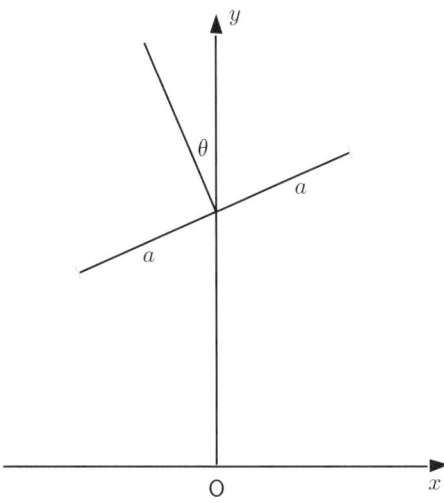

FIGURE 1 The x, y plane intersects the coin along a diameter of length $2a$. The normal to the side of the coin marked heads makes the angle θ with the positive y-axis.

that initially the coin is horizontal with heads up, and the toss gives it positive angular velocity ω. So that at $t = 0$, the initial spin conditions are:

$$\theta(0) = 0, \ \dot\theta(0) = \omega. \qquad (7)$$

Moreover, assuming no dissipation, as we do, the equation governing the rotational motion of the coin is just that of constant velocity.

$$\frac{d^2\theta(t)}{dt^2} = 0. \qquad (8)$$

The unique solution of (7) and (8) is:

$$\theta(t) = \omega t. \qquad (9)$$

Let t_s be the point in time at which the coin makes contact with the surface on which it lands, which we take to be the plane $y = 0$. Given the earlier stated assumption that the coin does not bounce at all, the coin will have a head up iff

$$2n\pi - \frac{\pi}{2} < \theta(t_s) < 2n\pi + \frac{\pi}{2}, \quad n = 0, 1, 2, \ldots. \qquad (10)$$

We now want to find t_s. First, we note that at any time t, the lowest point of the coin is at $y(t) - a|\sin\theta(t)|$. So t_s is the smallest positive root of the equation

$$y(t_s) - a|\sin\theta(t_s)| = 0. \tag{11}$$

We next want to find what Keller calls the *pre-image of heads,* i.e., to find the set of initial values of velocity u and angular velocity ω for which the coin ends with a head up. Let us call this set of points in the u, ω plane, H.

We look first at the endpoints defined by (10). These together with (9) yield—just for the endpoints—,

$$\omega t_s = (2n \pm \frac{1}{2})\pi, \tag{12}$$

and also at these endpoints $\sin\theta(t_s) = \pm 1$, so (11) in these cases simplifies to

$$y(t_s) - a = 0, \tag{13}$$

which combined with (8) yields

$$ut_s - \frac{gt_s^2}{2} = 0. \tag{14}$$

The importance of the endpoints is that they determine the boundaries of the region H. In particular, we examine solutions of (14) to determine t_s and then (12) to determine the boundaries. Equation (14) has two solutions:

$$t_s = 0, \qquad t_s = \frac{2u}{g}.$$

The first one yields only trivial results, so we use the second solution in (12) to obtain a relation in terms only of ω and u:

$$\omega = \frac{(2n \pm \frac{1}{2})\pi g}{2u}, \qquad n = 0, 1, 2, \ldots. \tag{15}$$

The relationship (15) is graphed in Figure 2 (after Keller 1986) for various values of n. As can be seen from (15), each curve is a hyperbola. On the axis $\omega = 0$, a head remains up throughout the toss, so the strip bordered by the axis belongs to H. The next strip is part of T, the complement of H, and, as is obvious, the alternation of strips being either part of H or T continues. From (15) we can infer that the strips are of equal vertical separation, namely, $\frac{\pi g}{2u}$, except for $n = 0$ for the lowest one where the vertical distance from the axis is $\frac{\pi g}{4u}$.

The critical observation is that, as the initial velocity u of the toss increases, the vertical separation decreases and tends to zero. This means that as u increases the alternation between H and T is generated by small changes in velocity.

As we shall show, essentially any mathematically acceptable probability distribution of u and ω, at time $t = 0$, will lead to the probability of heads being approximately 0.5. The mechanical process of tossing dominates the outcome. Small initial variations in u and ω, which are completely unavoidable, lead to the standard chance outcome. Put explicitly, the standard mechanical process of tossing a coin has a strong propensity to produce a head as outcome with probability 0.5.

To calculate P_H, the probability of heads, we assume an initial continuous probability density $p(u, \omega) > 0$, for $u > 0$ and $\omega > 0$. It is important to note that no features of

5.6 PROPENSITY REPRESENTATIONS OF PROBABILITY 217

symmetry of any kind are assumed for $p(u, \omega)$. We have at once

$$P_H = \int\int_H p(u,\omega) d\omega du. \tag{16}$$

It is obvious that (16) imposes no restriction on P_H. Given any value of P_H desired, i.e., any probability of heads, we can find a conditional density that will produce it in accordance with (16).

What can be proved is that as the velocity u increases, in the limit $P_H = \frac{1}{2}$. The actual rate of convergence will be sensitive to the given density $p(u, \omega)$.

THEOREM 4. (Representation Theorem).

$$\lim_{U \to \infty} P(H|u > U) = \frac{1}{2}.$$

Proof. We first write the conditional probability without taking the limit:

$$P(H|u > U) = \frac{\int_U^\infty \sum_{n=0}^\infty \int_{(2n-(1/2))\pi g/2u}^{(2n+(1/2))\pi g/2u} p(u,\omega) d\omega du}{\int_U^\infty \int_0^\infty p(u,\omega) d\omega du}. \tag{17}$$

(What has been done in the numerator is to integrate over each "slice" of H given by

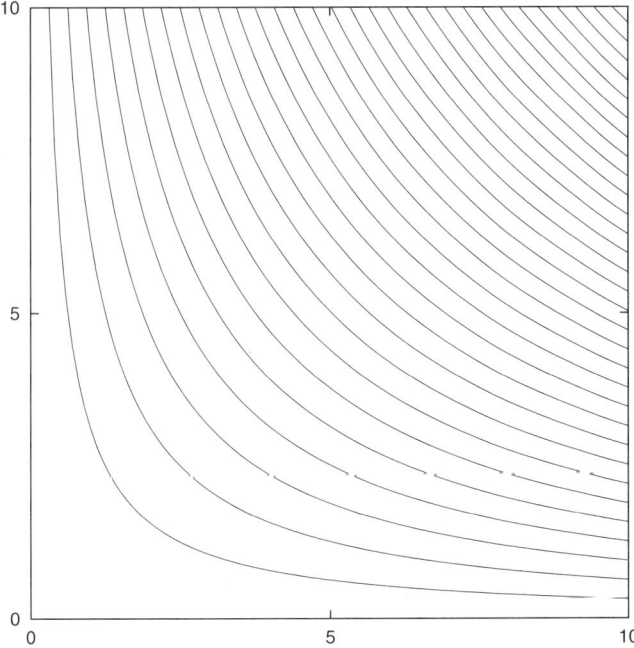

FIGURE 2 The curves which separate the sets H and T, the pre-images of heads and tails in the u, ω plane of initial conditions, are shown for various values of n, with the abscissa being $\frac{u}{g}$.

(15).) The set T is the complement of H, and so the boundaries of slices of T are:
$$\omega = \frac{(2n + 1 \pm \frac{1}{2})\pi g}{2u} \tag{18}$$
and we can write the denominator of (17) as:
$$\int_U^\infty \sum_{n=0}^\infty \int_{(2n-(1/2))\pi g/2u}^{(2n+(1/2))\pi g/2u} p(u, \omega) d\omega du + \int_U^\infty \sum_{n=0}^\infty \int_{(2n+(1/2))\pi g/2u}^{(2n+(3/2))\pi g/2u} p(u, \omega) d\omega du. \tag{19}$$

We next want to evaluate the numerator of the right-hand side of (17) as $U \to \infty$. From the definition of the Riemann integral, we can approximate each integral inside the summation side by the length of the interval, which is $\pi g/2u$, and the value of the integral at the midpoint of the interval, which is $p(u, \frac{n\pi g}{u})$.

$$\lim_{U \to \infty} \int_U^\infty \sum_{n=0}^\infty \int_{(2n-(1/2))\pi g/2u}^{(2n+(1/2))\pi g/2u} p(u, \omega) d\omega du = \lim_{U \to \infty} \int_U^\infty \sum_{n=0}^\infty p(u, \frac{n\pi g}{u}) \frac{\pi g}{2u} du. \tag{20}$$

And it is straightforward to show that as $U \to \infty$ the integral on the right converges to $\frac{1}{2} P(u > U)$. From this result, (17), (19), and (20), we have as desired:
$$\lim_{U \to \infty} P(H|u > U) = \frac{\lim_{U \to \infty} \frac{1}{2} P(u > U)}{\lim_{U \to \infty} \frac{1}{2} P(u > U) + \lim_{U \to \infty} \frac{1}{2} P(u > U)} = \frac{1}{2}.$$

Without pursuing the details, it is worth sketching a more realistic approximate analysis, based only on linear and angular velocities in the range of what would actually be observed in usual coin tossing. To do this, rather than take a physically unrealistic limit, we introduce more restrictions of the density $p(u, \omega)$:

(i) The density $p(u, \omega)$ is single peaked.
(ii) The intersection of $p(u, \omega)$ with any plane of fixed height parallel to the (u, ω) coordinate plane is convex.
(iii) The density $p(u, \omega)$ is positive only on a finite region of the u, ω plane. In particular there is a u^* and an ω^* such that $p(u, \omega) = 0$ if $u > u^*$ or $\omega > \omega^*$.
(iv) A further lower bound on the variance of u and ω is required, so that there is not too much concentration on a single small region of Figure 2, so as to produce too many heads in relation to tails, or vice versa.

With empirically reasonable choices for (i), (iii) and (iv), a pretty good bound, close to one-half, can be established for the probability of heads or tails. Some additional smoothing conditions on $p(u, \omega)$ might be made to facilitate computation; even a bivariate normal distribution might fit real data reasonably well.

Propensity for randomness in motion of three bodies. This fourth example is a special case of the three-body problem, certainly the most extensively studied problem in the history of mechanics. Our special case is this. There are two particles of equal mass m_1 and m_2 moving according to Newton's inverse-square law of gravitation in elliptic orbits relative to their common center of mass, which is at rest. The third particle has a nearly negligible mass, so it does not affect the motion of the other two particles, but they affect its motion. This third particle is moving along a line perpendicular to

5.6 PROPENSITY REPRESENTATIONS OF PROBABILITY

the plane of motion of the first two particles and intersecting the plane at the center of their mass–let this be the z axis. From symmetry considerations, we can see that the third particle will not move off the line. See Figure 3, after Moser (1973). The restricted problem is to describe the motion of the third particle.

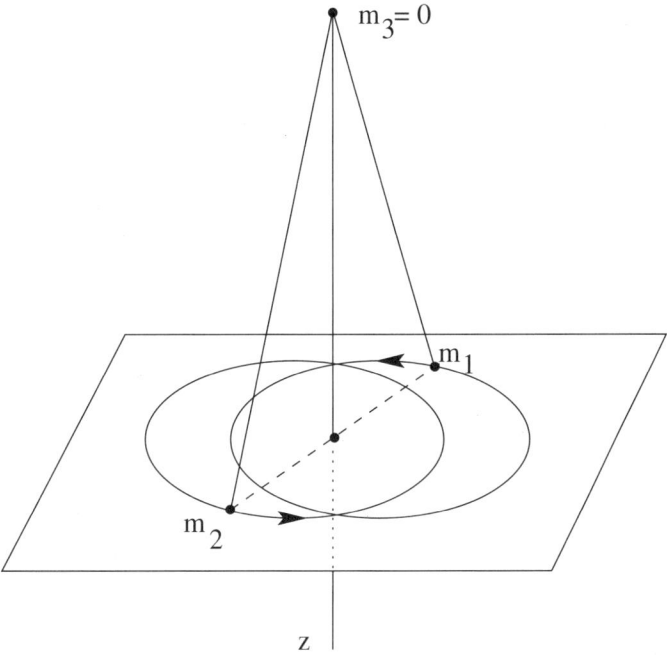

FIGURE 3   Restricted three-body problem.

To obtain a differential equation in simple form, we normalize the unit of time so that the temporal period of rotation of the two masses in the x, y plane is 2π, we take the unit of length to be such that the gravitational constant is one, and finally $m_1 = m_2 = \frac{1}{2}$, so that $m_1 + m_2 = 1$. The force on particle m_3, the particle of interest, from the mass of particle 1 is:

$$F_1 = \frac{m_1}{z^2 + r^2} \cdot \frac{(z, r)}{\sqrt{z^2 + r^2}},$$

where r is the distance in the x, y plane of particle 1 from the center of mass of the two-particle system m_1 and m_2, and this center is, of course, just the point $z = 0$ in the x, y plane. Note that $\frac{(z,r)}{\sqrt{z^2+r^2}}$ is the unit vector of direction of the force F_1. Similarly,

$$F_2 = \frac{m_2}{z^2 + r^2} \cdot \frac{(z, -r)}{\sqrt{z^2 + r^2}}.$$

So, simplifying, we obtain as the ordinary differential equation of the third particle

$$\frac{d^2 z}{dt^2} = -\frac{z}{(z^2 + r^2)^{3/2}}.$$

The analysis of this easily described situation is quite complicated and technical, but some of the results are simple to state in informal terms. Near the escape velocity for the third particle—the velocity at which it leaves and does not periodically return, the periodic motion is very irregular. In particular, the following remarkable theorem can be proved. Let $t_1, t_2 \ldots$ be the times at which the particle intersects the plane of motion of the other two particles. Let s_k be the largest integer equal to or less than the difference between t_{k+1} and t_k times a constant.[37] Variation in the s_k's obviously measures the irregularity in the periodic motion. The theorem, due to the Russian mathematicians Sitnikov (1960) and Alekseev (1969a,b), as formulated in Moser (1973), is this.

THEOREM 5. *Given that the eccentricity of the elliptic orbits is positive but not too large, there exists an integer, say α, such that any infinite sequence of terms s_k with $s_k \geq \alpha$, corresponds to a solution of the deterministic differential equation governing the motion of the third particle.*[38]

A corollary about random sequences immediately follows. Let s be any random (finite or infinite) sequence of heads and tails—for this purpose we can use any of the several variant definitions—Church, Kolmogorov, Martin-Löf, etc. We pick two integers greater than α to represent the random sequence—the lesser of the two representing heads, say, and the other tails. We then have:

COROLLARY 1. *Any random sequence of heads and tails corresponds to a solution of the deterministic differential equation governing the motion of the third particle.*

In other words, for each random sequence there exists a set of initial conditions that determines the corresponding solution. Notice that in essential ways the motion of the particle is completely unpredictable even though deterministic. This is a consequence at once of the associated sequence being random. It is important to notice the difference from the earlier coin-tossing case, for no distribution over initial conditions and thus no uncertainty about them is present in this three-body problem. No single trajectory in the coin-tossing case exhibits in itself such random behavior.

In this fourth case, propensity may be expressed as the tendency of the third particle to behave with great irregularity just below its escape velocity. In ordinary terms, we might want to say that the propensity is one for irregularity, but of course we can say more for certain solutions, namely, certain initial conditions lead to a propensity to produce random behavior, made explicit at the level of symbolic dynamics.[39]

Some further remarks on propensity. Galavotti's (1987) commentary on my views raises a number a fundamental questions about propensity that have been important in the philosophical literature. She raises too many issues for me to hope to reply in detail to all of them, but I think there are some points that I can take up explicitly that

---

[37] The constant is the reciprocal of the period of the motion of the two particles in the plane.

[38] The correspondence between a solution of the differential equation and a sequence of integers is the source of the term *symbolic dynamics*. The idea of such a correspondence originated with G. D. Birkhoff in the 1930s.

[39] For a recent review on chaotic motion in the solar system, based on research over the past decade showing that the orbits of the planets are inherently chaotic, see Murray and Holman (2001).

5.6 PROPENSITY REPRESENTATIONS OF PROBABILITY

will help clarify the discussion, or at least clarify my own position about the concept of propensity.

Propensity or objective probability. One of the first issues that Galavotti raises is whether my account is not really an account of objective chance rather than of propensity. I certainly answer negatively to this question. In my view, an objective theory of probability is just required to be objective as opposed to subjective, and no causal account of the kind implied in a propensity account is necessary. For example, purely logical theories of probability are intended to be objective, as opposed to subjective, as are also accounts based upon relative frequency or on a concept of complexity of finite sequences. It is certainly the view of the proponents of these various views that an objective theory of probability has been given. Moreover, the four extended examples considered certainly do not say anything about the relative frequency or complexity accounts of probability just mentioned. It seems to me that it is a virtue of the propensity theory to provide a particular causal view of probability, which will qualify as a special kind of objective probability.

This leaves open the question whether all objective accounts of probability should be encompassed by a propensity account. Philosophers such as Mellor (1971) seem to think so, but I hold a more pluralistic view. There is no reason in principle that accounts of probability or of randomness in terms of complexity can always be reduced to accounts in terms of propensity.

Question of details. Galavotti refers to the many detailed philosophical discussions of propensity by Popper, Mellor, Giere, and others. She rightly notes that my analysis does not go into the details of the views held by these various authors. I attempted to go into details in a different sense, namely, the mathematical and scientific details of specific examples that exhibit propensities. One purpose of the examples I selected was to illustrate the pluralism of the concept of propensity. It seems to me that Galavotti is asking for a more specific and more monolithic concept of propensity than I would be prepared to offer. I am skeptical that there is a technical general account of propensity, any more than there is a technical general account of the concept of disposition or that of tendency. The term *propensity* itself would not, in my view, occur ordinarily in any well worked out scientific theory. It might occur in the informal descriptions of experiments, but not at all in the statement of theories. This means that the concept differs from other general concepts that do occur in scientific theories, but it in no way challenges the usefulness of the concept. The same thing can be said about the concept of cause. The term *cause* never, or almost never, occurs in detailed scientific theories, but it certainly occurs in informal talk surrounding the theory and often in informal descriptions of experiments or of observational procedures. The fact that the term *cause* does not occur in the formulation of most scientific theories does not support the argument that the concept of cause is not important in science. It is just that in the formulation of theories we deal with more specific properties and concepts. The same is the case with propensity, which I take to be a way of talking about dispositional causal phenomena. Just as many different kinds of processes or things are causes, so many different kinds of processes or things are propensities.

To emphasize this point, let me just ring the changes on one of the natural ways to formulate talk about propensity in each of the four examples considered earlier: propensity to decay, propensity to respond, propensity for heads, and propensity for randomness. Notice that in the first two cases the propensity is referred to in relation to an event or an action, and in the last two cases to a standing phenomenon, but these grammatical differences are not as important as the various way in which the specific theories are formulated. The third and fourth examples are particularly interesting on this point, because they both are built in a general way within the framework of classical mechanics. On the other hand, the third example requires the introduction of an initial probability distribution on the initial conditions of vertical velocity and rate of rotation. Without this distribution expressing the variation in initial conditions the results certainly cannot be derived purely from physical considerations. The fourth example has quite a different character. No extraneous assumptions about probability are introduced. The random sequences exhibiting any probability desired are generated purely by the different objective initial conditions of the restricted three-body problem. I will return to this fourth example later in showing how it refutes many of the things that are said about the relation between propensity and indeterminism. The point for the moment is that the third and fourth examples illustrate that even within the framework of classical mechanics we can have very different talk of propensities in relation to probabilistic phenomena. The existence of so much pluralism, also exhibited in other ways in the first two examples, makes me skeptical of having any unified theory of propensity of any depth, just as I think we cannot have any fully satisfactory unified single theory of the general concept of cause. We can have, for example, a good general theory of probabilistic causality but not a detailed theory of causality that is meant to accurately reflect the relevant nuances of causal phenomena in diverse parts of science.

Single-case versus long-run frequencies. Various philosophers have differed on the application of propensities and have taken quite different views on whether propensity approaches to probability should apply to single-case phenomena or only to long-run frequencies. My general view is that this is a red herring and consequently I shall not say much about it. The most important point is that the problem is mainly one of statistical inference, not one of probability as such. I am happy to use propensity theory to make single-case predictions and also long-run frequency predictions, but how these predictions are to be tested is a matter of statistical methodology and not a matter of the foundations of probability. This is a summary way of dispensing with this issue and not, I recognize, fully satisfactory, but I want to avoid entering into the discussion of the methodological issues involved, which seem to me not special to propensity views of probability.

Propensities and indeterminism. From the standpoint of what has been said in the philosophical literature, Galavotti is certainly right in saying that propensities have often implied a commitment to indeterminism. Mellor (1971) begins his chapter on determinism and the laws of nature by saying "If propensities are ever displayed, deter-

5.6 PROPENSITY REPRESENTATIONS OF PROBABILITY

minism is false." The same view is held by Humphreys (1985), who says in his opening paragraph, "Although chance is undeniably a mysterious thing, one promising way to approach it is through the use of propensities—indeterministic dispositions possessed by systems in a particular environment ...". (I return later to other issues raised by Humphreys' article.) One reason for the inclusion of my fourth example was to give a particular detailed and sophisticated instance of the generation of random phenomena by a system that all would ordinarily classify as purely deterministic. Note that in the restricted three-body problem outlined in this example there is no question of collisions, which are sometimes used as the basis for introducing indeterminism in classical mechanics. In the ordinary sense, as reflected in the theory of differential equations, in the philosophical talk of determinism, etc., the three-body problem described is purely deterministic in character and yet randomness of any desired sort can be generated by appropriate initial conditions. The wedge that has sometimes been driven between propensity and determinism is illusory. There is a lot more to be said about indeterminism and the closely related concept of instability within the framework of determinism, but it would take us too far afield to develop the details here.

Propensities are not probabilities. In one of the better worked-out articles on propensities, Humphreys (1985) presents good arguments for why propensities cannot be probabilities. Humphreys makes the point that although propensities, as he puts it, are 'commonly held' to have the properties of probabilities, he thinks this is not the case, and in this I concur. He goes on to take this argument to be 'a reason for rejecting the current theory of probability as the correct theory of chance'. I strongly disagree with his conclusion and so I want to disentangle the basis of my agreements and disagreements with the various theses in his article, to which Galavotti refers in her commentary.

First he gives a formal argument in terms of Bayes' Theorem and inverse probabilities to show that propensities cannot have these properties. I will not go over the argument in detail, for I agree with it, but in my own case I never for a moment thought that propensities did have the properties of probability. It is my reason for rejecting this view initially but also leads to my rejection of Humphreys' conclusion.

The central point is based upon the nature of representation theorems, as exemplified in this chapter. It is standard that in proving a representation theorem, the objects out of which a representation is constructed do not themselves have the properties of the represented objects. The third and fourth examples in this section, which draw extensively on classical mechanics, illustrate the point. It is unusual, to say the least, in philosophical circles to assert that classical mechanics exemplifies the properties of probability. What we get in the coin-tossing example is a combination of an arbitrary initial probability distribution and the strong symmetry properties in a mechanical sense of a tossed coin. From these diverse ingredients we obtain a symmetric distribution for the actual tosses. The physical analysis that uses two simple differential equations of mechanics is far removed from any feature of probability. In the fourth case of the restricted three-body problem this strictly deterministic theory in itself certainly is not to be thought of as part of probability theory. It is one of the most surprising theorems in mechanics that we can represent in such a clearly deterministic way within such a purely deterministic framework any random sequence. But again, the well-known gravitational mechanical properties exhibited in the restricted three-body problem have not at all the properties of probability structures.

The first example is, as I said earlier, close to probability and therefore not a good example of how the representation of probability in terms of propensities is built up from ingredients that are themselves far from being probabilities. The second example is closer, because the intuitive notion of a response strength as in principle unbounded is not at all a feature of probability. The absence of constancy as, for instance, in a finite measure that always has a constant sum, is again a feature not characteristic of probability theory.

Other classical examples of representation theorems show the same sorts of things. Among the best known is the representation of thermodynamical properties by statistical aggregates of particles. The macroscopic thermodynamical properties of bodies, of gases, let us say, are represented by aggregates of particles for which the properties in question are not even defined, e.g., there is not a meaningful concept of temperature for a single particle.

So it is in terms of separating the represented objects from those ingredients out of which the representation is constructed that I reject Humphreys' conclusion that the current theory of probability should be rejected. It is the basis on which I argue for a classical conception of probability in quantum mechanics, but it is not appropriate here to enter into the quantum-mechanical example. In any case, in Humphrey's rejection of the standard theory of probability as an adequate theory of chance I do not think he has in mind the nonstandard features of probability theory characteristic of quantum mechanics.

In a related vein, there is much in Salmon's review article on propensities (1979) with which I agree. He certainly seems to concur with Humphreys that propensities are not probabilities. On the other hand, he does not go on to set forth a formal representation theory of propensity as I have attempted to do, and in my own judgment this is the most important missing ingredient needed to bring propensity theory to the same level of technical detail that is characteristic of relative-frequency theories or subjective theories of probability. The absence of a clear view about representation theorems is, I should also say, very much a missing ingredient of Mellor's book (1971), which has many useful and interesting things to say about propensities, but also, in that long Cambridge tradition of books on probability, is lacking a kind of formal clarity that a representation-theory view would bring to it.

Pluralism reemphasized. Finally, I want to comment on the last paragraphs of Galavotti's remarks. She compares the examples that I have given with her work with Costantini and Rosa on elementary-particle statistics. She remarks on their analysis of the phenomenological characteristics of elementary particle statistics in contrast to a propensity theory, which would presumably try to give an explanatory account of how and why the various types of statistics of Bose-Einstein and Fermi-Dirac, for example, arise. Unfortunately it seems to me that here she has missed the point of representation theorems and my claim for pluralism in the account of propensities. There is not going to be a deep general account of propensity, just as there will not be a deep general account of the physical properties of the world. We have many different subdisciplines in physics, each dealing in depth with particular properties of a class of phenomena. We do not naturally move in any easy way across the whole spectrum. The same is true of propensities. If we prove a representation theorem for given phenomena, it will be in terms of specific characteristics of those phenomena, and that particular account will not work well when transferred to another environment. This can already be seen in my four examples. It is a long way from response strengths to tossing a coin and even farther to the restricted three-body problem, yet in each case propensities of a certain kind form the ingredients out of which a representation theorem in terms of probability can be constructed.

There will be, I claim, no precise general account of propensity. There will be a pluralistic unfolding of many accounts of propensities for many different empirical situations. It is no more feasible to give a detailed general account of propensities than it is to give a detailed general theory of properties.

5.7 Theory of Subjective Probability

In the preceding sections I have mainly examined what might be termed the 'strong extensions' of the formal theory of probability. In each of these cases there has been at least one proposed method of uniquely determining the probability measure on the sample space, or the distribution of given random variables. We may use a term from Chapter 2 to characterize in a somewhat more technical way what these strong extensions attempt. Recall that a theory is categorical if any two models of the theory are isomorphic. Moreover, if the models of a theory have a reasonably simple structure, namely, a basic set constituting the domain of the model, and operations and relations defined on that domain, then the theory is said to be semicategorical if any two models whose domains have the same cardinality are isomorphic. The natural extension of these notions of categoricity to the present situation is to define $(\Omega, \Im)$-categoricity for probability spaces. We then say that the classical definition of Laplace, the relative-frequency theory, confirmation theory, and propensity theory provide four different theories of probability that are $(\Omega, \Im)$-categorical. I mean by this that once the structure of the sample space is determined and the algebra of events is fixed, then any one of these four theories uniquely (or at least up to a small set of parameters) determines the probability measure on the sample space. We have also examined in some detail the difficulties that arise for each of these $(\Omega, \Im)$-categorical theories. A common failing of all of them is clearly that they attempt to do too much. The many heterogeneous and diverse circumstances in which we want to apply probability concepts make it seem doubtful that any $(\Omega, \Im)$-categorical theory is possible.

Because we want to use probability concepts to talk about everything from the chance of drawing four aces in a hand of poker to the probability of rain tomorrow or the probability distribution of position in a quantum-mechanical experiment, it is hardly surprising that no simple categorical theory of probability can be found. The subjective theory of probability accepts this diversity of applications, and, in fact, utilizes it to argue that the many ways in which information must be processed to obtain a probability distribution do not admit of categorical codification. Consequently, two reasonable persons in approximately the same circumstances can hold differing beliefs about the probability of an event as yet unobserved. For example, according to the subjective theory of probability, two meteorologists can be presented with the same weather map and the same history of observations of basic meteorological variables such as surface temperature, air pressure, humidity, upper air pressures, wind, etc., and yet still differ in the numerical probability they assign to the forecast of rain tomorrow morning. I hasten to add, however, that the term 'subjective' can be misleading. It is not part of the subjective theory of probability to countenance every possible diversity in the assignment of subjective probabilities. It is a proper and important part of subjective theory to analyze, e.g., how classical relative-frequency data are to be incorporated into proper estimates of subjective probability. Such data are obviously important in any systematic estimate of probabilities, as we can see from examination of the scientific literature in which probabilities play a serious part. It is also obvious that the principles of symmetry naturally applied in the classical definition play an important role in the estimation of subjective probabilities whenever they are applicable.

The discussion of Bayes' Theorem in Section 1 has already provided an example of the sort of strong constraints to be placed on any subjective theory. The prior probability distributions selected by different investigators can differ widely without violating the subjective theory; but if these investigators agree on the method of obtaining further evidence, and if common observations are available to them, then these commonly accepted observations will usually force their beliefs to converge.

De Finetti's qualitative axioms. Let us turn now to a more systematic discussion of the major aspects of the subjective theory. For a more detailed treatment of many questions the reader is referred to the historically important article of de Finetti (1937/1964), which has been translated in Kyburg and Smokler (1964), and also to de Finetti's treatise (1974, 1975). The 1937 article of de Finetti's is one of the most important pieces of work in the subjective foundations of probability in the twentieth century. Probably the most influential work on these matters since 1950 is the book by Savage (1954). Savage extends de Finetti's ideas by paying greater attention to the behavioral aspects of decisions, and I shall have more to say later in this section about the relation between the subjective theory of probability and decision theory. Systematic works that are more philosophical and less mathematical are Jeffrey (1965) and Levi (1980).

Perhaps the best way to begin a systematic analysis of the subjective theory is by a consideration of de Finetti's axioms for qualitative probability. The spirit of these axioms is to place restraints on qualitative judgments of probability sufficient to prove a standard representation theorem, i.e., to guarantee the existence of a numerical probability measure in the standard sense. From this standpoint the axioms may be regarded as a contribution to the theory of measurement, with particular reference to comparative judgments of probability. The central question for such a set of axioms is how complicated the condition on the qualitative relation *more probable than* must be in order to obtain a numerical probability measure over events.

The intuitive idea of using a comparative qualitative relation is that individuals can

5.7 THEORY OF SUBJECTIVE PROBABILITY

realistically be expected to make such judgments in a direct way, as they cannot when the comparison is required to be quantitative. On most occasions I can say unequivocally whether I think it is more likely to rain or not in the next four hours at Stanford, but I cannot in the same direct way make a judgment of how *much* more likely it is not to rain than rain. Generalizing this example, it is a natural move to ask next what formal properties a qualitative comparative relation must have in order to be represented by a standard probability measure. (Later I review some of the empirical literature on whether people's qualitative judgments do have the requisite properties.)

We begin with the concept of a qualitative probability structure, the axioms for which are similar formally to those for the extensive structures of Section 3.3. The set-theoretical realizations of the theory are structures $(\Omega, \Im, \succeq)$ where Ω is a nonempty set, $\Im$ is a family of subsets of Ω, and the relation $\succeq$ is a binary relation on $\Im$. I initially follow the discussion given in Luce and Suppes (1965).

DEFINITION 1. *A structure* $\mathbf{\Omega} = (\Omega, \Im, \succeq)$ *is a* qualitative probability structure *if and only if the following axioms are satisfied for all A, B, and C in $\Im$:*

S1. $\Im$ is an algebra of sets on Ω;
S2. If $A \succeq B$ and $B \succeq C$, then $A \succeq C$;
S3. $A \succeq B$ or $B \succeq A$;
S4. If $A \cap C = \emptyset$ and $B \cap C = \emptyset$, then $A \succeq B$ if and only if $A \cup C \succeq B \cup C$;
S5. $A \succeq \emptyset$;
S6. Not $\emptyset \succeq \Omega$.

The first axiom on $\Im$ is the same as the first axiom of finitely additive probability spaces. Axioms S2 and S3 just assert that $\succeq$ is a weak ordering of the events in $\Im$. Axiom S4 formulates in qualitative terms the important and essential principle of additivity of mutually exclusive events. Axiom S5 says that any event is (weakly) more probable than the impossible event, and Axiom S6 that the certain event is strictly more probable than the impossible event. Defining the strict relation $\succ$ in the customary fashion:

$$A \succ B \text{ if and only if not } B \succeq A,$$

we may state the last axiom as: $\Omega \succ \emptyset$.

To give a somewhat deeper sense of the structure imposed by the axioms, we state some of the intuitively desirable and expected consequences of the axioms. It is convenient in the statement of some of the theorems also to use the (weakly) less probable relation, defined in the usual manner:

$$A \preceq B \text{ if and only if } B \succeq A.$$

The first theorem says that $\preceq$ is an extension of the subset relation.

THEOREM 1. *If $A \subseteq B$, then $A \preceq B$.*

Proof. Suppose, on the contrary, that not $A \preceq B$, i.e., that $A \succ B$. By hypothesis $A \subseteq B$, so there is a set C disjoint from A such that $A \cup C = B$. Then, because $A \cup \emptyset = A$, we have at once

$$A \cup \emptyset = A \succ B = A \cup C,$$

whence by contraposition of Axiom S4, $\emptyset \succ C$, which contradicts Axiom S5.   Q.E.D.

Some other elementary properties follow.

THEOREM 2.

(i) If $\emptyset \prec A$ and $A \cap B = \emptyset$, then $B \prec A \cup B$;

(ii) If $A \succeq B$, then $-B \succeq -A$;

(iii) If $A \succeq B$ and $C \succeq D$ and $A \cap C = \emptyset$, then $A \cup C \succeq B \cup D$.

(iv) If $A \cup B \succeq C \cup D$ and $C \cap D = \emptyset$, then $A \succeq C$ or $B \succeq D$.

(v) If $B \succeq -B$ and $-C \succeq C$, then $B \succeq C$.

Because it is relatively easy to prove that a qualitative probability structure has many of the expected properties, as reflected in the preceding theorems, it is natural to ask the deeper question of whether or not it has all of the properties necessary to guarantee the existence of a strictly agreeing numerical probability measure P such that for any events A and B in $\Im$

$$P(A) \geq P(B) \text{ if and only if } A \succeq B. \tag{1}$$

If Ω is an infinite set, it is moderately easy to show that the axioms of Definition 1 are not strong enough to guarantee the existence of such a probability measure. General arguments from the logical theory of models in terms of infinite models of arbitrary cardinality suffice; a counterexample is given in Savage (1954, p. 41). De Finetti (1951) stresses the desirability of obtaining an answer in the finite case. Kraft, Pratt, and Seidenberg (1959) show that the answer is also negative when Ω is finite; in fact, they find a counterexample for a set Ω having five elements, and, thus, 32 subsets. The gist of their counterexample is the following construction. Let $\Omega = \{a, b, c, d, e\}$, and let ϕ be a measure (not a probability measure) such that

$$\phi(a) = 4 - \epsilon$$
$$\phi(b) = 1 - \epsilon$$
$$\phi(c) = 2$$
$$\phi(d) = 3 - \epsilon$$
$$\phi(e) = 6,$$

and

$$0 < \epsilon < \frac{1}{3}.$$

Now order the 32 subsets of Ω according to this measure—the ordering being, of course, one that satisfies Definition 1. We then have the following strict inequalities in the ordering

$$\{a\} \succ \{b, d\} \quad \text{because } \phi(a) = 4 - \epsilon \succ 4 - 2\epsilon = \phi(b) + \phi(d),$$
$$\{c, d\} \succ \{a, b\} \quad \text{because } \phi(c) + \phi(d) = 5 - \epsilon \succ 5 - 2\epsilon = \phi(a) + \phi(b),$$
$$\{b, e\} \succ \{a, d\} \quad \text{because } \phi(b) + \phi(e) = 7 - \epsilon \succ 7 - 2\epsilon = \phi(a) + \phi(d).$$

5.7 THEORY OF SUBJECTIVE PROBABILITY

We see immediately also that any probability measure P that preserves these three inequalities implies that
$$\{c,e\} \succ \{a,b,d\},$$
as may be seen just by adding the three inequalities. In the case of ϕ
$$\phi(c) + \phi(e) = 8 > 8 - 3\epsilon = \phi(a) + \phi(b) + \phi(d).$$
However, no set A different from $\{c,e\}$ and $\{a,b,d\}$ has the property that
$$\phi(\{c,e\}) \geq \phi(A) \geq \phi(\{a,b,d\}).$$
Thus, we can modify the ordering induced by ϕ to the extent of setting
$$\{a,b,d\} \succ \{c,e\} \tag{2}$$
without changing any of the other inequalities. But no probability measure can preserve (2) as well as the three earlier inequalities, and so the modified ordering satisfies Definition 1, but cannot be represented by a probability measure.

Of course, it is apparent that by adding special structural assumptions to the axioms of Definition 1, such as Axiom 5 of Definition 1 of Section 3.3, it is possible to guarantee the existence of a strictly agreeing probability measure satisfying (1). In the finite case, for example, we can demand that all the atomic events be equiprobable, although this is admittedly a very strong requirement to impose.

Fortunately, a simple general solution of the finite case has been found by Scott (1964). (Necessary and sufficient conditions for the existence of a strictly agreeing probability measure in the finite case were formulated by Kraft, Pratt, and Seidenberg, but their multiplicative conditions are difficult to understand. Scott's treatment represents a real gain in clarity and simplicity.) The central idea of Scott's formulation is to impose an algebraic condition on the indicator (or characteristic) functions of the events. Recall that the indicator function of a set is just the function that assigns the value 1 to elements of the set and the value 0 to all elements outside the set. For simplicity of notation, if A is a set we shall denote by $\mathbf{A}^i$ its indicator function, which is a random variable. Thus, if A is an event
$$\mathbf{A}^i(\omega) = \begin{cases} 1 & \text{if } \omega \in A, \\ 0 & \text{otherwise.} \end{cases}$$
Scott's conditions are embodied in the following theorem, whose proof I do not give.

THEOREM 3. (Scott's Representation Theorem). *Let Ω be a finite set and $\succeq$ a binary relation on the subsets of Ω. Necessary and sufficient conditions that there exists a strictly agreeing probability measure P on Ω are the following: for all subsets A and B of Ω,*

(1) $A \succeq B$ or $B \succeq A$;
(2) $A \succeq \emptyset$;
(3) $\Omega \succ \emptyset$;
(4) For all subsets $A_0, \ldots, A_n$, $B_0, \ldots, B_n$ of Ω, if $A_i \succeq B_i$ for $0 \leq i < n$, and for all ω in Ω
$$\mathbf{A}_0^i(\omega) + \ldots + \mathbf{A}_n^i(\omega) = \mathbf{B}_0^i(\omega) + \ldots + \mathbf{B}_n^i(\omega),$$
then $A_n \preceq B_n$.

To illustrate the force of Scott's Condition (4), we may see how it implies transitivity. First, necessarily for any three indicator functions

$$\mathbf{A}^i + \mathbf{B}^i + \mathbf{C}^i = \mathbf{B}^i + \mathbf{C}^i + \mathbf{A}^i;$$

i.e., for all elements ω

$$\mathbf{A}^i(\omega) + \mathbf{B}^i(\omega) + \mathbf{C}^i(\omega) = \mathbf{B}^i(\omega) + \mathbf{C}^i(\omega) + \mathbf{A}^i(\omega).$$

By hypothesis, $A \succeq B$ and $B \succeq C$, whence by virtue of Condition (4),

$$C \preceq A,$$

and thus, by definition, $A \succeq C$, as desired. The algebraic equation of Condition (4) just requires that any element of Ω, i.e., any atomic event, belong to exactly the same number of A_i and B_i, for $0 \leq i \leq n$. Obviously, this algebraic condition cannot be formulated in the simple set language of Definition 1 and thus represents quite a strong condition.

General qualitative axioms. In the case that Ω is infinite, a number of strong structural conditions have been shown to be sufficient but not necessary. For example, de Finetti (1937/1964) and independently Koopman (1940a,b, 1941) use an axiom to the effect that there exist partitions of Ω into arbitrarily many events equivalent in probability. This axiom, together with those of Definition 1, is sufficient to prove the existence of a numerical probability measure. Related existential conditions are discussed in Savage (1954). A detailed review of these various conditions is to be found in Chapters 5 and 9 of Krantz et al. (1971).

However, as is shown in Suppes and Zanotti (1976), by going slightly beyond the indicator functions, simple necessary and sufficient conditions can be given for both the finite and infinite case. The move is from an algebra of events to the algebra $\Im^*$ of *extended* indicator functions relative to $\Im$. The algebra $\Im^*$ is just the smallest semigroup (under function addition) containing the indicator functions of all events in $\Im$. In other words, $\Im^*$ is the intersection of all sets with the property that if A is in $\Im$ then $\mathbf{A}^i$ is in $\Im^*$ and if $\mathbf{A}^*$ and $\mathbf{B}^*$ are in $\Im^*$, then $\mathbf{A}^* + \mathbf{B}^*$ is in $\Im^*$. It is easy to show that any function $\mathbf{A}^*$ in $\Im^*$ is an integer-valued function defined on Ω. It is the extension from indicator functions to integer-valued functions that justifies calling the elements of $\Im^*$ extended indicator functions, which, like indicator functions, are random variables.

The qualitative probability ordering must be extended from $\Im$ to $\Im^*$, and the intuitive justification of this extension must be considered. Let $\mathbf{A}^*$ and $\mathbf{B}^*$ be two extended indicator functions in $\Im^*$. Then, to have $\mathbf{A}^* \succeq \mathbf{B}^*$ is to have the expected value of $\mathbf{A}^*$ equal to or greater than the expected value of $\mathbf{B}^*$. The qualitative comparison is now not one about the probable occurrences of events, but about the expected value of certain restricted random variables. The indicator functions themselves form, of course, a still more restricted class of random variables, but qualitative comparison of their expected values is conceptually identical to qualitative comparison of the probable occurrences of events.

There is more than one way to think about the qualitative comparisons of the expected value of extended indicator functions, and so it is useful to consider several examples.

(i) Suppose Smith is considering two locations to fly to for a weekend vacation. Let $A_j, j = 1, 2$, be the event of sunny weather at location j and B_j be the event of warm weather at location j. The qualitative comparison Smith is interested in is the expected value of $\mathbf{A}_1^i + \mathbf{B}_1^i$ versus the expected value of $\mathbf{A}_2^i + \mathbf{B}_2^i$. It is natural to insist that the utility of the outcomes has been too simplified by the sums $\mathbf{A}_j^i + \mathbf{B}_j^i$. The proper response is that the expected values of the two functions are being compared as a matter of belief, not value or utility. Thus it would be quite natural to bet that the expected value of $\mathbf{A}_1^i + \mathbf{B}_1^i$ will be greater than that of $\mathbf{A}_2^i + \mathbf{B}_2^i$, no matter how one feels about the relative desirability of sunny versus warm weather. Put another way, within the context of decision theory, extended indicator functions are being used to construct the subjective probability measure, not the measurement of utility.

(ii) Consider a particular population of n individuals, numbered $1, \ldots, n$. Let A_j be the event of individual j going to Hawaii for a vacation this year, and let B_j be the event of individual j going to Acapulco. Then define

$$\mathbf{A}^* = \sum_{j=1}^n \mathbf{A}_j^i \text{ and } \mathbf{B}^* = \sum_{j=1}^n \mathbf{B}_j^i.$$

Obviously $\mathbf{A}^*$ and $\mathbf{B}^*$ are extended indicator functions—we have left implicit the underlying set Ω. It is meaningful and quite natural to qualitatively compare the expected values of $\mathbf{A}^*$ and $\mathbf{B}^*$. Presumably such comparisons are in fact of definite significance to travel agents, airlines, and the like.

We believe that such qualitative comparisons of expected value are natural in many other contexts as well. What the representation theorem below shows is that very simple necessary and sufficient conditions on the qualitative comparison of extended indicator functions guarantee existence of a strictly agreeing, finitely additive measure, whether the set Ω of possible outcomes is finite or infinite, i.e., $P(A) \geq P(B)$ iff $A \succeq B$.

The axioms are embodied in the definition of a qualitative algebra of extended indicator functions. Several points of notation need to be noted. First, $\mathbf{\Omega}^i$ and $\mathbf{\emptyset}^i$ are the indicator or characteristic functions of the set Ω of possible outcomes and the empty set $\emptyset$, respectively. Second, the notation $n\mathbf{A}^*$ for a function in $\Im^*$ is just the standard notation for the (functional) sum of $\mathbf{A}^*$ with itself n times. Third, the same notation is used for the ordering relation on $\Im$ and $\Im^*$, because the one on $\Im^*$ is an extension of the one on $\Im$: for A and B in $\Im$,

$$\mathbf{A}^i \succeq \mathbf{B}^i \text{ iff } A \succeq B.$$

Finally, the strict ordering relation $\succ$ is defined in the usual way: $\mathbf{A}^* \succ \mathbf{B}^*$ iff $\mathbf{A}^* \succeq \mathbf{B}^*$ and not $\mathbf{B}^* \succeq \mathbf{A}^*$.

DEFINITION 2. *Let Ω be a nonempty set, let $\Im$ be an algebra of sets on Ω, and let $\succeq$ be a binary relation on $\Im^*$, the algebra of extended indicator functions relative to $\Im$. Then the qualitative algebra $(\Omega, \Im, \succeq)$ is qualitatively satisfactory if and only if the following axioms are satisfied for every $\mathbf{A}^*$, $\mathbf{B}^*$, and $\mathbf{C}^*$ in $\Im^*$:*

Axiom 1. The relation $\succeq$ is a weak ordering of $\Im^*$;

Axiom 2. $\mathbf{\Omega}^i \succ \mathbf{\emptyset}^i$;

Axiom 3. $\mathbf{A}^* \succeq \emptyset^i$;

Axiom 4. $\mathbf{A}^* \succeq \mathbf{B}^*$ *iff* $\mathbf{A}^* + \mathbf{C}^* \succeq \mathbf{B}^* + \mathbf{C}^*$;

Axiom 5. *If* $\mathbf{A}^* \succ \mathbf{B}^*$ *then for every* $\mathbf{C}^*$ *and* $\mathbf{D}^*$ *in* $\Im^*$ *there is a positive integer* n *such that*
$$n\mathbf{A}^* + \mathbf{C}^* \succeq n\mathbf{B}^* + \mathbf{D}^*.$$

These axioms should seem familiar from the literature on qualitative probability. Note that Axiom 4 is the additivity axiom that closely resembles de Finetti's additivity axiom for events: *If $A \cap C = B \cap C = \emptyset$, then $A \succeq B$ iff $A \cup C \succeq B \cup C$.* As we move from events to extended indicator functions, functional addition replaces union of sets. What is formally of importance about this move is seen already in the exact formulation of Axiom 4. The additivity of the extended indicator functions is unconditional—there is no restriction corresponding to $A \cap C = B \cap C = \emptyset$. The absence of this restriction has far-reaching formal consequences in permitting us to apply without any real modification the general theory of extensive measurement. Axiom 5 has, in fact, the exact form of the Archimedean axiom used in Krantz et al. (1971, p. 73) in giving necessary and sufficient conditions for extensive measurement.

THEOREM 4. *Let Ω be a nonempty set, let $\Im$ be an algebra of sets on Ω, and let $\succeq$ be a binary relation on $\Im$. Then a necessary and sufficient condition that there exists a strictly agreeing probability measure on $\Im$ is that there be an extension of $\succeq$ from $\Im$ to $\Im^*$ such that the qualitative algebra of extended indicator functions $(\Omega, \Im^*, \succeq)$ is qualitatively satisfactory. Moreover, if $(\Omega, \Im^*, \succeq)$ is qualitatively satisfactory, then there is a unique strictly agreeing expectation function on $\Im^*$ and this expectation function generates a unique strictly agreeing probability measure on $\Im$.*

Proof. As already indicated, the main result used in the proof is from the theory of extensive measurement: necessary and sufficient conditions for existence of a numerical representation. In particular, let A be a nonempty set, $\succeq$ a binary relation on A, and $\circ$ a binary operation closed on A. Then there exists a numerical function ϕ on A unique up to a positive similarity transformation (i.e., multiplication by a positive real number) such that for a and b in A

(i) $\phi(a) \geq \phi(b)$ iff $a \succeq b$,

(ii) $\phi(a \circ b) = \phi(a) + \phi(b)$

if and only if the following four axioms are satisfied for all a, b, c, and d in A:

E1. The relation $\succeq$ is a weak ordering of A;

E2. $a \circ (b \circ c) \approx (a \circ b) \circ c$, where $\approx$ is the equivalence relation defined in terms of $\succeq$;

E3. $a \succeq b$ iff $a \circ c \succeq b \circ c$ iff $c \circ a \succeq c \circ b$;

E4. If $a \succ b$ then for any c and d in A there is a positive integer n such that $na \circ c \succeq nb \circ d$, where na is defined inductively.

It is easy to check that qualitatively satisfactory algebras of extended indicator functions as defined above satisfy these four axioms for extensive measurement structures. First, we note that functional addition is closed on $\Im^*$. Second, Axiom 1 is identical to

5.7 Theory of Subjective Probability

E1. Extensive Axiom E2 follows immediately from the associative property of numerical functional addition, that is, for any $\mathbf{A}^*$, $\mathbf{B}^*$, and $\mathbf{C}^*$ in $\Im^*$

$$\mathbf{A}^* + (\mathbf{B}^* + \mathbf{C}^*) = (\mathbf{A}^* + \mathbf{B}^*) + \mathbf{C}^*,$$

and so we have not just equivalence but identity. Axiom E3 follows from Axiom 4 and the fact that numerical functional addition is commutative. Finally, E4 follows from the essentially identical Axiom 5.

Thus, for any qualitatively satisfactory algebra $(\Omega, \Im^*, \succeq)$ we can infer there is a numerical function ϕ on $\Im^*$ such that for $\mathbf{A}^*$ and $\mathbf{B}^*$ in $\Im^*$

(i) $\phi(\mathbf{A}^*) \geq \phi(\mathbf{B}^*)$ iff $\mathbf{A}^* \succeq \mathbf{B}^*$,

(ii) $\phi(\mathbf{A}^* + \mathbf{B}^*) = \phi(\mathbf{A}^*) + \phi(\mathbf{B}^*)$,

(iii) $\min(\mathbf{A}^*) \leq \varphi(\mathbf{A}^*) \leq \max(\mathbf{A}^*)$,

since the min and max exist for any $\mathbf{A}^*$.

Second, since for every $\mathbf{A}^*$ in $\Im^*$

$$\mathbf{A}^* + \emptyset^i = \mathbf{A}^*,$$

we have at once that from (ii)

$$\phi(\emptyset^i) = 0.$$

Since $\mathbf{\Omega}^i$ is identically 1, to satisfy (iii)

$$\phi(\mathbf{\Omega}^i) = 1.$$

And thus ϕ is a standard (unique) expectation function E for extended indicator functions:

(i) $E(\emptyset^i) = 0$

(ii) $E(\mathbf{\Omega}^i) = 1$

(iii) $E(\mathbf{A}^* + \mathbf{B}^*) = E(\mathbf{A}^*) + E(\mathbf{B}^*)$.

But such an expectation function for $\Im^*$ defines a unique probability measure P on $\Im$ when it is restricted to the indicator functions in $\Im^*$, i.e., for A in $\Im$, we define

$$P(A) = E(\mathbf{A}^i).$$

Thus the axioms are sufficient, but it is also obvious that the only axioms, Axioms 2 and 3, that go beyond those for extensive structures are also necessary for a probabilistic representation. From the character of extended indicator functions, it is also clear that for each probability measure there is a unique extension of the qualitative ordering from $\Im$ to $\Im^*$.   Q.E.D.

The proof just given, even more than the statement of the theorem itself, shows what subset of random variables defined on a probability space suffices to determine the probability measure in a natural way. Our procedure has been to axiomatize in qualitative fashion the expectation of the extended indicator functions. There was no need to consider all random variables, and, on the other hand, the more restricted set of indicator functions raises the same axiomatic difficulties confronting the algebra of events.

Qualitative conditional probability. One of the more troublesome aspects of the qualitative theory of conditional probability is that $A|B$ is not an object—in particular it is not a new event composed somehow from events A and B. Thus the qualitative theory rests on a quaternary relation $A|B \succeq C|D$, which is read: event A given event B is at least as probable as event C given event D. There have been a number of attempts to axiomatize this quaternary relation (Koopman 1940a,b; Aczél 1961, 1966, p. 319; Luce 1968; Domotor 1969; Krantz et al., 1971; and Suppes 1973a). The only one of these axiomatizations to address the problem of giving necessary and sufficient conditions is the work of Domotor, which approaches the subject in the finite case in a style similar to that of Scott (1964).

By using indicator functions or, more generally, extended indicator functions, the difficulty of $A|B$ not being an object is eliminated, for $\mathbf{A}^i|B$ is just the indicator function of the set A restricted to the set B, i.e., $\mathbf{A}^i|B$ is a partial function whose domain is B. In similar fashion if $\mathbf{X}$ is an extended indicator function, $\mathbf{X}|A$ is that function restricted to the set A. The use of such partial functions requires care in formulating the algebra of functions in which we are interested, for functional addition $\mathbf{X}|A + \mathbf{Y}|B$ will not be well defined when $A \neq B$ but $A \cap B \neq \emptyset$. Thus, to be completely explicit we begin, as before, with a nonempty set Ω, the probability space, and an algebra $\Im$ of events, i.e., subsets of Ω, with it understood that $\Im$ is closed under union and complementation. Next we extend this algebra to the algebra $\Im^*$ of extended indicator functions, i.e., the smallest semigroup (under function addition) containing the indicator functions of all events in $\Im$. This latter algebra is now extended to include as well all partial functions on Ω that are extended indicator functions restricted to an event in $\Im$. We call this algebra of partial extended indicator functions $\Re\Im^*$, or, if complete explicitness is needed, $\Re\Im^*(\Omega)$. From this definition it is clear that if $\mathbf{X}|A$ and $\mathbf{Y}|B$ are in $\Re\Im^*$, then[40]

If $A = B$, $\mathbf{X}|A + \mathbf{Y}|B$ is in $\Re\Im^*$.

If $A \cap B = \emptyset$, $\mathbf{X}|A \cup \mathbf{Y}|B$ is in $\Re\Im^*$.

In the more general setting of decision theory or expected utility theory, there has been considerable discussion of the intuitive ability of a person to directly compare his preferences or expectations of two decision functions with different domains of restriction. Without reviewing this literature, we do want to state that we find no intuitive general difficulty in making such comparisons. Individual cases may present problems, but not necessarily because of different domains of definition. In fact, we believe comparisons of expectations under different conditions is a familiar aspect of ordinary experience. In the present setting the qualitative comparison of restricted expectations may be thought of as dealing only with beliefs and not utilities. The fundamental ordering relation is a weak ordering $\succeq$ of $\Re\Im^*$ with strict order $\succ$ and equivalence $\approx$ defined in the standard way.

Following Suppes and Zanotti (1982), we give axioms that are strong enough to prove that the probability measure constructed is unique when it is required to cover

---

[40] In this subsection the notation $\mathbf{A}^*$ and $\mathbf{B}^*$ for extended indicator functions is changed to $\mathbf{X}$ and $\mathbf{Y}$ to avoid any confusions in the conditional notation $\mathbf{X}|A$, etc.

5.7 Theory of Subjective Probability

expectation of random variables. It is worth saying something more about this problem of uniqueness. The earlier papers mentioned have all concentrated on the existence of a probability distribution, but from the standpoint of a satisfactory theory it seems obvious for many different reasons that one wants a unique distribution. For example, if we go beyond properties of order and have uniqueness only up to a convex polyhedron of distributions, as is the case with Scott's axioms for finite probability spaces, we are not able to deal with a composite hypothesis in a natural way, because the addition of the probabilities is not meaningful.

DEFINITION 3. *Let Ω be a nonempty set, let $\Re\Im^*(\Omega)$ be an algebra of partial extended indicator functions, and let $\succeq$ be a binary relation on $\Re\Im^*$. Then the structure $(\Omega, \Re\Im^*, \succeq)$ is a partial qualitative expectation structure if and only if the following axioms are satisfied for every $\mathbf{X}, \mathbf{Y}, \mathbf{X}_1, \mathbf{X}_2, \mathbf{Y}_1$ and $\mathbf{Y}_2$ in $\Im^*$ and every A, B and C in $\Im$ with $A, B \succ \emptyset$:*

Axiom 1. The relation $\succeq$ is a weak ordering of $\Re\Im^*$;

Axiom 2. $\Omega^i \succ \emptyset^i$;

Axiom 3. $\Omega^i|A \succeq \mathbf{C}^i|B \succeq \emptyset^i|A$;

Axiom 4a. If $\mathbf{X}_1|A \succeq \mathbf{Y}_1|B$ and $\mathbf{X}_2|A \succeq \mathbf{Y}_2|B$ then

$$\mathbf{X}_1|A + \mathbf{X}_2|A \succeq \mathbf{Y}_1|B + \mathbf{Y}_2|B;$$

Axiom 4b. If $\mathbf{X}_1|A \preceq \mathbf{Y}_1|B$ and $\mathbf{X}_1|A + \mathbf{X}_2|A \succeq \mathbf{Y}_1|B + \mathbf{Y}_2|B$ then

$$\mathbf{X}_2|A \succeq \mathbf{Y}_2|B;$$

Axiom 5. If $A \subseteq B$ then

$$\mathbf{X}|A \succeq \mathbf{Y}|A \text{ iff } \mathbf{X} \cdot \mathbf{A}^i|B \succeq \mathbf{Y} \cdot \mathbf{A}^i|B;$$

Axiom 6. (Archimedean). If $\mathbf{X}|A \succ \mathbf{Y}|B$ then for every $\mathbf{Z}$ in $\Im^*$ there is a positive integer n such that

$$n\mathbf{X}|A \succeq n\mathbf{Y}|B + \mathbf{Z}|B.$$

The axioms are simple in character and their relation to the axioms of Definition 2 is apparent. The first three axioms are very similar. Axiom 4, the axiom of addition, must be relativized to the restricted set. Notice that we have a different restriction on the two sides of the inequality.

I have been unable to show whether or not it is possible to replace the two parts of Axiom 4 by the following weaker and more natural axiom. If $\mathbf{X}_2|A \approx \mathbf{Y}_2|B$, then $\mathbf{X}_1|A \succeq \mathbf{Y}_1|B$ iff $\mathbf{X}_1|A + \mathbf{X}_2|A \succeq \mathbf{Y}_1|B + \mathbf{Y}_2|B$. The really new axiom is Axiom 5. In terms of events and numerical probability, this axiom corresponds to the following: If $A \subseteq B$, then $P(C|A) \geq P(D|A)$ iff $P(C \cap A|B) \geq P(D \cap A|B)$. Note that in the axiom itself, function multiplication replaces intersection of events. (Closure of $\Im^*$ under function multiplication is easily proved.) Axiom 6 is the familiar and necessary Archimedean axiom.

We now state and prove the main theorem. In the theorem we refer to a strictly agreeing expectation function on $\Re\Im^*(\Omega)$. From standard probability theory and conditional expected utility theory, it is evident that the properties of this expectation should be the following for $A, B \succ \emptyset$:

(i) $E(\mathbf{X}|A) \geq E(\mathbf{Y}|B)$ iff $\mathbf{X}|A \succeq \mathbf{Y}|B$,
(ii) $E(\mathbf{X}|A + \mathbf{Y}|A) = E(\mathbf{X}|A) + E(\mathbf{Y}|A)$,
(iii) $E(\mathbf{X} \cdot \mathbf{A}^i|B) = E(\mathbf{X}|A)E(\mathbf{A}^i|B)$ if $A \subseteq B$,
(iv) $E(\emptyset^i|A) = 0$ and $E(\mathbf{\Omega}^i|A) = 1$.

Using primarily (iii), it is then easy to prove the following property, which occurs in the earlier axiomatic literature mentioned above:

$$E(\mathbf{X}|A \cup \mathbf{Y}|B) = E(\mathbf{X}|A)E(\mathbf{A}^i|A \cup B) + E(\mathbf{Y}|B)E(\mathbf{B}^i|A \cup B),$$

for $A \cap B = \emptyset$.

THEOREM 5. *Let Ω be a nonempty set, let $\Im$ be an algebra of sets on Ω, and let $\succeq$ be a binary relation on $\Im \times \Im$. Then a necessary and sufficient condition that there is a strictly agreeing conditional probability measure on $\Im \times \Im$ is that there is an extension $\succeq^*$ of $\succeq$ from $\Im \times \Im$ to $\Re\Im^*(\Omega)$ such that the structure $(\Omega, \Re\Im^*(\Omega), \succeq^*)$ is a partial qualitative expectation structure. Moreover, if $(\Omega, \Re\Im^*(\Omega), \succeq^*)$ is a partial qualitative expectation structure, then there is a unique strictly agreeing conditional expectation function on $\Re\Im^*(\Omega)$ and this expectation generates a unique strictly agreeing conditional probability measure on $\Im \times \Im$.*

The proof is given in Suppes and Zanotti (1982), and I follow it here.

Proof. For every $\mathbf{X}|A$, with $A \succ \emptyset$, we define the set

$$S(\mathbf{X}|A) = \{\tfrac{m}{n} : m\mathbf{\Omega}^i|A \succeq n\mathbf{X}|A\}.$$

(We note that it is easy to prove from the axioms that $\mathbf{\Omega}^i \approx \mathbf{\Omega}^i|A$, and thus for general purposes we can write: $m\mathbf{\Omega}^i \succeq n\mathbf{X}|A$.) Given this definition, on the basis of the reduction in the proof of Theorem 4 of Axioms 1-4 and 6 to a known necessary and sufficient condition for extensive measurement (Krantz et al. 1971, Ch. 3), we know first that the greatest lower bound of $S(\mathbf{X}|A)$ exists. Following the proof in Krantz et al., we use this to define the expectation of $\mathbf{X}$ given A:

$$E(\mathbf{X}|A) = g.l.b.\{\frac{m}{n} : m\mathbf{\Omega}^i \succeq n\mathbf{X}|A\}. \tag{3}$$

It then follows from these earlier results that the function E (for fixed A) is unique and:

$$E(\mathbf{X}|A) \geq E(\mathbf{Y}|A) \text{ iff } \mathbf{X}|A \succeq \mathbf{Y}|A. \tag{4}$$

$$E(\mathbf{X}|A + \mathbf{Y}|A) = E(\mathbf{X}|A) + E(\mathbf{Y}|A). \tag{5}$$

$$E(\emptyset^i|A) = 0 \text{ and } E(\mathbf{\Omega}^i|A) = 1. \tag{6}$$

The crucial step is now to extend the results to the relation between given events A and B.

We first prove the preservation of order by the expectation function. For the first half of the proof, assume

$$\mathbf{X}|A \succeq \mathbf{Y}|B, \tag{7}$$

and suppose, on the contrary, that

$$E(\mathbf{Y}|B) > E(\mathbf{X}|A). \tag{8}$$

5.7 Theory of Subjective Probability

Then there must exist natural numbers m and n such that
$$E(\mathbf{Y}|B) > \frac{m}{n} > E(\mathbf{X}|A), \tag{9}$$
and so from the definition of the function E, we have
$$m\mathbf{\Omega}^i \prec n\mathbf{Y}|B, \tag{10}$$
and
$$m\mathbf{\Omega}^i \succeq n\mathbf{X}|A, \tag{11}$$
whence
$$n\mathbf{Y}|B \succ n\mathbf{X}|A, \tag{12}$$
but from (7) and Axiom 4a we have by a simple induction
$$n\mathbf{X}|A \succeq n\mathbf{Y}|B, \tag{13}$$
which contradicts (12), and thus the supposition (8) is false.

Assume now
$$E(\mathbf{X}|A) \geq E(\mathbf{Y}|B), \tag{14}$$
and suppose
$$\mathbf{Y}|B \succ \mathbf{X}|A. \tag{15}$$
Now if $E(\mathbf{X}|A) > E(\mathbf{Y}|B)$, by the kind of argument just given we can show at once that
$$\mathbf{X}|A \succ \mathbf{Y}|B, \tag{16}$$
which contradicts (15). On the other hand, if
$$E(\mathbf{X}|A) = E(\mathbf{Y}|B), \tag{17}$$
then we can argue as follows. By virtue of (15) and Axiom 6, there is an n such that
$$n\mathbf{Y}|B \succeq (n+1)\mathbf{X}|A, \tag{18}$$
whence by the earlier argument
$$E(n\mathbf{Y}|B) \geq E((n+1)\mathbf{X}|A), \tag{19}$$
and by (5)
$$nE(\mathbf{Y}|B) \geq (n+1)E(\mathbf{X}|A), \tag{20}$$
and so by (17) and (20)
$$E(\mathbf{Y}|B) \leq 0, \tag{21}$$
but from (4)–(6) it follows easily that
$$E(\mathbf{Y}|B) \geq 0, \tag{22}$$
whence
$$E(\mathbf{Y}|B) = 0. \tag{23}$$
But then, using again (4)–(6), we obtain
$$\mathbf{Y}|B \approx \emptyset^i|B, \tag{24}$$
and by virtue of Axiom 3
$$\mathbf{X}|A \succeq \emptyset^i|B, \tag{25}$$

whence from (24) and (25) by transitivity

$$\mathbf{X}|A \succeq \mathbf{Y}|B, \tag{26}$$

contradicting (15). We have thus now shown that

$$E(\mathbf{X}|A) \geq E(\mathbf{Y}|B) \text{ iff } \mathbf{X}|A \succeq \mathbf{Y}|B. \tag{27}$$

Finally, we need to prove that for $A \succ \emptyset$ and $A \subseteq B$

$$E(\mathbf{X} \cdot \mathbf{A}^i | B) = E(\mathbf{X}|A) E(\mathbf{A}^i | B). \tag{28}$$

We first note that by putting $m\mathbf{\Omega}^i$ for $\mathbf{X}$ and $n\mathbf{X}$ for $\mathbf{Y}$ in Axiom 5, we obtain

$$m\mathbf{\Omega}^i \geq n\mathbf{X}|A \text{ iff } m\mathbf{A}^i|B \geq n\mathbf{X} \cdot \mathbf{A}^i|B. \tag{29}$$

It follows directly from (29) that

$$\{\frac{m}{n} : m\mathbf{\Omega}^i \succeq n\mathbf{X}|A\} = \{\frac{m}{n} : m\mathbf{A}^i|B \succeq n\mathbf{X} \cdot \mathbf{A}^i|B\} \tag{30}$$

whence their greatest lower bounds are the same, and we have

$$E(\mathbf{X}|A) = E'_{\mathbf{A}^i|B}(\mathbf{X} \cdot \mathbf{A}^i | B), \tag{31}$$

where E' is the measurement function that has $\mathbf{A}^i|B$ as a unit, that is,

$$E'_{\mathbf{A}^i|B}(\mathbf{A}^i|B) = 1.$$

As is familiar in the theory of extensive measurement, there exists a positive real number c such that for every $\mathbf{X}$

$$cE'_{\mathbf{A}^i|B}(\mathbf{X} \cdot \mathbf{A}^i|B) = E(\mathbf{X} \cdot \mathbf{A}^i|B). \tag{32}$$

Now by (31) and taking $\mathbf{X} = \mathbf{\Omega}^i$

$$cE(\mathbf{\Omega}^i|A) = E(\mathbf{\Omega}^i \cdot \mathbf{A}^i|B),$$

but $E(\mathbf{\Omega}^i|A) = 1$, so

$$c = E(\mathbf{\Omega}^i \cdot \mathbf{A}^i|B) = E(\mathbf{A}^i|B). \tag{33}$$

Combining (31), (32) and (33) we obtain (28) as desired.

The uniqueness of the expectation function follows from (6) and the earlier results of Theorem 4 about unconditional probability.

For $A \succ \emptyset$, we then define for every B in $\mathfrak{F}$,

$$P(B|A) = E(\mathbf{B}^i|A),$$

and it is trivial to show the function P is a conditional probability measure on $\mathfrak{F}$, which establishes the sufficiency of the axioms. The necessity of each of the axioms is easily checked.

Historical background on qualitative axioms. The qualitative axioms of de Finetti (1937/1964) were, as already indicated, not the first ones.[41] Various incomplete sets of such axioms were given much earlier by Keynes (1921) and Jeffreys and Wrinch (1919) in the framework of logical theories of probability. Undoubtedly the historically

---

[41] A particularly early source of much detailed discussion of the use of probability qualitatively in ordinary affairs is to be found in various sections of Hume's *Treatise* (1739), especially Section XIII, Part III, Bk I, entitled "Of Unphilosophical Probability".

5.7 THEORY OF SUBJECTIVE PROBABILITY

most important predecessor of de Finetti, within the framework of subjective theories of probability, was Ramsey's (1931) "Truth and Probability", published posthumously, but written in 1926. Ramsey's basic approach differs from de Finetti's in that he begins with actions and expected utility, or, in terms now familiarly used, both beliefs and desires, not beliefs alone. He describes this general program well in the following two paragraphs.

> In order therefore to construct a theory of quantities of belief which shall be both general and more exact, I propose to take as a basis a general psychological theory, which is now universally discarded, but nevertheless comes, I think, fairly close to the truth in the sort of cases with which we are most concerned. I mean the theory that we act in a way we think most likely to realize the objects of our desires, so that a person's actions are completely determined by his desires and opinions. This theory cannot be made adequate to all the facts, but it seems to me a useful approximation to the truth particularly in the case of our self-conscious or professional life, and it is presupposed in a great deal of our thought. It is a simple theory and one which many psychologists would obviously like to preserve by introducing unconscious desires and unconscious opinions in order to bring it more into harmony with the facts. How far such fictions can achieve the required result I do not attempt to judge: I only claim for what follows approximate truth, or truth in relation to this artificial system of psychology, which like Newtonian mechanics can, I think, still be profitably used even though it is known to be false.
>
> It must be observed that this theory is not to be identified with the psychology of the Utilitarians, in which pleasure had a dominating position. The theory I propose to adopt is that we seek things which we want, which may be our own or other people's pleasure, or anything else whatever, and our actions are such as we think most likely to realize these goods. But this is not a precise statement, for a precise statement of the theory can only be made after we have introduced the notion of quantity of belief.
>
> (Ramsey 1931, p. 173)

As Ramsey himself says, his ideas are only sketched in the article cited—the only one he wrote on the topic. But even this sketch is important, for he outlines a simple clear way to find an 'ethically neutral proposition p believed to degree $\frac{1}{2}$'—his first axiom. He next introduces an ordering relation for preference over "worlds", as he puts it, amplified by further axioms to measure value, although his axioms are only casually formulated. He then proposes to use the measure of value to measure by the usual method of betting ratios the quantity, or probability, of belief.[42]

Ramsey's ideas were later used directly as a basis of a detailed finitistic axiomatization of subjective probability and utility by Davidson and Suppes (1956). Slightly later these ideas were then used in experiments concerned to measure individuals' personal probability and utility (Davidson, Suppes and Siegel 1957).

But the essential point to keep in mind is that de Finetti developed his ideas in great detail, both mathematically and philosophically, in many publications, in contrast to Ramsey, whose death at 26 abruptly ended any further work. The much greater subsequent influence of de Finetti is hardly surprising.

---

[42] As de Finetti says in a footnote (p. 102) added to the English translation of his 1937 article, he did not know about Ramsey's work before 1937, two years after his lectures, on which the article is based, were given in Paris.

De Finetti's representation theorem. A central problem for subjectivists such as de Finetti is how to deal with the standard, even if artificial, problem of estimating the objectively unknown probability p of observing heads with a fair toss of a possibly biased coin. The relative-frequency approach is just to estimate p from an infinite sequence of trials. (The artificiality comes from the use of an infinite sequence, which it is not possible to produce or observe in finite time.) De Finetti's response is that, assuming the same artificial use of an infinite sequence, this is not a problem for subjectivists.[43] The concept he uses, already mentioned several times, is that of exchangeability, which we now define for infinite sequences. A sequence of $0, 1$ random variables $\mathbf{X}_1, \mathbf{X}_2, \ldots, \mathbf{X}_n, \ldots$ on a probability space is *exchangeable* if for every n and every permutation σ of the first n integers, $\mathbf{X}_1, \mathbf{X}_2, \ldots \mathbf{X}_n$ has the same probability distribution as $\mathbf{X}_{\sigma 1}, \mathbf{X}_{\sigma 2}, \ldots \mathbf{X}_{\sigma n}$. The other concept we need to state in the theorem is that of a mixture of probabilities. Here we simplify slightly and write the mixture as a probability density f on the interval $[0, 1]$. So f has the properties:

(i) $f(p) \geq 0$, for p in $[0, 1]$,
(ii) $\int_0^1 f(x)dx = 1$.

Now it is easy to show that any mixture f of independent sequences $\mathbf{X}_1, \mathbf{X}_2, \ldots \mathbf{X}_n \ldots$ is exchangeable. De Finetti proved the hard part, the converse.

THEOREM 6. (de Finetti 1937/1964). *A sequence of random variables* $\mathbf{X}_1, \mathbf{X}_2, \ldots \mathbf{X}_n$ *...taking only the values 0 and 1 is exchangeable if and only if there exists a mixture density f on $[0,1]$ such that the probability that for any sample of n trials with m successes, $0 \leq m \leq n$, the probability of the sample is*

$$\int_0^1 p^m(1-p)^{n-m}f(p)dp.$$

Another appealing formulation is that in a sample of n trials, the probability of the finite sequence of heads and tails observed depends just on two numbers, n, the length of the sample and m, the number of heads. This is an invariance statement corresponding to the formulation given at the end of Section 1. Note also the important fact that given p, the outcomes of the sample are independent. De Finetti's (1937/1964) careful and detailed philosophical discussion of the theorem is highly recommended. The proof is also presented in detail by de Finetti, as well as the generalization to random quantities, de Finetti's term for general random variables. I cannot do better than end this discussion with one of de Finetti's own summarizing paragraphs:

> The result at which we have arrived gives us the looked-for answer, which is very simple and very satisfactory: the nebulous and unsatisfactory definition of "independent events with fixed but unknown probability" should be replaced by that of "exchangeable events". This answer furnishes a condition which applies directly to the evaluations of the probabilities of individual events and does not run up against any of the difficulties that the subjectivistic considerations propose to eliminate. It constitutes a very natural and very clear condition, of a purely qualitative character, reducing to the demand that certain events be judged equally probable, or, more precisely, that all the combinations

---

[43]Explicit results on exchangeability for finite sequences may be found in Diaconis (1977) and Diaconis and Freedman (1980).

of n events $E_{i_1}, \ldots, E_{i_n}$ have always the same probability, whatever be the choice or the order of the E_i. The same simple condition of "symmetry" in relation to our judgments of probability defines exchangeable random quantities, and can define, in general, exchangeable random elements in any space whatever. It leads in all cases to the same practical conclusion: a rich enough experience leads us always to consider as probable future frequencies or distributions close to those which have been observed.

(de Finetti 1937/1964, p. 142)

Defense of objective priors. The concept of an objective prior probability occurs already in logical theories of probability in several different forms discussed in Section 5. But there is another version, sometimes called an objective Bayesian position, that holds only objective priors based on prior solid empirical and scientific results should be used. The most straightforward approach to this has been that of E. T. Jaynes, a physicist, whose principal papers were collected in a volume edited by Roger Rosenkrantz (1983). Jaynes argues forcibly in many papers for using the principle of maximum entropy as the objective prior. Especially in Jaynes (1978), he shows how this principle certainly makes sense in statistical mechanics and leads to direct derivations of classical results by Maxwell, Boltzmann and Gibbs.

Restricting ourselves to a discrete density on a finite set, the principle of maximum entropy is to choose the objective prior so as to maximize the entropy H of the distribution, i.e., maximize

$$H = -\sum_{i=1}^{n} p_i \log p_i,$$

subject to whatever objective constraints are known. The standard argument is that this distribution is the prior with the maximum uncertainty. The proof, outlined in Section 4.5, that entropy is a complete invariant for Bernoulli or Markov processes, helps support that claim, although the ergodic results summarized in that section are not ordinarily discussed in the context of objective priors.[44]

---

[44] The help comes in the following form. Maximizing entropy will maximize uncertainty, at least for processes that are Markov—a Bernoulli process is a special case of a Markov process, because for such processes, those which have the same entropy have isomorphic uncertainty structures and conversely, as explained in Section 4.5. For the more familiar standard arguments as to why entropy is a good measure of uncertainty, see Rosenkrantz (1977, pp. 13ff). I quote his summary on p. 13. "That entropy is a satisfactory measure of uncertainty is further attested by the following of its properties:

(i) $H(p_1, \ldots, p_m) = H(p_1, \ldots, p_m, 0)$, the entropy is wholly determined by the alternatives which are assigned a nonzero probability.

(ii) When all the p_i are equal, $H(p_1, \ldots, p_m)$ is increasing in m, the number of equiprobable alternatives.

(iii) $H(p_1, \ldots, p_m) = 0$, a minimum, when some $p_i = 1$.

(iv) $H(p_1, \ldots, p_m) = \log m$, a maximum, when each $p_i = 1/m$.

(v) Any averaging of the p_i (i.e., any flattening of the distribution) increases H.

(vi) H is non-negative.

(vii) $H(p_1, \ldots, p_m)$ is invariant under every permutation of the indices $1, \ldots, m$.

(viii) $H(p_1, \ldots, p_m)$ is continuous in its arguments."

A second argument is familiar from statistical mechanics. The maximum-entropy distribution is the one to choose because, under the null hypothesis that no other constraints are present besides those already taken account, most of the possible distributions will be close to it. Here is a familiar example from Jaynes (1979) that illustrates this point and also shows how the principle works.

Suppose a die is tossed $N = 1000$ times. If the die were "true", the expected number of spots up should be 3.5. But we are told the correct expectation for the slightly biased die is 4.5. So

$$\sum_{i=1}^{6} i f_i = 4.5.$$

The result, derived in detail in Jaynes (1978) is: $(f_1, \ldots, f_6) = (0.05, 0.08, 0.11, 0.17, 0.24, 0.35)$. The bias toward the faces with more spots is evident. The computation, even of this simple example, is nontrivial, a point sometimes stated in criticism of the maximum entropy principle. A good discussion of this and other points may be found in Rosenkrantz's (1977) book on these and related topics, especially Chapter 3.

The objective Bayesian viewpoint, defended most clearly and effectively by Jaynes in the context of many physical examples, has not won out over the personal subjectivism of de Finetti. Perhaps the main difficulty of objective approaches, emphasized by de Finetti, arises from attempts to apply them when it is not appropriate. The subjective Bayesian always has a prior, even if he cannot explain or justify it. This is an attitude well suited to any complex and subtle domain of investigation or application, such as forecasting the weather. But, all the same, some of Jaynes' applications in statistical mechanics are very appealing, special though they may be. Moreover, as de Finetti attests, "objectivity has its rightful, important place".[45]

General issues. I now want to turn to a number of general issues that arise in evaluating the subjective view of probability.

Use of symmetries. A natural first question is to ask how subjective theory utilizes the symmetries that are such a natural part of the classical, Laplacean definition of probability. If we think in terms of Bayes' Theorem the answer seems apparent. The symmetries that we all accept naturally in discussing games of chance are incorporated immediately in the subjective theory as prior probabilities. Thus, for example, if I wish to determine the probability of getting an ace in the next round of cards dealt face up in a hand of stud poker, given the information that one ace is already showing on the board, I use as prior probabilities the natural principles of symmetry for games of chance, which are a part of the classical definition. Of course, if I wish to introduce refined corrections I could do so, particularly corrections arising from the fact that in ordinary shuffling, the permutation groups introduced are groups of relatively small finite order, and, therefore, there is information carry-over from one hand to another, in spite of the fact that a rather thorough shuffling may have been given the cards

---

[45] This brief quotation is taken from de Finetti (1975, p. 201), but his lengthy philosophical analysis on this and the following pages is very pertinent to the conceptual problems of inductive inference.

5.7 THEORY OF SUBJECTIVE PROBABILITY

by the dealer. These second-order refinements with respect to shuffling are simply an indication of the kind of natural corrections that arise in the subjective theory and that would be hard to deal with in principle within the framework of the classical definition of probability. On the other hand, I emphasize that the principles of symmetry used in the classical definition are a natural part of the prior probabilities of an experienced card player. The extent to which these symmetries are compelling is a point I shall return to later.

Use of relative frequencies. It should also be clear that the proper place for the use of relative-frequency data in the subjective theory is in the computation of posterior probabilities. The example given earlier of the application of Bayes' Theorem provides an artificial but simple instance in which such use is made of relative-frequency data. It is clear what is required in order to get convergence of opinion between observers whose initial opinions differ. The observers must agree on the method of sampling, and, of course, they must also agree on the observations that result from this sampling. Under very weak restrictions, no matter how much their initial opinions differ, they can be brought arbitrarily close to convergence on the basis of a sufficient amount of sampled observations. The obvious requirement is that the individual observations be approximately independent, or at least exchangeable. If, for example, the observations are temporally strongly dependent, then many observations will count for no more than a single observation. Since these matters have already been considered, we will not discuss them again in detail.

Reflection upon the conditions under which convergence of beliefs will take place also throws light on the many situations in which no such convergence occurs. The radically differing opinions of men and women about religion, economics, and politics are excellent examples of areas in which there is a lack of convergence; no doubt a main source of this divergence is the lack of agreement on what is to count as evidence, or, put another way, inability to agree on an experiment, or some other regime of collecting data, which may be summarized in an agreed-upon likelihood function.

Problem of the forcing character of information. As already indicated, it is an important aspect of the subjective theory to emphasize that equally reasonable persons may hold rather different views about the probability of the same event. The ordinary use of the word *rational* seems to go beyond what is envisaged in the subjective theory of probability. Let us consider one or two examples of how this difference in usage may be expressed.

The first kind of example deals with the nonverbal processing of information by different individuals. One man is consistently more successful than another in predicting tomorrow's weather. At least before the advent of powerful mathematical methods of predicting weather, which are now just beginning to be a serious forecasting instrument, it was the common observation of experienced meteorologists that there was a great difference in the ability of meteorologists with a common training and background, and with a common set of observations in front of them, to predict successfully tomorrow's weather in a given part of the world. As far as I can see, in terms of the standard subjective theory as expressed, for example, by de Finetti, there is no very clear way of stating that on a single occasion the better predictor is in some sense more rational in his processing of information than the other man; yet in common usage we would be very inclined to say this. It is a stock episode in novels, and a common experience in real life for many people, for someone to denigrate the intelligence or rationality of individuals who continue to hold naive beliefs about other people's behavior in the face of much contrary, even though perhaps subtle, evidence.

But successive predictions can be studied like any other empirical phenomena, and there is a large literature on evaluating the performance of forecasters, an important practical topic in many arenas of experience. Examination of quantitative methods of evaluation of subjective, as well as objective, forecasts lies outside the scope of this chapter. The *Journal of Forecasting* is entirely devoted to the subject. See also, for example, Makridakis et al. (1984) and Dawid (1986).

Contrary to the tenor of many of de Finetti's remarks, it seems fair to say that the subjective theory of probability provides necessary but not sufficient conditions of rationality.

Bayesian probabilities and the problem of concept formation. An important point revolving around the notion of *mistaken* belief is involved in the analysis of how information is processed. In common usage, a belief is often said to be mistaken or irrational when later information shows the belief to be false. According to the subjective theory of probability, and much sensible common usage in addition, this view of mistaken beliefs is itself a mistake. A belief is not shown to be mistaken on the basis of subsequent evidence not available at the time the belief was held. Proper changes in belief are reflected in the change from a prior to a posterior probability on the basis of new information. The important point for subjective theory is that the overall probability measure does not itself change, but rather we pass from a prior to a posterior, conditional probability. Applications of Bayes' Theorem that we have considered demonstrate this well enough. The following quotation from de Finetti (1937/1964, p. 146) illustrates this point beautifully.

> Whatever be the influence of observation on predictions of the future, it never implies and never signifies that we *correct* the primitive evaluation of the probability $P(E_{n+1})$ after it has been *disproved* by experience and substitute for it another $P^*(E_{n+1})$ which conforms to that experience and is therefore probably *closer to the real probability;* on the contrary, it manifests itself solely in the sense that when experience teaches us the result A on the first n trials, our judgment will be expressed by the probability $P(E_{n+1})$ no longer, but by the probability $P(E_{n+1}|A)$, i.e., that which our initial opinion would already attribute to the event E_{n+1} considered as conditioned on the outcome A. Nothing of this initial opinion is repudiated or corrected; it is not the function P which has been modified (replaced by another P^*), but rather the argument E_{n+1} which has been replaced by $E_{n+1}|A$, and this is just to remain faithful to our original opinion (as manifested in the choice of the function P) and coherent in our judgment that our predictions vary when a change takes place in the known circumstances.

5.7 THEORY OF SUBJECTIVE PROBABILITY

In spite of the appeal of what de Finetti says, there seems to be a wide class of cases in which the principles he affirms have dubious application. I have in mind all those cases in which a genuinely new concept is brought to bear on a subject. I do not mean necessarily the creation of a new scientific concept, but rather any situation in which an individual applies a concept that he was not previously using in his analysis of the data in front of him.[46]

Problem of unknown probabilities. Another feature of the subjective theory of probability that is in conflict with common usage of probability notions is the view that there are no unknown probabilities. If someone asks me what is the probability of rain in the Fiji Islands tomorrow, my natural inclination is to say, 'I don't know', rather than to try to give a probability estimate. If another person asks me what I think the probability is that Stanford University will have an enrollment of at least 50,000 students 500 years from now, I am naturally inclined simply to say, 'I haven't the faintest idea what the probability or likelihood of this event is'. De Finetti insists on the point that a person always has an opinion, and, therefore, a probability estimate about such matters, but it seems to me that there is no inherent necessity of such a view. It is easy to see one source of it. The requirement that one always have a probability estimate of any event, no matter how poor one's information about the circumstances in which that event might occur may be, arises from a direct extension of two-valued logic. Any statement is either true or false, and, correspondingly, any statement or event must always have a definite probability for each person interrogated. From a formal standpoint it would seem awkward to have a logic consisting of any real number between 0 and 1, together with the quite disparate value, 'I don't know'.

A little later we shall examine the view that one can always elicit a subjective probability for events about which the individual has very little background information by asking what sort of bet he will make concerning the event. Without anticipating that discussion, I still would like to insist that it does not really seem to be a proper part of the subjective theory to require an assignment of a probability to every imaginable event. In the same spirit with which we reply to a question about the truth of a statement by saying that we simply don't know, we may also reply in the same fashion to a request for an estimate of a probability.

Decisions and the measurement of subjective probability. It is commonplace to remark that a man's actions or decisions, and not his words, are the true mark of his beliefs. As a reflection of this commonly accepted principle, there has been considerable discussion of how one may measure subjective probabilities on the basis of decisions actually made. This is a complex subject, and I shall not attempt to give it a fully detailed treatment.

The classical response in terms of subjective probability is that we may find out the

---

[46] I. J. Good (1983) emphasizes this point well in many passages, indexed under the heading 'evolving probabilities'. An even more important point is that conditionalization on new information, fundamental though it may be, as de Finetti emphasizes, does not replace the need for an explicit development of Bayesian statistics. An excellent and quite accessible introduction is Lindley (1971).

subjective probabilities a man truly holds by asking him to place wagers. For example, if he thinks the probability of rain tomorrow is really $\frac{1}{2}$, then he will be willing to place an even-money bet on this occurrence. If he thinks that the probability of snow tomorrow has a subjective probability of 0.1, then he will bet against snow at odds of 1:9. It is also clear how this same procedure may be used to test precise statements. For example, if a man says the probability of rain tomorrow is at least $\frac{1}{2}$, then presumably he will accept any bet that provides odds at least this favorable to him.

Unfortunately, there is a central difficulty with this method of measuring subjective probability. Almost all people will change the odds at which they will accept a bet if the amount of money varies. For example, the man who will accept an even-money bet on the probability of rain tomorrow with the understanding that he wins the bet if in fact it does rain, will not accept an even-money bet if the amount of money involved moves from a dollar on each side to a hundred dollars on each side. Many people who will casually bet a dollar will not, in the same circumstances and at the same odds, be willing to bet a hundred dollars, and certainly not a thousand dollars. The man who will accept an even-money bet on its raining tomorrow will perhaps be willing to accept odds of two to one in his favor only if the wager is of the order of a hundred dollars, while he will accept only still more favorable odds for a bet involving a larger sum of money. What then are we to say is his true estimate of the probability of rain tomorrow if we use this method of wagers to make the bet?

In spite of this criticism, there have been a number of interesting empirical studies of the measurement of subjective probability using the simple scheme we have just described. A brief diversion which describes some of these empirical results may be of interest. Again I follow the discussion in Luce and Suppes (1965).

The first attempt to measure subjective probability experimentally was apparently by Preston and Baratta (1948). Subjects were run in groups of two or more, and they used play money to bid for gambles in a simple auction game. The successful bidder was permitted to roll a set of dice after the bidding was completed. The probability of winning with the dice corresponded exactly to the probability stated on the card presented for auction. For example, on a given play, subjects might bid for a prize of 250 points with probability of 0.25 of winning. If for this gamble the average successful bid was 50, then the authors computed the psychological probability to be $\frac{50}{250} = .20$. Using this method of computation, perhaps their most interesting conclusion was that objective probabilities less than 0.20 are systematically overestimated, and objective probabilities greater than 0.20 are systematically underestimated.

Two other conclusions are noteworthy. First, by comparing the data of sophisticated subjects with those of subjects who were relatively naive about the facts of probability, they concluded that the underestimating and overestimating effects just described exist in both kinds of subjects. Second, an increase in the number of players, and, therefore, an increase in the competition among them, tended to increase the underestimation of high probabilities and the overestimation of low probabilities, which is a somewhat surprising result.

A ready-made, real-life comparison of subjective and objective probabilities is provided by the betting on horses under the parimutuel system as compared with the

5.7 THEORY OF SUBJECTIVE PROBABILITY

objective probabilities of winning, as determined a posteriori after the races. The total amount of money bet on a horse divided by the net total bet on all horses in a race determines the odds, and, thus, the collective subjective probability that the horse will win. Griffith (1949) examined data from 1,386 races run in 1947; his results are in remarkable qualitative agreement with those obtained by Preston and Baratta. Low objective probabilities are systematically overestimated by the subjective probabilities, and high objective probabilities are systematically underestimated. For Griffith's data, objective and subjective probability are equal at 0.16, which is close to the 0.20 value obtained by Preston and Baratta. Griffith remarks in a final footnote that the same computations were carried through for all races run at the same tracks in August, 1934, and essentially the same results were obtained. In this case the indifference point fell at 0.18 rather than at 0.16. The invariance of the results from the depression economy of 1934 to the relative affluence of 1947 increases their significance.

A still more extensive study very similar to Griffith's has been made by McGlothlin (1956) of 9,605 races run from 1947 to 1953, mostly on California tracks. As one would expect, the general character of his results agrees closely with those of the earlier study. The objective and subjective probability curves intersect between 0.15 and 0.22. McGlothlin also found some interesting tendencies during the course of the usual eight races on a given day. First, there was an increasing tendency to overestimate low probabilities, and this phenomenon was particularly striking for the final race of the day. There seems to be evidence that on the last race of the day many bettors are uninterested in odds that are not sufficient to recoup their losses from the earlier races. This phenomenon is so striking that bets placed on low-odds horses had a net positive expectation even after subtracting the tracks' take, which was at the time 13 percent in California.

Toda (1951, 1958) proposed a two-person game method for measuring subjective probability that is very similar to the auction procedure of Preston and Baratta. One of its more interesting applications has been made by Shuford (1959) to obtain extensive measurements of subjective probability of both elementary and compound events. Subjects rolled a 20-face die (with the integers 1 to 20 on the faces) twice to select the row and column of a 20×20 matrix of vertical and horizontal bars. The subjects, 64 airmen in training at an airbase, were run in pairs. The sequence of events on a given trial was as follows. The matrix of horizontal and vertical bars was projected on a screen. Subject A wrote down his bid x that a horizontal bar, say, would be selected ($0 \leq x \leq 10$); subject B decided to 'buy' or 'sell' A's bid; the die was rolled by the experimenter to decide the bet, that is, which element of the projected matrix was selected; and, finally, the subjects scored themselves for the play. It can be shown that A's optimal (minimax) strategy in this game is to set x equal to 10 times his subjective probability of the favorable outcome.

Two games were played with each pair of subjects. In the first game the payoff was determined by the occurrence of a horizontal or vertical bar, as the case might be, in the position of the matrix selected by the two rolls of the 20-face die. In the second game, the bet paid off if two successive selections of elements of the matrix made with the die resulted in two bars of the same type.

Confirming the earlier findings of Preston and Baratta, Griffith, and McGlothlin, a fairly large fraction of the subjects overestimated low probabilities and underestimated high ones. This was true for both the elementary and the compound events. On the other hand, Shuford found that the subjective probability estimates of a number of the subjects were fit quite well by a linear function of objective probability, although the slope and intercept of this function varied from one subject to another. His findings about the estimation of compound events are particularly interesting. A majority of the subjects approximated the correct rule, that is, they estimated the probability of the compound event as approximately the square of the probability of the elementary event. That the application of the correct rule was so common is surprising, because when the subjects were asked at the end of the series of trials what rule they had used, only two stated the correct one.

It is apparent from the studies cited that it is possible empirically to measure subjective probabilities, and, also, that the obvious method of doing this is in terms of the kind of wagers individuals will make. On the other hand, recall the objections offered above to this method of measurement, because of the lack of invariance of the odds at which wagers will be made when the amounts of money involved are changed. They have prompted a deeper theoretical development in terms of the concept of expected utility, which I will not explore here. For references see Savage (1954), Luce and Suppes (1965), Fishburn (1970), Krantz et al. (1971, Ch. 8) and Luce (2000). Savage's axioms are discussed in the next subsection, but not his basic representation theorem.

Inexact measurement of belief: upper and lower probabilities.[47] Almost everyone who has thought about the problems of measuring beliefs in the tradition of subjective probability or Bayesian statistical procedures concedes some uneasiness with the problem of always asking for the next decimal of accuracy in the prior estimation of a probability or of asking for the parameter of a distribution that determines the probabilities of events. On the other hand, the formal theories that have been developed for rational decision-making under uncertainty by Ramsey (1931), de Finetti (1931, 1937/1964), Koopman (1940a,b), Savage (1954) and subsequent authors have almost uniformly tended to yield a result that guarantees a unique probability distribution on states of nature or whatever other collection of entities is used for the expression of prior beliefs. I examine some of these standard theories and address the question of how we can best criticize the claims they make. Among other points, I consider the claim that the idealizations expressed in the axioms can be regarded as theories of pure rationality.

Because the standard theories mentioned earlier reach essentially the same formal results, namely, the existence of a unique probability distribution on states of nature, criticisms of one will pretty much apply to criticisms of the lot. For this reason, it may pay to concentrate on Savage's (1954) axioms, because of their familiarity to a wide audience and because they have been much discussed in the literature. I emphasize, however, that what I have to say about Savage's axioms will apply essentially without change to other standard theories.

Because Savage's axioms are rather complicated from a formal standpoint, I shall not

---

[47] Much of this subsection is drawn from Suppes (1974d).

5.7 THEORY OF SUBJECTIVE PROBABILITY

TABLE 1   Decision matrix for buying bread.

	d_1 buy 700 loaves	d_2 buy 800 loaves	d_3 buy 900 loaves
s_1–rain	$21.00	$19.00	$17.00
s_2–no rain	$21.00	$24.00	$26.50

state them explicitly here, but shall try to describe their intuitive content. The axioms are about preference among decisions, where decisions are mappings or functions from the set of states of nature to the set of consequences. To illustrate these ideas, let me use an example I have used before (Suppes 1956).

A certain independent distributor of bread must place his order for a given day by ten o'clock of the preceding evening. His sales to independent grocers are affected by whether or not it is raining at the time of delivery, for if it is raining, the grocers tend to buy less on the accumulated evidence that they have fewer customers. On a rainy day the maximum the distributor can sell is 700 loaves; on such a day he makes less money if he has ordered more than 700 loaves. On the other hand, when the weather is fair, he can sell about 900 loaves. If the simplifying assumption is made that the consequences to him of a given decision with a given state of nature (s_1–rain or s_2–no rain) may be summarized simply in terms of his net profits, the situation facing him is represented in Table 1. The distributor's problem is to make a decision.

Clearly, if he knows for certain that it is going to rain, he should make decision d_1, and if he knows for certain that it is not going to rain, he should make decision d_3. The point of Savage's theory, expanded to more general and more complex situations, is to place axioms on choices or preferences among the decisions in such a way that anyone who satisfies the axioms will be maximizing expected utility. This means that the way in which he satisfies the axioms will generate a subjective probability distribution about his beliefs concerning the true state of nature and a utility function on the set of consequences such that the expectation of a given decision is defined in a straightforward way with respect to the subjective probability distribution on states of nature and the utility function on the set of consequences. As one would expect, Savage demands, in fact in his first axiom, that the preference among decisions be transitive and that given any two decisions one is at least weakly preferred to the other. Axiom 2 extends this ordering assumption to having the same property hold when the domain of definition of decisions is restricted to a given set of states of nature; for example, the decision-maker might know that the true state of nature lies in some subset of the whole set. Axiom 3 asserts that knowledge of an event cannot change preferences among consequences, where preferences among consequences are defined in terms of preferences among decisions. Axiom 4 requires that given any two sets of states of nature, that is, any two events, one is at least as probable as the other, that is, qualitative probability among events is strongly connected. Axiom 5 excludes the trivial case in which all consequences are equivalent in utility and, thus, every decision is equivalent to every other. Axiom 6 says essentially that if event A is less probable than event B (A and B are subsets of the same set of states of nature), then there is a partition of the states of nature such that the union of each element of the partition with A is less probable than B. As is well known, this axiom of Savage's is closely related to the axiom of de Finetti and Koopman, which requires the existence of a partition of the states of nature into arbitrarily many events that are equivalent in probability. Finally, his last axiom, Axiom 7, is a formulation of the sure-thing principle.

My first major claim is that some of Savage's axioms do not in any direct sense represent axioms of rationality that should be satisfied by any ideally rational person but, rather, they represent structural assumptions about the environment that may or may not be satisfied in given applications.

Many years ago, at the time of the Third Berkeley Symposium (1955), I introduced the distinction between structure axioms and rationality axioms in the theory of decision-making (Suppes 1956). Intuitively, a structure axiom as opposed to a rationality axiom is existential in character. In the case of Savage's seven postulates, two (5 and 6) are structure axioms, because they are existential in character.

Savage defended his strong Axiom 6 by holding it applicable if there is a coin that a decision-maker believes is fair for any finite sequence of tosses. There are however, several objections to this argument. First of all, if it is taken seriously then one ought to redo the entire foundation and simply build it around Bernoulli sequences with $p = 0.5$ and get arbitrarily close approximations to the probability of any desired event. More importantly, without radical changes in human thinking, it is simply not natural on the part of human beings to think of finite sequences of tosses of a coin in evaluating likelihoods or probabilities, qualitative or quantitative, of significant events with which they are concerned.

Consider the case of a patient's deciding whether to follow a surgeon's advice to have major surgery. The surgeon, let us suppose, has evaluated the pros and cons of the operation, and the patient is now faced with the critical decision of whether to take the risk of major surgery with at least a positive probability of death, or whether to take the risk of having no surgery and suffering the consequences of the continuing disease. I find it very unlikely and psychologically very unrealistic to believe that thinking about finite sequences of tosses of a fair coin will be of any help in making a rational decision on the part of the patient.

On the other hand, other axioms like those on the ordering of preferences or qualitative probability seem reasonable in this framework and are not difficult to accept. But the important point is this. In a case in which uncertainty has a central role, in practice, decisions are made without any attempt to reach the state of having a quantitative probability estimate of the alternatives or, if you like, a computed expected utility.

It is, in fact, my conviction that we usually deal with restricted situations in which the set of decisions open to us is small and in which the events that we consider relevant are small in number. The kind of enlarged decision framework provided by standard theories is precisely the source of the uneasiness alluded to in the first sentence of this subsection. Intuitively we all move away from the idea of estimating probabilities with arbitrary refinement. We move away as well from the introduction of an elaborate mechanism of randomization in order to have a sufficiently large decision space. Indeed, given the Bayesian attitude towards randomization, there is an air of paradox about the introduction *à la* Savage of finite sequences of tosses of a fair coin.

Another way of putting the matter, it seems to me, is that there is a strong intuitive feeling that a decision-maker is not irrational simply because a wide range of decision possibilities or events is not available to him. It is not a part of rationality to require that the decision-maker enlarge his decision space, for example, by adding a coin that may be tossed any finite number of times. I feel that the intrinsic theory of rationality should be prepared to deal with a given set of states of nature and a given set of decision functions, and it is the responsibility of the formal theory of belief or decision to provide a theory of how to deal with these restricted situations without introducing strong structural assumptions.

A technical way of phrasing what I am saying about axioms of pure rationality is the following. For the moment, to keep the technical apparatus simple, let us restrict ourselves to a basic set S of states of nature and a binary ordering relation of qualitative probability on subsets of S, with the usual Boolean operations of union, intersection and complementation having their intuitive meaning in terms of events. I then say that an axiom about such structures is an axiom of pure rationality only if it is closed under submodels. Technically, closure under submodels means that if the axiom is satisfied for a pair $(S, \succeq)$ then it is satisfied for any nonempty subset of S with the binary relation $\succeq$ restricted to the power set of the subset, i.e., restricted to the set of all subsets of the given subset. (Of course, the operations of union, intersection and complementation are closed in the power set of the subset.) Using this technical definition, we can easily see that of Savage's seven axioms, five of them satisfy this restriction, and the two already mentioned as structure axioms do not.

Let me try to make somewhat more explicit the intuition which is behind the requirement that axioms of pure rationality should satisfy the condition of closure under submodels. One kind of application of the condition is close to the axiom on the independence of irrelevant alternatives in the theory of choice. This axiom says that if we express a preference among candidates for office, for example, and if one candidate is removed from the list due to death or for other reasons, then our ordering of preferences among the remaining candidates should be unchanged. This axiom satisfies closure under submodels. The core idea is that existential requirements that reach out and make special requirements on the environment do not represent demands of pure rationality but rather structural demands on the environment, and such existential demands are ruled out by the condition of closure under submodels.

A different, but closely related, way of defining axioms of pure rationality is that such an axiom must be a logical consequence of the existence of the intended numerical representation. This criterion, which I shall call the *criterion of representational consequence*, can be taken as both necessary and sufficient, whereas the criterion of closure under submodels is obviously not sufficient. On the other hand, the extrinsic character of the criterion of representational consequence can be regarded as unsatisfactory. It is useful for identifying axioms that are not necessary for the intended representation and thus smuggle in some unwanted arbitrary structural assumption. As should be clear, Savage's Axioms 5 and 6 do such smuggling.

I am quite willing to grant the point that axioms of rationality of a more restricted kind could be considered. One could argue that we need special axioms of rationality for special situations, and that we should embark on a taxonomy of situations providing appropriate axioms for each of the major classes of the taxonomy. In the present primitive state of analysis, however, it seems desirable to begin with a sharp distinction between rationality and structure axioms and to have the concept of pure rationality universal in character.

Returning now to my criticisms of Savage's theory, it is easy to give finite or infinite models of Savage's five axioms of rationality for which there exists no numerical representation in terms of utility and subjective probability. In the language I am using here, Savage's axioms of pure rationality are insufficient for establishing the existence of representing numerical utility and subjective probability functions.

Moreover, we may show that no finite list of additional elementary axioms of a universal character will be sufficient to guarantee the existence of appropriate numerical functions. By *elementary axioms* I mean axioms that can be expressed within first-order logic. As was explained in Chapter 2, first-order logic essentially consists of the conceptual apparatus of sentential connectives, one level of variables and quantifiers for these variables, together with nonlogical predicates, operation symbols and individual constants. Thus, for example, the standard axioms for groups or for ordered algebraic fields are elementary, but the least upper-bound axiom for the field of real numbers is not. It is possible to formulate Savage's Axiom 5 in an elementary way, but not his Axiom 6.

In the case of infinite models, the insufficiency of elementary axioms, without restriction to their being of a universal character, follows from the upward Löwenheim-Skolem-Tarski theorem, plus some weak general assumptions. This theorem asserts that if a set of elementary axioms has an infinite model (i.e., a model whose domain is an infinite set, as is the case for Savage's theory), then it has a model of every infinite cardinality. Under quite general assumptions, e.g., on the ordering relation of preference or greater subjective probability, it is impossible to map the models of high infinite cardinality into the real numbers, and thus no numerical representation exists.

In the case of finite models, the methods of Scott and Suppes (1958), discussed in Section 2.2, apply to show that no finite set of universal elementary axioms will suffice. The system consisting of Savage's five axioms of pure rationality has finite models, but by the methods indicated we can show there is no finite elementary extension by means of universal axioms of rationality that will be strong enough to lead to the standard numerical representation.

Qualitative axioms for upper and lower probabilities. I introduce purely in terms of belief or subjective probability what I consider the appropriate finitistic analogue of Savage's axioms. These constitute an extension of de Finetti's qualitative axioms, stated in Definition 1, and lead to simple approximate measurement of belief in arbitrary events. The axioms require something that I partly criticized earlier, namely, the existence of some standard set of events whose probability is known exactly. They would, for example, be satisfied by tossing a fair coin n times for some fixed n. They do not require that n be indefinitely large and therefore n may be looked upon as more realistic. I give

5.7 THEORY OF SUBJECTIVE PROBABILITY

the axioms here in spite of my feeling that, from the standpoint of a serious decision like that on surgery mentioned earlier, they may be unsatisfactory. They do provide a combination of de Finetti's ideas and a *finite* version of the standard structural axiom on infinite partitions.

The concept of upper and lower probabilities seems to be rather recent in the literature, but it is obviously closely related to the classical concepts of inner and outer measure, which were introduced by Caratheodory and others at the end of the nineteenth century and the beginning of the twentieth century. Koopman (1940b) explicitly introduces upper and lower probabilities but does nothing with them from a conceptual standpoint. He uses them as a technical device, as in the case of upper and lower measures in mathematical analysis, to define probabilities. The first explicit conceptual discussions seem to be quite recent (Smith, 1961; Good, 1962). Smith especially enters into many of the important conceptual considerations, and Good states a number of the quantitative properties it seems natural to impose on upper and lower probabilities. Applications to problems of statistical inference are to be found in Dempster (1967). However, so far as I know, no other simple axiomatic treatment starting from purely qualitative axioms does yet exist in the literature, and the axioms given below represent such an effort. It is apparent that they are not the most general axioms possible, but they do provide a simple and rather elegant qualitative base.

From a formal standpoint, the basic structures to which the axioms apply are quadruples $(\Omega, \Im, \mathcal{S}, \succeq)$, where Ω is a nonempty set, $\Im$ is an algebra of subsets of Ω, that is, $\Im$ is a nonempty family of subsets of Ω and is closed under union and complementation, $\mathcal{S}$ is a similar algebra of sets, intuitively the events that are used for standard measurements, and I shall refer to the events in $\mathcal{S}$ as *standard* events S, T, etc. The relation $\succeq$ is the familiar ordering relation on $\Im$. I use familiar abbreviations for equivalence and strict ordering in terms of the weak ordering relation. (As has been mentioned earlier, a weak ordering is transitive and strongly connected, i.e., for any events A and B, either $A \succeq B$ or $B \succeq A$.)

DEFINITION 4. *A structure* $\mathbf{\Omega} = (\Omega, \Im, \mathcal{S}, \succeq)$ *is a finite approximate measurement structure for beliefs if and only if* Ω *is a nonempty set,* $\Im$ *and* $\mathcal{S}$ *are algebras of sets on* Ω, *and the following axioms are satisfied for every* A, B *and* C *in* $\Im$ *and every* S *and* T *in* $\mathcal{S}$:

Axiom 1. The relation $\succeq$ is a weak ordering of $\Im$;

Axiom 2. If $A \cap C = \emptyset$ and $B \cap C = \emptyset$ then $A \succeq B$ if and only if $A \cup C \succeq B \cup C$;

Axiom 3. $A \succeq \emptyset$;

Axiom 4. $\Omega \succ \emptyset$;

Axiom 5. $\mathcal{S}$ is a finite subset of $\Im$;

Axiom 6. If $S \neq \emptyset$ then $S \succ \emptyset$;

Axiom 7. If $S \succeq T$ then there is a V in $\mathcal{S}$ such that $S \approx T \cup V$.

In comparing Axioms 3 and 6, note that A is an arbitrary element of the general algebra $\Im$, but event S (referred to in Axiom 6) is an arbitrary element of the subalgebra $\mathcal{S}$. Also in Axiom 7, S and T are standard events in the subalgebra $\mathcal{S}$, not arbitrary events in the general algebra. Axioms 1–4 are just the familiar de Finetti axioms without any change. Because all the standard events (finite in number) are also events (Axiom 5), Axioms 1–4 hold for standard events as well as arbitrary events. Axiom 6 guarantees that every minimal element of the subalgebra $\mathcal{S}$ has positive qualitative probability. Technically a minimal element of $\mathcal{S}$ is any event A in $\mathcal{S}$ such that $A \neq \emptyset$, and it is not the case that there is a nonempty B in $\mathcal{S}$ such that B is a proper subset of A. A *minimal open interval* (S, S') of $\mathcal{S}$ is such that $S \prec S'$ and $S' - S$ is equivalent to a minimal element of $\mathcal{S}$. Axiom 7 is the main structural axiom, which holds only for the subalgebra and not for the general algebra; it formulates an extremely simple solvability condition for standard events. It was stated in this form in Suppes (1969a, p. 6) but in this earlier case for the general algebra $\mathfrak{F}$.

In stating the representation and uniqueness theorem for structures satisfying Definition 4, in addition to an ordinary probability measure on the standard events, I shall use upper and lower probabilities to express the inexact measurement of arbitrary events. A good discussion of the quantitative properties one expects of such upper and lower probabilities is found in Good (1962). All of his properties are not needed here because he dealt with conditional probabilities. The following properties are fundamental, where $P_*(A)$ is the lower probability of an event A and $P^*(A)$ is the upper probability (for every A and B in $\mathfrak{F}$):

I. $P_*(A) \geq 0$.

II. $P_*(\Omega) = P^*(\Omega) = 1$.

III. If $A \cap B = \emptyset$ then $P_*(A) + P_*(B) \leq P_*(A \cup B) \leq P_*(A) + P^*(B) \leq P^*(A \cup B) \leq P^*(A) + P^*(B)$.

Condition (I) corresponds to Good's Axiom D2 and (III) to his Axiom D3.

For standard events $P(S) = P_*(S) = P^*(S)$. For an arbitrary event A not equivalent in qualitative probability to a standard event, I think of its "true" probability as lying in the open interval $(P_*(A), P^*(A))$.

Originally I included as a fourth property

$$P_*(A) + P^*(\neg A) = 1,$$

where $\neg A$ is the complement of A, but Mario Zanotti pointed out to me that this property follows from (II) and (III) by the following argument:

$$1 = P_*(\Omega) = P_*(A \cup \neg A) \leq P_*(A) + P^*(\neg A) \leq P^*(A \cup \neg A) = P^*(\Omega) = 1.$$

In the fourth part of Theorem 7, I define a certain relation and state it is a semiorder with an implication from the semiorder relation holding to an inequality for upper and lower probabilities. Semiorders have been fairly widely discussed in the literature as a generalization of simple orders, first introduced by Duncan Luce. I use here the axioms given by Scott and Suppes (1958). A structure $(A, *\!\succ)$ where A is a nonempty set and $*\!\succ$ is a binary relation on A is a *semiorder* if and only if for all $a, b, c, d \in A$:

Axiom 1. Not $a *\!\succ a$;

Axiom 2. If $a *\!\succ b$ and $c *\!\succ d$ then either $a *\!\succ d$ or $c *\!\succ b$;

5.7 THEORY OF SUBJECTIVE PROBABILITY

Axiom 3. If $a \mathbin{*\!\succ} b$ and $b \mathbin{*\!\succ} c$ then either $a \mathbin{*\!\succ} d$ or $d \mathbin{*\!\succ} c$.[48]

THEOREM 7. *Let $\Omega = (\Omega, \Im, \mathcal{S}, \succeq)$ be a finite approximate measurement structure for beliefs. Then*

(i) *there exists a probability measure P on $\mathcal{S}$ such that for any two standard events S and T*
$$S \succeq T \text{ if and only if } P(S) \geq P(T),$$

(ii) *the measure P is unique and assigns the same positive probability to each minimal event of $\mathcal{S}$,*

(iii) *if we define P_* and P^* as follows:*

 (a) *for any event A in $\Im$ equivalent to some standard event S,*
 $$P_*(A) = P^*(A) = P(S),$$

 (b) *for any A in $\Im$ not equivalent to some standard event S, but lying in the minimal open interval (S, S') for standard events S and S'*
 $$P_*(A) = P(S) \text{ and } P^*(A) = P(S'),$$

 then P_* and P^* satisfy conditions (I)–(III) for upper and lower probabilities on $\Im$, and

 (c) *if n is the number of minimal elements in $\mathcal{S}$ then for every A in $\Im$*
 $$P^*(A) - P_*(A) \leq 1/n,$$

(iv) *if we define for A and B in $\Im$*
$$A \mathbin{*\!\succ} B \text{ if and only if } \exists S \text{ in } \mathcal{S} \text{ such that } A \succ S \succ B,$$

then $\mathbin{*\!\succ}$ is a semiorder on $\Im$, if $A \mathbin{*\!\succ} B$ then $P_*(A) \geq P^*(B)$, and if $P_*(A) \geq P^*(B)$ then $A \succeq B$.

Proof. Parts (i) and (ii) follow from the proof given in Suppes (1969a, pp. 7–8) once it is observed that the subalgebra $\mathcal{S}$ is isomorphic to a finite algebra of sets with the minimal events of $\mathcal{S}$ corresponding to unit sets, i.e., atomic events.

As to part (iii), conditions (I) and (II) for upper and lower probabilities are verified immediately. To verify condition (III) it will be sufficient to assume that neither A nor B is equivalent to a standard event, for if either is, the argument given here is simplified, and if both are, (III) follows at once from properties of the standard measure P. So we may assume that A is in a minimal interval (S, S') and B in a minimal interval (T, T'), i.e., $S \prec A \prec S'$ and $T \prec B \prec T'$. Since by hypothesis of (III), $A \cap B = \emptyset$, $T \preceq \neg S$ for if $T \succ \neg S$, we would have $A \cup B \succ S \cup \neg S$, which is impossible. Now it is easily checked that for standard events if $T \preceq \neg S$ then $\exists T^*$ in $\mathcal{S}$ such that $T^* \approx T$ and $T^* \subseteq \neg S$. So we have
$$P_*(A) + P_*(B) \leq P(S) + P(T^*) = P(S \cup T^*) \leq P_*(A \cup B),$$
with the last inequality following from $S \cup T^* \prec A \cup B$, which is itself a direct consequence of $S \prec A, T^* \prec B, A \cap B = \emptyset$ and Axiom 2. For the next step, if $\exists T^{**}$ in $\mathcal{S}$ such that

---

[48] In the article cited it is shown that, when A is a finite set, then the following representation theorem can be proved for semiorders. There is a numerical function φ defined on A such that for all a and b in A
$$\varphi(a) > \varphi(b) + 1 \text{ iff } a \mathbin{*\!\succ} b.$$
However, this representation does not necessarily hold if A is an infinite set, so it cannot be used in clause (iv) of Theorem 7 below.

$T^{**} \approx T'$ and $T^{**} \subseteq \neg S'$, then $A \cup B \prec S' \cup T^{**}$ and let $A \cup B$ be in the minimal closed interval $[V, V']$, i.e., $V \preceq A \cup B \preceq V'$. Then it is easy to show that $V \preceq S \cup T^{**}$, whence

$$P_*(A \cup B) = P(V) \leq P(S \cup T^{**}) = P(S) + P(T^{**}) = P_*(A) + P^*(B)$$

and since $S \cup T^* \prec A \cup B$, and $V \preceq S \cup T^{**}$, either $A \cup B \preceq S \cup T^{**}$ or $A \cup B \preceq S' \cup T^{**}$. In either case

$$\begin{aligned} P_*(A) + P^*(B) &= P(S \cup T^{**}) \leq P^*(A \cup B) \leq P(S' \cup T^{**}) \\ &= P(S') + P(T^{**}) = P^*(A) + P^*(B). \end{aligned}$$

On the other hand, if there were no T^{**} such that $T^{**} \approx T'$ and $T^{**} \subseteq \neg S'$, then $T' \succ \neg S'$, so that $S \cup T^* = S \cup \neg S$, and consequently $A \cup B \approx S \cup T^*$, so that $A \succeq S$ or $B \succeq T^*$ contrary to hypothesis, which completes the proof of (III).

Proof of (c) of part (iii) follows at once from (ii) and the earlier parts of (iii). Proof of (iv) is also straightforward and will be omitted.

In my opening remarks I mentioned the embarrassing problem of being asked for the next decimal of a subjective probability. Without claiming to have met all such problems, the results embodied in Theorem 7 show that the axioms of Definition 4 provide a basis for a better answer. If there are n minimal standard events, then the probabilities of the 2^n standard events are known exactly as rational numbers of the form m/n, with $0 \leq m \leq n$, and further questions about precision are mistaken. The upper and lower probabilities of all other events are defined in terms of the probabilities of the 2^n standard events, and so the upper and lower probabilities are also known exactly as rational numbers of the same form m/n.

Finally, I note explicitly that there is no need in Definition 4 to require that the sample space Ω be finite. The only essential requirement is that the set $\mathcal{S}$ of standard events be finite. The algebra $\mathfrak{I}$ could even have a cardinality greater than that of the continuum and thus the order relation $\succeq$ on $\mathfrak{I}$ might not be representable numerically, and yet the upper and lower probabilities for all events in $\mathfrak{I}$ would exist and be defined as in the theorem.

5.8 Epilogue: Pragmatism about Probability

This long survey of the main conceptual ways of representing probability might misleadingly suggest that when scientists use probability concepts, they choose their favorite representation to serve as a foundation. As in the case of most other foundational issues of a similar character, the scientific practice is rather different. Perhaps the two subjects in which we most naturally might expect to find a foundational commitment to some one representation of probability are statistical mechanics and quantum mechanics, but this is far from being the case. I will not try to document in detail what I think the literature since Maxwell shows for statistical mechanics or the literature since Heisenberg and Dirac show for quantum mechanics. I will just give a few illustrations.

Early statistical mechanics. In the early days, for instance, Maxwell's famous papers of 1860 and 1867 on the dynamical theory of gases, there was a variety of probability calculations, but no mention of how probability should be interpreted. From 1860 until the end of the century there evolved an ever more sophisticated analysis on

5.8 Epilogue: Pragmatism about Probability

how probability concepts could be used in statistical physics. Such well-known figures as Boltzmann, Gibbs, Loschmidt, Kirchhoff, Planck, Poincaré and Zermelo argued in great detail about the foundations of statistical physics, but not about the foundations of probability. A good discussion of these matters and extensive references are to be found in Brush (1976).

Quantum mechanics. An even more austere attitude toward probability dominates the early papers on quantum mechanics. The famous 1925 paper by Heisenberg and the two papers published early in 1926 by Dirac do not mention probability at all. (Reprints of these and other early papers on quantum mechanics are to be found in van der Waerden (1968).) In his well-known treatise on quantum mechanics, Dirac (1947, pp. 47–48) has a clear discussion of computing mean values of observables and the probability of an observable having a given value, but not a word is said about how probability is to be interpreted.

Fock. A philosophically deeper and more interesting discussion is to be found in the text by the well-known Russian quantum physicist V. A. Fock (1931/1978), but his analysis does not fit into the various major interpretations of probability discussed earlier in this chapter. He emphasizes in a careful way two points. The first is that probability in quantum mechanics is an expression of potential possibilities, and the second is that the probabilistic difficulties of understanding the Heisenberg uncertainty relations disappear if we fully admit the dual wave-corpuscular nature of the electron (p. 93). This brief summary does not give a rich or even fair summary of what Fock has to say, so let me give some extensive quotations, which may not be well known to some philosophers of science. Here is a good passage about potential possibility:

> If we take the act of interaction between an atomic object and a measuring instrument as the source of our judgements about the object's properties and if in studying phenomena we allow for the concept of relativity with respect to the means of observation, we are introducing a substantially new element into the description of the atomic object and its state and behaviour, that is, the idea of probability and thereby the idea of potential possibility. The need to consider the concept of probability as a substantial element of description rather than a sign of incompleteness of our knowledge follows from the fact that for given external conditions the result of the object's interaction with the instrument is not, generally speaking, predetermined uniquely but only has a certain probability of occurring. With a fixed initial state of the object and with given external conditions a series of such interactions results in a statistics that corresponds to a certain probability distribution. This probability distribution reflects the potential possibilities that exist in the given conditions.
>
> (Fock 1931/1978, p. 19)

And here is a good passage on statistical ensembles and the uncertainty relations:

> In the first years of development of quantum mechanics, in the early attempts to find a *statistical (probabilistic) interpretation,* physicists were still bound by the notion of the electron being a classical mass point. Even when de Broglie's idea on the wave nature of matter emerged, waves of matter were at times interpreted as something that carries the

mass points. Later, when Heisenberg's relations appeared, they were interpreted as *inaccuracy relations,* and not as uncertainty relations. For instance, it was thought that the electron had definite position and velocity but that there was no possibility of determining either. The square of the modulus of the wave function was interpreted as probability density for a particle—irrespective of the conditions of the actual experiment—to have given coordinates (the coordinates were thought of as definite). A similar interpretation was given to the square of the modulus of the wave function in momentum space. Both probabilities (in ordinary space and in momentum space) were considered simultaneously as the probability of a certain compound event, specifically that the particle has definite values of coordinates and momemtum. The actual impossibility, expressed by Heisenberg's relations, of their simultaneous measurement therefore appeared as a paradox or caprice of nature, according to which not everything existing is cognizable.

All these difficulties vanish if we fully admit the dual wave-corpuscular nature of the electron, establish its essence, and grasp what the quantum mechanical probabilities refer to and what statistical ensemble they belong to.

First, let us try to give a general definition of a *statistical ensemble.* We assume an unlimited set of elements having various features, which make it possible to sort these elements and to observe the frequency of occurrence of an element with a given feature. If for this there exists a definite probability (that is, for each element of the set), the set constitutes a statistical ensemble.

In quantum mechanics, as in classical physics, the only sets that can be considered are those whose elements have definite values of the parameters (features) according to which sorting can be done. This implies that the elements of a statistical ensemble must be described in a classical language, and that a quantum object cannot be an element of a statistical ensemble even if a wave function can be ascribed to the object.

The elements of statistical ensembles considered in quantum mechanics are not the microobjects themselves but the results of experiments with them, a definite experimental arrangement corresponding to a definite ensemble. These results are described classically and thus can serve as a basis for sorting the elements of the ensemble. Since for different quantities the probability distribution arising from a given wave function correspond to different experimental arrangements, they belong to different ensembles.

(Fock 1931/1978, pp. 93–94)

Slightly later Fock does say this about relative frequency.

The probability of this or that behaviour of an object with a given initial state is determined by the internal properties of the object and by the nature of the external conditions; it is a number characterizing the potential possibilities of this or that behaviour of the object. And the probability manifests itself in the frequency of occurrence of a given behaviour of the object; the relative frequency is its numerical measure. The probability thus belongs, in essence, to the individual object (and not to an ensemble of objects) and characterizes its potential possibilities. At the same time, to determine its numerical value from experiment one must have the statistics of the realization of these possibilities, so that the experiment must be repeated many times. It is clear from this that the probabilistic character of the quantum theory does not exclude the fact that the theory is based on the properties of an individual object.

5.8 Epilogue: Pragmatism about Probability

> To summarize we can say that the purpose of the main concept of quantum mechanics, the concept of a state described by a wave function, is to analyze objectively all potential possibilities inherent in the micro-object. This determines the probabilistic nature of the theory.
> (Fock 1931/1978, pp. 94–95)

In this last passage Fock makes passing reference to relative frequency, but his manner of dealing with the concept is standard and superficial, in contrast to his probing remarks on other matters. In contrast, let us look at the statements of two distinguished mathematicians whose work on the foundations of quantum mechanics is well-known.

Weyl. First Weyl (1928/1931) has this to say about relative frequencies. It is meant to hold for all experimental science, not just quantum mechanics or even physics.

> The significance of probabilities for experimental science is that they determine *the relative frequency of occurrence in a series of repeated observations*. According to classical physics it is in principle possible to create conditions under which every quantity associated with a given physical system assumes an arbitrarily sharply defined value which is exactly reproducible whenever these conditions are the same. *Quantum physics denies this possibility.*
> (Weyl 1928/1931, p. 75)

Weyl continues on the next page with a markedly Kantian remark about natural science.

> Natural science is of a constructive character. The concepts with which it deals are not qualities or attributes which can be obtained from the objective world by direct cognition. They can only be determined by an indirect methodology, by observing their reaction with other bodies, and their implicit definition is consequently conditioned by definite laws of nature governing reactions. Consider, for example, the introduction of the Galilean concept of mass, which essentially amounts to the following indirect definition: "Every body possesses a momentum, that is, a vector mv having the same direction as its velocity v; the scalar factor m is called its mass. The momentum of a closed system is conserved, that is, the sum of the momenta of a number of reacting bodies is the same before the reaction as after it." On applying this law to the observed collision phenomena data are obtainable which allow a determination of the relative masses of the various bodies. But scientists have long held the opinion that such *constructive concepts were nevertheless intrinsic attributes of the "Ding an sich"*, even when the manipulations necessary for their determination were not carried out. *In quantum theory we are confronted with a fundamental limitation to this metaphysical standpoint.* (Weyl 1928/1931, p. 76)

Finally, I include a quote from Weyl that is very much in the spirit of relative-frequency problems with the reference class, but made just in the context of the problem of homogeneity in quantum mechanics.

> In general the conditions under which an experiment is performed will not even guarantee that all the individuals constituting the system under observation are in the same "state", as represented in the quantum theory by a ray in system space. This is, for example, the case when we only take care that all the atoms are in the quantum state (n, l) without undertaking to separate them, with respect to m by means of the *Stern-Gerlach* effect. In order to apply quantum mechanics it is therefore necessary to set up a criterion which will enable us to determine whether the given conditions are sufficient to insure such a "*pure state*." We say that the conditions $\Im'$ effect a greater homogeneity than the conditions $\Im$

if (1) every quantity which has a sharp, reproducible value under $\mathfrak{S}$ has the same definite value under $\mathfrak{S}'$ and if (2) there exists a quantity which is strictly determinate under $\mathfrak{S}'$ but not under $\mathfrak{S}$. The desired criterion is obviously this: *The conditions $\mathfrak{S}$ guarantee a pure state if it is impossible to produce a further increase in homogeneity.* (This maximum of homogeneity was obtained in classical physics only when all quantities associated with the system had definite values.) (Weyl 1928/1931, pp. 77–78)

von Neumann. Even today in many ways the most detailed discussion of probability in quantum mechanics is to be found in von Neumann's (1932/1955) classic treatise. Most of the analysis of a probabilistic character focuses on two problems. The first is the contrast between causal and noncausal processes. The time evolution of a state according to the time-dependent Schrödinger equation is deterministic and causal in character. On the other hand, the measurement of a system in a state with a discrete spectrum can undergo a change to any one of several possible states. Because of its probabilistic nature this measurement process in quantum mechanics is noncausal. As is evident, in making this distinction, von Neumann is using the strictly deterministic causal stance of classical physics. The second problem he addresses is closely related to this one, namely, the possibility of deterministic hidden variables that will convert the probabilistic measurement process into a classical deterministic process. He gives a proof, now regarded as not quite satisfactory, that there can be no such hidden variables. A survey of more recent results on hidden variables for quantum mechanics is given in Chapter 7, and so will not be pursued here.

A third problem that has attracted more attention in statistical mechanics than in quantum mechanics is that of the ergodic character of quantum processes, to which von Neumann contributed but which he only mentions in the book on quantum mechanics, probably because he did not want to add another layer of concepts to present ergodic ideas in any detail. An elementary introduction to ergodic processes is given here in the last section of Chapter 4 on invariance, but applications to quantum mechanics are not touched upon.

In his final chapter, which is on the measurement process, von Neumann has one of the most famous passages on the subjectivity of observation in physics to be found anywhere in the large literature on the foundations of quantum mechanics. I quote a substantial part of it.

> First, it is inherently entirely correct that the measurement or the related process of the subjective perception is a new entity relative to the physical environment and is not reducible to the latter. Indeed, subjective perception leads us into the intellectual inner life of the individual, which is extra-observational by its very nature (since it must be taken for granted by any conceivable observation or experiment). (Cf. the discussion above.) Nevertheless, it is a fundamental requirement of the scientific viewpoint—the so-called principle of the psycho-physical parallelism—that it must be possible so to describe the extra-physical process of the subjective perception as if it were in reality in the physical world—i.e., to assign to its parts equivalent physical processes in the objective environment, in ordinary space. (Of course, in this correlating procedure there arises the frequent necessity of localizing some of these processes at points which lie within the portion of space occupied by our own bodies. But this does not alter the

5.8 EPILOGUE: PRAGMATISM ABOUT PROBABILITY

fact of their belonging to the "world about us," the objective environment referred to above.) In a simple example, these concepts might be applied about as follows: We wish to measure a temperature. If we want, we can pursue this process numerically until we have the temperature of the environment of the mercury container of the thermometer, and then say: this temperature is measured by the thermometer. But we can carry the calculation further, and from the properties of the mercury, which can be explained in kinetic and molecular terms, we can calculate its heating, expansion, and the resultant length of the mercury column, and then say: this length is seen by the observer. Going still further, and taking the light source into consideration, we could find out the reflection of the light quanta on the opaque mercury column, and the path of the remaining light quanta into the eye of the observer, their refraction in the eye lens, and the formation of an image on the retina, and then we would say: this image is registered by the retina of the observer. And were our physiological knowledge more precise than it is today, we could go still further, tracing the chemical reactions which produce the impression of this image on the retina, in the optic nerve tract and in the brain, and then in the end say: these chemical changes of his brain cells are perceived by the observer. But in any case, no matter how far we calculate—to the mercury vessel, to the scale of the thermometer, to the retina, or into the brain, at some time we must say: and this is perceived by the observer. That is, we must always divide the world into two parts, the one being the observed system, the other the observer. In the former, we can follow up all physical processes (in principle at least) arbitrarily precisely. In the latter, this is meaningless. The boundary between the two is arbitrary to a very large extent. In particular we saw in the four different possibilities in the example above, that the observer in this sense needs not to become identified with the body of the actual observer: In one instance in the above example, we included even the thermometer in it, while in another instance, even the eyes and optic nerve tract were not included. That this boundary can be pushed arbitrarily deeply into the interior of the body of the actual observer is the content of the principle of the psycho-physical parallelism—but this does not change the fact that in each method of description the boundary must be put somewhere, if the method is not to proceed vacuously, i.e., if a comparison with experiment is to be possible. Indeed experience only makes statements of this type: an observer has made a certain subjective observation; and never any like this: a physical quantity has a certain value.

(von Neumann 1932/1955, pp. 418–420)

It is obvious that the emphasis by von Neumann on subjective perception resonates with the basic ideas of subjective probability, although the wider literature on subjective probability is not discussed. Secondly, his undeveloped remarks about psychophysical parallelism suggest a more radical subjectivism than is characteristic of typical Bayesian views of probability.

Pragmatism in physics. This survey of the attitude toward probability in physics could easily be extended. But pragmatism, in the sense of using the formal methods of probability, without much attention to foundational questions of interpretation or representation of probability, wholly dominates the use of probability in physics. There is wide agreement on one point by everyone, including von Neumann. That is the idea of confirming or testing quantum mechanics by the relative frequency of observed phenomena in repeated experiments of the same design. The statement quoted above from p. 75

of Fock's work would be endorsed by almost every physicist. But this statement about experiments would be acceptable to almost every view of the interpretation of probability, as well, from the most extreme Bayesian to the most extreme propensity view. Moreover, the characterization of randomness, the bête noire of the relative-frequency interpretation of probability, is seldom if ever mentioned in the physics literature. More generally, in a recent article (Suppes 1998) I tried to make a broader case for the pragmatic exuberance of most physics, now and in the past.[49]

Statistical practice. The previous paragraphs have emphasized the pragmatic character of the attitude toward the nature of probability in physics. Surprisingly enough, in view of the many conceptual disputes between Bayesians and non-Bayesians within statistics itself, a much more explicit attitude about pragmatism in statistical practice can be found among statisticians concerned with detailed problems of some intricacy. I shall not try to survey the literature on this point, but rather quote from one splendid example, far from physics and quantum mechanics, but from a statistical standpoint more sophisticated than almost any quantum mechanical example I can think of.

The example I have in mind is the statistical analysis of the *Federalist Papers* to determine whether the author of certain disputed parts was Hamilton or Madison. The *Papers* constitute one of the most famous series of political documents written about the time of the American Revolution. They are still much read and the disputed authorship is a continuing problem of historical interest. The wonderfully detailed statistical study by Mosteller and Wallace (1964/1984) settles the question about as definitively as it ever will be, even with continued improvements in statistical and computational methods in the future. Here is the skeptical and pragmatic attitude they express in the preface to the first edition.

> Some may feel that we should make more definite pronouncements about the comparative value of different methods of inference—especially Bayesian versus classical. Obviously we are favorably disposed toward the Bayesian approach, but in the field of data analysis at least, the techniques are in their infancy. As these techniques evolve, their merits and demerits may be easier to judge.
>
> Even though individual statisticians may claim generally to follow the Bayesian school or the classical school, no one has an adequate rule for deciding what school is being represented at a given moment. When we have thought we were at our most Bayesian, classicists have told us that we were utterly classical; and when we have thought ourselves to be giving classical treatment, Bayesians have told us that the ideas are not in the classical lexicon. So we cannot pretend to speak for anyone but ourselves.
>
> (Mosteller and Wallace 1964/1984, p. ix)

---

[49]That such a pragmatic attitude dominates physics and should dominate economics is well-argued for in Cartwright (1999). On the other hand, the excellent book of T. L. Fine (1973), which covers, from a different angle, many of the sections of this chapter, is focused on the continued need for deeper satisfactory representations of probability. The conflicts in viewpoint represented by the pragmatism of physicists and Fine's plea for greater and more specific depth of interpretation and practice are central to the philosophical foundations of scientific methods. For reasons that cannot be spelled out in detail here, I predict that the physicists' minimalist program of interpretation will continue to dominate the use of probability concepts in scientific theories, which is why I have included this final section of an already overly long chapter.

5.8 Epilogue: Pragmatism about Probability

Bayesian concepts and methods have been extensively developed since Mosteller and Wallace first published this passage in 1964, but Bayesian methods are in fact still used relatively rarely in scientific papers, in spite of such excellent recent texts as Rouanet et al. (1998). But this is not because scientists are devoted to an objective rather than subjective view of probability—the "classical school" of statisticians, in Mosteller and Wallace's terminology, holding the objective view of probability. No, it is because by and large scientists in most disciplines remain indifferent to the conceptual foundations of probability and pragmatically apply statistical concepts without any foundational anxiety, just computing the statistical tests someone has taught them to use. But indifference, on the other hand, sometimes disappears when statistical analysis is required of some process or phenomenon with strong practical implications.

Other forces of a purely scientific nature can also create a change in attitude, which is most likely to occur throughout this century. Statistics itself is now in the midst of a revolution much more far-reaching in its implications than the conflict between objective and subjective views of probability that dominated the last half of the twentieth century. This revolution has been started by the demand for new statistical methods for analyzing very large data bases, now prevalent in almost every area of science. The new methods already developed are intensely computational in nature, and will, no doubt, continue to be so. There is much about this new context and focus that is more like that of physics than the scientific sources of data associated with the preceding fifty years of statistics. It is not unreasonable to believe that the pragmatic attitude so well exemplified by physicists, and also by some of the best statistical practice, will be the dominant one toward the nature of probability for quite some time. From a foundational standpoint, I cannot help observing how much the qualitative axiomatic methods used in Section 6 for analyzing objective propensity interpretations and those used in Section 7 for subjective interpretations of probability are essentially the same, and, in retrospect, I see could be improved by more unification of the formal results.

This leads to the kind of pragmatism I am happy to endorse. Uniform formal methods are accompanied by a rich variety of interpretations or representations of the concepts used. Probability is too rich and diversified in application to be restricted to one single overweening representation. The pragmatic context of use will sometimes fix the representation chosen, but more often, a deeper pragmatic attitude will dominate and no explicit choice of representation will be made, as is well illustrated statistically in the work of Mosteller and Wallace (1964/1984) referred to earlier, and more recently, in an avalanche of papers on data mining, adaptive statistics, boosting, neural networks and a variety of other approaches to large-scale data analysis.[50]

---

[50] For a good not-too-technical overview of these recent developments, see Hastie, Tibshirani and Friedman (2001).

6

Representations of Space and Time

Well before Aristotle's famous extended and complicated argument that the world is eternal, without beginning or end, and there cannot be more than one world (*On the Heavens,* Book I), varying representations of space and time were a central topic in Greek philosophy. This emphasis continued for many centuries. Perhaps the last major philosopher to consider representations of space and time as central to his philosophy in the detailed context of the physics of his time was Kant in the *Critique of Pure Reason* and the *Metaphysical Foundations of Natural Science.* In any case, detailed conceptual analysis of space and time is no longer a central topic for most philosophers, but a flourishing activity for those with specialized knowledge and interest in the foundations of physics. The focus of the first three sections of this chapter is of this latter sort. The first section states some necessary geometric preliminaries. The second one assumes a standard analytic geometric representation used in classical physics and concentrates on studying the invariant properties of this representation, but only kinematical properties of classical mechanics are needed to prove the invariance. The third section provides a similar analysis of the space-time of special relativity. The analysis reflects the importance of questions of invariance in classical and relativistic physics; some fairly extensive historical remarks are made at the end of both sections. The strong simplifying assumption used, and not built up from more primitive physical concepts, is that both classical and restricted relativistic space-time structures are special cases of affine structures, which are defined in the first section.

This affine unification has several justifications. First, many properties of both the classical and the space-time of special relativity are invariant under the general affine group of transformations—for example, the linearity of inertial paths and conservation of their being parallel. More importantly, the group of Galilean transformations of classical physics and the group of Lorentz transformations of special relativity are both subgroups of the affine group. These common features are often not given much notice in philosophical discussions of space and time.

In the fourth section, the focus moves from physics to psychology, in particular, to the question of how we can decide if visual space is Euclidean. Because there has been much less consideration of spatial perceptual phenomena by philosophers, the treatment is much more detailed. As might be expected, the results are less definitive in character. Continuing in this line of thought, the fifth section examines a variety of philosophical and experimental claims about the nature of visual space. The psychological experiments of Foley and Wagner, discussed in some detail in this section, provide, in Section 6, the framework for some restricted axioms of congruence. These axioms extend the purely affine axioms of Definition 1, but avoid the paradoxes for any space of constant curvature, not just the Euclidean one, posed by the Foley and Wagner experiments. The seventh section concludes the examination of visual space with reflection on three major problems for any spatial theory of visual phenomena.

In the last section, the eighth one, I turn briefly to a related topic in the philosophy of mathematics, finitism in geometry. Conceptually what is considered is the finitary nature of geometric constructions throughout a very long tradition going back to the earliest times of ancient Greek geometry, and the kind of problems that can be solved by such methods.

As can be seen, various sections of this chapter move from physics to psychology and finally to geometry itself. This seems like a bit of a mixture and it is. The second and third sections on invariance in physics could have been placed at the beginning of the next chapter, Chapter 7, on representations in mechanics. But I kept them here for three reasons. First, no dynamical assumptions are used formally in these two sections, so the use of mechanical ideas is pretty thin. Second, the derivation of the Lorentz transformations is more relevant to electromagnetic theory, especially historically, than to mechanics. Third, I wanted to emphasize in this chapter the many different ways of thinking about the representations of space and time. Different disciplines concerned with spatial and temporal phenomena pose problems that require different approaches.

6.1 Geometric Preliminaries

The viewpoint toward space-time developed in this chapter is that standard geometric concepts and objects shall be assumed and not developed in detail. What is needed in fact are the concepts of an affine and of a Euclidean space, and the geometric relations such as betweenness in terms of which such spaces are constructed.

To keep the initial focus on the special axioms needed for classical or Minkowski space-time, standard and familiar axioms for affine and Euclidean spaces will be reviewed only briefly. The representation and invariance theorems will be formulated for later use.

Ordered affine spaces are axiomatized, in one standard version, in terms of a nonempty set A of points and the ternary relation B of betweenness. (I also used the standard brief notation $a|b|c$ for $B(a, b, c)$.) In the coordinate representation of such spaces, points are ordered n-tuples of real numbers forming the standard n-dimensional Cartesian space R^n. Let $\mathbf{x}, \mathbf{y}$, and $\mathbf{z}$ be any three such Cartesian points. Then, in the intended interpretation, the Cartesian relation of betweenness $B_R(\mathbf{x}, \mathbf{y}, \mathbf{z})$ holds iff there exists an $\alpha \in [0, 1]$ such that

$$\mathbf{y} = \alpha \mathbf{x} + (1 - \alpha)\mathbf{z}.$$

The structure (R^n, B_R) is the n-dimensional affine Cartesian space.

We need the following definitions. For $a \neq b$, a *line ab* is the set of all points c such that $a|b|c$, $b|c|a$ or $c|a|b$, and abc is a *triangle* if the points a, b and c are noncollinear, i.e., do not lie on a line. (Note that here a triangle is just an ordered triple of three noncollinear points. With some abuse of notation, the same notation is used for the plane consisting of the set of points defined by such a triple of points.)

Let a, b, c be noncollinear; then the plane abc is defined as follows:

$$abc = \{d \mid (\exists e)(\exists f)(e \neq f \text{ and } e, f \in ab \cup ac \cup bc \\ \text{ and } d, e, f \text{ are collinear})\}.$$

(By requiring $e|d|f$ rather than just collinearity, we define the triangular region abc.) For 3-spaces, we use a *tetrahedron* $abcd$ with the four points noncoplanar. Following the usual terminology, a, b, c, and d are the *vertices* of the tetrahedron, the segments ab, bc, ac, ad, bd, and cd are the *edges,* and the four triangular regions abc, bcd, cda, and dab are the *faces*. The 3-space $abcd$ is then the set of all points collinear with pairs of points in one or two faces of the tetrahedron $abcd$.

We proceed in the same way for the 4-space, a concept that we need for the spaces of classical and special relativity. A quintuple $abcde$ is a *simplex* if and only if the five points do not all lie in a common 3-space. The *cells* of the simplex are the five tetrahedral regions determined by four points of the simplex. The 4-space $abcde$ is then the set of points collinear with pairs of points on one or two cells of the simplex.

DEFINITION 1. *A structure* $\mathfrak{A} = (A, B)$ *is an* affine space *if and only if the following axioms are satisfied for* $a, b, c, d, e, f, g, a', b'$ *and* c' *in A:*

1. *If* $a|b|a$, *then* $a = b$.
2. *If* $a|b|c$, *then* $c|b|a$.
3. *If* $a|b|c$ *and* $b|d|c$, *then* $a|b|d$.
4. *If* $a|b|c$ *and* $b|c|d$ *and* $b \neq c$, *then* $a|b|d$.
5. (*Connectivity*) *If* $a|b|c$, $a|b|d$, *and* $a \neq b$, *then* $b|c|d$ *or* $b|d|c$.
6. (*Extension*) *There exists* f *in* A *such that* $f \neq b$ *and* $a|b|f$.
7. (*Pasch's Axiom*) *If* abc *is a triangle*, $b|c|d$, *and* $c|e|a$, *then there is on line* de *a point* f *such that* $a|f|b$.
8. (*Axiom of Completeness*) *For every partition of a line into two nonempty sets Y and Z such that*

    (i) *no point b of Y lies between any a and c of Z and*

    (ii) *no point b' of Z lies between any a' and c' of Y,*

    there is a point b of $Y \cup Z$ such that for every a in Y and c in Z, b lies between a and c.

9. (*Dimensionality*). *Moreover, if all points of $\mathfrak{A}$ lie on a plane, but not on a line, $\mathfrak{A}$ is* two-dimensional; *if all points of $\mathfrak{A}$ lie in a 3-space but not on a plane, $\mathfrak{A}$ is* three-dimensional; *if all points of $\mathfrak{A}$ lie in a 4-space but not in a 3-space, $\mathfrak{A}$ is* four-dimensional.[1]

The formulation just in terms of betweenness is rather complicated, and so I have avoided it by using the concepts defined before Definition 1.[2]

---

[1] Axiom 9 is in the form of a definition, but it gives the forms of axioms needed to specify the dimensionality. For example, if someone wants to focus on two-dimensional affine space, the axiom of dimensionality is: all points of $\mathfrak{A}$ lie on a plane, but not on a line.

[2] I remark that an entirely different, but equivalent, approach to affine geometry is to begin with a projective space, say three-dimensional as an example, and then remove a plane, labeled the 'plane

THEOREM 1. *Every n-dimensional affine space (A, B) for $n = 2, 3, 4$, is isomorphic to the n-dimensional affine Cartesian space (R^n, B_R), i.e., there is a one-one function f mapping A onto R^n such that for any points a, b and c in A*

$$B(a, b, c) \text{ iff } B_R(f(a), f(b), f(c)).$$

Moreover, any two n-dimensional affine spaces are isomorphic.

In other words, the theorem asserts the categorical character of affine spaces of a given dimension. (The elementary axioms for finite affine constructions introduced later in Section 8 are noncategorical.)

Of equal importance, although neglected by some, is the companion theorem on the invariance or uniqueness of the representation. In standard geometric terms, the invariance is simply uniqueness up to the group of affine transformations. An *affine transformation* is any linear transformation φ of n-dimensional Cartesian space such that for every vector $\mathbf{x}$[3]

$$\varphi(\mathbf{x}) = \mathbf{Ax} + \mathbf{b},$$

where $\mathbf{A}$ is a nonsingular $n \times n$ matrix and $\mathbf{b}$ is any n-dimensional vector, that is, any n-tuple of real numbers. Intuitively, the matrix $\mathbf{A}$ represents a rotation together with stretches in different amounts along the n Cartesian coordinates. The vector $\mathbf{b}$ represents a translation of the origin of the Cartesian space. Two familiar facts about such transformations are these: they preserve collinearity, and they transform parallel lines into parallel lines. It is also apparent that the set of affine transformations is a group under functional composition.

THEOREM 2. *The representation of an n-dimensional affine space in terms of the n-dimensional affine Cartesian space over the field of real numbers is invariant under the group of affine transformations. In particular, if f and f' are two functions mapping the given affine space isomorphically onto R^n, then there exists an affine transformation φ such that $f' = \varphi \circ f$; also if f is a function so mapping the given affine space onto R^n and φ is any affine transformation, then $\varphi \circ f$ is also such an isomorphic mapping.*

Both Galilean and Lorentz transformations are special kinds of affine transformations, as will be spelled out later.

To obtain Euclidean spaces from affine ones we need add only the quaternary relation of congruence or equidistance: $ab \approx cd$ iff points a and b are the same distance apart as are points c and d. In other terms, the line segment with endpoints a and b is congruent to the line segment with endpoints c and d. The interpretation of equidistance in the Cartesian space R^n is familiar: for any vectors $\mathbf{x}, \mathbf{y}, \mathbf{u}$ and $\mathbf{v}$ of R^n the relation $\mathbf{xy} \approx_R \mathbf{uv}$ holds iff

$$\sum_{i=1}^{n}(x_i - y_i)^2 = \sum_{i=1}^{n}(u_i - v_i)^2.$$

I omit the familiar axioms of congruence for Euclidean spaces, which need to be added

---

at infinity'. The resulting space is affine. In the two-dimensional case, a line at infinity is removed to obtain the affine plane.

[3] A change in the conventions of notation of Chapter 5 is made in the first three sections of this chapter. Bold-face capital letters such as $\mathbf{X}$ or $\mathbf{A}$ no longer denote random variables, but rather matrices, and now boldface lower-case letters such as $\mathbf{x}$ and $\mathbf{y}$ denote vectors.

to the affine axioms of Definition 1. They extend the concept of affine congruence, which holds just for parallel segments. Implicitly assuming the congruence axioms, we then have the standard representation:

THEOREM 3. *Every n-dimensional Euclidean space $(A, B, \approx)$ is isomorphic to the n-dimensional Euclidean Cartesian space $(R^n, B_R, \approx_R)$, i.e., there is a one-one function f mapping A onto R^n such that*

(i) *for any points a, b and c in A*

$$B(a,b,c) \text{ iff } B_R(f(a), f(b), f(c)),$$

and

(ii) *for any points a, b, c and d in A*

$$ab \approx cd \text{ iff } \sum_{i=1}^{n}(f_i(a) - f_i(b))^2 = \sum_{i=1}^{n}(f_i(c) - f_i(d))^2,$$

where $f_i(a)$ is the ith coordinate of the Cartesian vector $f(a)$. Moreover, any two n-dimensional Euclidean spaces are isomorphic.

Again, of equal importance is the invariance theorem that is a companion to Theorem 3. An affine transformation $\varphi(\mathbf{x}) = \mathbf{Ax} + \mathbf{b}$, as defined above, is a *similarity* transformation iff the matrix $\mathbf{A}$ is a similarity matrix, i.e., there is an orthogonal matrix $\mathbf{A}'$ and a positive real number δ such that $\mathbf{A} = \delta \mathbf{A}'$. In other words, a similarity transformation permits only a uniform stretch of units in all directions, whereas a general affine transformation imposes no such restriction.

THEOREM 4. *The representation of an n-dimensional Euclidean space in terms of the n-dimensional Euclidean Cartesian space over the field of real numbers is invariant under the group of similarity transformations.*

In contrast to the earlier statement, it is obvious, but worth noting, that an arbitrarily selected Galilean or Lorentz transformation will not be a general similarity transformation in its spatial part, because the unit of measurement for spatial distance is held constant. This will become clear in the next two sections.

6.2 Classical Space-time

Rather than give a purely qualitative geometric formulation, as may be found in Suppes et al. (1989, Ch. 13) with references to the history of such qualitative formulations, it is also physically natural to introduce frames of reference, one-dimensional temporal ones and three-dimensional spatial ones, with Cartesian, i.e., orthogonal, axes.

Formally, a space-time measurement frame f is a pair of functions (s, r) mapping each space-time point a to a spatial vector $(x_1, x_2, x_3,)$ in Cartesian space $(R^3, B_R, \approx_R)$, where spatially $s_i(a) = x_i$ and temporally $r(a) = t$. These space and time functions expressing measurements relative to a frame of reference are the same for classical and special relativistic physics. As we go from the classical to the relativistic case, the laboratory or observatory methods of measurement do not change, but the fundamental invariants do. In the axioms given below we refer to inertial frames, but nothing beyond affineness and the appropriate invariance are assumed for these frames.[4] It is a consequence, not an assumption back of the invariance axioms, that any two inertial frames are related by a constant relative velocity.

DEFINITION 1. *A structure* $\mathfrak{A} = (A, B, \mathcal{F})$ *is a four-dimensional classical space-time structure with inertial frames iff the following axioms are satisfied:*

Axiom 1. The structure (A, B) is a four-dimensional affine space.

Axiom 2. $\mathcal{F}$ is a nonempty collection of (inertial) frames $f = (s, r)$ that are isomorphic mappings of (A, B) onto the affine Cartesian space (R^4, B_R).

Axiom 3. (Temporal invariance). For any two points a and b of A, and for any two frames $f = (s, r)$ and $f' = (s', r')$ in $\mathcal{F}$, the temporal distance is invariant, i.e.,

$$|r(a) - r(b)| = |r'(a) - r'(b)|. \qquad (I)$$

Axiom 4. (Spatial Invariance). For any two points a and b of A, and for any two frames (s, r) and (s', r') in $\mathcal{F}$, the spatial distance is invariant, i.e.,

$$\sum_{i=1}^{3}(s_i(a) - s_i(b))^2 = \sum_{i=1}^{3}(s'_i(a) - s'_i(b))^2. \qquad (II)$$

Axiom 5. (Closure of Invariant Frames). If $f = (s, r)$ is in $\mathcal{F}$ and if φ is an affine transformation of (R^4, B_R) such that f and $\varphi \circ f$ satisfy (I) and (II), then $\varphi \circ f$ is in $\mathcal{F}$.

The intuitive meaning of each of the axioms should be clear, but some comments may be useful. Axiom 1, which in the analysis given here is common to both classical and relativistic space-time, expresses the affine character of the spaces. The flatness or linearity of both kinds of spaces is obviously one of their most important shared characteristics. The assumption of an affine four-dimensional space-time structure is not always used in foundational analyses of the space and time of classical physics. Often a separate analysis is given for space and for time, but the unification used here is natural to make clear just how classical and restricted relativistic space-time rest on common assumptions and how they differ. From a qualitative geometric standpoint, the affine framework axiomatized just in terms of betweenness must be supplemented by additional qualitative primitives if a completely qualitative approach to physical space and time is to be developed. A concept of congruence for the spatial distance between simultaneous point events is the natural addition for classical space-time. However, physicists almost always avoid this qualitative geometric approach, even in foundational discussions. The use of frames here is much closer to physical ways of thinking. Another justification for assuming a common four-dimensional affine space for both classical and

---

[4]From a physical standpoint, it would be desirable to postulate the existence of one fixed frame 'at rest with respect to the fixed stars', as the classical phrase goes. But this introduction would obviously disturb the purely formal character of the axiomatic characterization given in Definition 1. Still, the intuition is physically important that an inertial frame is one moving with a constant velocity with respect to some reference point that has a massive or eternal character, absolute space for Newton, a triad of fixed, i.e., distant, stars for some, the center of mass of the solar system for others, or finally, in more grandiose terms, the center of mass of the universe, the last three of which are apparently empirically very similar, in the sense of moving relative to each other with a nearly constant velocity.

6.2 CLASSICAL SPACE-TIME

relativistic space-time is that a common invariant of both is the linearity of inertial paths of particles, i.e., the space-time trajectories of particles acted on by a null resultant force.

A separate representation theorem for classical space-time structures is not required, for the representing functions are just the inertial frames postulated in Axiom 2 of Definition 1. What is needed is the Galilean invariance theorem.

The invariance theorem for classical space-time structures is a specialization of Theorem 6.1.2 on the invariance of the representation of affine spaces. Recall that the uniqueness was up to an arbitrary affine transformation. We now put additional conditions on the matrix $\mathbf{A}$ of such transformations. We say that a four-dimensional matrix $\mathbf{A}$ is a *Galilean matrix with respect to a frame f in $\mathcal{F}$* if and only if there exist a three-dimensional vector $\mathbf{u}$, an orthogonal matrix $\mathbf{E}$ of order 3, and $\delta = \pm 1$ such that

$$\mathbf{A} = \begin{pmatrix} \mathbf{E} & 0 \\ 0 & \delta \end{pmatrix} \begin{pmatrix} \mathbf{I} & -\mathbf{u}^* \\ -\mathbf{u} & 1 \end{pmatrix}$$

where $\mathbf{I}$ is the identity matrix of order 3.

The interpretation of the constants is as follows. The number δ represents a possible transformation in the direction of time—if $\delta = 1$ there is no change and if $\delta = -1$ the direction is reversed; $\mathbf{u}$ is the relative velocity of the new inertial frame of reference with respect to the old one; and the matrix $\mathbf{E}$ represents a rotation of the spatial coordinates. The vector $\mathbf{b}$ of the affine transformation also has a simple interpretation. The first three coordinates represent a translation of the origin of the spatial coordinates of f, and the fourth coordinate a translation of the origin of the one-dimensional coordinate system for measurement of time of f. We are thus assuming that the same measurement scales of space and time are used by all the different frames in $\mathcal{F}$, as is implicit in Axioms 3 and 4, but a translation of spatial or temporal position is permitted in moving from one frame to another.

We say that an affine transformation is *Galilean* if and only if its four-dimensional matrix is Galilean. Using these concepts we then may state the following invariance theorem for classical space and time.

THEOREM 1. *The representation of a classical space-time structure with inertial frames in terms of the four-dimensional affine Cartesian space over the field of real numbers is invariant under the group of Galilean transformations.*

Proof. Let f and f' be two frames in $\mathcal{F}$. We analyze the affine transformation φ such that for every four-dimensional Cartesian vector $\mathbf{x}$ of f, $\varphi(\mathbf{x}) = \mathbf{x}'$. By virtue of Theorem 6.1.2 there is a nonsingular matrix $\mathbf{A}$ of order 4 and a four-dimensional $\mathbf{b}$ such that for every four-dimensional Cartesian vector $\mathbf{x}$ of f

$$\varphi(\mathbf{x}) = \mathbf{A}\mathbf{x} + \mathbf{b}.$$

The proof reduces to showing that $\mathbf{A}$ is a Galilean matrix. Since $\mathbf{A}$ is nonsingular, we may decompose $\mathbf{A}$ as follows:

$$\mathbf{A} = \begin{pmatrix} \mathbf{D} & \mathbf{e}^* \\ \mathbf{h} & g \end{pmatrix},$$

where $\mathbf{D}$ is a 3×3 nonsingular matrix acting on the three spatial coordinates x_1, x_2, x_3 of any vector $\mathbf{x}$ of f, $\mathbf{h}$ is a row vector, $\mathbf{e}^*$ is a column vector and $g \neq 0$ is a real number.

First, suppose **D** is not an orthogonal matrix. Then it is easy to show there exist vectors **x** and **y** such that the spatial distance between **x** and **y** is not preserved by φ, contrary to Axiom 4, i.e.

$$\sum (x_i - y_i)^2 \neq \sum (\varphi(x_i) - \varphi(y_i))^2.$$

By a similar argument we can show that by virtue of Axiom 3, $g = \delta = \pm 1$.

Some general theorems about classical mechanics are to be found in Section 7.1, but they necessarily go beyond the kinematical framework of the present section.

Historical remarks. The formal invariance axioms of Definition 1, or even the first axiom on affine spaces providing the general setting, will not be intuitively obvious to readers unfamiliar with the foundations of classical mechanics, where these ideas originate, and have since been abstracted away from their mechanical context to become geometric in character. The only remnant of classical physics left, but an essential one of course, is that the first three coordinates of a frame's location of a space-time point are spatial coordinates, in the sense of physical space, and the fourth coordinate location of a space-time point in physical time.

The discovery of this invariance is relatively late. It certainly does not fit in with Aristotle's physics, which holds that the center of the universe is the center of the Earth. And, more importantly, from observation it is inferred that the natural uniform motion of the heavens, i.e., the planets and the stars, is circular.

Here is Ptolemy's cogent summary of these views:

> The general preliminary discussion covers the following topics: the heaven is spherical in shape, and moves as a sphere; the Earth too is sensibly spherical in shape, when taken as a whole; in position it lies in the middle of the heavens very much like its centre; in size and distance it has the ratio of a point to the sphere of the fixed stars; and it has no motion from place to place. ... It is plausible to suppose that the ancients got their first notions on these topics from the following kind of observations. They saw that the sun, moon and other stars were carried from east to west along circles which were always parallel to each other, that they began to rise up from below the Earth itself, as it were, gradually got up high, then kept on going round in similar fashion and getting lower, until, falling to Earth, so to speak, they vanished completely, then, after remaining invisible for some time, again rose afresh and set; and [they saw] that the periods of these [motions], and also the places of rising and setting, were, on the whole, fixed and the same.

> What chiefly led them to the concept of a sphere was the revolution of the ever-visible stars, which was observed to be circular, and always taking place about one centre, the same [for all]. For by necessity that point became [for them] the pole of the heavenly sphere: those stars which were closer to it revolved on smaller circles, those that were farther away described circles ever greater in proportion to their distance, until one reaches the distance of the stars which become invisible. In the case of these, too, they saw that those near the ever-visible stars remained invisible for a short time, while those farther away remained invisible for a long time, again in proportion [to their distance]. The result was that in the beginning they got to the aforementioned notion soley from such considerations; but from then on, in their subsequent investigation, they found that

6.2 CLASSICAL SPACE-TIME

everything else accorded with it, since absolutely all phenomena are in contradiction to the alternative notions which have been propounded.

For if one were to suppose that the stars' motion takes place in a straight line towards infinity, as some people have thought, what device could one conceive of which would cause each of them to appear to begin their motion from the same starting-point every day? How could the stars turn back if their motion is towards infinity? Or, if they did turn back, how could this not be obvious? [On such a hypothesis], they must gradually diminish in size until they disappear, whereas, on the contrary, they are seen to be greater at the very moment of their disappearance, at which time they are gradually obstructed and cut off, as it were, by the Earth's surface. ...

To sum up, if one assumes any motion whatever, except spherical, for the heavenly bodies, it necessarily follows that their distances, measured from the Earth upwards, must vary, wherever and however one supposes the Earth itself to be situated. Hence the sizes and mutual distances of the stars must appear to vary for the same observers during the course of each revolution, since at one time they must be at a greater distance, at another at a lesser. Yet we see that no such variation occurs.

(Ptolemy, *Almagest*, pp. 38–39)

The point to be emphasized above all is that this conclusion about uniform circular motion was not from a priori arguments,[5] but an inference from astronomical observations recorded over many hundreds of years. Given the great success and comparative accuracy it is no wonder that it took many more hundreds of years to dislodge these Aristotelian conclusions.

Moreover, it is not surprising that the development of new ideas came from earthly phenomena, not astronomical observations. Out of the long development of impetus concepts in medieval mechanics, Benedetti (1585/1959) was perhaps the first to insist on a rectilinear path exclusively for a freely moving body. Here is what he says about a body thrown from a rotating sling:

Indeed the circular motion, as the body rotates, makes it so that the body by natural inclination (*naturalis inclinatio*) and under the force of impetus already initiated pursues a rectilinear path (*recta iter peragere*).

(Benedetti, in Clagett, *The Science of Mechanics in the Middle Ages,* p. 664)

Only a little later Galileo's own progress toward conceiving inertial motion as rectilinear was slow and complicated.[6]

The rapid developments in the seventeenth century cannot be surveyed in detail here,

---

[5]The well-known physicist, David Bohm (1965, p. 5) supports this mistaken a priori view wholeheartedly in the following passage, taken from his textbook on special relativity:

A part of Aristotle's doctrine was that bodies in the Heavens (such as planets), being more perfect than Earthly matter, should move in an orbit which expresses the perfection of their natures. Since the circle was considered to be the most perfect geometrical figure, it was concluded that a planet must move in a circle around the Earth. When observations failed to disclose perfect circularity, this discrepancy was accommodated by the introduction of "epicycles," or of "circles within circles." In this way, the Ptolemaic theory was developed, which was able to "adjust" to any orbit whatsoever, by bringing in many epicycles in a very complicated way.

[6]See, for example, the discussion of Clagett against the background of medieval mechanics (Clagett 1959, pp. 666–671).

but an early, clear formulation almost equivalent to Newton's first law is to be found in Descartes' *Principles of Philosophy,* Book II, (1644/1983, pp. 59–60), Propositions 37 and 39:

> 37. The first law of nature: that each thing, as far as is in its power, always remains in the same state; and that consequently, when it is once moved, it always continues to move.
>
> 39. The second law of nature: that all movement is, of itself, along straight lines; and consequently, bodies which are moving in a circle always tend to move away from the center of the circle which they are describing.

It was this clear rectilinear concept of inertial motion, along with the rejection of a fixed, unmoved center point, that was needed to clear the way for the classical notion of invariance.[7]

Undoutedly one reason for Descartes emphasizing the relativity of motion was so that in Part III of the *Principles of Philosophy* he could use the Copernican system, but still not require that the Earth moves. Here is Proposition 19 of Book III (1644/1983, p. 91).

> 19. That I deny the motion of the Earth more carefully than Copernicus and more truthfully than Tycho.[8]

Descartes was an adept fence-straddler on difficult points of conflict between science and Catholic theology.

Although Newton introduces the concept of absolute rest in the opening pages of his *Principia* (1687/1946, p. 9), in his prudent, careful way he asserts 'that absolute space cannot be determined from the position of bodies in our regions'. He goes on to elaborate the uses and need for a concept of relative motion as well as that of true motion.

But, it is still a long time before invariance in classical physics is fully recognized and used. In Mach's (1883/1942) important but flawed foundational study *The Science of Mechanics, A Critical and Historical Account of Its Development,* there is no systematic consideration of any questions of invariance in, I believe, any of its nine editions. This is also true of a work of a very different character, E. T. Whittaker's (1904/1937) well-known treatise on analytic dynamics.

In truth, it is rather like an induction backward in time. Einstein's theory of special relativity had to be introduced first with an emphasis on invariants coming to the fore as a natural way to explain the differences from classical physics. It was only in this late context that the affine assumptions and the invariants of Definition 1 were fully and systematically formulated, including the explicit identification of the Galilean transformations.[9]

---

[7] Descartes rejects the unmoved center in the discussion of Proposition 13 of Book II, and in other places as well.

[8] 'Tycho' refers to Tycho Brahe (1546–1601), famous for his excellent systematic astronomical observations, and of great importance for Kepler's theoretical work.

[9] According to Abraham Pais (1982, p. 140), the term *Galilean* was first introduced by Frank (1909, p. 382).

6.3 Axioms for Special Relativity

The purpose of this section is to characterize the invariance properties of special relativity. As in the case of the previous section, the geometric points of the space are interpreted as four-dimensional point-events. The desired invariance result is given in terms of the group of Lorentz transformations.

As in the classical case, linearity is introduced by means of the ternary relation of betweenness. The approach used here is to keep in the foreground the question of congruence of segments, i.e., equidistance. Of course, the simple classical formulation of the separate invariance of spatial and temporal measurements of distance is not possible. However, as was pointed out by Robb (1936) many years ago, there is a meaningful notion of congruence in special relativity for segments of inertial lines, i.e., possible paths of particles moving uniformly. This congruence is that of the *proper time* of inertial segments. There is also a separate notion of congruence for segments of separation or spatial lines, i.e., lines determined by two points that determine neither an inertial line nor an optical line. In contrast, there is no natural notion of congruence for segments of optical lines, i.e., lines representing optical paths.

The intuition back of these ideas is most easily explained by introducing the usual coordinate representation for special relativity. The four real coordinates (x_1, x_2, x_3, t) of a point-event are measured relative to an inertial space-time frame f of reference, just as in classical physics. As in the preceding section, the coordinates x_1, x_2 and x_3 are the three orthogonal spatial coordinates and t is the time coordinate of f. It is often notationally convenient to replace t by x_4. The relativistic distance, or *proper time*, between two points $x = (x_1, x_2, x_3, x_4)$ and $y = (y_1, y_2, y_3, y_4)$ is given by the indefinite quadratic metric, relative to frame $f = (s, r)$, as:

$$I_f(xy) = \sqrt{c^2(x_4 - y_4)^2 - \sum_{i=1}^{3}(x_i - y_i)^2} \qquad (1)$$

where c is the numerical velocity of light. It is simpler to use the square of the relativistic distance.[10] It is easy to see that $I_f^2(xy) > 0$ if x and y lie on an inertial line, $I^2(xy) < 0$ if x and y lie on a separation line, and $I^2(xy) = 0$ if x and y lie on an optical line. In these terms, two *inertial segments* xy and uv, i.e., segments of inertial lines, are congruent iff

$$I_f^2(xy) = I_f^2(uv) \qquad (2)$$

and a similar condition holds for two segments of separation lines. On the other hand, this equality does not work at all for optical lines, because the relativistic length of any optical segment is zero.

Given an affine space-time structure, invariance of the speed of light in a vacuum, c, and invariance of proper time of inertial segments for different inertial frames is a sufficient condition to prove that inertial frames must be related by Lorentz transforma-

---

[10]Equation (1) is convenient for computation, but conceptually it is natural to divide by c, so that in any given coordinate system f the units of proper time are temporal, i.e.,

$$\text{Proper time}_f(x, y) = \frac{1}{c} I_f(x, y).$$

tions. The restriction to inertial segments is physically natural, for such segments are more susceptible to direct measurement than are segments of separation lines.

The axioms for restricted relativistic space-time require only a change in the invariance axioms of classical space-time. The separate measurement of space and time relative to an inertial frame is unchanged.

We now add to the structure $(A, B, \mathcal{F})$ the invariant constant c for the speed of light.

DEFINITION 1. *A structure $\mathfrak{A} = (A, B, \mathcal{F}, c)$ is a (four-dimensional) restricted relativistic space-time structure with inertial frames if and only if the following axioms are satisfied:*

Axiom 1. Same affine axiom as Definition 1, Section 2.

Axiom 2. Same frame axiom as Definition 1, Section 2.

Axiom 3. (Invariance of speed of light). The positive real number c, representing the measurement of the speed of light in a vacuum, is invariant, i.e., constant, for every frame f in $\mathcal{F}$.

Axiom 4. (Invariance of proper time of inertial segments). For any two points a and b of A such that for some frame f of $\mathcal{F}$, $I_f^2(ab) > 0$, the proper time of the segment ab is invariant in $\mathcal{F}$, i.e., for any two frames f and f' in $\mathcal{F}$,

$$I_f(ab) = I_{f'}(ab). \tag{III}$$

Axiom 5. (Closure of Invariant Frames). If f is in $\mathcal{F}$ and if φ is any affine transformation of (R^4, B_R) such that f and $\varphi \circ f$ satisfy (III), then $\varphi \circ f$ is in $\mathcal{F}$.

We next need to define formally the concept of a Lorentz matrix and then that of a Lorentz transformation.

DEFINITION 2. *A matrix $\mathbf{A}$ (of order 4) is a Lorentz matrix if and only if there exist real numbers β, δ, a three-dimensional vector $\mathbf{u}$, and an orthogonal matrix $\mathbf{E}$ of order 3 such that*

$$\beta^2 \left(1 - \frac{\mathbf{u}^2}{c^2}\right) = 1$$

$$\delta^2 = 1$$

$$\mathbf{A} = \begin{pmatrix} \mathbf{E} & \mathbf{0} \\ \mathbf{0} & \delta \end{pmatrix} \begin{pmatrix} \mathbf{I} + \frac{\beta-1}{\mathbf{u}^2}\mathbf{u}^*\mathbf{u} & -\frac{\beta\mathbf{u}^*}{c^2} \\ -\beta\mathbf{u} & \beta \end{pmatrix}.$$

(In this definition and elsewhere, if $\mathbf{A}$ is a matrix, $\mathbf{A}^*$ is its transpose, and vectors like $\mathbf{u}$ are vectors—thus $\mathbf{u}^*$ is a column vector.[11]) The physical interpretation of the various quantities in Definition 2 should be obvious. The number β is the *Lorentz contraction factor*. When $\delta = -1$, we have a reversal of the direction of time. The matrix $\mathbf{E}$ represents a *rotation* of the spatial coordinates of f, or a rotation followed by a reflection. The vector $\mathbf{u}$ is the *relative velocity* of the two frames of reference. For future reference it may be noted that every Lorentz matrix is nonsingular.

DEFINITION 3. *A Lorentz transformation is a one-one function φ mapping R_4 onto itself such that there is a Lorentz matrix $\mathbf{A}$ and a four-dimensional vector $\mathbf{b}$ with respect*

---

[11] For simplicity of notation I sometimes omit the transpose * on column vectors.

6.3 AXIOMS FOR SPECIAL RELATIVITY

to a frame f in $\mathcal{F}$ so that for all $\mathbf{x}$ in R_4

$$\varphi(\mathbf{x}) = \mathbf{xA} + \mathbf{b}.$$

The physical interpretation of the vector $\mathbf{b}$ is clear. Its first three coordinates represent a translation of the origin of the spatial coordinates, and its last coordinate a translation of the time origin. Definition 3 makes it clear that every Lorentz transformation is a nonsingular affine transformation of R_4, a fact which we shall use in several contexts.

In the proof of the invariance theorem it is convenient to use a Lemma about Lorentz matrices, which is proved in Rubin and Suppes (1954) and is simply a matter of direct computation.

LEMMA 19. *A matrix $\mathbf{A}$ (of order 4) is a Lorentz matrix if and only if*

$$\mathbf{A} \begin{pmatrix} \mathbf{I} & 0 \\ 0 & -c^2 \end{pmatrix} \mathbf{A}^* = \begin{pmatrix} \mathbf{I} & 0 \\ 0 & -c^2 \end{pmatrix}.$$

We now state the invariance theorem:

THEOREM 1. *The representation of a restricted relativistic space-time structure with inertial frames in terms of the affine four-dimensional Cartesian space over the field of real numbers is invariant under the group of Lorentz transformations.*

Proof. Let f, f' be two frames in $\mathcal{F}$. As before, for a in A, $f(a) = \mathbf{x}$, $f_1(a) = x_1$, $f'(a) = \mathbf{x}'$, etc. We consider the transformation φ such that for every a in A, $\varphi(\mathbf{x}) = \mathbf{x}'$. By virtue of Axioms 1 and 2 and Theorem 1.2 there is a nonsingular matrix $\mathbf{A}$ (of order 4) and a four-dimensional vector $\mathbf{b}$ such that for every $\mathbf{x}$ in R^4

$$\varphi(\mathbf{x}) = \mathbf{xA} + \mathbf{b}.$$

The proof reduces to showing that $\mathbf{A}$ is a Lorentz matrix.

Let

$$\mathbf{A} = \begin{pmatrix} \mathbf{D} & \mathbf{e}^* \\ \mathbf{f} & g \end{pmatrix}. \tag{3}$$

And let α be a light line (in f) such that for any two distinct points $\mathbf{x}$ and $\mathbf{y}$ of α if $\mathbf{x} = (Z_1, t_1)$ and $\mathbf{y} = (Z_2, t_2)$, then

$$\frac{Z_1 - Z_2}{t_1 - t_2} = W. \tag{4}$$

Clearly by Axiom 3, $|W| = c$. Now let

$$W' = \frac{Z'_1 - Z'_2}{t'_1 - t'_2}. \tag{5}$$

From (3), (4) and (5) we have:

$$W' = \frac{(Z_1 - Z_2)\mathbf{D} + (t_1 - t_2)\mathbf{f}}{(Z_1 - Z_2)\mathbf{e}^* + (t_1 - t_2)g}. \tag{6}$$

Dividing all terms on the right of (6) by $t_1 - t_2$, and using (4), we obtain:

$$W' = \frac{W\mathbf{D} + \mathbf{f}}{W\mathbf{e}^* + g}. \tag{7}$$

From Axiom 3 again

$$|W'| = c. \tag{8}$$

Since $|W'| = c$, we have by squaring (7):
$$\frac{W\mathbf{DD}^*W^* + 2W\mathbf{Df}^* + |\mathbf{f}|^2}{(W\mathbf{e}^* + g)^2} = c^2, \tag{9}$$
and consequently
$$W(\mathbf{DD}^* - c^2\mathbf{e}^*\mathbf{e})W^* + 2W(\mathbf{Df}^* - c^2\mathbf{e}^*g) + |\mathbf{f}|^2 - c^2g^2 = 0. \tag{10}$$
Since (10) holds for an arbitrary light line, we may replace W by $-W$, and obtain (10) again. We thus infer:
$$W(\mathbf{Df}^* - c^2\mathbf{e}^*g) = 0,$$
but the direction of W is arbitrary, whence
$$\mathbf{Df}^* - c^2\mathbf{e}^*g = 0. \tag{11}$$

Now let $\mathbf{x} = (0,0,0,0)$ and $\mathbf{y} = (0,0,0,1)$. Then
$$I_f^2(\mathbf{xy}) = c^2.$$
But it is easily seen from (3) that
$$I_{f'}^2(\mathbf{xy}) = c^2g^2 - |\mathbf{f}|^2,$$
and thus by our fundamental invariance axiom (Axiom 4)
$$c^2g^2 - |\mathbf{f}|^2 = c^2. \tag{12}$$
From (10), (11), (12) and the fact that $|W|^2 = c^2$, we infer:
$$W(\mathbf{DD}^* - c^2\mathbf{e}^*\mathbf{e})W^* = |W|^2,$$
and because the direction of W is arbitrary, we conclude:
$$\mathbf{DD}^* - c^2\mathbf{e}^*\mathbf{e} = \mathbf{I}, \tag{13}$$
where $\mathbf{I}$ is the identity matrix.

Now by direct computation on the basis of (3),
$$\mathbf{A} \begin{pmatrix} \mathbf{I} & 0 \\ 0 & -c^2 \end{pmatrix} \mathbf{A}^* = \begin{pmatrix} \mathbf{DD}^* - c^2\mathbf{e}^*\mathbf{e} & \mathbf{Df}^* - c^2\mathbf{e}^*g \\ (\mathbf{Df}^* - c^2\mathbf{e}^*g)^* & \mathbf{ff}^* - c^2g^2 \end{pmatrix}. \tag{14}$$
From (11), (12), (13) and (14) we arrive finally at the result:
$$\mathbf{A} \begin{pmatrix} \mathbf{I} & 0 \\ 0 & -c^2 \end{pmatrix} \mathbf{A}^* = \begin{pmatrix} \mathbf{I} & 0 \\ 0 & -c^2 \end{pmatrix},$$
and thus by virtue of Lemma 1, $\mathbf{A}$ is a Lorentz matrix. Q.E.D.

Historical remarks. As already stated, explicit questions of invariance were made prominent in the earliest development of the special theory of relativity, quite the opposite of the corresponding case for classical physics.

The predecessors of Einstein are Voigt, FitzGerald, Lorentz and Poincaré, none of whom explicitly formulated a relativistic invariance principle. In fact, much of the effort was still implicitly motivated by search for an ether absolutely at rest, as the basis of a single preferred reference frame. Voigt (1887) was apparently the first actually to publish a version of the Lorentz transformations, but his concern was the Doppler shift.[12]

---
[12] Here and in what follows up to the publication of Einstein (1905/1923) I have used the excellent historical analysis in Pais (1982).

6.3 AXIOMS FOR SPECIAL RELATIVITY

The Irish physicist G. F. FitzGerald (1889), much impressed by the famous Michelson-Morley (1887) experiment on the null result on the motion of the Earth in the ether, was the first to propose the contraction factor $\sqrt{1 - v^2/c^2}$ in the length of a body in the direction of its motion with velocity v relative to the measurement frame of reference. But for FitzGerald the change was real, as he put it "...the length of material bodies changes, according as they are moving through the aether...". Without knowledge of FitzGerald's article, Lorentz (1892) also proposed such a contraction factor. Then Lorentz (1895) derived a form of Lorentz transformations useful in terrestrial optical experiments, as part of his development of electromagnetic theory. Poincaré (1898) discusses with wonderful clarity, but with only programmatic remarks, the obvious operational difficulties of judging two distant events simultaneous.

Publication of Einstein (1905/1923) is the decisive event, with a clear statement of invariance, the crucial step in the move from Maxwell's electromagnetic theory to the kinematical special theory of relativity. Here is the English translation (1923) of the first two paragraphs of this justly famous article.

> It is known that Maxwell's electrodynamics—as usually understood at the present time—when applied to moving bodies, leads to asymmetries which do not appear to be inherent in the phenomena. Take, for example, the reciprocal electrodynamic action of a magnet and a conductor. The observable phenomenon here depends only on the relative motion of the conductor and the magnet, whereas the customary view draws a sharp distinction between the two cases in which either the one or the other of these bodies is in motion. For if the magnet is in motion and the conductor at rest, there arises in the neighbourhood of the magnet an electric field with a certain definite energy, producing a current at the places where parts of the conductor are situated. But if the magnet is stationary and the conductor in motion, no electric field arises in the neighbourhood of the magnet. In the conductor, however, we find an electromotive force, to which in itself there is no corresponding energy, but which gives rise—assuming equality of relative motion in the two cases discussed—to electric currents of the same path and intensity as those produced by the electric forces in the former case.
>
> Examples of this sort, together with the unsuccessful attempts to discover any motion of the Earth relatively to the "light medium," suggest that the phenomena of electrodynamics as well as of mechanics possess no properties corresponding to the idea of absolute rest. They suggest rather that, as has already been shown to the first order of small quantities, the same laws of electrodynamics and optics will be valid for all frames of reference for which the equations of mechanics hold good.[13] We will raise this conjecture (the purport of which will hereafter be called the "Principle of Relativity") to the status of a postulate, and also introduce another postulate, which is only apparently irreconcilable with the former, namely, that light is always propagated in empty space with a definite velocity c which is independent of the state of motion of the emitting body. These two postulates suffice for the attainment of a simple and consistent theory of the electrodynamics of moving bodies based on Maxwell's theory for stationary bodies. The introduction of a "luminiferous ether" will prove to be superfluous inasmuch as the view here to be developed will not require an "absolutely stationary space" provided with special properties, nor assign a velocity-vector to a point of the empty space in which electromagnetic processes take place. (Einstein 1905/1923, pp. 37–38)

---

[13] The reference here is to the inertial frames of classical physics, which are, of course, the same for special relativity.

The first paragraph of Einstein's article is focused on defining simultaneity of distant events, hinted at by Poincaré (1898) earlier, but only here stated exactly and boldly used. I give an equivalent affine geometric formulation. Let a be any space-time, i.e., affine, point. Let L be an inertial line through a; here L serves as the time axis of the inertial frame with respect to which simultaneity is defined. Any affine line is *orthogonal* to L if the two lines form the diagonals of an optical parallelogram, i.e., an affine parallelogram whose sides are segments of optical lines, as defined earlier.

Then
$$\mathcal{S}(a, L) = \{b : \text{line } ab \text{ is orthogonal to } L\} \cup \{a\}$$
is the set of points simultaneous to a, relative to L. Note that the definition of simultaneity must be relative to a given frame of reference. The absolute simultaneity of classical physics is given up.

In qualitative axiomatizations of the geometric structure of special relativity it is natural to postulate that for every affine space-time point a and any inertial line through a, the set $\mathcal{S}(a, L)$ is a three-dimensional Euclidean space with respect to the affine betweenness relation and a congruence relation $\approx$ satisfying the standard Euclidean congruence axioms.

Section 2 of this famous 1905 article explicitly states the two principles Einstein uses:

> The following reflexions are based on the principle of relativity and on the principle of the constancy of the velocity of light. These two principles we define as follows:
>
> 1. The laws by which the states of physical systems undergo change are not affected, whether these changes of state be referred to the one or the other of two systems of coordinates in uniform translatory motion.
> 2. Any ray of light moves in the "stationary" system of coordinates with the determined velocity c, whether the ray be emitted by a stationary or by a moving body.
>
> (Einstein 1905/1923, p. 41)

These principles are then used to show the relativity of lengths and time.

In Section 3 a form of the Lorentz transformations is derived, but, as noted earlier, without awareness of Lorentz's earlier paper. Section 4 considers the physical meaning of the transformations derived. Here Einstein introduces the proper time τ of a moving body, i.e., the time 'when at rest relatively to the moving system,' and derives the equation

$$\tau = t\sqrt{1 - \frac{v^2}{c^2}}.$$

In Section 5 the relativistic composition of velocities is derived.

The remainder of the article is devoted to electromagnetic theory. In particular, the covariance[14] of Maxwell's equations under Lorentz transformations is derived.

Einstein submitted his paper to the *Annalen der Physik* on June 30, 1905. Poincaré's article (1905) was submitted on June 5, 1905 and his second, published in 1906, was completed in July 1905. Of course, neither author was aware of the other's publication at the time. In the second paper Poincaré gave a completely detailed proof of the covariance of Maxwell's equations. But curiously enough, Poincaré apparently did not fully grasp

---

[14] The relation between covariance and invariance was analyzed in Section 4.4.

the sense of special relativity. In later lectures given at Göttingen in 1909, published in 1910, he still introduced, beyond the two principles used by Einstein in Section 2 of the 1905 article and quoted above, as a third hypothesis the physical reality of the FitzGerald-Lorentz contraction, a clear confusion about the essential nature of special relativity.

Later qualitative axiomatic approaches. In the case of special relativity, the primitive binary relation *after* of temporal order is a sufficient structural basis for the entire geometry. This was shown many years ago by Robb (1911, 1914, 1921, 1928, 1930, 1936). The most extensive exposition of his ideas is to be found in his 1936 book. A detailed discussion and some extensions of Robb's work can be found in Winnie (1977). Alexandrov (1950, 1975) showed that invariance of the numerical representation of Robb's single binary relation is sufficient to imply the Lorentz transformations. This was also shown later and independently by Zeeman (1964, 1967). Surprisingly, this natural question was not discussed at all by Robb.

Since the time of Robb's early work, a number of papers have been published on qualitative axiomatic approaches to the geometry of special relativity. An early approach was that of Wilson and Lewis (1912), who gave an incomplete axiomatization of the two-dimensional case with an affine emphasis on parallel transformations. Reichenbach (1924) gave a philosophically interesting and intuitively appealing axiomatic discussion, but it was more in the spirit of physics than of an explicit and precise geometrical axiomatization. He emphasized the analysis of simultaneity and philosophical issues of conventionalism. Walker (1948, 1959), building on the physical approach to relativity of the British physicist E. A. Milne (1948), took as undefined, in addition to space-time points, particles as certain classes of such points (intuitively the classes are particle paths) and Robb's binary relation of *after*, with one particle being distinguished as the particle-observer. In addition, he assumed a primitive relation of signal correspondence between the space-time points of any two particles. Perhaps the most unsatisfactory feature of Walker's approach is that the signal-mappings are complex functions that, at one stroke, do work that should be done by a painstaking buildup of more elementary and more intuitively characterizable operations or relations. An independent but related approach to that of Walker was given by Hudgin (1973), who began with timelike geodesics and signal correspondence represented by a communication function. Like Walker, Hudgin used the physical ideas of Milne.

Domotor (1972) published a detailed discussion of Robb and provided another viewpoint on Robb's approach by giving a vector-space characterization. Latzer (1972) showed that it is possible to take a binary relation of light signaling as primitive. He did not give a new axiomatic analysis but showed how Robb's ordering relation can be defined in terms of his symmetric light relation. From a purely geometrical standpoint, there is a good argument for adopting such a symmetric relation because there is no nonarbitrary way of distinguishing a direction of time on purely geometrical grounds. It actually follows from much earlier work of Alexandrov (1950, 1975) that Latzer's binary primitive is adequate.

One of the most detailed recent efforts, roughly comparable to that of Robb, is the work of Schutz (1973), which is in the spirit of the work of Walker. Schutz's axiom system goes beyond Walker's in two respects. First, Walker did not state axioms strong enough to characterize completely the space-time of restricted relativity; and, second, more than half of the axioms have a form that is not closely related to any of those given by Walker, though in several cases they have a rather strong appeal in terms of physical intuition. As Schutz indicated, the most important immediate predecessor of his book is the article by Szekeres (1968), whose approach resembled that of Walker, but Szekeres treated both particles and light signals as objects rather than treating light signals as formally being characterized by a binary relation of signaling. Later, Schutz (1979) simplified his axioms and cast them in better form, but with essentially the same approach.

Also important is the recent work of Mundy (1986a, 1986b). In (1986b) Mundy gave a detailed analysis of the physical content of Minkowski geometry. He also included a survey of other axiomatizations more oriented toward physics, such as that of Mehlberg (1935, 1937). Mundy stressed in his axiomatization the constancy of the speed of light as the physical core of Minkowski geometry.

In (1986a) Mundy gave an optical axiomatization that is close in spirit to that of Robb, although the single binary primitive is close to Latzer's (1972). Mundy's formulation is certainly an improvement on Robb's and is the best set of axioms thus far that depends on a single binary relation.

In a more recent work, Goldblatt (1987) emphasizes, probably for the first time in the literature, the central role that can be played by the concept of orthogonality generalized to relativistic space-time. In a Euclidean vector space with the standard inner product, two vectors $\mathbf{x}$ and $\mathbf{y}$ are orthogonal iff $\mathbf{x} \cdot \mathbf{y} = 0$. Using the indefinite quadratic metric $I^2(\mathbf{x}, \mathbf{y})$, we obtain the following: when two distinct lines are orthogonal, one must be timelike (an inertial line) and the other spacelike (a separation line); an optical line is orthogonal to itself. Axiomatically, Goldblatt adds to the affine axioms based on betweenness, additional axioms based on the quaternary relation of ab being orthogonal to cd.

6.4 How to Decide if Visual Space is Euclidean[15]

Philosophers of past times have claimed that the answer to the question of whether visual space is Euclidean can be found by a priori or purely philosophical methods. Today such a view is presumably held only by a few redoutable Kantians. It would be generally agreed that one way or another the answer is surely empirical, but it might be empirical for indirect reasons. It could be decided by physical arguments that physical space is Euclidean and then by conceptual arguments about perception that necessarily visual space must be Euclidean. To some extent this must be the view of many laymen who accept that to a high degree of approximation physical space is Euclidean, and therefore automatically hold the view that visual space is Euclidean.

I begin with the question, How do we test the proposition that visual space is Euclidean? This section is devoted to this problem of methodology and includes a brief

---

[15]Much of the material in this section and the next is taken from Suppes (1977a).

overview of the hierarchy of geometries relevant to visual phenomena. The next section reviews a number of answers that have been given to the question. I examine both philosophical and psychological claims. Section 7 is devoted to central issues raised by the variety of answers that have been given.

What would seem to be, in many ways, the most natural mathematical approach to the question has also been the method most used experimentally. It consists of considering a finite set of points. Experimentally, the points are approximated by small point sources of light of low illumination intensity, displayed in a darkened room. The intuitive idea of the setting is to make only a finite number of point-light sources visible and to make these light sources of sufficiently low intensity to exclude illumination of the surroundings. The second step is to ask the person making visual judgments to state whether certain geometric relations hold between the points. For example, are points a and b the same distance from each other as points c and d? (Hereafter in this discussion I shall refer to points but it should be understood that I have in mind the physical realization in terms of point-light sources.) Another kind of question might be, Is the angle formed by points $a\ b\ c$ congruent or equal in measure to the angle formed by points $d\ e\ f$?

Another approach to such judgments is not to ask whether given points have a certain relation, but rather to permit the individual making the judgments to manipulate some of the points. For example, first fix points a, b and c and then ask him to adjust d so that the distance between c and d is the same as the distance between a and b. Although the formulation I am giving of these questions sounds as if they might be metric in character, they are ordinarily of a qualitative nature—for example, that of congruence of segments, which I formulate as having the same distance. No metric requirements are imposed upon the individuals making such judgments. For instance, no one would naturally ask subjects in the experiments relevant to our question to set the distance between two points to be approximately 1.3 meters or to determine an angle of, say, 21 degrees.

Once such judgments are obtained, whether on the basis of fixed relations or by adjusting the position of points, the formal or mathematical question to ask is whether the finite relational structure can be embedded in a two- or three-dimensional Euclidean space. The dimensionality depends upon the character of the experiment. In many cases the points will be restricted to a plane and therefore embedding in two dimensions is required; in other cases embedding in three dimensions is appropriate. By a *finite relational structure* I mean a relational structure whose domain is finite. To give a simple example, suppose that A is the finite set of points and the judgments we have asked for are judgments of equidistance of points. Let E be the quaternary relation of equidistance. Then to say that the finite relational structure $\mathfrak{A} = (A, E)$ can be embedded in three-dimensional Euclidean space is to say that there exists a function φ defined on A such that φ maps A into the set of triples of real numbers and such that for every $a, b, c,$ and d in A the following relation holds:

$$ab\ E\ cd \quad \text{iff} \quad \sum_{i=1}^{3}(\varphi_i(a) - \varphi_i(b))^2 = \sum_{i=1}^{3}(\varphi_i(c) - \varphi_i(d))^2,$$

where $\varphi_i(a)$ is the ith coordinate of $\varphi(a)$. Note that the mapping into triples of real numbers is just mapping visual points into a Cartesian representation of three-dimensional Euclidean space.

In principle, it is straightforward to answer the question raised by this embedding procedure. So that, given a set of data from an individual's visual judgments of equidistance between points, we can determine in a definite and constructive mathematical manner whether such isomorphic embedding is possible.

Immediately, however, a problem arises. This problem can be grasped by considering the analogous physical situation. Suppose we are making observations of the stars and want to test a similar proposition, or some more complex proposition of celestial mechanics. We are faced with the problem recognized early in the history of astronomy, and also in the history of geodetic surveys, that the data are bound not to fit the theoretical model exactly. The classical way of putting this is that errors of measurement arise, and our problem is to determine if the model fits the data within the limits of the errors of measurement. In examining data on the advancement of the perihelion of Mercury, which is one of the important tests of Einstein's general theory of relativity, the most tedious and difficult aspect of the data analysis is to determine whether the theory and the observations are in agreement within the estimated error of measurement.

Laplace (1799/1966), for example, used such methods with unparalleled success. He would examine data from some particular aspect of the solar system, for example, irregularities in the motion of Jupiter and Saturn, and would then raise the question of whether these observed irregularities were due to errors of measurement or to the existence of 'constant' causes. When the irregularities were too great to be accounted for by errors of measurement, he then searched for a constant cause to explain the deviations from the simpler model of the phenomena. In the case mentioned, the irregularities in the motion of Jupiter and Saturn, he was able to explain them as being due to the mutual gravitational attraction of the two planets, which had been ignored in the simple theory of their motion. But Laplace's situation is different from the present one in the following important respect. The data he was examining were already rendered in quantitative form and there was no question of having a numerical representation. Our problem is that we start from qualitative judgments and we are faced with the problem of simultaneously assigning a measurement and determining the error of that measurement. Because of the complexity and subtlety of the statistical questions concerning errors of measurement in the present setting, for purposes of simplification I shall ignore them, but it is absolutely essential to recognize that they must be dealt with in any detailed analysis of experimental data.

Returning to the formal problem of embedding qualitative relations among a finite set of points into a given space, it is surprising to find that the results of the kinds that are needed in the present context are not really present in the mathematical literature on geometry. There is a very large literature on finite geometries; for example, Dembowski (1968) contains over 1200 references. Moreover, the tradition of considering finite geometries goes back at least to the beginning of this century. Construction of such geometries by Veblen and others was a fruitful source of models for proving independence of axioms, etc.[16] On the other hand, the literature that culminates in Dembowski's magisterial survey consists almost entirely of projective and affine geometries that have a relatively weak structure. From a mathematical standpoint, such structures have been of considerable interest in connection with a variety of problems in abstract algebra. The corresponding theory of finite geometries of a stronger type, for example, finite Euclidean, finite elliptic, or finite hyperbolic geometries, is scarcely developed at all. As a result, the experimental literature does not deal directly with such finite geometries, although they are a natural extension of the weaker finite geometries on the one hand and finite measurement structures on the other. (For more details, see Section 8.)

A second basic methodological approach to the geometric character of visual space is to assume that a standard metric representation already exists and then to examine which kind of space best fits the data. An excellent example of this methodology is to be found in various publications of Foley (1964, 1972). Foley shows experimentally that the size-distance invariance hypothesis, which asserts that the perceived size-distance ratio is equal to the physical size-distance ratio, is grossly incorrect. At the same time he also shows that perceived visual angles are about ten percent greater than physical angles. These studies are conducted on the assumption that a variety of more primitive and elementary axioms are satisfied. In contrast, Luneburg (1948) assumes that the perceived visual angle equals the physical angle, that is, that the transformation between the two is conformal, but what is back of the use of this assumption is a whole variety of assumptions that both physical space and visual space are homogeneous spaces of constant curvature, that is, are Riemannian spaces, and essentially Luneburg does not propose to test in any serious way the many consequences implied by this very rich assumption of having a homogeneous space with constant curvature. In other words, in this second approach there is no serious attempt to provide tests that will show if all of the axioms that hold for a given type of space are satisfied.

A third approach is to go back to the well-known Helmholtz-Lie problem on the nature of space and to replace finiteness by questions of continuity and motion. In a famous lecture of 1854, Riemann (1866-1867) discussed the hypotheses on which the foundations of geometry lie. More than a decade later, Helmholtz (1868) responded in a paper entitled "Über die Tatsachen, die der Geometrie zu Grunde liegen". The basic argument of Helmholtz's paper was that, although arbitrary Riemannian spaces are conceivable, actual physical space has, as an essential feature, the free mobility of rigid bodies. From a mathematical standpoint, such motions are characterized in metric geometry as transformations of a space onto itself that preserve distances. Such transformations are called *isometries*, meaning metrically isomorphic. Because of the extensive mathematical development of the topic (for modern review, see Busemann 1955, Section 48, or Freudenthal 1965), an excellent body of formal results is available to facilitate use of this viewpoint in the investigation of the character of visual space. Under various axiomatizations of the Helmholtz-Lie approach, it can be proved that the only spaces satisfying the axioms are the following three kinds of elementary spaces: Euclidean, hyperbolic, and elliptic.

---

[16]See Veblen (1904) and Veblen and Young (1910, 1918).

From a philosophical standpoint, it is important to recognize that considerations of continuity and motion are probably more fundamental in the analysis of the nature of visual space than the mathematically more elementary properties of finite spaces. Unfortunately, I am not able to report any experimental literature that uses the Helmholtz-Lie approach as a way of investigating the nature of visual space. Although it is implicit in some of the results reported below, it would be difficult to interpret the experimental results as satisfying an axiom of free mobility. Let me be clear on this point. Some of the experimental investigations lead to the result that visual space cannot be elementary in the sense just defined, but these investigations do not explicitly use the kind of approach to motion suggested by the rich mathematical developments that have followed in response to the Helmholtz-Lie problem, i.e., the study of spaces that permit the free mobility of rigid bodies.

A fourth approach, which lies outside the main body of the literature to be considered in this chapter, is the approach through picture grammars and the analysis of perceptual scenes. Its growing literature has been in response especially to problems of pattern recognition that center on construction of computer programs and peripheral devices that have rudimentary perceptual capacities. Although this approach has a formal character quite different from the others considered and it has not been used to address directly the question about the Euclidean character of visual space, it should be mentioned because it does provide an approach that in many respects is very natural psychologically and that is in certain aspects more closely connected to the psychology of perception than most of the classical geometric approaches that have been used thus far in the analysis of visual space. (An elementary introduction and references to the earlier literature are to be found in Suppes and Rottmayer (1974); an encyclopedic review is given by Fu (1974). The literature of the last 25 years is too large to cite single survey references. Important recent statistical trends are discussed in Attias (1999), Breiman (1999) and Friedman, Hastie and Tibshirani (2000).)

A typical picture grammar has the following character. Finite line segments or finite curves of a given length and with a given orientation are concatenated together as basic elements to form geometric figures of greater complexity. A typical problem in the literature of pattern recognition is to provide such a concatenation (not necessarily one dimensional) so as to construct handwritten characters, or, as a specialized example that has received a fair amount of attention, to recognize handwritten mathematical symbols. These approaches are often labelled picture grammars, because they adopt the approach used in mathematical linguistics for writing phrase-structure grammars to generate linguistic utterances. Picture grammars can in fact be characterized as context free, context sensitive, etc., depending upon the exact character of the rules of production. What is missing is the question, Can the set of figures generated by the picture grammars be embedded in Euclidean space or other metric spaces of an elementary character? This question would seem to have some conceptual interest from the standpoint of the theory of perception. It is clearly not of the same importance for the theory of pattern recognition. Picture grammars base perception on a set of primitive concepts that seem much more natural than the more abstract concepts familiar in classical geometry. Moreover, extensive results on the behavior of single neurons in the

6.4 How to Decide if Visual Space is Euclidean

brain support that, at least partially, specialized neurons in the visual systems of many animals detect specific features of the kind used in picture grammars.

The hierarchy of geometries. Those who have declared that visual space is not Euclidean have usually had a well-defined alternative in mind. The most popular candidates have been claims that visual space is either elliptic or hyperbolic, although some more radical theses are implicit in some of the experimental work.

How the various geometries are to be related hierarchically is not entirely a simple matter, for by different methods of specialization one may be obtained from another. A reasonably natural hierarchy for purposes of talking about visual space is shown in Figure 1. In the figure, I have also referred to geometries rather than to spaces, although from a certain conceptual standpoint the latter is preferable. I have held to the language of *geometries* in deference to tradition in the literature on visual space. The weakest geometry considered here is either projective geometry on the left-hand side at the top of the figure or ordered geometry at the right. There are various natural primitive concepts for projective geometry. Fundamental in any case is the concept of incidence and, once order is introduced, the concept of separation, introduced in Chapter 3. In contrast, ordered geometry is based upon the single ternary relation of betweenness holding for three points in the fashion standard for Euclidean geometry, but of course axioms based only upon betweenness are weaker than those required for hyperbolic or Euclidean geometry. Without entering into technical details, elliptic geometry of the plane is obtained from projective geometry by defining it as the geometry corresponding to the group of projective collineations that leave an imaginary ellipse invariant in the projective plane. Although elliptic geometry has been important in the consideration of visual space, as we shall see later, the details of elliptic geometry are complicated and subtle, and as far as I know have not actually been adequately studied in detail in relation to any serious body of experimental data.

Turning now to the right-hand side of Figure 1, affine geometry is obtained from ordered geometry by adding Euclid's axiom that, given a line and a point external to the line, there is at most one line (i) through the point, (ii) in the plane formed by the point and the line, and (iii) that does not meet the line. Going in the other direction from ordered geometry in Figure 1, we obtain absolute geometry by adding the concept of congruence of segments, which is just the notion of equidistance mentioned earlier. We add Euclid's axiom to absolute geometry to obtain Euclidean geometry and we add the negation of Euclid's axiom to absolute geometry to obtain hyperbolic geometry. These are the only two extensions of absolute geometry. Given the fundamental character

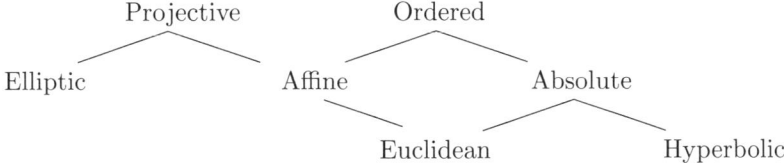

FIGURE 1   Hierarchy of geometries.

of absolute geometry in relation to the claims often made that visual space is either Euclidean or hyperbolic, it is somewhat surprising that there has been no more detailed investigation experimentally of whether the axioms of absolute geometry hold for visual space.

There is another way of organizing the hierarchy of geometries in terms of metric spaces. Recall that a *metric space* is a pair (A, d) such that A is a nonempty set, d is a real-valued function defined on the Cartesian product $A \times A$, such that for all a, b, c in A, (i) $d(a, a) = 0$ and if $a \neq b, d(a, b) > 0$; (ii) $d(a, b) = d(b, a)$; and (iii) $d(a, b) + d(b, c) \geq d(a, c)$. The elements of the set A are called *points*. The first axiom asserts that distances are positive, except for the distance between identical points, which is zero. The second axiom asserts that distance is symmetric; that is, it is a function only of the unordered pair of points, not a function of their order. The third axiom is the triangle inequality.

Most of the metric spaces important for the theory of perception have the property that any two points can be joined by a segment. Such spaces are called metric spaces with additive segments. These spaces are naturally divided into two broad subclasses, affine metrics and coordinate-free metrics. By further specialization of each of these subclasses we are led naturally to the Euclidean, hyperbolic, and spherical spaces, as well as to generalizations of the Euclidean metric in terms of what are called Minkowski metrics. An important subclass of the coordinate-free metrics is the family of Riemannian metrics. It may be shown that the only spaces that are Riemannian and affine metric are either Euclidean or hyperbolic. We shall not use these concepts in detail, but it is important to mention that this alternative hierarchy of metric spaces is as natural to use as the more classical hierarchy exhibited in Figure 1. All of the concepts I have introduced in this brief survey of the hierarchy of geometries are familiar in the mathematical literature of geometry. More details are to be found in Suppes et al. (1989, Chs. 12 & 13).

6.5 The Nature of Visual Space: Experimental and Philosophical Answers

My main purpose in this section is to provide a survey of the answers that have been given to the nature of visual space. The natural place to begin is with Euclid's *Optics*, the oldest extant treatise on mathematical optics. As stated in Section 2.4, it is important to emphasize that Euclid's *Optics* is really a theory of vision, not a treatise on physical optics.

Euclid's restriction to monocular vision is one that we shall meet repeatedly in this survey. However, it should be noted that he proves several propositions involving more than one eye; for example, "If the distance between the eyes is greater than the diameter of the sphere, more than the hemispheres will be seen". Euclid is not restricted to some simple geometric optics but is indeed concerned with the theory of vision, as is evident from the proposition that "if an arc of a circle is placed on the same plane as the eye, the arc appears to be a straight line". This kind of proposition is a precursor of later theories—for example, that of Thomas Reid (1764/1967)—which emphasize the non-Euclidean character of visual space.

6.5 VISUAL SPACE: EXPERIMENTAL AND PHILOSOPHICAL ANSWERS

I skip rapidly through the period after Euclid to the eighteenth century, not because there are not matters of interest in this long intervening period but because there do not seem to be salient changes of opinion about the character of visual space, or at least if there are, they are not known to me. I looked, for example, at the recent translation by David C. Lindberg (1970) of the thirteenth-century treatise *Perspectiva Communis* of John Pecham and found nothing to report in the present context, although the treatise itself and Lindberg's comments on it are full of interesting matter of great importance concerning other questions in optics, as, for example, theories about the causes of light.

Newton's *Opticks* (1704/1931) is in marked contrast to Euclid's. The initial definitions do not make any mention of the eye until Axiom VIII, and then in very restrained fashion. Almost without exception, the propositions of Newton's optics are concerned with geometric and especially physical properties of light. Only in several of the Queries at the end are there any conjectures about the mechanisms of the eye, and these conjectures do not bear on the topic at hand.

Five years after the publication of the first edition of Newton's *Opticks*, Berkeley's *An Essay towards a New Theory of Vision* (1709/1901) appeared in 1709. Berkeley does not have much of interest to say about the geometry of visual space, except in a negative way. He makes the point that distance cannot be seen directly and, in fact, seems to categorize the perception of distance as a matter of tactile, rather than visual, sensation, because the muscular convergence of the eyes is tactile in character. He emphatically makes the point that we are not able geometrically to observe or compute the optical angle generated by a remote point as a vertex, with sides pointing toward the centers of the two eyes. Here is what he says about the perception of optical angles. "Since therefore those angles and lines are not themselves perceived by sight, it follows, ...that the mind does not by them judge the distance of objects" (# 13). What he says about distance he also says about magnitude not being directly perceived visually. In this passage (# 53), he is especially negative about trying to use the geometry of the visual world as a basis for visual perception.

It is clear from these and other passages that for Berkeley visual space is not Euclidean, because there is no proper perception of distance or magnitude; at least, visual space is not a three-dimensional Euclidean space. What he seems to say is also ambiguous as to whether one should argue that it is at least a two-dimensional Euclidean space. My own inclination is to judge that his views on this are more negative than positive. Perhaps a sound negative argument can be made up from his insistence on there being a minimum visible. As he puts it, "It is certain sensible extension is not infinitely divisible. There is a minimum tangible, and a minimum visible, beyond which sense cannot perceive. This everyone's experience will inform him" (# 54).

In fact, toward the end of the essay, Berkeley makes it clear that even two-dimensional geometry is not a proper part of visual space or, as we might say, the visual field. As he says in the final paragraph of the essay, "By this time, I suppose, it is clear that neither abstract nor visible extension makes the object of geometry".

Of much greater interest is Thomas Reid's *Inquiry into the Human Mind*, first published in 1764 (1764/1967). Chapter 6 deals with seeing, and Section 9 is the celebrated one entitled 'Of the geometry of visibles'. It is sometimes said that this section is a proper precursor of non-Euclidean geometry, but if so, it must be regarded as an implicit precursor, because the geometry explicitly discussed by Reid as the geometry of visibles is wholly formulated in terms of spherical geometry, which had of course been recognized as a proper part of geometry since ancient times. The viewpoint of Reid's development is clearly set forth at the beginning of the section: "Supposing the eye placed in the centre of a sphere, every great circle of the sphere will have the same appearance to the eye as if it was a straight line; for the curvature of the circle being turned directly toward the eye, is not perceived by it. And, for the same reason, any line which is drawn in the plane of a great circle of the sphere, whether it be in reality straight or curve, will appear to the eye." It is important to note that Reid's geometry of visibles is a geometry of monocular vision. He mentions in other places binocular vision, but the detailed geometric development is restricted to the geometry of a single eye. The important contrast between Berkeley and Reid is that Reid develops in some detail the geometry in a straightforward, informal, mathematical fashion. No such comparable development occurs in Berkeley.

Daniels (1972) has argued vigorously that Reid's geometry of visibles is not simply a use of spherical geometry but is an introduction by Reid of a double elliptic space. A similar argument is made by Angell (1974). I am sympathetic with these arguments, but it seems to me that they go too far, and for a fairly straightforward reason not discussed by either Daniels or Angell. Let us recall how elliptic geometry was created by Felix Klein at the end of the nineteenth century. He recognized that a natural geometry, very similar to Euclidean geometry or hyperbolic geometry, could be obtained from spherical geometry by identifying antipodal points as a single point. The difficulty with spherical geometry as a geometry having a development closely parallel to that of Euclidean geometry is that two great circles, which correspond to lines, have two points, not one point, of intersection. However, by identifying the two antipodal points as a single point, a fair number of standard Euclidean postulates remain valid. It is quite clear that no such identification of antipodal points was made by Reid, for he says quite clearly in the fifth of his propositions, "Any two right lines being produced will meet in two points, and mutually bisect each other". This property of meeting in two points rather than one is what keeps his geometry of visibles from being a proper elliptic geometry and forces us to continue to think of it in terms of the spherical model used directly by Reid himself.

In spite of the extensive empirical and theoretical work of Helmholtz on vision, he does not have a great deal to say that directly bears on this question, and I move along to experiments and relevant psychological theory in the twentieth century. The initial stopping point is Blumenfeld (1913). (See Table 1 for the chronology.)

Blumenfeld was among the first to perform a specific experiment to show that, in one sense, phenomenological visual judgments do not satisfy all Euclidean properties. Blumenfeld performed experiments with so-called parallel and equidistance alleys. In a darkened room the subject sits at a table, looking straight ahead, and he is asked to adjust two rows of point sources of light placed on either side of the normal plane, i.e., the vertical plane that bisects the horizontal segment joining the centers of the two eyes. The two furthest lights are fixed and are placed symmetrically and equidistant

TABLE 1   Is Visual Space Euclidean?

Name	Claim	Answer
Euclid (300 B.C.)	Theory of perspective	Yes
Reid (1764), Daniels (1972), Angell (1974)	Geometry of visibles is spherical	No
Blumenfeld (1913)	Parallel alleys not equal to equidistance alleys	No
Luneburg (1947, 1948, 1950)	Visual space is hyperbolic	No
Blank (1953, 1957, 1958a,b, 1961)	Essentially the same as Luneburg	No
Hardy et al. (1953)	Essentially the same as Luneburg	No
Zajaczkowska (1956)	Positive results on experimental test of Luneburg theory	No
Schelling (1956)	Hyperbolic relative to given fixation point	No
Gogel (1956a,b, 1963, 1964a,b, 1965)	Equidistance tendency evidence for contextual geometry	No
Foley (1964, 1965, 1966, 1969, 1972, 1978)	Visual space is nonhomogeneous	No but
Indow (1967, 1968, 1974a,b, 1975)	MDS methods yield good Euclidean fit	Not sure
Indow et al. (1962a,b, 1963)	Close to Indow	Not sure
Nishikawa (1967)	Close to Indow	Not sure
Matsushima and Noguchi (1967)	Close to Indow	Not sure
Grünbaum (1963)	Questions the theory of Luneburg	Yes
Strawson (1966)	Phenomenal geometry is Euclidean	Yes

from the normal plane. The subject is then asked to arrange the other lights so that they form a parallel alley extending toward him from the fixed lights. His task is to arrange the lights so that he perceives them as being straight and parallel to each other in his visual space. This is the task for construction of a parallel alley. The second task is to construct a distance alley. In this case, all the lights except the two fixed lights are turned off and a pair of lights is presented, which are adjusted as being at the same physical distance apart as the fixed lights—the kind of equidistance judgments discussed earlier. That pair of lights is then turned off and another pair of lights closer to him is presented for adjustment, and so forth. The physical configurations do not coincide, but in Euclidean geometry straight lines are parallel if and only if they are equidistant from each other along any mutual perpendiculars. The discrepancies observed in Blumenfeld's experiment are taken to be evidence that visual space is not Euclidean. In both the parallel-alley and equidistance-alley judgments the lines diverge as you move away from the subject, but the angle of divergence tends to be greater in the case of parallel than in the case of equidistance alleys. The divergence of the alleys as one moves away from the subject has been taken by Luneburg to support his hypothesis that visual space is hyperbolic.

In fact, Luneburg, in several publications in the late forties, has been by far the strongest supporter of the view that visual space is hyperbolic. He, in conjunction with his collaborators, has set forth a detailed mathematical theory of binocular vision and at the same time has generated a series of experimental investigations to test the basic tenants of the theory. In many respects, Luneburg's article (1947) remains the best detailed mathematical treatment of the theory of binocular vision. Without extensive discussion, Luneburg restricts himself to Riemannian geometries of constant curvature in order to preserve rigid motions, that is, free mobility of rigid bodies. Luneburg develops, in a coordinate system natural for binocular vision, the theory of Riemannian spaces of constant curvature in a quite satisfactory form, although an explicit axiomatic treatment is missing. In orthogonal sensory coordinates the line element ds can be represented in terms of sensory coordinates α, β and γ (these Greek letters are not the ones used by Luneburg) by:

$$ds^2 = \frac{d\alpha^2 + d\beta^2 + d\gamma^2}{\left[(1 + \frac{1}{4}K(\alpha^2 + \beta^2 + \gamma^2))\right]^2}, \tag{1}$$

where

$K = 0$ for Euclidean space,
$K < 0$ for hyperbolic space,
$K > 0$ for elliptic space.

The best more recent treatment is Indow (1979).

It is important to recognize that the Luneburg approach is strictly a psychophysical approach to visual space. It assumes a transformation of physical Euclidean space to yield a psychological one. In this sense it is not a fundamental qualitative approach to the axioms of visual space. Luneberg nowhere examines with any care or explicitness the more general primitive assumptions that lead to the proof that visual space is a Riemannian space of constant curvature.

6.5 VISUAL SPACE: EXPERIMENTAL AND PHILOSOPHICAL ANSWERS

After these general developments he turns to the detailed arguments for the view that the appropriate space of constant curvature for visual space is hyperbolic ($K < 0$ in equation (1)). It is not possible to enter into the details of Luneburg's argument here, but he bases it on three main considerations, all of which have had a great deal of attention in the theory of vision: first, the data arising from the frontal-plane horopter where curves which appear as straight are physically curved (data on these phenomena go back to the time before Helmholtz); second, the kind of alley phenomena concerning judgments of parallelness mentioned earlier; and, third, accounting for judgments of distorted rooms in which appropriate perspective lines are drawn and which consequently appear as rectangular or regular. In this last case, Luneburg draws on some unpublished classic and spectacular demonstrations by A. Ames, Jr. For related work, see Ames (1946). One of the difficulties of this field is that the kind of detailed mathematical and quantitative arguments presented by Luneburg in connection with these three typical kinds of problems are not always satisfactorily analyzed in the later literature. Rather, new data of a different sort are often presented to show that different phenomena argue against Luneburg's hypothesis that visual space is hyperbolic. Still, the literature in both directions is substantial. The numerous references given in the rest of this chapter are by no means complete in scope.

Luneburg died in 1949, but a number of his former students and collaborators have continued his work and provided additional experimental support as well as additional mathematically based arguments in favor of his views. I refer especially to Blank (1953, 1957, 1958a, 1958b, 1961) and Hardy et al. (1953), although this is not an exhaustive list. Another positive experimental test was provided by Zajaczkowska (1956).

Schelling (1956) agrees with Luneburg but makes an important point of modification, namely, the metrics of negative curvature—that is, of the hyperbolic spaces that Luneburg argues for—are essentially momentary metrics. At a given instant the eye has a certain fixation point, and relative to this fixation point Luneburg's theory is, according to Schelling, probably approximately correct, but the applicability of the theory is severely restricted because the eyes are normally moving about continuously and the points of fixation are continually changing. This fundamental fact of change must be taken account of in any fully adequate theory.

Gogel (1956a,b, 1963, 1964a,b, 1965) has studied what is called the equidistance tendency, or what in the context of this chapter we might term the Berkeley tendency. Remember that Berkeley held that distance from the observer was not a visual idea at all, but derived from the tactile sense. Without entering into a precise analysis of Berkeley's views, Gogel has provided an important body of evidence that when other cues are missing there is a strong tendency to view objects as being at the same distance from the observer. These careful and meticulous studies of Gogel are important for establishing not only the equidistance tendency but also its sensitivity to individual variation, on the one hand, and to the presence of additional visual cues on the other. The equidistance tendency is certainly present as a central effect, but any detailed theory of visual space has a bewildering complexity of contextual and individual differences to account for, and it seems to me that Gogel's experiments are essentially decisive on this point. In the papers referred to, Gogel does not give a sharp answer to the question about the character of visual space. But I have listed him in Table 1 because it seems to me that the impact of his studies is to argue strongly for skepticism about fixing the geometry of visual space very far up in the standard hierarchy. Rather, his studies support the point that the full geometry is strongly contextual in character and therefore quite deviant from the classical hierarchy.

A number of interesting experimental studies of the geometry of visual space have been conducted by John Foley. In Foley (1964) an experiment using finite configurations of small point sources of light was conducted to test the Desarguesian property of visual space.[17] (Of course, the property was tested on the assumption that a number of other axioms were valid for visual space.) The results confirmed the Desarguesian property for most observers but not for all. In Foley (1966), perceived equidistance was studied as a function of viewing distance. Like most of Foley's experiments, this was conducted in the horizontal eye-level plane. The locus of perceived equidistance was determined at distances of 1.2, 2.2, 3.2, and 4.2 meters from the observer. As in other Foley experiments, the stimuli were small, point-like light sources viewed in complete darkness. The observer's head was held fixed but his eyes were permitted to move freely. There were five lights, one in the normal plane, which was fixed, and four variable lights on each side of the normal plane at angles of 12 degrees and 24 degrees with respect to the normal plane. The locus of perceived equidistance was found to be concave toward the observer at all distances. Perhaps most importantly, the locus was found to vary with viewing distance, which indicates that the visual space does not depend on the spatial distribution of retinal stimulation alone. Again, there is here a direct argument for a contextual geometry and results are not consistent with Luneburg's theory. The equidistance judgments were of the following sort. A subject was instructed to set each of the lights, except the fixed light, in the normal plane to be at the same distance from himself as the fixed light. Thus, it should appear to him that the lights lie on a circle, with himself as observer at the center. The important point is that for none of the ten subjects in the experiment did the judgments of the locus for equidistance lie on the Vieth-Mueller horopter or circle mentioned earlier as one of the supporting arguments for Luneburg's theory. Also important for the fundamental geometry of visual space is the fact that the loci determined by the observers were not symmetric about the normal plane.

Here is another beautiful experiment by Foley (1972). The situation is shown in Figure 2. The instructions to the subject are to make judgments of perpendicularity ($\perp$) and congruence ($\approx$) as follows, with the subject given point A as fixed in the depth axis directly in front of his position as observer O:

---

[17] First, in explaining Desargues' proposition, we need two definitions. Two triangles are *perspective from a point* if and only if there is a one-to-one correspondence between the vertices of the two triangles such that the three lines passing through the three pairs of corresponding vertices meet in a common point, the perspective point. Two triangles are *perspective from a line* if and only if there is a one-to-one correspondence between the lines forming the sides of the two triangles such that the three pairs of corresponding lines forming the sides of the two triangles meet at points that themselves lie on a common line. The Desarguesian property or proposition is that two triangles that are perspective from a point are also perspective from a line. In projective geometry, planes can be either Desarguesian or not, in three-dimensional projective geometry the property is provable from other standard axioms.

1. Find point B so that $AB \perp OB$ & $AB \approx OB$.
2. Find point C so that $OC \perp OB$ & $OC \approx OB$.
3. Judge the relative lengths of OA & BC.

The results are that the 24 subjects in 40 of 48 trials judged BC significantly longer than OA. This judgment that BC is longer than OA contradicts properties of any space of constant curvature, whether that curvature is positive, negative, or zero.

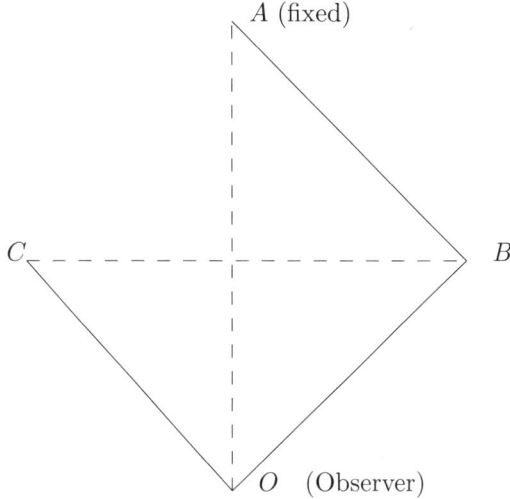

FIGURE 2   Foley experiment.

Foley's (1972) study shows experimentally that, on the one hand, the size-distance invariance hypothesis is incorrect, and that in fact the ratio of perceived frontal extent to perceived egocentric distance greatly exceeds the physical ratio, while, on the other hand, perceived visual angles are quite close to physical ones. These results, together with other standard assumptions, are inconsistent with the Luneburg theory that visual space is hyperbolic. Foley describes the third experiment in this paper in the following way:

> How can it be that in the primary visual space reports of perceived size-distance ratio are not related to reports of perceived visual angle in a Euclidean way? One possibility is that the two kinds of judgments are in part the product of different and independent perceptual processes .... The results are consistent with the hypothesis that the two kinds of judgments are the product of independent processes. They also show that no one geometric model can be appropriate to all stimulus situations, and they suggest that the geometry may approach Euclidean geometry with the introduction of cues to distance.   (Foley 1972, p. 328)

Again, there is in Foley's analysis a strong case for a contextual geometry. A number of other detailed experimental studies of Foley that have not been referenced here build a case for the same general contextual view, which I discuss in more detail below.

A number of detailed investigations on the geometry of visual space have been conducted by Tarow Indow (1967, 1968, 1974a,b, 1975) and other Japanese investigators closely associated with him (Indow et al. 1962a, 1962b, 1963; Matsushima and Noguchi 1967; Nishikawa 1967). They have found, for example, that multidimensional scaling methods (MDS), which have been intensively developed in psychology, in many cases yield extremely good fits to Euclidean space (Indow 1982). Indow has experimentally tested the Luneburg theory based upon the kind of alley experiments that go back to Blumenfeld (1913). As might be expected, he duplicates the result that the equidistance alleys always lie outside the parallel alleys, which, under the other assumptions that are standard, implies that the curvature of the space is negative and therefore it must be hyperbolic. But Indow (1974a,b) properly challenges the simplicity of the Luneburg assumptions, especially the constancy of curvature. It is in this context that he has also tried the alternative approach of determining how well multidimensional scaling will fit a Euclidean metric. As he emphasizes, the Luneburg approach is fundamentally based upon differential geometry as a method of characterizing Riemannian spaces with constant curvature, but for visual judgments it is probably more appropriate to depend upon judgments in the large, and therefore upon a different conceptual basis for visual geometry. Throughout his writings, Indow recognizes the complexity and difficulty of reaching any simple answer to give the proper characterization of visual space. The wealth of detail in his articles and those of his collaborators is commended to the reader who wants to pursue these matters in greater depth.

I myself have learned more about the theory of visual space from Indow and his colleagues than from anyone else. I was initially enormously skeptical of the Luneburg ideas and I came to realize they could be converted into a realistic program, just because of the extraordinarily careful experiments performed by Indow and his colleagues. In this case the fact that the program has not turned out to be more satisfactory than it is, is not because of the weakness of the experiments, but in fact because of their very strength. They have given us confidence to recognize that there are fundamental things wrong with the Luneburg approach to visual space. Above all, as Indow and his colleagues have brought out on several occasions, there is a complete lack of parametric stability once we assume that a space of constant curvature, for example, negative curvature in the hyperbolic case, is a reasonable hypothesis. When we estimate the curvature we find that, even for the same experiments, the results are not stable from day to day for a given subject, and certainly when we transfer from one kind of experiment, for example, the classical alley experiments, to judgments of a different kind, there is little transfer at all of the parametric values estimated in one situation to the next.

Careful mathematical analysis of several different features of visual space relevant to the preceding review has been given by Jan Drösler in a series of articles (1966, 1979a,b,c, 1987, 1988, and 1992). The topics range from multidimensional scaling in Cayley-Klein geometries (1979c) to the psychophysical function of binocular space perception (1988), which is especially pertinent to the more detailed development of the psychophysical ideas of Luneburg in a richer psychological setting.

Another experiment (Wagner 1985) dealt with perceptual judgments about distances and other measures among 13 white stakes in an outdoor viewing area. In physical coordinates let x equal measurement along the depth axis and y along the frontal axis and let perceptual judgments of distance be shown by primes (see Figure 3 for depth and frontal axis). Then the general result of Wagner is that $x' \approx 0.5y'$ if physically $x = y$. Notice how drastic the foreshortening is along the depth axis. This result of Wagner's is not anomalous or peculiar but represents a large number of different perception experiments showing dramatic foreshortening along the depth axis.

In his important book on the philosophy of space and time, Grünbaum (1963) rejects the Luneburg theory and affirms that, in order to yield the right kinds of perceptual judgments, visual space must be Euclidean. His argument is rather brief and I shall not examine it in any detail. It would be my own view that he has not given proper weight to the detailed experimental studies or to the details of the various theoretical proposals that have been made.

I close this survey by returning to a philosophical response to the question, that of Strawson (1966) in his book on Kant's *Critique of Pure Reason*. From the standpoint of the large psychological literature I have surveyed, it is astounding to find Strawson asserting as a necessary proposition that phenomenal geometry is Euclidean. The following quotation states the matter bluntly:

> With certain reservations and qualifications, to be considered later, it seems that Euclidean geometry may also be interpreted as a body of unfalsifiable propositions about phenomenal straight lines, triangles, circles, etc.; as a body of a priori propositions about spatial appearances of these kinds and hence, of course, as a theory whose application is restricted to such appearances.

(Strawson 1966, p. 286)

The astounding feature of Strawson's view is the absence of any consideration that phenomenal geometry could be other than Euclidean and that it surely must be a matter, one way or another, of empirical investigation to determine what is the case. The qualifications he gives later do not bear on this matter, but pertain rather to questions of idealization and of the nature of constructions, etc. The absence of any attempt to deal in any fashion whatsoever with the large theoretical and experimental literature on the nature of visual space is surprising.

6.6 Partial Axioms for the Foley and Wagner Experiments[18]

I am not able to give a fully satisfactory geometric analysis of the strongly supported experimental facts found in these experiments but do think there are some things of interest to be said that will help clarify the foundational situation. I have divided the axioms that I propose into various groups.

Affine plane. These I take to be the standard axioms of Definition 1.1. We could of course, reduce them and not require the whole plane but that is not important here. We can take as primitives, either betweenness, or parallelness and some concept like that of a midpoint algebra. Again the decision is not critical for considerations here. We now add judgments of perceived congruence ($\approx$).

Three distinguished points. Intuitively, let o_1 be the center of the left eye, o_2 the center of the right eye and o the bisector of the segment o_1o_2. Explicitly, they satisfy the following two axioms.

---

[18]This section draws on the material in Suppes (1995).

2a. The three points are collinear and distinct;

2b. $o_1 o \approx o o_2$.

Affine axioms of congruence.

3a. Opposite sides of any parallelogram are congruent.

3b. If $aa \approx bc$ then $b = c$.

3c. $ab \approx ba$.

3d. If $ab \approx cd$ & $ab \approx ef$ then $cd \approx ef$.

3e. If $a|b|c$ & $a'|b'|c', ab \parallel a'b'$ & $ab \approx a'b'$ & $bc \approx b'c'$ then $ac \approx a'c'$ (this is a familiar and weak affine additivity axiom).

The *frontal axis* is the line containing o_1, o, and o_2 and the *depth axis* is the half line through o such that for any point a on the axis $o_1 a \approx o_2 a$. (Notice that we cannot characterize the depth axis in terms of a general notion of perpendicularity for that is not available, and in fact will not be available within the framework of these axioms. The depth axis is only a half line because a subject cannot see directly behind the frontal axis.)

3f. *First special congruence axiom.* If $a \neq c, a$ and c are on the frontal axis and b is on the depth axis, and $ao \approx oc$, then $ab \approx bc$ (see Figure 3).

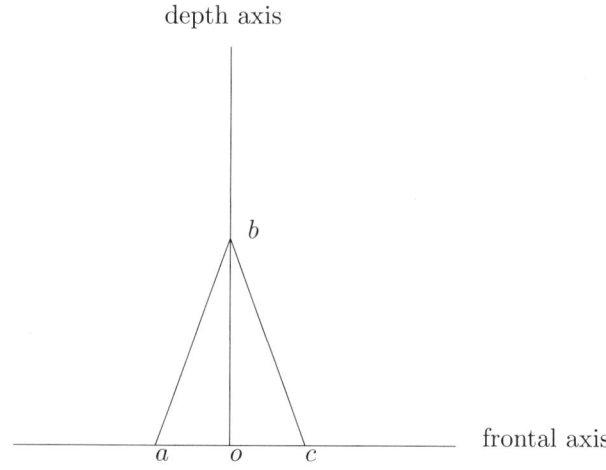

FIGURE 3   Congruence Axiom 3f.

3g. *Second special congruence axiom.* If $a \neq c, a, c$ on the frontal axis, $ao \approx oc$, ab and $cd \parallel$ to depth axis, $ab \approx cd$, then $ob \approx od$ (see Figure 4).

The last two special axioms of congruence extend affine congruence to congruence of segments that are not parallel, but only in the case where the segments are symmetric about the depth axis as is seen from Figures 3 and 4. This means that we have a weak extension of affine congruence. An extension that is far too weak even to give us the axioms of congruence for absolute spaces (see Suppes et al. 1989, Ch. 13). (As already remarked, absolute geometry can be thought of this way. Drop the Euclidean axiom that through a given point a exterior to a line α there is at most one line through a

6.6 FOLEY AND WAGNER EXPERIMENTS

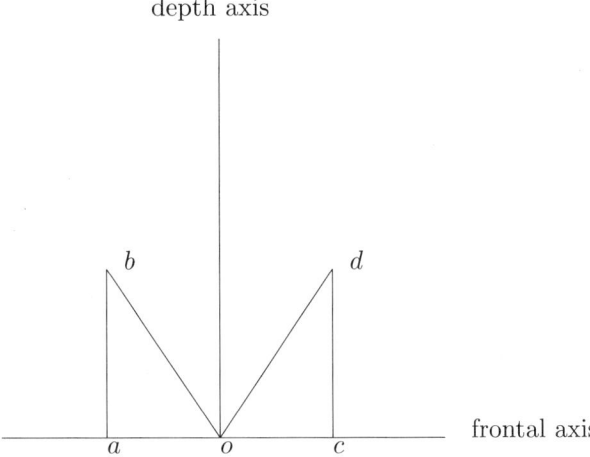

FIGURE 4   Congruence Axiom 3g.

that is parallel to a and lies in the plane formed by a and α. Adding this axiom to the axioms of absolute geometry gives us Euclidean geometry, as is obvious. What is much more interesting is that adding the negation of this axiom to those of absolute geometry gives us hyperbolic geometry.) We can prove the following theorem.

THEOREM 1. *Let the halfspace consisting of all points on the same side of the frontal axis as the depth axis be designated the frontal half plane:*

(1). The frontal halfplane is isomorphic under a real-valued function φ to a two-dimensional affine halfplane over the field of real numbers, with the x-axis the depth axis and the y-axis the frontal axis. Moreover, congruence as measured by the Euclidean metric is satisfied when line segments are congruent, i.e., if $ab \approx cd$ then $\sum_{i=1}^{2}(\varphi_i(a) - \varphi_i(b))^2 = \sum_{i=1}^{2}(\varphi_i(c) - \varphi_i(d))^2$.

(2). The only affine transformations possible are those consisting of stretches α of the frontal axis and stretches β of the depth axis with $\alpha, \beta > 0$.

The proof of (1) is obvious from familiar results of classical geometry. The proof of (2) follows from observing that the affine transformations described are the only ones that preserve the symmetry around the depth axis required by the two special congruence axioms.

It is clear that the results of Theorem 1 are quite weak. The theorem is not contradicted by the Foley and Wagner experiments, but this is not surprising, for the apparatus of betweenness plus symmetric congruence about the depth axis cannot describe the results of either experiment. If we were to add a concept of perpendicularity, as required by the Foley procedure, then we would essentially get a Euclidean halfplane, and the resulting structure would contradict the Foley results.

Correspondingly, we cannot describe the Wagner psychophysical results of extensive perceptual foreshortening along the depth axis, without adding some psychophysical assumptions radically different from those of Luneburg. Of course, the Foley results also are best interpreted as perceptual foreshortening along the depth axis. The natural conclusion is that we cannot consistently describe visual geometric relations in any space close to the specificity and simplicity of structure of a space of constant curvature.

Even the weak affine structure of Theorem 1 is too strong, and probably should be replaced by a standard version of absolute geometry, but with the congruence axioms weakened as given above. Such an absolute geometry can be extended to hyperbolic geometry, but the affine structure cannot. What we seem to end up with is a variety of fragments of geometric structures to describe different experiments. A hyperbolic fragment perhaps for alley experiments, a fragment outside the standard ones for the Foley experiments, etc.

The goal of having a unified structure of visual space adequate to account for all the important experimental results now seems mistaken. Only a pluralistic and fragmentary approach seems possible.

6.7 Three Conceptual Problems About Visual Space

In this section, I center my remarks around three clusters of issues. The first is concerned with the contextual character of visual geometry, the second with problems of distance perception and motion, and the third with the problem of characterizing the nature of the objects of visual space.

Contextual geometry. A wide variety of experiments and ordinary experience as well testify to the highly contextual character of visual space. The presence or absence of 'extraneous' points can sharply affect perceptual judgments. The whole range of visual illusions, which I have not discussed here, also provides a broad body of evidence for the surprising strength of these contextual effects.

As far as I can tell, no one has tried seriously to take account of these contextual effects from the standpoint of the axiomatic foundations of visual geometry. In a way it is not surprising, for the implications for the axiomatic foundations are, from the ordinary standpoint, horrendous. Let us take a simple example to illustrate the point.

In ordinary Euclidean geometry, three points form an isosceles triangle just when two sides of the triangle are of the same length. Suppose now that Euclidean geometry had the much more complicated aspect that whether a triangle were isosceles or not depended not simply on the configuration of the three points but also on whether there was a distinguished point lying just outside the triangle alongside one of the dual sides. This asymmetry may well make the visual triangle no longer appear isosceles. This is but one simple instance of a combinatorial nightmare of contextual effects that can easily be imagined and, without much imagination or experimental skill, verified as being real. For a bewildering, diverse collection of examples of visual illusions and a history of the subject, see Coren and Girgus (1978).

What are we to say about such effects? It seems to me the most important thing is to recognize that perceptual geometry is not really the same as classical geometry at all, but in terms of the kinds of judgments we are making it is much closer to physics. Consider, for example, the corresponding situation with bodies that attract each other by gravitation. The introduction of a third body makes all the difference to the motions of the two original bodies and it would be considered bizarre for the situation to be

6.7 Conceptual Problems About Visual Space

otherwise. This also applies to electromagnetic forces, mechanical forces of impact, etc. Contextual effects are the order of the day in physics, and the relevant physical theories are built to take account of such effects.

Note that physical theories depend upon distinguished objects located in particular places in space and time. Space-time itself is a continuum of undistinguished points, and it is characteristic of the axiomatic foundations of classical geometry that there are no distinguished points in the space. But it is just a feature of perception that we are always dealing with distinguished points which are analogous to physical objects, not geometric points. Given this viewpoint, we are as free to say that we have contextual effects in visual geometry as we are to make a similar claim in general relativity due to the presence of large masses in a given region.

Interestingly enough, there is some evidence that as we increase the visual cues, that is, as we fill up the visual field with an increasingly complex context of visual imagery, the visual space becomes more and more Euclidean. It is possible that we have here the exact opposite of the situation that exists in general relativity. In the case of perception it may be that spaces consisting of a very small number of visible points may be easily made to deviate from any standard geometry.

The geometric viewpoint can be brought into close alignment with the physical one, when the isomorphic embedding of finite sets of points in some standard geometry is taken as the appropriate analysis of the nature of visual space. This approach was mentioned earlier and is implicit in some of the experimental literature discussed. It has not sufficiently been brought to the surface, and the full range of qualitative axioms that must be satisfied for the embedding of a finite collection of points in a unique way in a given space, whether Euclidean, hyperbolic, elliptic, or whatever, needs more explicit and detailed attention.

It also seems satisfactory to avoid the problems of contextual effects in initial study of this kind by deliberately introducing symmetries and also certain additional assumptions such as quite special relations of a fixed kind to the observer. The many different experimental studies, and the kind of mathematical analysis that has arisen out of the Luneburg tradition, suggest that a good many positive and almost definitive results can be achieved under special restrictive assumptions. It seems that making these results as definitive as possible, admitting at the same time their specialized character and accepting the fact that the general situation is contextual in character, is an appropriate research strategy. It also seems likely that for these special situations one can give a definitely negative answer to the question, Is visual space Euclidean?, and respond that, to high approximation, in many special situations it is hyperbolic and possibly in certain others elliptic in character. This restricted answer is certainly negative. A general answer at the present time does not seem available as to how to characterize the geometry in a fully satisfactory way that takes account of the contextual effects that are characteristic of visual illusions, equidistance tendencies, etc.

Distance perception and motion. As indicated earlier in the brief discussion of the Helmholtz-Lie problem, most of the work surveyed in the preceding section has not taken sufficient account of problems of motion. There is an excellent survey article of Foley (1978) on distance perception which indicates that eye motion during the initial stage of focusing on an object is especially critical in obtaining information about perceptual distance. Philosophical traditions in perception have tended to ignore the complicated problems of motion of the eyes or head as an integral part of visual perception, but the most elementary considerations are sufficient to demonstrate their fundamental importance. It was a fundamental insight of Luneburg to recognize that it is important to characterize invariance properties of motions of the eyes and head that compensate each other. The deeper aspects of scanning that determine the character of the visual field have not really been studied in a thoroughly mathematical and quantitative fashion. There is little doubt in my mind that this is the area most important for future developments in the theory of visual space. We should, I would assume, end up with a kinematics of visual perception replacing the geometry of visual perception. For example, Lamb (1919) proves that under Donders' law, which asserts that the position of the eyeball is completely determined by the primary position and the visual axis aligned to the fixation point, it is not possible for every physically straight line segment to be seen as straight. This kinematical theorem of Lamb's, which is set forth in detail in Roberts and Suppes (1967), provides a strong kinematical argument against the Euclidean character of visual space. I cite it here simply as an example of the kind of results that one should expect to obtain in a more thoroughly developed kinematics of visual perception.

Objects of visual space. Throughout the analysis given in this chapter the exact characterization of what are to be considered as the objects of visual space has not been settled in any precise or definitive way. This ambiguity has been deliberate, because the wide range of literature to which I have referred does not have a settled account of what are to be regarded as visual objects. The range of views is extreme—from Berkeley, who scarcely even wants to admit a geometry of pure visual space, to those who hold that visual space is simply a standard Euclidean space and there is little real distinction between visual objects and physical objects. In building up the subject axiomatically and systematically, clearly some commitments are needed, and yet it seems that one can have an intelligible discussion of the range of literature considered here without having to fix upon a precise characterization, because there is broad agreement on the look of things in the field of vision. From the standpoint of the geometry of visual space, we can even permit such wide disagreement as to whether the objects are two dimensional or three dimensional in order to discuss the character of the geometry. Thomas Reid would lean strongly toward the two-dimensional character of visual space. Foley would hold that visual space is three dimensional; note, however, that most of his experiments have been restricted to two dimensions. At the very least, under several different natural characterizations of the objects of visual space, it is apparent that strong claims can be made that visual space is not Euclidean, and this is a conclusion of some philosophical interest.

6.8 Finitism in Geometry[19]

Developments in the foundations of mathematics in the twentieth century have been dominated by a concern to provide an adequate conceptual basis for classical analysis, as well as classical number theory. The attitude of Frege, Hilbert, Russell and Brouwer, along with others who have followed them, have had a continual focus on providing a proper conceptual foundation for classical parts of pure mathematics. What is remarkable about this history is how little attention has been paid to finitism in the application of mathematics. Attention to various applied traditions would have supported a much more radical finitism than was possible, as long as the viewpoint was that of providing an adequate basis for classical elementary number theory. Even the weakest systems of recursive arithmetic have closure conditions that lead to constructions far more complex than are needed in any standard applications. The passions for closure and for completeness are in themselves wonderful intellectual passions, but they run counter to the drive for solutions of applied problems by limited means, ones that are computationally feasible. This line of thought easily leads to finitism in the applications of mathematics to the real world.

Those whose focus is pure mathematics might immediately respond to my emphasis on finitism by saying that there is already a considerable emphasis on finitism in pure geometry, and by then citing the extensive and deep research on finite geometries in the last half century. It is also natural to think of this work as an extension in many ways of Felix Klein's Erlangen program, that is, the approach to geometry that emphasizes, above all, the structure of the group of automorphisms of the geometry.[20] The study of automorphisms is also an important way of studying finite geometries, for given a finite geometry, the group of automorphisms is a finite group. Also given a finite group, a natural problem is to determine its finite geometry. As noted earlier, a survey up until the mid-sixties of the twentieth century is to be found in Dembowski (1968). In the last thirty years there has also been a substantial body of work, and a number of open problems of a purely mathematical sort have been solved, especially in the relation of finite geometries to finite simple groups.

However, this large body of research is not the topic of this section. I have in mind more elementary mathematical problems, ones that are focused on applications. What are the applications I have in mind? I begin with the mathematical methods used by architects since ancient times. As early as the sixth century BC, architects were applying geometry to detailed questions of construction. It was not that they were simply mechanically using the elementary geometry familiar from the philosophical and mathematical tradition of that time, but they were using detailed geometric ideas to provide visual illusions. These methods of providing for visual illusions led to what are called in the architectural literature, then and now, 'refinements'. A principal example is *entasis,* which is a geometric construction that makes the vertical outline of marble columns convex rather than linear. Here is what Vitruvius in the first century BC has to say about this construction.

---

[19]The material in this section is drawn from Suppes (2001a).
[20]See Section 4.3.

> These proportionate enlargements are made in the thickness of columns on account of the different heights to which the eye has to climb. For the eye is always in search of beauty, and if we do not gratify its desire for pleasure by a proportionate enlargement in these measures, and thus make compensation for ocular deception, a clumsy and awkward appearance will be presented to the beholder. With regard to the enlargment made at the middle of columns, which among the Greeks is called ἔντασις, at the end of the book a figure and calculation will be subjoined, showing how an agreeable and appropriate effect may be produced by it. (Vitruvius, 1960 edition, p. 86)

Entasis is, of course, a famous example, but there are many more; the most extensive extant written record is Vitruvius' Roman treatise, written in the time of Augustus. The detailed applications to be found in Vitruvius are not exceptional but standard. It is a justified lament that the book on architecture written by Ictinus, architect of the Parthenon, is now lost. The important point in the present context is that, long before Ictinus and long after Vitruvius, architects were trained in geometry and how to apply it in a detailed way to the construction of buildings. As I like to put it, these architects were not proving theorems, but doing constructions (*problems* in the language of Euclid). The constructions were not all just simple and obvious; in fact, the lost drawing of Vitruvius' diagram of the construction of entasis was more complicated in its mathematical content than what can be found in most well-known architectural treatises today. (But certainly not more so than the mathematics back of modern computer programs for architectural design.) More than a thousand years after Vitruvius, in Palladio's famous four books of architecture (1570/1965), the Greek tradition of proportion, rather than direct numerical computation, continued to dominate the detailed thinking of architects, as may be seen in many places in Palladio's exposition. In various ways, the computer resources now available are encouraging once again the emphasis on proportion.

Beyond entasis and other visual illusions, the most important mathematical or geometric discovery already used by architects of the fifth century was the discovery of perspective, traditionally attributed to the painter Agatharchus, who designed the background set for the staging of a tragedy by Aeschylus and wrote a book about scene painting, as a consequence of this experience. Mathematical investigation of the theory of perspective began as early as Democritus and Anaxagoras. This is what Vitruvius had to say about this early work on perspective, writing half a millenium later in Rome.[21]

> Agatharcus, in Athens, when Aeschylus was bringing out a tragedy, painted a scene, and left a commentary about it. This led Democritus and Anaxagoras to write on the same subject, showing how, given a centre in a definite place, the lines should naturally correspond with due regard to the point of sight and the divergence of the visual rays, so that by this deception a faithful representation of the appearance of buildings might be given in painted scenery, and so that, though all is drawn on a vertical flat facade, some parts may seem to be withdrawing into the background, and others to be standing out in front. (Vitruvius, 1960 edition, p. 198)

---

[21] According to Alexander Jones (2000), the only two writers on mathematical optics in antiquity who studied problems of perspective were Ptolemy in Book VII of his *Geography* and Pappus in Book VI of his *Collection*.

6.8 FINITISM IN GEOMETRY

What is important about this variety of applications of geometry, including, of course, the important case of perspective, is that the methods and results are highly finitistic in character, just as are, in fact, most, if not all, the results of Greek geometry.

So, if we look at the applications of geometry in this long and important architectural tradition, which includes the thorough development of the mathematical theory of perspective and also of projective geometry, one is naturally struck by the tight interaction between the finitistic character of the mathematics and the familiar applications. The applications of numerical methods in ancient times have the same finitistic character, but I do not try to consider the details here.

At a certain point in the nineteenth century, nonfinitistic methods entered classical analysis and were extended, in some sense, to geometry, due to several different forces at work. I would stress especially the demand for representation theorems in complete form, because the representations of geometric spaces came to be formulated so as to be isomorphic to a given Cartesian product of the real number system, the product itself depending upon the dimension of the space. Such a search for representation theorems is not at all a part of the Greek tradition in much earlier times. One way of putting the matter, in terms of many different discussions of the foundations of mathematics, is that such representation theorems demand a representation of the actual infinite. It is this representation of the actual infinite, or, to put it in more pedantic terms, sets of infinite cardinality, which are mapped from the geometric space onto a Cartesian product of the set of real numbers isomorphically preserving the structure of the geometric space. The stronger sense of nonfinitistic methods involving the axiom of choice, or its equivalent, is a still further remove from the finitism of applications I am focusing on here.

It is a separate question to what extent the parts of classical analysis actually used in applications of mathematics in physics and elsewhere can avoid commitments not only to any results using the axiom of choice, but, also, any sets of an actually infinite character.[22] My own conviction is that one can go the entire distance, or certainly almost the entire distance, in a purely finitistic way, but this is not what I focus on here.

Quantifier-free axioms and constructions. One requirement of very strict constructive methods is that the axioms of the theory in question should be quantifier-free and thus avoid all purely existential assumptions. Such existential demands are replaced by specific constructions, which are either primitive or definable, in a quantifier-free way, in terms of the given primitive constructions. Often, in traditional formulations of geometry, existential axioms are innocuous, in the sense that they can easily be replaced by essentially equivalent quantifier-free axioms which use a fixed finite number of given constants.

---

[22]The foundational effort leading to the arithmetization of analysis in the nineteenth century forced even the most elementary concepts to have as extensions infinite sets, e.g., the representation of a given rational number as the equivalence class of all fractions equal in the usual sense to a representative fraction. Thus, the rational number $[\frac{1}{2}] = \{\frac{p}{q} : 2p = q \ \& \ q \neq 0\}$. But such unintuitive abstractions never played a serious part in applications and do not in proving standard theorems of analysis.

Temporal sequence. An aspect of applications ignored in standard axiomatic formulations of constructions is the temporal sequence in which any actual constructions are performed. The usual mathematical notation for operations or relations does not provide a place for indicating temporal order. In probability theory, however, such temporal matters are of great importance in developing the theory or applications of stochastic processes. Whether time is treated as continuous or discrete, the time of possible occurrence of events or random variables is indicated by subscripts, t, t', t_1, etc., for continuous time and $n, n', n+1$, etc., for discrete time. When the time of possible occurrence is not stipulated as part of the theoretical framework, then, in principle, a new random variable is required to describe this uncertainty.

Such probabilistic considerations lie outside the framework of geometry being considered here, although they could naturally be included if the actual fact of errors in real constructions were incorporated. Such matters are of undoubted importance in applications, but, only for the sake of simplicity, will be ignored here. This is because of the much greater complexity of the axioms required to include from the beginning the conceptual apparatus for dealing with the inevitable errors of approximation in real constructions. For examples of the complexity of such axioms, see Suppes et al. (1989, Ch. 16).

What we can easily use in the notation for constructions are subscripts to show temporal sequence. This provides an easy way of showing whether or not two constructions require the same number of steps. Here is a simple example from affine geometry using only the two constructions of finding the midpoint, i.e., bisecting a line segment, and doubling, in one of two directions, a line segment. In customary fashion, line segments are referred to in the intuitive discussion, but only the two points that are endpoints of a segment are required formally. In this example I formulate the axioms of construction informally.

Problem. *Given three noncollinear points α_0, β_0 and γ_0, construct a parallelogram with adjacent sides $\alpha_0\beta_0$ and $\alpha_0\gamma_0$.*

1. Construct the midpoint a_1 of $\beta_0\gamma_0$.
2. Construct the point a_2 that doubles the segment $\alpha_0 a_1$ in the direction of a_1. The figure $\alpha_0\gamma_0 a_2\beta_0$ is the desired parallelogram (see Figure 5).

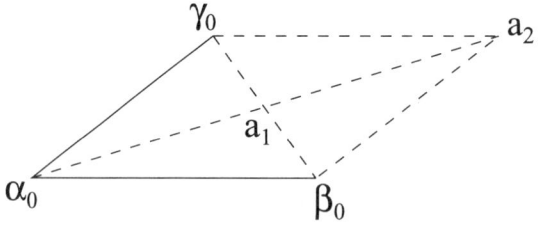

FIGURE 5   Constructed parallelogram.

Affine Axioms. I now turn to axioms for the affine plane constructions that can be made by bisecting and doubling the distances between pairs of points. The axioms are

6.8 Finitism in Geometry

quantifier-free, as would be expected from my earlier remarks. I have, in stating the axioms, dropped the notation for temporal order of construction, but the appropriate subscripts can be added, if desired, when presenting constructions, in an obvious manner. The subscripts are less useful in stating the axioms.

A few elementary definitions and theorems follow the axioms, but no proofs are given here, except the representation theorem for introducing coordinates. (More theorems and proofs are to be found in Suppes (2000), although in a slightly different notation.) To avoid closure conditions on operations, the bisection construction is formally expressed as a relation $B(ab, c)$, with the intuitive meaning that c is the constructed point bisecting segment ab. Correspondingly, $D(ab, c)$ is the doubling construction in the direction b. Intuitively, the position of c is shown in Figure 6. Also needed as a primitive ternary relation is that of linearity. Intuitively $L(abc)$ if a, b and c all lie on a line.

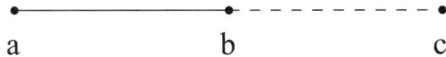

FIGURE 6   The doubling construction.

First, the axioms involving just linearity.

L1. If $a = b$, $a = c$ or $b = c$, then $L(abc)$.

L2. If $a \neq b$, $L(abp), L(abq)$ and $L(abr)$, then $L(pqr)$.

Next, the axioms just involving bisection. Note that the first axiom asserts that when points a and b, not necessarily distinct, are given, then the constructed point is unique.

B1. If $B(ab, c)$ and $B(ab, c)'$ then $c = c'$. (*Uniqueness*)

B2. $B(aa, a)$. (*Idempotency*)

B3. If $B(ab, c)$ then $B(ba, c)$. (*Commutativity*)

B4. If $B(ab, c)$, $B(de, f)$, $B(ad, g)$, $B(be, h)$, $B(cf, i)$ and $B(gh, j)$, then $i = j$. (*Bicommutativity*)

B5. If $B(ab, c)$ and $B(ab', c)$ then $b = b'$. (*Cancellation*)

Axiom B4 illustrates well the simplification of understanding introduced by using algebraic-operation notation. This axiom then has the much simpler form:

$$(a \oplus b) \oplus (d \oplus e) = (a \oplus d) \oplus (b \oplus e).$$

As should be clear from what I said already, I use the awkward relation-style notation to make the restriction to finite configurations transparent later.

The four axioms for the doubling construction are the following.

D1. If $D(ab, c)$ and $D(ab, c')$, then $c = c'$. (*Uniqueness*)

D2. If $D(ab, c)$ and $D(ba, c)$, then $a = b$. (*Antisymmetry*)

D3. If $D(ab, c)$ and $D(ab', c)$, then $b = b'$. (*Left cancellation*)

D4. If $D(ab, c)$ and $D(a'b, c)$ then $a = a'$. (*Right cancellation*)

Finally, there are three axioms using more than one relation.

BD. If $D(ab, c)$ then $B(ac, b)$. (*Reduction*)

LB. *If $B(ab,c)$ then $L(abc)$.* (*Linearity of Bisection*)

LBL. *If $B(ab,d)$, $B(bc,e)$, $B(ac,f)$ and $L(def)$ then $L(abc)$.* (*Linearity of Midpoints*)

I introduce two conditional definitions to reintroduce the simplifying operational notation.

Definition 1. If $B(ab,c)$ then $a \oplus b = c$.

Definition 2. If $D(ab,c)$ then $a * b = c$.

In other words, $\oplus$ is the conditional operation of bisecting and $*$ of doubling. In the following theorems and definitions, which use the operational notation, the conditions required by Definitions 1 and 2 to employ the operations $\oplus$ and $*$ are assumed satisfied in all cases, but without explicit statement.

Theorems. First, I summarize in one theorem the elementary properties of collinearity.

THEOREM 1. (Collinearity).

(i) If $L(abc)$ then L holds for any permutation of abc.

(ii) $L(aba)$.

(iii) If $a \neq b$, $L(abc)$ and $L(abd)$ then $L(acd)$.

(iv) If $p \neq q$, $L(abp)$, $L(abq)$ and $L(pqr)$ then $L(abr)$.

Szmielew (1983) points out that (i), (ii) and (iii) of Theorem 1 are equivalent to Axioms L2 and L3—actually a weaker form of (i), namely, if $L(abc)$ then $L(bac)$.

THEOREM 2. $(a \oplus b) \oplus c = (a \oplus c) \oplus (b \oplus c)$ (*Self-distributivity*).

THEOREM 3. If $a \oplus b = a$ then $a = b$.

The next theorem shows that reduction and linearity hold for doubling.

THEOREM 4. $a * (a \oplus b) = b$ and $L(ab(a*b))$.

Any three noncollinear points 'form' a triangle, so we may define the ternary relation T of triangularity as the negation of L.

DEFINITION 1. $T(abc)$ iff it is not the case $L(abc)$.

In the next definition, the quaternary relation P defined has the intuitive meaning that four points standing in this relation form a parallelogram (thus 'P' for 'parallelogram').

DEFINITION 2. $P(abcd)$ iff $T(abc)$ and $a \oplus c = b \oplus d$.

This definition characterizes parallelograms, which do the work 'locally' of parallel lines in standard nonconstructive affine geometry. The triangularity or nonlinearity condition eliminates degeneracy. The important condition is that a convex quadrilateral $abcd$ (see Figure 7) is a parallelogram if and only if the midpoints of the diagonals ac and bd coincide, which is a familiar property of parallelograms, but also sufficient for the definition. Concave quadrilaterals violate the midpoint condition.

I now define informally what a construction is in terms of the concepts already introduced. Given three noncollinear points α_0, β_0 and γ_0, a *construction* is a finite sequence of terms $(C_1, \ldots, C_n)$, with each term $C_k, k = 1, \ldots, n$ being an ordered triple:

6.8 FINITISM IN GEOMETRY

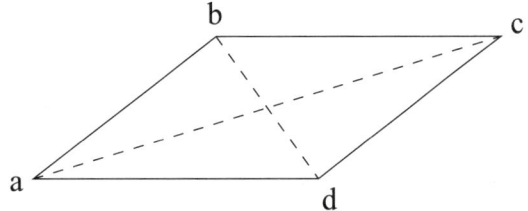

FIGURE 7   Intersection of diagonals of a parallelogram.

(i) whose first member is a point a_k provided a_k is neither $\alpha_0, \beta_0, \gamma_0$ nor a previously constructed point;[23]

(ii) whose second member is a pair of points (a_i, a_j) which are either $\alpha_0, \beta_0, \gamma_0$ or points previously constructed in the sequence, i.e., $i, j < k$;

(iii) and whose third member is either 'B' standing for use of bisection or 'D' for doubling such that the operation shown is applied to the pair of points (a_i, a_j) to construct a_k.

Thus, the earlier construction of a parallelogram with adjacent sides $\alpha_0\beta_0$ and $\alpha_0\gamma_0$ is the sequence:

$$((a_1, \beta_0\gamma_0, B), (a_2, \alpha_0 a_1, D)).$$

As already remarked, a construction requires the construction of a new point at every step. The constructive method of verifying that a point has not been previously introduced is by introducing coordinates inductively at each step to make a simple numerical test possible. In many applications this theoretical coordinate check is ordinarily replaced by a quick visual inspection of the construction at each stage.

Analytic representation theorem.   Here is the finitistic representation theorem.

THEOREM 5. *Given three noncollinear points $\alpha_0, \beta_0, \gamma_0$ and a construction $C = (C_1, \ldots, C_n)$ based on these three points, then the pair (φ_1, φ_2) of rational numerical functions as defined below provide an analytic representation of the points $(a_1, \ldots, a_k)$ constructed by C, where:*

(i) $\varphi_1(\alpha_0) = 0, \varphi_2(\alpha_0) = 0, \varphi_1(\beta_0) = 1, \varphi_2(\beta_0) = 0, \varphi_1(\gamma_0) = 0, \varphi_2(\gamma_0) = 1$,

(ii) $\varphi_i(a \oplus b) = \frac{\varphi_i(a) + \varphi_i(b)}{2}, i = 1, 2$, (*bisecting*),

(iii) $\varphi_i(a * b) = 2\varphi_i(b) - \varphi_i(a), i = 1, 2$, (*doubling*).

Outline of Proof. By induction: For $n = 1$, if $C_1 = (a_1, \alpha_0\beta_0, B)$, then

$$\varphi_1(a_1) = \frac{0+1}{2} = \frac{1}{2},$$
$$\varphi_2(a_1) = \frac{0+0}{2} = 0.$$

And similarly for the remaining 11 possible cases of C_1.

---

[23]The condition that a_k is not previously given or constructed is not really needed. Such a condition is not standard, for example, in the usual formal recursive definitions of proofs. On the other hand, it is an intuitively natural restriction.

If the representation holds for $1, \ldots, n-1$, then, although there are many possible C_n's, for the given construction, C_n is unique, with the coordinates inductively defined up to $n-1$ being used to check that the point a_n constructed is new. So if the general term is of the form $C_n = (a_n, a_i a_j, B), i, j < n$,

$$\varphi_1(a_n) = \frac{\varphi_1(a_i) + \varphi_1(a_j)}{2},$$
$$\varphi_2(a_n) = \frac{\varphi_2(a_i) + \varphi_2(a_j)}{2}.$$

If the general form is $C_n = (a_n, a_i a_j, D), i, j < n$, then

$$\varphi_1(a_n) = 2\varphi_1(a_j) - \varphi_1(a_i),$$
$$\varphi_2(a_n) = 2\varphi_2(a_j) - \varphi_2(a_i).$$

As is evident, the method of inductively assigning numerical coordinates is obvious and so is the proof, which is how it should be for the standard geometric constructions, as opposed to the necessarily lengthy and complicated proofs for the affine, projective or Euclidean representation of the entire real plane or three-dimensional space. A rigorous example of such a proof is developed over many pages in Borsuk and Szmielew (1960).

Analytic invariance theorem. There is also a finitistic affine invariance theorem that holds for the introduction of coordinates, which I state without proof.

THEOREM 6. (Invariance Theorem). *Let C be a construction based on three non-collinear points α_0, β_0 and γ_0, and let (φ_1, φ_2) be coordinate functions as defined above. Let*

$$\varphi'_1 = a\varphi_1 + b, \ a \neq 0$$
$$\varphi'_2 = c\varphi_2 + d, \ c \neq 0,$$

then (φ'_1, φ'_2), is also a coordinate representation of C with the origin possibly changed (if $b \neq 0$ or $d \neq 0$) and with affine changes of scale unless $a = 1$ or $c = 1$, where now $\varphi'_1(\alpha_0) = b, \varphi'_2(\alpha_0) = d, \varphi_1(\beta_0) = a + b, \varphi_2(\beta_0) = d, \varphi_1(\gamma_0) = b$ and $\varphi_2(\gamma_0) = c + d$.

It should be obvious that for each particular construction something more is needed, namely, the kind of proof, so familiar in Euclid, that the construction actually carries out what was intended as the construction. Constructions are like proofs in that each step can be valid, but what is constructed, like what is proved, is not for what was given as the problem or theorem. So accompanying each construction there should be a matching proof, which can be either direct or use the analytic representation.

Second, we in general want for constructions an independence-of-path theorem. If we are given a finite configuration of points we can prove the configuration has certain properties independent of the particular construction used to generate it.

None of these additional proof requirements for particular constructions take us outside the finitistic framework already described. In Suppes (2000) an additional trapezoid construction is introduced to permit the easy, finitistic construction of any rational number.

6.8 FINITISM IN GEOMETRY

Algebraic problems. The emphasis on introducing coordinates via constructions simplifies drastically the analytic representation theorem. A purely algebraic approach with a general representation for finite models of the axioms, but with the order of construction of points not given, requires additional algebraic identities as axioms and a more complicated representation theorem. Examples of the sorts of identities that are needed are the following, which is not complete:

$$\begin{aligned}
(a * b) * b &= a, \\
(a * b) \oplus (b * a) &= a \oplus b, \\
(a * b) * (a \oplus b) &= b * a.
\end{aligned}$$

It is easy to check that the intended coordinate representation requires that these be satisfied. It is also apparent that these and similar identities represent redundant steps in constructions, which are eliminated by the requirement that each point constructed be a new one. These identities show well enough why the restricted nature of constructions drastically simplifies the theory, which should be the aim of any theory intended to be used in a variety of applications, as has historically been the case for the classical geometric constructions that preceded by many centuries the theory of real numbers.

Final remark. The main philosophical point that I have tried to illustrate in this section is that the foundations of many parts of applied mathematics can be given a simple finitistic formulation, and, if desired, a finitistic numerical representation. Some efforts in this direction for the more complicated case of applications of analysis are to be found in my joint articles with Rolando Chuaqui (1990, 1995) or Richard Sommer (1996, 1997), and Suppes and Chuaqui (1993) as well, listed in the references.

A different way to make the point is to examine empirically the mathematics used in most published articles and textbooks of physics. Most of what is standardly taught or used in articles is surprisingly constructive in character, and very finitistic in practice. Scarcely any proofs by induction will be found, and almost never an $\epsilon - \delta$ proof about continuity or convergence. The widespread use of Taylor's theorem by physicists requires only the first few terms of the expansion. For the elementary derivation of some standard differential equations that appear often in physics, see Suppes and Chuaqui (1993). But the strongest new support for finitistic mathematical methods in science is the now nearly universal use of finitistic discrete methods in almost all complex problems, because of their implementation on digital computers, a situation which seems unlikely to change in the near future.

7

Representations in Mechanics

In this chapter I survey some of the many topics on representation and invariance that have played a prominent role in mechanics. The first section focuses on classical particle mechanics. The concept of a real affine space introduced in this chapter as the space-time structure of classical physics is, of course, in a sense easy to see, a vector formulation of the affine space of classical space-time introduced in Section 6.1. The second section focuses on quantum mechanics, but really only on the philosophically interesting question of hidden variables, for which various representation theorems are given. It is worth emphasizing that nonlocal quantum entanglement of particles and the accompanying nonexistence of classical common causes, otherwise known as hidden variables, is the most persistent foundational puzzle of standard quantum mechanics. The last section, the third one, analyzes two important but different senses of reversibility for causal chains. Temporal reversibility of physical systems is the most important natural concept of invariance for the time domain. Philosophical reflections on the meaning of reversibility have often not been sensitive to the fundamental distinction between weak and strong reversibility, a distinction which is more familiar in the theory of stochastic processes than in the philosophy of science.

7.1 Classical Particle Mechanics

As another example of a substantive theory, I have selected classical particle mechanics for several reasons. In the first place, the intuitive content of this theory is widely familiar. Secondly, the general mathematical framework required is relatively simple and straightforward, but at the same time, the theory is rich enough to generate a large number of significant problems, both in terms of its philosophical foundations and in terms of its systematic applications to the real world.

Assumed mathematical concepts. Before turning to mechanics proper, it will be desirable to review some of the mathematical, especially geometric, notions that will be used in the statement of the axioms and in subsequent discussion of the significance and meaning of these axioms. To begin with, I review quickly some notions introduced in earlier chapters. The closed interval $[a, b]$ is the set of all real numbers x such that $a \leq x \leq b$. The open interval (a, b) is the interval which is the set of all numbers x such that $a < x < b$. Correspondingly, $(a, b]$ is a half-open, half-closed interval, and so forth. The interval $(-\infty, a)$ is the set of all real numbers x such that $x < a$, the interval (a, ∞) is the set of all real numbers x such that $a < x$, and the interval $(-\infty, \infty)$ is the set Re of all real numbers. We assume as given all the familiar operations on real numbers and their properties.

In particular, we take as given the field of real numbers, which is just the given structure $(Re, +, \cdot, 0, 1)$, where $+$ is the operation of addition on real numbers, $\cdot$ is the operation of multiplication, 0 is the number zero, and 1 is the number one.

We now turn to vector spaces over the field of real numbers. A vector space has the set-theoretical structure $(V, +, \cdot, \mathbf{0})$, where V is a nonempty set, $+$ is a binary operation on V (vector addition), $\cdot$ is a function from $Re \times V$ into V (multiplication of a vector by a real number to yield another vector), and $\mathbf{0}$ is the null vector. The axioms for vector spaces are given in the following definition.

DEFINITION 1. *A structure* $\mathfrak{V} = (V, +, \cdot, \mathbf{0})$ *is a vector space over* Re *iff the following axioms hold for all* $\mathbf{x}, \mathbf{y}, \mathbf{z}$ *in* V *and all* a, b *in Re:*

1. $(\mathbf{x} + \mathbf{y}) + \mathbf{z} = \mathbf{x} + (\mathbf{y} + \mathbf{z})$
2. $\mathbf{x} + \mathbf{y} = \mathbf{y} + \mathbf{x}$
3. $\mathbf{x} + \mathbf{0} = \mathbf{x}$
4. *There exists* $\mathbf{u}$ *in* V *such that* $\mathbf{x} + \mathbf{u} = \mathbf{0}$
5. $a(\mathbf{x} + \mathbf{y}) = a\mathbf{x} + a\mathbf{y}$
6. $(a + b)\mathbf{x} = a\mathbf{x} + b\mathbf{x}$
7. $(ab)\mathbf{x} = a(b\mathbf{x})$
8. $1\mathbf{x} = \mathbf{x}$.

The notation of Axioms 5-8 is informal. In a style that is widely used, beginning with elementary arithmetic, juxtaposition is used to denote multiplication, and the explicit symbol '·' is omitted. In the spirit of Definition 3.1 for groups it would have been natural to introduce as primitive the unary operation $-\mathbf{x}$ and have written Axiom 4 as: $\mathbf{x} + -\mathbf{x} = \mathbf{0}$. In deference to most standard axiomatizations of vector spaces we did not do this, but hereafter we shall treat this additive inverse notation as properly defined and use it as desired. Of course, to treat $-\mathbf{x}$ as well defined, we must also prove that the vector $\mathbf{u}$ of Axiom 4 not only exists but for each $\mathbf{x}$ is unique. But this is easily done. We can establish the following identities to show that we cannot have $\mathbf{u} \neq \mathbf{u}'$ and $\mathbf{x} + \mathbf{u} = \mathbf{x} + \mathbf{u}' = \mathbf{0}$.

$$\begin{aligned} \mathbf{u}' &= \mathbf{0} + \mathbf{u}' & \text{by Axioms 2, 3} \\ &= (\mathbf{x} + \mathbf{u}) + \mathbf{u}' & \text{by hypothesis on } \mathbf{u} \\ &= (\mathbf{x} + \mathbf{u}') + \mathbf{u} & \text{by Axioms 1, 2} \\ &= \mathbf{0} + \mathbf{u} & \text{by hypothesis on } \mathbf{u}' \\ &= \mathbf{u} & \text{by Axioms 2, 3.} \end{aligned}$$

A succinct way of reformulating Axioms 1-4 is that $(V, +)$ is a commutative group, or, in the sense of Definition 2.3.1, $(V, +, \mathbf{0}, -)$ is a commutative group. In linear algebra a more general form of Definition 1 would ordinarily be given. A vector space would be defined over an arbitrary field, and not simply the given field of real numbers,

7.1 CLASSICAL PARTICLE MECHANICS

but for purposes of mechanics the restriction to the field of real numbers is a natural one. Instead of saying $\mathfrak{V}$ is a vector space over Re, we shall often say that $\mathfrak{V}$ is a *real vector space*.

The most familiar model of a vector space is Re^3, i.e., the model in which a vector is an ordered triple of real numbers, the familiar three-dimensional Cartesian space of analytic geometry. This Cartesian model was used extensively in Chapter 6, but the more abstract view taken here could also have been used. However, I believe the more concrete Cartesian model is better, especially for the extensive discussion of visual space. Within physics the usual interpretation of vectors is that they represent the magnitude and direction of some physical quantity such as force, with respect to perpendicular spatial coordinate axes.

Binary operations on vectors, or vectors and real numbers, are defined in a natural way. Thus if $\mathbf{x} = (x_1, x_2, x_3)$ is a vector and a is a real number

$$a\mathbf{x} = a(x_1, x_2, x_3) = (ax_1, ax_2, ax_3) = (x_1 a, x_2 a, x_3 a) = (x_1, x_2, x_3)a = \mathbf{x}a.$$

If $\mathbf{x} = (x_1, x_2, x_3)$ and $\mathbf{y} = (y_1, y_2, y_3)$ are vectors then

$$\mathbf{x} + \mathbf{y} = (x_1, x_2, x_3) + (y_1, y_2, y_3) = (x_1 + y_1, x_2 + y_2, x_3 + y_3)$$

and

$$-\mathbf{x} = -(x_1, x_2, x_3) = (-x_1, -x_2, -x_3).$$

Also the vector $\mathbf{0} = (0, 0, 0)$. It is easy to check that the structure $(Re^3, +, \cdot, \mathbf{0})$ as just defined is a vector space over Re.

We next define the *scalar product*—also called *inner product*—of two vectors, which is a mapping or function from $V \times V$ to Re. The *scalar product* $(\mathbf{x}, \mathbf{y})$ of any two vectors has the following defining properties:

1. (*Bilinearity*)

    $(\mathbf{x} + \mathbf{y}, \mathbf{z}) = (\mathbf{x}, \mathbf{z}) + (\mathbf{y}, \mathbf{z})$
    $(\mathbf{x}, \mathbf{y} + \mathbf{z}) = (\mathbf{x}, \mathbf{y}) + (\mathbf{x}, \mathbf{z})$
    $(a\mathbf{x}, \mathbf{y}) = a(\mathbf{x}, \mathbf{y}) = (\mathbf{x}, a\mathbf{y})$

2. (*Symmetry*)
    $(\mathbf{x}, \mathbf{y}) = (\mathbf{y}, \mathbf{x})$

3. (*Positive definite*)
    If $\mathbf{x} \neq \mathbf{0}$ then $(\mathbf{x}, \mathbf{x}) > 0$.

Moreover, the *Euclidean norm* $|\mathbf{x}|$ is the non-negative-valued function defined by

$$|\mathbf{x}| = (\mathbf{x}, \mathbf{x})^{\frac{1}{2}}$$

and the *Euclidean metric* $d(\mathbf{x}, \mathbf{y})$ is the non-negative function defined by

$$d(\mathbf{x}, \mathbf{y}) = |\mathbf{x} - \mathbf{y}| = (\mathbf{x} - \mathbf{y}, \mathbf{x} - \mathbf{y})^{\frac{1}{2}}.$$

The scalar product has the geometric interpretation:

$$(\mathbf{x}, \mathbf{y}) = |\mathbf{x}||\mathbf{y}| \cos(\mathbf{x}, \mathbf{y}),$$

where $\cos(\mathbf{x}, \mathbf{y})$ is the cosine of the angle between the vectors $\mathbf{x}$ and $\mathbf{y}$. In the

Cartesian model of a vector space already discussed, the scalar product is:
$$((x_1, x_2, x_3), (y_1, y_2, y_3)) = x_1 y_1 + x_2 y_2 + x_3 y_3.$$

For expressing Newton's third law in simple vectorial form we also need to state the defining properties of the *vector product*—also called *exterior* product—of two vectors, which is a mapping from $V \times V$ to V. The *vector product* $[\mathbf{x}, \mathbf{y}]$ of any two vectors has the following defining properties:

1. (*Bilinearity*)
$$[\mathbf{x} + \mathbf{y}, \mathbf{z}] = [\mathbf{x}, \mathbf{z}] + [\mathbf{y}, \mathbf{z}]$$
$$[\mathbf{x}, \mathbf{y} + \mathbf{z}] = [\mathbf{x}, \mathbf{y}] + [\mathbf{x}, \mathbf{z}]$$
$$[a\mathbf{x}, \mathbf{y}] = a[\mathbf{x}, \mathbf{y}] = [\mathbf{x}, a\mathbf{y}]$$

2. (*Skew symmetry*)
$$[\mathbf{x}, \mathbf{y}] = -[\mathbf{y}, \mathbf{x}]$$

3. (*Jacobi identity*)
$$[[\mathbf{x}, \mathbf{y}], \mathbf{z}] + [[\mathbf{y}, \mathbf{z}], \mathbf{x}] + [[\mathbf{z}, \mathbf{x}], \mathbf{y}] = \mathbf{0}$$

4. (*Orthogonality*)
$$[[\mathbf{x}, \mathbf{y}], \mathbf{x}] = \mathbf{0}.$$

In the Cartesian model of a vector space discussed earlier, the vector product has the following characterization:
$$[(x_1, x_2, x_3), (y_1, y_2, y_3)] = (x_2 y_3 - x_3 y_2, x_3 y_1 - x_1 y_3, x_1 y_2 - x_2 y_1).$$

There are certain standard vector concepts we also need in this section. First, vectors $\mathbf{x}_1, \ldots, \mathbf{x}_n \in V$ are *linearly independent* iff for any real numbers $a_1, \ldots, a_n$ if $\sum_n a_i \mathbf{x}_i = 0$ then all $\mathbf{x}_i = 0$. The vector space $\mathfrak{V}$ has *dimension* n over Re iff n is the maximal number of linearly independent vectors of $\mathfrak{V}$. If $\mathfrak{V}$ has dimension n then a *basis* of $\mathfrak{V}$ is any set of n linearly independent vectors of $\mathfrak{V}$. Although classical physical space is 3-dimensional, larger dimensional vector spaces are needed in mechanics.

The next concept needed is that of an affine space to represent the positions of particles. In prior axiomatizations of classical mechanics (McKinsey, Sugar, and Suppes 1953; Suppes 1957/1999) we represented position by a vector in Re^3, but a more coordinate-free representation is desirable. Already a vector space is an improvement in this respect, but because a vector space has a natural origin it is not entirely satisfactory either. An arbitrary affine space seems the appropriate framework for representing classical space and time, and therefore the positions of particles[1].

There are several different but essentially equivalent ways of axiomatizing the concept of a real affine space, i.e., the concept of an affine space associated with a vector space over the field of real numbers. Let A be an arbitrary nonempty set which we shall

---
[1] As the reader can easily determine, this analysis is closely related to that in Section 6.1, where affine spaces are first characterized from the standpoint of synthetic geometry, using really just the ternary relation of betweenness as primitive.

7.1 CLASSICAL PARTICLE MECHANICS

endow with an affine structure. The elements of A we call *points*. One way to so endow A is by the *action* of the assumed real vector space $\mathfrak{V}$ on A, more precisely, by the action of the additive group of $\mathfrak{V}$ on A. Such an action is a function from $V \times A$ to A, which we also denote by the symbol of addition, and which satisfies the following three axioms:

(1) If $\mathbf{x}, \mathbf{y}$ are in V and P is in A then
$$(\mathbf{x} + \mathbf{y}) + P = \mathbf{x} + (\mathbf{y} + P).$$

(2) $\mathbf{0} + P = P$, where $\mathbf{0}$ is the zero vector in V.

(3) For every ordered pair of points (P, Q) of A, there is exactly one vector $\mathbf{x}$ in V such that
$$\mathbf{x} + P = Q.$$

Especially because of this last equation, the additive group of V is often referred to as the group of *parallel displacements* of V.

Although this approach and its associated terminology is popular (see, for example, Snapper & Troyer 1971, and Arnold 1978), it seems unnatural, especially in notation, because it is unusual to use '+' for a binary operation combining different types of objects, and addition is not defined for two points P and Q. A slightly different approach emphasizes that the meaningful operation on points is that of difference, which is a mapping from $A \times A$ to V. It is customary not to write the difference as $Q - P$ but rather as $\underline{PQ}$ (or sometimes with an arrow on top of 'PQ' to indicate the vector runs from P to Q). Both the notations $Q - P$ and $\underline{PQ}$ will be used here. In any case, it is this difference approach that I shall mainly use in subsequent developments, but the additive displacement operation is easily defined in terms of difference:

$$\mathbf{x} + P = Q \text{ iff } Q - P = \mathbf{x}.$$

DEFINITION 2. *A structure* $\mathfrak{A} = (A, \mathfrak{V})$ *is a real affine space iff A is a nonempty set, $\mathfrak{V}$ is a real vector space, and the following axioms are satisfied:*

1. For any point P in A and vector $\mathbf{x}$ in $\mathfrak{V}$ there is exactly one point Q in A such that $\underline{PQ} = \mathbf{x}$.

2. For any points P, Q and R in A
$$\underline{PQ} + \underline{QR} = \underline{PR}.^2$$

The second axiom is the important vector law of addition for points.

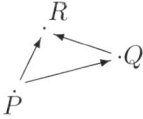

It will be useful to prove a few elementary facts about real affine spaces. First, in Axiom 2 we set $Q = P$ and obtain
$$\underline{PP} + \underline{PR} = \underline{PR},$$

---

[2]Contrast this definition of affine spaces with that in Section 6.1. Here much more is assumed in the definition, namely, a real vector space. I leave it to the reader to construct the appropriate isomorphism between the structures satisfying these two different definitions.

whence $\underline{PP} = \mathbf{0}$ for every point P in A—note that $\mathbf{0}$ is the zero vector of $\mathfrak{V}$. Setting now $R = P$ in Axiom 2 we have

$$\underline{PQ} + \underline{QP} = \underline{PP},$$

and so

$$\underline{PQ} = -\underline{QP}.$$

We leave it to the reader to show that $\underline{P_1Q_1} = \underline{P_2Q_2}$ implies that $\underline{P_1P_2} = \underline{Q_1Q_2}$, which is a form of the parallelogram law. The *dimension* of the real affine space $(A, \mathfrak{V})$ is just the dimension of the real vector space $\mathfrak{V}$. To distinguish dimensions we sometimes write $\mathfrak{V}(3)$ or $\mathfrak{V}(4)$, with the three-dimensional vector space $\mathfrak{V}(3)$ ordinarily being a subspace of the four-dimensional vector space $\mathfrak{V}(4)$. When $\mathfrak{V}$ has an inner product and Euclidean norm as defined earlier, $(A, \mathfrak{V})$ is said to be a *Euclidean space*, and the *distance* $d(P,Q)$ between two points P and Q is just the Euclidean norm of the vector $\underline{PQ}$, i.e.,

$$d(P,Q) = |\underline{PQ}| = (\underline{PQ}, \underline{PQ})^{\frac{1}{2}}.$$

It is easy to show that the distance between points has the properties of a metric space:

1. $d(P,Q) \geq 0$,
2. $d(P,Q) = 0$ iff $P = Q$,
3. $d(P,Q) = d(Q,P)$,
4. $d(P,Q) \leq d(P,R) + d(R,Q)$.

Space-time structure. The next important structure to define is that of a *classical space-time structure*.

DEFINITION 3. *Let $\mathfrak{A} = (A, \mathfrak{V})$ be a four-dimensional real affine space, let τ be a mapping from A to Re, and let $\boldsymbol{\alpha}$ be a mapping from the range of τ to A. Then $(\mathfrak{A}, \tau, \boldsymbol{\alpha})$ is a* classical space-time structure *iff for every point P in A, $\{Q | \tau(Q) - \tau(P) = 0\}$ is a three-dimensional Euclidean space with the distance function defined earlier on the three-dimensional real vector space $\mathfrak{V}(3)$, and $\boldsymbol{\alpha}(\tau(P))$ is a point in this three-dimensional Euclidean space, i.e.,*

$$\tau(\boldsymbol{\alpha}(\tau(P))) - \tau(P) = \mathbf{0}.$$

As should be apparent, the mapping τ is the clock time-function that gives the time of occurrence of any space-time point-event in A. The set $\{Q | \tau(Q) - \tau(P) = 0\}$ is the set of all points or, more vividly, point-events, in A that are simultaneous with P. For each t in the range of τ, we shall use the notation $\mathfrak{A}(t)$ for the three-dimensional affine subspace of $\mathfrak{A}$ that consists of all point-events occurring at time t. As is immediate from Definition 3, $\mathfrak{A}(t)$ is Euclidean. Note that $\mathfrak{A}$ itself is not a Euclidean space, but only affine, for no objective meaning can be assigned to the comparison of distances between pairs of nonsimultaneous events, a matter we explored in more detail in the previous chapter, when we considered the representation theorem for classical structures. In the present context this means the vector space associated with the four-dimensional affine space does not have an inner product.

7.1 Classical Particle Mechanics

The function $\boldsymbol{\alpha}$ from the range of τ, i.e., measured instants of time, to A has an obvious and important physical meaning. For each instant of time t, the point $\boldsymbol{\alpha}(t)$ of A is the origin of the fixed inertial frame of reference to which motions are referred. A more invariant formulation could be given by generalizing $\boldsymbol{\alpha}$ to the family of all inertial frames of reference. This is done for special relativity explicitly in Hermes (1938) and in Suppes (1959), but for present purposes it introduces too much apparatus and takes us too far from the traditional notation of mechanics toward which we converge later in this section.

The meaning of the formal requirement on $\boldsymbol{\alpha}$ is apparent. The origin of the inertial frame of reference must be temporally present at all times.

In stating the axioms for classical mechanics we assume the concept of a position function $\mathbf{s}$, from an interval T of real numbers to a real vector space with given inner product, being differentiable or twice differentiable at every point of T, and therefore on T. Similarly, we assume it is understood what is meant when the first and second derivatives of $\mathbf{s}$ with respect to T, written as $\frac{d}{dt}\mathbf{s}$ and $\frac{d^2}{dt^2}\mathbf{s}$, or as $D\mathbf{s}$ and $D^2\mathbf{s}$, or as $\dot{\mathbf{s}}$ and $\ddot{\mathbf{s}}$, are referred to. (Following familiar practice in mathematics and physics, more than one notation for derivatives will be used; when the notation D or D^2 is used, the variable of differentiation will be obvious.)

It is easy to show that the usual product rule for differentiation holds for the scalar and vector products. For explicitness, let $\mathbf{x}(t)$ and $\mathbf{y}(t)$ be differentiable vector-valued functions. Then

$$\text{scalar product:} \quad \frac{d}{dt}(\mathbf{x}(t), \mathbf{y}(t)) = (\frac{d}{dt}\mathbf{x}(t), \mathbf{y}(t)) + (\mathbf{x}(t), \frac{d}{dt}\mathbf{y}(t))$$

and

$$\text{vector product:} \quad \frac{d}{dt}[\mathbf{x}(t), \mathbf{y}(t)] = [\frac{d}{dt}\mathbf{x}(t), \mathbf{y}(t)] + [\mathbf{x}(t), \frac{d}{dt}\mathbf{y}(t)].$$

We also need the concept of absolute convergence for a denumerable sum of vectors, i.e., an infinite series of vectors. The requirement is just that the numerical series $\sum_{i=1}^{\infty} |\mathbf{x}_i|$ is absolutely convergent, where $|\mathbf{x}_i|$ is the Euclidean norm.

Primitive notions. Our next task is to characterize the primitive notions of the axiomatization of mechanics to be given here. The axiomatization is with respect to a given classical space-time structure. Let T be the range of τ. We use five primitive notions: a set P of particles; a position function $\mathbf{s}$ from the Cartesian product $P \times T$ to A, the given affine space; a mass function m from the set P of particles to Re; a force function $\mathbf{f}$ for *internal* forces which is a function from the Cartesian product of the set of particles with itself and the time interval T, i.e., $P \times P \times T$, to the three-dimensional vector space $\mathfrak{V}(3)$; and an *external* force function $\mathbf{g}$ from the Cartesian product of the set of particles, the set T and the set of positive integers, i.e., $P \times T \times \underline{I}^+$ to $\mathfrak{V}(3)$. The positive integers enter into the definition of the external force function in order to provide a method of enumerating the forces.

To make these notions completely definite, I state in a slightly different way their intended interpretation. If $p \in P$ and $t \in T$, then $\mathbf{s}(p,t)$ is physically interpreted as the position of particle p at time t. Of course, what is observable is the vector $\mathbf{s}(p,t) - \boldsymbol{\alpha}(t)$, not $\mathbf{s}(p,t)$ simpliciter. For each p in P we will often write its position function as $\mathbf{s}_p$. For p in P, $m(p)$ is interpreted as the numerical value of the mass of the particle p. If p and q are in P and t is in T, then $\mathbf{f}(p,q,t)$ is the force which particle q exerts upon particle p at time t. Such forces are the internal forces of the mechanical system. If p is in P, t is in T, and n is a positive integer then $\mathbf{g}(p,t,n)$ is the nth external force acting on particle p at time t.

In characterizing the primitive notions of mechanics we use the customary assumption that the primitive functions and relations are not defined over any sets larger than those mentioned in the axioms. Thus, for example, $m(p)$ is defined only when p is in P; similarly, $\mathbf{s}(p,t)$ is defined only when p is in P and t is in T, and so forth. It follows from these considerations that the primitive P is actually definable in terms of the domain of the mass function m. However, it is also customary in the usual mathematical formulations of axioms not to eliminate such primitives just because they are definable, and so we do not insist on formulating the axioms in terms of a set of independent primitive notions.

The axioms. We are now ready to formulate the axioms for systems of particle mechanics. The axioms assume no other parts of physics and only those mathematical concepts that have been explicitly discussed. Thus the axioms do provide the core of a purely set-theoretical definition of the predicate 'is a system of classical particle mechanics'. The axioms themselves are divided along classical lines into kinematical and dynamical axioms. The intuitive idea behind this distinction is that the kinematical axioms are those required just to describe the motion of particles. The dynamical axioms enter into any considerations concerned with the causes of motion. We expand on this distinction further, after the axioms are explicitly stated.

DEFINITION 4. *A structure* $\mathfrak{P} = (P, \mathbf{s}, m, \mathbf{f}, \mathbf{g})$ *is a system of (classical) particle mechanics with respect to the classical space-time structure* $(\mathfrak{A}, \tau, \boldsymbol{\alpha})$ *iff the following axioms are satisfied:*

Kinematical Axioms

1. *The set T, the range of τ, is an interval of real numbers.*

2. *The set P is finite and nonempty.*

3. *For every p in P and t in T, $\mathbf{s}_p(t)$ is in $\mathfrak{A}(t)$.*

4. *For every p in P and t in T, the vector function $\mathbf{s}_p(t) - \boldsymbol{\alpha}(t)$ is twice differentiable at t.*

Dynamical Axioms

5. *For p in P, $m(p)$ is a positive real number.*

6. *For p and q in P and t in T*
$$\mathbf{f}(p,q,t) = -\mathbf{f}(q,p,t).$$

7. *For p and q in P and t in T*
$$[\mathbf{s}(p,t) - \mathbf{s}(q,t), \mathbf{f}(p,q,t) - \mathbf{f}(q,p,t)] = \mathbf{0}.$$

7.1 CLASSICAL PARTICLE MECHANICS

8. For p in P and t in T, the series
$$\sum_{n=1}^{\infty} \mathbf{g}(p,t,n) \text{ is absolutely convergent.}$$

9. For p in P and t in T,
$$m(p)D^2(\mathbf{s}_p(t) - \boldsymbol{\alpha}(t)) = \sum_{q \in P} \mathbf{f}(p,q,t) + \sum_{n=1}^{\infty} \mathbf{g}(p,t,n).$$

The coordinate-free formulation and the explicit introduction of the affine space $\mathfrak{A}$ are the main differences between the above axioms and those to be found in McKinsey, Sugar, and Suppes (1953) and Suppes (1957/1999). The explicit introduction of the affine space makes the axioms more complicated, but is, in my judgment, a decided improvement over the earlier formulation mentioned in making more fully explicit the physical content of mechanics. In the earlier formulations a Cartesian coordinate system was assumed from the beginning, and the affine character of space-time was not made evident. What now seems particularly important is to introduce the point of reference $\boldsymbol{\alpha}(t)$, i.e., the origin of the 'fixed' frame of reference.[3] Absolute position in the affine space $\mathfrak{A}(t)$ has no empirical meaning, only relative position, which is as matters should be. Second, by placing positions in the affine space and forces in the associated vector space the physical difference is emphasized, as it was not in the earlier work.

There are several further intuitive and conceptual remarks to be made about the axioms. I first comment on the kinematical axioms. In the case of Axiom 1 we are willing to take as a measurement of time any interval of real numbers. It is certainly not necessary for many applications, in fact, most practical applications, that this interval be the entire set of real numbers. It should also be noted that already in this axiom we see an element of mathematical convenience entering, quite independently of any questions about empirical constraints. From an empirical standpoint it would certainly be impossible to distinguish between Axiom 1 and the axiom that the set T of elapsed times is the set of algebraic numbers or even the set of rational numbers in the theoretically desirable full interval of real numbers, although this change would require modification of other axioms, especially Axiom 9. The full interval of real numbers is useful for purposes of applying mathematical analysis, in particular, the standard apparatus of the calculus and differential equations, but it is important to recognize that this apparatus is justified in terms of the kind of computational results it produces, not because the differentiability of the position function can be verified directly in observable empirical data, or even by any quite indirect argument from empirical data.

The requirement expressed in Axiom 2 that the set of particles is finite agrees with the usual formulations of particle mechanics. If this requirement were rejected then it would be necessary to put in some kind of convergence condition on the internal forces in order for the expression of the second law of motion in Axiom 9 to be satisfactory. Moreover, we might wish to add additional constraints that would require the total mass

---

[3]From an intuitive and informal standpoint this is done very nicely in Appell's classic treatise (1941, Tome I, p. 100) by assuming that $\boldsymbol{\alpha}(t)$ is the center of mass of the solar system.

of the system, and also perhaps the kinetic energy of the system, to be finite.[4]

Concerning Axiom 4 it should be noted that a slightly weaker requirement is imposed by an axiom of Hamel (1958), which requires merely that the position function be piecewise twice differentiable (a function is piecewise differentiable if with the exception of a finite number of points it is differentiable at all points of any finite interval on which it is defined). The piecewise restriction is convenient for the idealized analysis of the effects of impact. It is, for example, a reasonable approximation to assume that when two rigid bodies collide, at the point of collision their position functions are not differentiable although they are continuous. For the conceptual purposes of the present discussion, this slightly more realistic assumption of Hamel's introduces complications that need not be gone into, and, therefore, the more restricted and simpler Axiom 4 has been used.

As indicated, the first four axioms constitute the kinematical axioms for systems of particle mechanics. It is apparent that these axioms are quite weak in their general form, but it is to be emphasized that particular problems can be extraordinarily difficult. Perhaps the two most beautiful classical examples are the Ptolemaic analysis of the motion of the planets and Kepler's derivation of his three kinematical laws for the motion of the planets. The detailed results embodied in Ptolemy's *Almagest* concerning the motion of the planets represent the most significant work in empirical science in ancient times. Ancient astronomy refutes the claim that the Greek heritage in science was nonempirical in character. In any case I mention these important historical kinematical results to emphasize that the development of mechanics over centuries has entailed much more than the discovery of the dynamical or causal laws. Without the kinematical regularities abstracted from refractory astronomical observations for more than 2,000 years before Newton, it is doubtful he could have been so successful in formulating and solving the two-body gravitational problem, his greatest achievement.[5]

The first dynamical axiom, Axiom 5, asserts the standard requirement that the mass of a particle be positive. Various generalizations of this axiom exist in the literature,

---

[4]In general terms, the move from assuming a finite set of particles to an infinite set is characteristic of continuum mechanics. And, as might be expected, the axioms are much more complicated, as can be seen by perusing the important papers of Walter Noll, reprinted in Noll (1974). Noll is one of the pre-eminent contributors to the modern foundations of continuum mechanics. Of course, there is an important denumerably infinite case as well, namely, idealized versions of statistical mechanics. For many excellent insights on the history of both continuum and statistical mechanics, Truesdell (1968) is much to be recommended for its lively style and unvarnished opinions. From a purely epistemological standpoint I would argue strongly for the number of particles always being finite, but the number, if it is merely finite, is so large that asymptotic methods assuming an infinity are all that are practical for obtaining concrete results. But my view, so easy to state, in some sense represents a philosophical prejudice that has a distinguished pedigree. The field approach to matter, which dominates many aspects of modern physics, does not yield pride of fundamental place to the particle view of matter. Noll (1955) shows that under rather general conditions the mean quantities of statistical mechanics,– mean velocity, mean flow of energy, for instance–, satisfy the field equations for continuous materials. So, we cannot theoretically distinguish between the particle or molecular view of a fluid or solid and the field view, as far as the most important physical quantities are concerned.

[5]That Newton's results for earthly finite bodies moving in a fluid, an important part of continuum or fluid mechanics, were much less impressive can partly be explained by the absence of an earlier extensive development of the purely descriptive, noncausal theory. In any case, for a good discussion and appraisal of Newton's work in fluid mechanics, see Truesdell (1968, Ch. 3).

7.1 CLASSICAL PARTICLE MECHANICS

for example, generalizations that permit massses to be either zero or negative or also generalizations that permit mass to vary with time, but it does not seem pertinent to examine these technical variations here.

Axioms 6 and 7 together provide an exact formulation of Newton's Third Law of Motion. Axiom 6 corresponds to what Hamel (1958, p. 25) calls the first complete reaction principle. It requires that the force that particle q exerts on particle p be numerically equal and opposite to the force that particle p exerts on particle q. Axiom 7, which Hamel calls the second complete reaction principle, requires that the line of action of the forces be parallel to the line joining the positions of the two particles. (We leave it as an exercise for the reader to show that it follows from the definition of the vector product that Axiom 7 does indeed require that the line of action of the forces be parallel to the line joining the positions of particles p and q.)

Axiom 8 requires that the sum of the external forces acting on a particle be absolutely convergent. The reason for requiring absolute convergence rather than simple convergence is that the motion of the particle should be independent of the order of naming the particles applied to it. In the present context, absolute convergence is a principle of symmetry or invariance that has a strong physical basis. The arbitrary ordering assigned by number-names to external forces acting on a particle has no obvious physical significance.

Finally, Axiom 9 is the formulation in the present context of Newton's Second Law. There are two other axioms that are often discussed in considering the foundations of particle mechanics. The first is simply the expression of Newton's First Law. I have omitted this as an axiom because it is a trivial consequence of the present axioms (see Theorem 1 below). The second example is an axiom of impenetrability, which might be formulated as follows. For any two distinct particles p and q of P and any time t in T, $\mathbf{s}(p,t)$ is not equal to $\mathbf{s}(q,t)$. Such an axiom of impenetrability has been omitted primarily because one of the standard interpretations of systems of particle mechanics is in terms of the particles' being the centers of mass of rigid bodies, and it is easy to think of physical examples in which the centers of mass of distinct bodies may at certain moments coincide, for example, when a ball is thrown through a ring, a bullet is fired through the hole of a doughnut, a person is sitting at a certain point in an airplane. On the other hand, it should be noted that the axiom of impenetrability played an important role in the eighteenth-century discussions of the theory of matter by Euler and others.[6]

Two theorems—one on determinism. As mentioned earlier, the first theorem is a formulation of Newton's First Law. In this and subsequent theorems it is understood that the systems of mechanics are with respect to a given classical space-time structure $(A, \boldsymbol{\tau}, \boldsymbol{\alpha})$.

---

[6]Euler discusses impenetrability in many different places, but perhaps his best philosophical discussion is in the first two letters (21 October 1760 and 25 October 1760) of the Second Part of his *Lettres de Euler a une Princesse d'Allemagne, sur divers sujets de physique et de philosophie* (1842), where he also criticizes Descartes' theory of matter as pure extension.

THEOREM 1. *Let* $\mathfrak{P} = (P, \mathbf{s}, m, \mathbf{f}, \mathbf{g})$ *be a system of particle mechanics, and let* p *be a member of* P *such that, for all* t *in* T, *the range of* τ,

$$\sum_{q \in P} \mathbf{f}(p, q, t) + \sum_{n=1}^{\infty} \mathbf{g}(p, t, n) = 0.$$

Then there are vectors $\mathbf{a}$ *and* $\mathbf{b}$ *such that, for all* t *in* T,

$$\mathbf{s}(p, t) - \boldsymbol{\alpha}(t) = \mathbf{a} + \mathbf{b}t.$$

The proof of this theorem follows at once from Axioms 4, 5, and 9.

The next theorem is a classic expression of the fact that particle mechanics is a thoroughly deterministic science. The theorem says that the whole history of a system is determined by P, T, m, $\mathbf{f}$, $\mathbf{g}$, and appropriate initial conditions. In particular, if we know the masses and the forces acting on a collection of particles, as well as their positions and velocities at a given moment of time, we know the entire history of the system. The proof rests upon elementary considerations concerning the uniqueness of solutions of the kind of differential equation exemplified by Newton's Second Law and is thus omitted here.

THEOREM 2. *Let* $\mathfrak{P} = (P, \mathbf{s}, m, \mathbf{f}, \mathbf{g})$ *and* $\mathfrak{P}' = (P, \mathbf{s}', m, \mathbf{f}, \mathbf{g})$ *be two systems of particle mechanics such that for some* t *in* T *and every* p *in* P

$$\mathbf{s}(p, t) - \boldsymbol{\alpha}(t) = \mathbf{s}'(p, t) - \boldsymbol{\alpha}(t)$$

and

$$D(\mathbf{s}_p(t) - \boldsymbol{\alpha}(t)) = D(\mathbf{s}'_p(t) - \boldsymbol{\alpha}(t)).$$

Then for every t *in* T *and every* p *in* P

$$\mathbf{s}(p, t) - \boldsymbol{\alpha}(t) = \mathbf{s}'(p, t) - \boldsymbol{\alpha}(t).$$

Because the trajectory or path of a particle is fully determined by the forces acting on it and "initial" conditions of position and velocity at some time t, it can easily be mistakenly thought that we can often or usually directly solve the vectorial equation of motion—three equations in Cartesian coordinates—to obtain the trajectory of the particle as an *explicit function* of t, built up from a finite number of known mathematical functions. Unfortunately, this is far from being the case. Even for physical situations that have a simple formulation and consequently an easy derivation of the equations of motion, it is likely that no closed solution of the sort just described can be found. The most celebrated example is the three-body problem: to determine the motion of three particles acted upon only by their mutual forces of gravitational attraction. In this case as in others it is always possible to express the solution in power series, but such infinite series are not at all satisfactory for studying either the qualitative or long-term behavior of a system. The complexity of the three-body problem has already been exhibited for an important special case in Section 5.6.[7] The glory of mechanics is the ingenuity that has been used in the solution of various special cases, of which the most famous historical example is Newton's solution of the two-body problem for gravitational attraction.

---

[7] For a readable account of some recent surprising progress on some new special solutions of the three-body problem, see Montgomery (2001).

7.1 CLASSICAL PARTICLE MECHANICS

Momentum and angular momentum. The remainder of this section on classical mechanics is oriented toward some general theorems on momentum and angular momentum, which also require the explicit introduction of several of the most significant mechanical concepts not required for statement of the axioms.

To have a more compact notation for the position vector of a particle at time t, we shall use the familiar notation $\mathbf{q}_i(t)$, where now we also replace 'p' for the particle p, by the numerical index i, so:

$$\mathbf{q}_i(t) = \mathbf{s}(i, t) - \boldsymbol{\alpha}(t),$$

and where there is little possibility of confusion, we shall write $\mathbf{q}_i$ rather than $\mathbf{q}_i(t)$, in order to conform more closely to the standard notation of physics. We also need a notation for the *resultant* force on a particle i, where N is the number of particles:

$$\mathbf{F}_i(t) = \sum_{j=1}^{N} \mathbf{f}(i,j,t) + \sum_{n=1}^{\infty} \mathbf{g}(i,t,n),$$

and again we shall often write $\mathbf{F}_i$ rather than $\mathbf{F}_i(t)$. Second, we need a notation for the resultant *external* force on a particle i:

$$\mathbf{G}_i(t) = \sum_{n=1}^{\infty} \mathbf{g}(i,t,n).$$

The (*linear*) *momentum* of particle i is $m_i \mathbf{q}_i$. From this definition it follows that *the rate of change of momentum of a particle is equal to the resultant force on the particle.* In symbols,

$$\frac{d}{dt}(m_i \dot{\mathbf{q}}_i) = \mathbf{F}_i.$$

In this form the law is often referred to as the *principle of linear momentum*. The application of the principle to a system yields a significant result. First we define the momentum, $\mathbf{P}(t)$, of a system of particles by:

$$\mathbf{P}(t) = \sum_{i=1}^{N} m_i \dot{\mathbf{q}}_i(t)$$

where N is the number of particles.

THEOREM 3. *The rate of change of the momentum of a system of particles is equal to the sum of the external resultant forces on the particles, i.e.,*

$$\frac{d}{dt}\mathbf{P}(t) = \sum_{i=1}^{N} \mathbf{G}_i(t).$$

The proof is simple. It depends particularly on Axiom 6, the principle of equal and opposite internal forces, which is the basis of their cancellation in determining the rate of change of $\mathbf{P}(t)$.

We now extend Theorem 3 to the center of mass of a system. Let

$$\underline{m} = \sum_{i=1}^{N} m_i,$$

i.e., let $\underline{m}$ be the total mass of the system. The center of mass, $\mathbf{c}(t)$, of a system is defined as follows. Let $\mathbf{b}(t)$ be any affine point in $\mathfrak{A}(t)$. The (*linear*) *moment* $\mathbf{L}(t)$

about $\mathbf{b}(t)$ of the system $\mathfrak{P}$ of particles is defined as:

$$\mathbf{L}(t) = \sum_{i=1}^{N} m_i(\mathbf{s}_i(t) - \mathbf{b}(t)).$$

Then the *center of mass* of $\mathfrak{P}$ is defined as the point $\mathbf{c}(t)$ such that the (linear) moment about $\mathbf{c}(t)$ of $\mathfrak{P}$ is zero. Of course, for the definition to be a proper one, we must show that such a point $c(t)$ exists and is unique. Consider the vector

$$\frac{\sum m_i(\mathbf{s}_i(t) - \boldsymbol{\alpha}(t))}{m}$$

and the affine point $\boldsymbol{\alpha}(t)$. By Definition 2 of affine spaces, a unique point $\mathbf{c}(t)$ exists such that

$$\frac{\sum m_i(\mathbf{s}_i(t) - \boldsymbol{\alpha}(t))}{m} = \mathbf{c}(t) - \boldsymbol{\alpha}(t),$$

and so

$$\sum m_i(\mathbf{s}_i(t) - \boldsymbol{\alpha}(t)) = \underline{m}(\mathbf{c}(t) - \boldsymbol{\alpha}(t)).$$

Then

$$\sum m_i(\mathbf{s}_i(t) - \mathbf{c}(t)) = \underline{m}(\boldsymbol{\alpha}(t) - \boldsymbol{\alpha}(t)) = 0, \tag{1}$$

and since for any three affine points P, Q and R

$$\underline{PQ} + \underline{QR} = \underline{PR},$$

i.e.,

$$\underline{QR} = \underline{PR} - \underline{PQ},$$

or in difference notation

$$Q - R = (P - R) - (P - Q), \tag{2}$$

we infer from (1) and (2)

$$\sum m_i(\mathbf{s}_i(t) - \mathbf{c}(t)) = 0.$$

so that the center of mass as defined always exists and is unique.

We leave to the reader proof of the following classical representation.

THEOREM 4. *The center of mass of a system of particles moves like a particle that has mass equal to the total mass of the system and that is acted on by a force equal to the resultant external force of the system.*

We now turn to angular momentum. The angular momentum $\mathbf{M}(t)$ of a particle relative to a point $\mathbf{b}(t)$ is the moment of the (linear) momentum relative to $\mathbf{b}(t)$, i.e., the vector product of the position vector of the particle i relative to $\mathbf{b}(t)$ and its momentum. In symbols

$$\mathbf{M}_i(t) = [\mathbf{s}_i(t) - \mathbf{b}(t), m_i D(\mathbf{s}_i(t) - \mathbf{b}(t))].$$

In case $\mathbf{b}(t)$ is chosen to be $\boldsymbol{\alpha}(t)$, we have the familiar formula

$$\mathbf{M}_i(t) = [\mathbf{q}_i, m_i \dot{\mathbf{q}}_i].$$

The angular momentum of a system of particles relative to a point is just

$$\mathbf{M}(t) = \sum_{i=1}^{N} \mathbf{M}_i(t).$$

7.1 Classical Particle Mechanics

The *torque,* or *moment* of a force $\mathbf{F}$ on a particle i about a point $\mathbf{b}(t)$ is just

$$[\mathbf{s}(i,t) - \mathbf{b}(t), \mathbf{F}].$$

THEOREM 5. *The rate of change of the angular momentum of a system of particles about a point is equal to the sum of the moments of the external forces acting on the particles of the system. In symbols,*

$$\frac{d\mathbf{M}(t)}{dt} = \sum_{i=1}^{N} [\mathbf{s}(i,t) - \mathbf{b}(t), \mathbf{G}_i].$$

Proof. From the definition of $\mathbf{M}(t)$

$$\frac{d\mathbf{M}(t)}{dt} = \sum_{i=1}^{N} ([D(\mathbf{s}_i(t) - \mathbf{b}(t)), m_i D(\mathbf{s}_i(t) - \mathbf{b}(t))] + [\mathbf{s}(t) - \mathbf{b}(t), m_i D^2(\mathbf{s}_i(t) - \mathbf{b}(t))]).$$

The first vector product is zero, and the second term of the second vector product can be replaced by the sum of forces—using directly Axiom 9, i.e., Newton's second law. So

$$\frac{d\mathbf{M}(t)}{dt} = \sum_{i=1}^{N} [\mathbf{s}_i(t) - \mathbf{b}(t), \mathbf{F}_i + \mathbf{G}_i],$$

but the moments of the internal forces cancel, for $\mathbf{f}_{ij} = -\mathbf{f}_{ji}$ and thus

$$[\mathbf{s}_i(t) - \mathbf{b}(t), \mathbf{f}_{ij}] + [\mathbf{s}_j(t) - \mathbf{b}(t), \mathbf{f}_{ji}] = [\mathbf{s}_i(t) - \mathbf{s}_j(t), \mathbf{f}_{ij}] = 0,$$

by Axiom 7, which proves the theorem.

Laws of conservation. If there are no resultant external forces on any particles, i.e., if for all particles i and times t, $\mathbf{G}_i(t)$ is zero, then the system of particles is said to be *closed*. For closed systems we have as immediate consequences of Theorems 3 and 5, two important laws of conservation:

THEOREM 6. *For closed systems of particles, momentum* $\mathbf{P}(t)$ *and angular momentum* $\mathbf{M}(t)$ *are conserved, i.e., both* $\mathbf{P}(t)$ *and* $\mathbf{M}(t)$ *are constants.*

Some form of the law of conservation of momentum goes back at least to Descartes' *Principles of Philosophy* (1644), but the account is, from our vantage point, enormously confused. Descartes defined the quantity of motion as the product of the size of a body and its velocity as a scalar. Conservation is formulated in this way (II, 36):[8]

> That God is the primary cause of motion; and that He always maintains an equal quantity of it in the universe.

That Descartes did have in mind size and not mass is supported by statements in the following articles: II, 36, 40, 43, 47-52; IV, 199, 203. The classical controversy between Cartesians and Lebnizians about the correct definition of quantity of motion was not really settled until the eighteenth century. Modern usage is close to that of Euler, as in many other related matters of mathematics and mechanics.

---

[8]This reference is to Part II, Article 36, of the *Principles*. Any quotations are from the translation of Miller and Miller (1983).

Newton did informally state the law of conservation of momentum in a correct fashion for bodies acted upon only by forces of mutual attraction. In the Introduction to Section XI of Book I of the *Principia,* he says this:

> And if there be more bodies, which either are attracted by one body, which is attracted by them again, or which all attract each other mutually, these bodies will be so moved among themselves, that their common centre of gravity will either be at rest, or move uniformly forwards in a right line.

Conservation of angular momentum is a more recent concept. Although the moment of momentum, or angular momentum, was used in various ways, e.g., in the theory of rigid bodies, in the eighteenth century, according to Truesdell (1968), Theorem 5 was probably first proved by Poisson in 1833 in the second edition of his *Traité de Mécanique.*

These conservation laws have important applications in a great variety of problems, sometimes in order to reduce the number of independent variables, in other cases to provide a general law that can be used to derive a great deal of information about a system, even though a full dynamical specification of the forces may not be feasible.

The first kind of application can be illustrated by the classical reduction of independent variables in the three-body problem, already mentioned as the most celebrated problem of classical mechanics. In Cartesian coordinates we begin with 18 independent variables—nine of position and nine of velocity. The conservation of momentum means that the center of mass of the system moves with uniform velocity, and this translates into six constants for the center's motion, thus reducing the number of variables to 12. The conservation of angular momentum imposes three more equations and thus a further reduction of the number of variables to nine. Finally, the conservation of energy, discussed below, leads to one more equation and thus a reduction to a total of eight variables. Bruns (1887) showed that these ten conditions of conservation are the only algebraic constraints possible on the three-body problem. Note in summary that they come from the three laws of conservation of classical mechanics.

Fortunately these conservation laws generalize to relativistic (nonquantum) mechanics and classical quantum mechanics. An important application in which the full dynamical system is not specified is the Compton effect (1923). Einstein had pointed out earlier that photons should have a definite momentum in spite of their zero rest mass. Compton argued that the law of conservation of momentum should hold for collisions between photons and electrons. Therefore the scattering of photons by electrons should lead to a transfer of momentum to the electrons, as well as a reduction in frequency from the scattering of the photons. Compton showed experimentally that the reduction in frequency did occur in the scattering of x-rays (the Compton effect) and the transfer of momentum to electrons doing the scattering could be observed. What is methodologically important about this example is that no assumption about the quantitative nature of the interacting forces was made.

The philosophical idea of conservation is implicit in a number of places in pre-modern physics. The Aristotelian theory of the motion of the heavens as something permanent and unchanging in its character represented one important idea of constancy or conservation. Another is Aristotle's concept of matter as the substratum of that which is changing. The scholastic dictum on this is:

7.1 CLASSICAL PARTICLE MECHANICS

Matter cannot be generated nor can it be corrupted, since all that is generated, is generated from matter, and all that perishes, perishes into matter.[9]

Apparently the actual term *conservation* was first used (in French) by Leibniz (1692) in his controversies concerning *vis viva* and its proper quantitative definition.[10] At least since the middle of the nineteenth century the laws of conservation have been rightly regarded as among the most universal and significant laws of physics. Not only are they powerful in application, but they also satisfy a deep philosophical need to identify that which is unchanging and permanent in the world. The conservation laws of momentum, angular momentum, and energy are perhaps the first great global laws of physics which can be applied to an isolated physical system without knowing many things about its detailed dynamical structure.[11]

Embedding in closed systems. I consider now a theorem asserting that any system of particle mechanics can be embedded in a closed system. To formulate our embedding results we need the notion of two systems of mechanics being *equivalent*. This concept of equivalence is somewhat weaker than the strictest notion of isomorphism we might define for systems of particle mechanics. It is weaker in the sense that we do not require that the structure of the individual forces be the same but just that the resultant forces be identical.

DEFINITION 5. *Two systems of particle mechanics* $\mathfrak{P} = (P, \mathbf{s}, m, \mathbf{f}, \mathbf{g})$ *and* $\mathfrak{P}' = (P', \mathbf{s}', m', \mathbf{f}', \mathbf{g}')$ *are* equivalent *if and only if*

$$P = P'$$
$$\mathbf{s} = \mathbf{s}'$$
$$m = m'.$$

Notice that this notion of equivalence is one of several concepts which are both weaker and stronger than the "natural" notion of isomorphism; it is weaker in that two equivalent systems do not have the same structure of individual forces, but it is stronger in that two equivalent systems must be kinematically identical and identical in their mass functions.

We also need to define the notion of one system of particle mechanics being a subsystem of another. This definition follows the discussion already given for the concept of subsystem in a simpler algebraic setting.

DEFINITION 6. *Let* $\mathfrak{P} = (P, \mathbf{s}, m, \mathbf{f}, \mathbf{g})$ *be a system of particle mechanics, let* P' *be a nonempty subset of* P, *let* $\mathbf{s}'$ *and* m' *be the functions* $\mathbf{s}$ *and* m *with their first arguments restricted to* P', *let* $\mathbf{f}'$ *be the function* $\mathbf{f}$ *with its first two arguments*

---

[9] Cited in Jammer (1961, p. 41). The Latin is: "Materia non est generabilis nec corruptibilis, quia omne quod generatur, generatur ex materia, et quod corrumpitur, corrumpitur in materiam." The dictum is attributed by some to Thomas Aquinas. Also, see Chapter IX of Book I of Aristotle's *Physics* 192 a30 for a very similar statement.

[10] See Leibniz's Werke, edited by Pertz, Mathematics VI, p. 217.

[11] The relation of conservation laws to invariance is discussed briefly at the end of Section 4.1, in connection with Noether's (1918) famous theorem. The theorem on conservation of energy is stated later, as Theorem 8, because the concept of a *conservative* system, a condition stronger than that for a closed system, must first be introduced.

restricted to P', and let $\mathbf{g}'$ be the function such that for every p in P' and t in T

$$\mathbf{g}'(p,t,1) = \sum_{q \in P-P'} \mathbf{f}(p,q,t),$$

and for every i

$$\mathbf{g}'(p,t,i+1) = \mathbf{g}(p,t,i).$$

Then $\mathfrak{P}' = (P', \mathbf{s}', m', \mathbf{f}', \mathbf{g}')$ is a subsystem of $\mathfrak{P}$.

The embedding representation theorem is then just the following simple result.

THEOREM 7. *Every system of particle mechanics is equivalent to a subsystem of a closed system of particle mechanics.*

The proof of this theorem may be found in Suppes (1957/1999, pp. 302-304) and therefore will be omitted here. The interpretation raises a rather interesting problem from the standpoint of physics. Embedding theorems of this sort are standard in mathematics, but they have a more restricted significance in physics, for the simple reason that the larger closed system in which we embed a given system consists of the initial system augmented by imagined or purely conceptualized particles and not by real physical particles. In other words, when we embed a system in a closed system we are not guaranteeing the actual existence of the additional particles in the larger closed system. The conceptual point of the theorem, however, remains the same as it would be generally in mathematics. What we have shown is that any sorts of motion that may be produced in an arbitrary system of particle mechanics with arbitrary external forces acting on it can be reproduced and represented in an enlarged closed system of particle mechanics. So the theorem has significance in terms of conceptual arguments about the adequacy of mechanics to explain the origin of all the motions that may be found in actual objects in the universe. What the theorem shows is that we may always give an account of the motion of a system by conceptualizing it in terms of internal forces satisfying Newton's Third Law of equal and opposite forces.

Conservative systems. Of great importance, in ways that will at least be partially pointed out, is the representation of the forces acting on the particle of a system by a potential energy function. The concept was introduced by Lagrange in 1773 (*Oeuvres*, VI, p. 335), and the name *potential* is due to the British mathematician and scientist George Green (1828).

DEFINITION 7. *Let $\mathfrak{P} = (P, T, m, \mathbf{s}, \mathbf{f}, \mathbf{g})$ be a system of particle mechanics with n particles. Then the system is conservative iff there exists a differentiable function U (the potential energy function of $\mathfrak{P}$) from E^{3n} to R, such that for $1 \leq i \leq n$ and t in T*

$$\mathbf{F}_i(t) = -\frac{\partial U}{\partial \mathbf{q}_i}(\mathbf{q}_1(t), \ldots, \mathbf{q}_n(t)), \quad i = 1, \ldots, n.$$

As the definition makes clear, the resultant forces in a conservative system are represented by the partial derivatives of the potential energy function. The definition has been given in this form to emphasize this point, but the traditional formulation is to put $m_i \ddot{\mathbf{q}}_i$ on the left-hand side of the defining equation, rather than $F_i(t)$. Within

7.1 Classical Particle Mechanics

the present framework, where we start from forces and Newton's second law, the two formulations are obviously equivalent.

Here are some simple but important examples of conservative systems.

Falling stone. The problem is one-dimensional, with $x(t)$ the height of the stone above the surface of the earth at time t. The equation of motion is

$$m\ddot{x} = -gm,$$

where m is the mass of the stone and g is the gravitational constant, with g being approximately 9.8 meters/sec. Then

$$U(x) = gmx.$$

Harmonic oscillator. This simple mechanical system consisting of a single particle has many diverse applications in both classical and quantum mechanics. Classical instances are these: simple pendulum executing small oscillations, small extensions of the sort of spring that is used in a spring balance. The one-dimensional equation of motion is

$$m\ddot{x} = -m\alpha^2 x,$$

where m is the mass of the particle and α is a physical constant varying from one kind of application to another. Here

$$U(x) = \frac{m\alpha^2 x^2}{2},$$

and we can write the equation of motion as

$$m\ddot{x} = \frac{\partial U(x)}{\partial x}.$$

Three-body problem. This celebrated problem concerning three particles of masses m_1, m_2 and m_3 with the only forces being those of mutual gravitational attraction has the following potential energy function:

$$U(\mathbf{q}_1, \mathbf{q}_2, \mathbf{q}_3) = -\frac{m_1 m_2}{|\mathbf{q}_1 - \mathbf{q}_2|} - \frac{m_2 m_3}{|\mathbf{q}_2 - \mathbf{q}_3|} - \frac{m_3 m_1}{|\mathbf{q}_3 - \mathbf{q}_1|}.$$

We now want to prove an important theorem about conservation of total energy in a conservative system. First, for any system of n particles the kinetic energy $T(t)$ as a function of time is defined by:

$$T(t) = \frac{1}{2} \sum_{i=1}^{n} m_i \dot{\mathbf{q}}_i^2.$$

This quantity has a long and complicated history in mechanics. It was essentially first introduced by Leibniz (*Acta erud.*, 1695). He called the mass of a particle multiplied by the square of its velocity the *vis viva*. The total energy E of a conservative system with potential energy function U is:

$$E = T + U,$$

where the arguments of the functions have been omitted. More explicitly

$$E(t) = T(t) + U(\mathbf{q}_i(t), \ldots, \mathbf{q}_n(t)).$$

THEOREM 8. (Law of Conservation of Energy). *The total energy of a conservative system is constant in time.*

Proof. Using the definitions given above of T and U, it is easy to show that
$$\frac{d}{dt}(T+U) = 0.$$

The following theorem relates closed systems–no external forces–, and conservative systems. The proof is omitted.

THEOREM 9. *If a system of particles is closed and the internal forces of interaction depend only on the distances between the particles, i.e., for all particles i and j and times t*
$$\mathbf{f}(i,j,t) = \mathbf{f}(|\mathbf{q}_i - \mathbf{q}_j|),$$
then the system is conservative, i.e., it has a potential energy function.

7.2 Representation Theorems for Hidden Variables in Quantum Mechanics

The literature on hidden variables in quantum mechanics is now enormous. This section covers mainly the part dealing with probabilistic representation theorems for hidden variables, even when the hidden variables may be deterministic. Fortunately, this body of results can be understood without an extensive knowledge of quantum mechanics, which is not developed *ab initio* here.[12]

First, we state, and sketch the proof, of the fundamental theorem of the collection we consider: there is a factoring hidden variable, i.e., formally a common cause, for a finite set of finite or continuous observables, i.e., random variables in the language of probability theory, if and only if the observables have a joint probability distribution. The physically important aspect of this theorem is that under very general conditions the existence of a hidden variable can be reduced completely to the relationship between the observables alone, namely, the problem of determining whether or not they have a joint probability distribution compatible with the given data, e.g., means, variances and correlations of the observables.

We emphasize that although most of the literature is restricted to no more than second-order moments such as covariances and correlations, there is no necessity to make such a restriction. It is in fact violated in the third-order or fourth-order moment that arises in the well-known Greenberger, Horne and Zeilinger (1989) three- and four-particle configurations providing new Gedanken experiments on hidden variables. For our probabilistic proof of an abstract GHZ result, see Theorem 7, and for related inequalities, see Theorem 8.

As is familiar, Bell's results on hidden variables were mostly restricted to ±1 observables, such as spin or polarization. But there is nothing essential about this restriction. Our general results cover any finite or continuous observables. At the end we give various results on hidden variables for Gaussian observables and formulate as the final

---

[12] Many of the results given are taken from joint work with Acacio de Barros and Gary Oas (Suppes, de Barros and Oas 1998; de Barros and Suppes 2000).

7.2 Hidden Variables in Quantum Mechanics

theorem a nonlinear inequality that is necessary and sufficient for three Gaussian random variables to have a joint distribution compatible with their given means, variances and correlations.

Factorization. In the literature on hidden variables, the principle of factorization is sometimes baptized as a principle of locality. The terminology is not really critical, but the meaning is. We have in mind a quite general principle for random variables, continuous or discrete, which is the following. Let $\mathbf{X}_1, \ldots, \mathbf{X}_n$ be random variables, then a necessary and sufficient condition that there is a random variable $\boldsymbol{\lambda}$, which is intended to be the hidden variable, such that $\mathbf{X}_1 \ldots, \mathbf{X}_n$ are conditionally independent given $\boldsymbol{\lambda}$, is that there exists a joint probability distribution of $\mathbf{X}_1, \ldots, \mathbf{X}_n$, without consideration of $\boldsymbol{\lambda}$. This is our first theorem, which is the general fundamental theorem relating hidden variables and joint probability distributions of observable random variables.

THEOREM 1. (Suppes and Zanotti 1981; Holland and Rosenbaum 1986). *Let n random variables $X_1, \ldots, X_n$, finite or continous, be given. Then there exists a hidden variable $\boldsymbol{\lambda}$ such that there is a joint probability distribution F of $(\mathbf{X}_1, \ldots, \mathbf{X}_n, \boldsymbol{\lambda})$ with the properties*

(i) $F(x_1, \ldots, x_n \mid \lambda) = P(\mathbf{X}_1 \leq x_1, \ldots, \mathbf{X}_n \leq x_n \mid \boldsymbol{\lambda} = \lambda)$

(ii) *Conditional independence holds, i.e., for all $x_1, \ldots, x_n, \lambda$,*

$$F(x_1, \ldots, x_n \mid \lambda) = \prod_{j=1}^{n} F_j(x_j \mid \lambda),$$

if and only if there is a joint probability distribution of $\mathbf{X}_1, \ldots, \mathbf{X}_n$. Moreover, $\boldsymbol{\lambda}$ may be constructed so as to be deterministic, i.e., the conditional variance given $\boldsymbol{\lambda}$ of each $\mathbf{X}_i$ is zero.

To be completely explicit in the notation

$$F_j(x_j \mid \lambda) = P(\mathbf{X}_j \leq x_j \mid \boldsymbol{\lambda} = \lambda). \tag{1}$$

Idea of the proof. Consider three ± 1 random variables $\mathbf{X}$, $\mathbf{Y}$ and $\mathbf{Z}$. There are 8 possible joint outcomes $(\pm 1, \pm 1, \pm 1)$. Let p_{ijk} be the probability of outcome (i, j, k). Assign this probability to the value λ_{ijk} of the hidden variable $\boldsymbol{\lambda}$ we construct. Then the probability of the quadruple (i, j, k, λ_{ijk}) is just p_{ijk} and the conditional probabilities are deterministic, i.e.,

$$P(\mathbf{X} = i, \mathbf{Y} = j, \mathbf{Z} = k \mid \lambda_{ijk}) = 1,$$

and factorization is immediate, i.e.,

$$P(\mathbf{X} = i, \mathbf{Y} = j, \mathbf{Z} = k \mid \lambda_{ijk}) = P(\mathbf{X} = i \mid \lambda_{ijk}) P(\mathbf{Y} = j \mid \lambda_{ijk}) P(\mathbf{Z} = k \mid \lambda_{ijk}).$$

Extending this line of argument to the general case proves the joint probability distribution of the observables is sufficient for existence of the factoring hidden variable, i.e., common cause. From the formulation of Theorem 1 necessity is obvious, since the joint distribution of $(\mathbf{X}_1, \ldots, \mathbf{X}_n)$ is a marginal distribution of the larger distribution $(\mathbf{X}_1 \ldots, \mathbf{X}_n, \boldsymbol{\lambda})$.

It is apparent that the construction of $\boldsymbol{\lambda}$ is purely mathematical. It has in itself no physical content. In fact, the proof itself is very simple. All the real mathematical difficulties are to be found in giving workable criteria for observables to have a joint probability distribution. As we remark in more detail later, we still do not have good criteria in the form of inequalities for necessary and possibly sufficient conditions for a joint distribution of three random variables with $m > 2$ finite values, as in higher spin cases.

When additional physical assumptions are imposed on the hidden variable $\boldsymbol{\lambda}$, then the physical content of $\boldsymbol{\lambda}$ goes beyond the joint distribution of the observables. A simple example is embodied in the following theorem about two random variables. We impose an additional condition of symmetry on the conditional expectations, and then a hidden variable exists only if the correlation of the two observables is non-negative, a strong additional restriction on the joint distribution. The proof of this representation theorem is found in the article cited with its statement.

THEOREM 2. (Suppes and Zanotti 1980). *Let* $\mathbf{X}$ *and* $\mathbf{Y}$ *be two-valued random variables, for definiteness with possible values* 1 *and* −1, *and with positive variances, i.e.,* $\sigma(\mathbf{X})$, $\sigma(\mathbf{Y}) > 0$. *In addition, let* $\mathbf{X}$ *and* $\mathbf{Y}$ *be exchangeable, i.e.,*

$$P(\mathbf{X}=1, \mathbf{Y}=-1) = P(\mathbf{X}=-1, \mathbf{Y}=1).$$

Then a necessary and sufficient condition that there exist a hidden variable $\boldsymbol{\lambda}$ *such that*

$$E(\mathbf{XY} \mid \boldsymbol{\lambda} = \lambda) = E(\mathbf{X} \mid \boldsymbol{\lambda} = \lambda) E(\mathbf{Y} \mid \boldsymbol{\lambda} = \lambda)$$

and

$$E(\mathbf{X} \mid \boldsymbol{\lambda} = \lambda) = E(\mathbf{Y} \mid \boldsymbol{\lambda} = \lambda)$$

for every value λ *(except possibly on a set of measure zero) is that the correlation of* $\mathbf{X}$ *and* $\mathbf{Y}$ *be non-negative.*

Often, in physics, as in the present section, we are interested only in the means, variances and covariances—what is called the second-order probability theory, because we consider only second-order moments. We say that a hidden variable $\boldsymbol{\lambda}$ satisfies the *Second-Order Factorization Condition* with respect to the random variables $\mathbf{X}_1, \ldots, \mathbf{X}_n$ whose two first moments exist if and only if

(a) $E(\mathbf{X}_1 \cdots \mathbf{X}_n | \boldsymbol{\lambda}) = E(\mathbf{X}_1|\boldsymbol{\lambda}) \cdots E(\mathbf{X}_n|\boldsymbol{\lambda})$,
(b) $E(\mathbf{X}_1^2 \cdots \mathbf{X}_n^2 | \boldsymbol{\lambda}) = E(\mathbf{X}_1^2|\boldsymbol{\lambda}) \cdots E(\mathbf{X}_n^2|\boldsymbol{\lambda})$.

We then have as an immediate consequence of Theorem 1 the following representation.

THEOREM 3. *Let n random variables discrete or continuous be given. If there is a joint probability distribution of* $\mathbf{X}_1, \ldots, \mathbf{X}_n$, *then there is a deterministic hidden variable* $\boldsymbol{\lambda}$ *such that* $\boldsymbol{\lambda}$ *satisfies the Second-Order Factorization Condition with respect to* $\mathbf{X}_1, \ldots, \mathbf{X}_n$.

The informal statement of Theorems 1 and 3, which we call the *Factorization Theorems*, is that the necessary and sufficient condition for the existence of a factorizing hidden variable $\boldsymbol{\lambda}$ is just the existence of a joint probability distribution of the given random variables $\mathbf{X}_i$. In our view, the condition of factorizability is often too strong a condition for hidden variables. A striking example is that gravitational phenomena in classical mechanics satisfy locality but not factorizability.

7.2 HIDDEN VARIABLES IN QUANTUM MECHANICS

Locality. The next systematic concept we want to discuss is locality. We mean by locality what we think John Bell meant by locality in the following quotation from his well-known 1964 paper (Bell 1964).

> It is the requirement of locality, or more precisely that the result of a measurement on one system be unaffected by operations on a distant system with which it has interacted in the past, that creates the essential difficulty.... The vital assumption is that the result B for particle 2 does not depend on the setting **a**, of the magnet for particle 1, nor A on **b**.

Although Theorems 1 and 3 are stated at an abstract level without any reference to space-time or other physical considerations, there is an implicit hypothesis of locality in their statements. To make the locality hypothesis explicit, we need to use additional concepts. For each random variable $\mathbf{X}_i$, we introduce a vector M_i of parameters for the local apparatus (in space-time) used to measure the values of random variable $\mathbf{X}_i$.

DEFINITION 1. (Locality Condition I)

$$E(\mathbf{X}_i^k | M_i, M_j, \boldsymbol{\lambda}) = E(\mathbf{X}_i^k | M_i, \boldsymbol{\lambda}),$$

where $k = 1, 2$, corresponding to the first two moments of $\mathbf{X}_i$, $i \neq j$, and $1 \leq i, j \leq n$.

Note that we consider only M_j on the supposition that in a given experimental run, only the correlation of $\mathbf{X}_i$ with $\mathbf{X}_j$ is being studied. Extension to more variables is obvious. In many experiments the direction of the measuring apparatus is the most important parameter that is a component of M_i.

DEFINITION 2. (Locality Condition II) *The distribution of $\boldsymbol{\lambda}$ is independent of the parameter values M_i and M_j, i.e., for all functions g for which the expectation $E(g(\boldsymbol{\lambda}))$ and $E(g(\boldsymbol{\lambda})|M_i, M_j)$ are finite,*

$$E(g(\boldsymbol{\lambda})) = E(g(\boldsymbol{\lambda})|M_i, M_j).$$

Here we follow Suppes (1976). In terms of Theorem 3, locality in the sense of Condition I is required to satisfy the hypothesis of a fixed mean and variance for each $\mathbf{X}_i$. If experimental observation of $\mathbf{X}_i$ when coupled with $\mathbf{X}_j$ were different from what was observed when coupled with $\mathbf{X}_{j'}$, then the hypothesis of constant means and variances would be violated. The restriction of Locality Condition II must be satisfied in the construction of $\boldsymbol{\lambda}$ and it is easy to check that it is.

We embody these remarks in Theorem 4.

THEOREM 4. *Let n random variables $\mathbf{X}_1, \ldots, \mathbf{X}_n$ be given satisfying the hypothesis of Theorem 3. Let M_i be the vector of local parameters for measuring $\mathbf{X}_i$, and let each $\mathbf{X}_i$ satisfy Locality Condition I. Then there is a hidden variable $\boldsymbol{\lambda}$ satisfying Locality Condition II and the Second-Order Factorization Condition if there is a joint probability distribution of $\mathbf{X}_1, \ldots, \mathbf{X}_n$.*

The next theorem states two conditions equivalent to an inequality condition given in Suppes and Zanotti (1981) for three random variables having just two values.

THEOREM 5. *Let three random variables $\mathbf{X}, \mathbf{Y}$ and $\mathbf{Z}$ be given with values ± 1 satisfying the symmetry condition $E(\mathbf{X}) = E(\mathbf{Y}) = E(\mathbf{Z}) = 0$ and with covariances $E(\mathbf{XY}), E(\mathbf{YZ})$ and $E(\mathbf{XZ})$ given. Then the following three conditions are equivalent:*

(i) There is a hidden variable λ satisfying Locality Condition II and equation (a) of the Second-Order Factorization Condition holds.

(ii) There is a joint probability distribution of the random variables **X**, **Y**, and **Z** compatible with the given means and covariances.

(iii) The random variables **X**, **Y** and **Z** satisfy the following inequalities:

$$-1 \leq E(\mathbf{XY}) + E(\mathbf{YZ}) + E(\mathbf{XZ}) \leq 1 + 2\mathrm{Min}(E(\mathbf{XY}), E(\mathbf{YZ}), E(\mathbf{XZ})).$$

There are several remarks to be made about this theorem, especially the inequalities given in (iii). A first point is how these inequalities relate to Bell's well-known inequality (Bell 1964):

$$1 + E(\mathbf{YZ}) \geq |E(\mathbf{XY}) - E(\mathbf{XZ})|. \tag{2}$$

Bell's inequality is in fact neither necessary nor sufficient for the existence of a joint probability distribution of the random variables **X**, **Y** and **Z** with values ± 1 and expectations equal to zero. That it is not sufficient is easily seen from letting all three covariances equal $-\frac{1}{2}$. Then Bell's inequality is satisfied, for

$$1 - \frac{1}{2} \geq |-\frac{1}{2} - (-\frac{1}{2})|,$$

i.e.,

$$\frac{1}{2} \geq 0,$$

but, as is clear from (iii) there can be no joint distribution with the three covariances equal to $-\frac{1}{2}$, for

$$-\frac{1}{2} + -\frac{1}{2} + -\frac{1}{2} < -1.$$

Secondly, Bell's inequality is not necessary. Let $E(\mathbf{XY}) = \frac{1}{2}$, $E(\mathbf{XZ}) = -\frac{1}{2}$, and $E(\mathbf{YZ}) = -\frac{1}{2}$, then (2) is violated, because

$$1 - \frac{1}{2} < |\frac{1}{2} - (-\frac{1}{2})|,$$

but (iii) is satisfied, and so there is a joint distribution:

$$-1 \leq \frac{1}{2} - \frac{1}{2} - \frac{1}{2} \leq 1 + 2\mathrm{Min}(\frac{1}{2}, -\frac{1}{2}, -\frac{1}{2}),$$

i.e.,

$$-1 \leq -\frac{1}{2} \leq 0.$$

Bell derived his inequality for certain cases satisfied by a local hidden-variable theory, but violated by the quantum mechanical covariance equal to $-\cos\theta_{ij}$. In particular, let $\theta_{\mathbf{XY}} = 30°, \theta_{\mathbf{XZ}} = 60°, \theta_{\mathbf{YZ}} = 30°$, so, geometrically **Y** bisects **X** and **Z**. Then

$$\left|-\frac{1}{2} - \left(-\frac{\sqrt{3}}{2}\right)\right| > 1 - \frac{\sqrt{3}}{2}.$$

The second remark is that (iii) is not necessary for three-valued random variables with expectations equal to zero. Let the three values be $1, 0, -1$. Here is a counterexample where each of the three correlations is $-\frac{1}{2}$, and thus with a sum equal to $-\frac{3}{2}$, violating (iii). Note that for ± 1 variables with expectations zero, covariances and correlations are equal, because the variances are 1. In general, this is not the case, and it is in particular

7.2 Hidden Variables in Quantum Mechanics

not the case for our counterexample. Except for the special case mentioned, inequalities should be written in terms of correlations rather than covariances to have the proper generality (see, e.g., Theorem 10 below).

There is a joint probability distribution with the following values. Let $p(x, y, z)$ be the probability of a given triple of values, e.g., $(1, -1, 0)$. Then, of course, we must have for all x, y and z

$$p(x, y, z) \geq 0 \text{ and } \sum_{x,y,z} p(x, y, z) = 1,$$

where x, y and z each have the three values $1, 0, -1$. So, let

$$p(-1, 0, 1) = p(1, -1, 0) = p(0, 1, -1) = p(1, 0, -1) = p(-1, 1, 0) = p(0, -1, 1) = \frac{1}{6}$$

and the other 21 $p(x, y, z) = 0$. Then it is easy to show that in this model $E(\mathbf{X}) = E(\mathbf{Y}) = E(\mathbf{Z}) = 0$, $\text{Var}(\mathbf{X}) = \text{Var}(\mathbf{Y}) = \text{Var}(\mathbf{Z}) = \frac{2}{3}$, and $\text{Cov}(\mathbf{XY}) = \text{Cov}(\mathbf{YZ}) = \text{Cov}(\mathbf{XZ}) = -\frac{1}{3}$, so that the correlations

$$\rho(\mathbf{XY}) = \rho(\mathbf{YZ}) = \rho(\mathbf{XZ}) = -\frac{1}{2}.$$

The sum of the three correlations is then $-\frac{3}{2}$.

It is a somewhat depressing mathematical fact that even for three random variables with n-values and with expectations equal to zero, a separate investigation seems to be needed for each n to find necessary and sufficient conditions to have a joint probability distribution compatible with given means, variances and covariances. A more general recursive result would be highly desirable, but seems not to be known. Such results are pertinent to the study of multi-valued spin phenomena.

The next well-known theorem states two conditions equivalent to Bell's Inequalities for random variables with just two values. This form of the inequalities is due to Clauser, Horne, Shimony, and Holt(1969), referred to as CHSH. The equivalence of (ii) and (iii) is due to Fine (1982).

THEOREM 6. (Bell's Inequalities). *Let n random variables be given satisfying the locality hypothesis of Theorem 4. Let $n = 4$, the number of random variables, let each $\mathbf{X}_i$ be discrete with values ± 1, let the symmetry condition $E(\mathbf{X}_i) = 0$, $i = 1, \ldots, 4$ be satisfied, let $\mathbf{X}_1 = \mathbf{A}$, $\mathbf{X}_2 = \mathbf{A}'$, $\mathbf{X}_3 = \mathbf{B}$, $\mathbf{X}_4 = \mathbf{B}'$, with the covariances $E(\mathbf{AB})$, $E(\mathbf{AB}')$, $E(\mathbf{A}'\mathbf{B})$ and $E(\mathbf{A}'\mathbf{B}')$ given. Then the following three conditions are equivalent.*

(i) *There is a hidden variable λ satisfying Locality Condition II and equation (a) of the Second-Order Factorization Condition holds.*

(ii) *There is a joint probability distribution of the random variables $\mathbf{A}$, $\mathbf{A}'$, $\mathbf{B}$ and $\mathbf{B}'$ compatible with the given means and covariances.*

(iii) *The random variables $\mathbf{A}$, $\mathbf{A}'$, $\mathbf{B}$ and $\mathbf{B}'$ satisfy Bell's inequalities in the CHSH form*

$$-2 \leq E(\mathbf{AB}) + E(\mathbf{AB}') + E(\mathbf{A}'\mathbf{B}) - E(\mathbf{A}'\mathbf{B}') \leq 2$$
$$-2 \leq E(\mathbf{AB}) + E(\mathbf{AB}') - E(\mathbf{A}'\mathbf{B}) + E(\mathbf{A}'\mathbf{B}') \leq 2$$
$$-2 \leq E(\mathbf{AB}) - E(\mathbf{AB}') + E(\mathbf{A}'\mathbf{B}) + E(\mathbf{A}'\mathbf{B}') \leq 2$$
$$-2 \leq -E(\mathbf{AB}) + E(\mathbf{AB}') + E(\mathbf{A}'\mathbf{B}) + E(\mathbf{A}'\mathbf{B}') \leq 2.$$

It will now be shown that the CHSH inequalities remain valid for three-valued random variables (spin-1 particles). Consider a spin-1 particle with the 3-state observables, $A(a,\lambda) = +1, 0, -1$, $B(b,\lambda) = +1, 0, -1$. λ is a hidden variable having a normalized probability density, $\rho(\lambda)$. The expectation of these observables is defined as,

$$E(a,b) = \int AB\rho(\lambda)d\lambda. \quad (3)$$

We have suppressed the variable dependence on A and B for clarity. Consider the following difference,

$$|E(a,b) - E(a,b')| = \left|\int A[B - B']\rho(\lambda)d\lambda\right|. \quad (4)$$

Since the density $\rho > 0$ and $|A| = 1, 0$ we have the following inequality,

$$|E(a,b) - E(a,b')| \leq \int |A[B - B']|\rho(\lambda)d\lambda, \quad (5)$$

$$\leq \int |[B - B']|\rho(\lambda)d\lambda. \quad (6)$$

Similarly we have the following inequality,

$$|E(a',b) + E(a',b')| = \left|\int A'[B + B']\rho(\lambda)d\lambda\right|, \quad (7)$$

$$\leq \int |[B + B']|\rho(\lambda)d\lambda. \quad (8)$$

Adding the two expressions we arrive at the following inequality,

$$|E(a,b) - E(a,b')| + |E(a',b) + E(a',b')|$$
$$= \int (|[B - B']| + |[B + B']|)\rho(\lambda)d\lambda. \quad (9)$$

The term in square brackets is equal to 2 in all cases except when B and B' are both equal to zero; there the right-hand side vanishes. With this and the normalization condition for the hidden variable density, we have the same inequality as the spin-$\frac{1}{2}$ CHSH inequality,

$$|E(a,b) - E(a,b')| + |E(a',b) + E(a',b')| \leq 2. \quad (10)$$

Note that we could create a stronger inequality by adding the function $2(|E(a,b)| - 1)(|E(a,b')| - 1)$ to the left-hand side.

For higher spins we can proceed analogously and derive the following inequality which must be satisfied for spin-j particles,

$$|E(a,b) - E(a,b')| + |E(a',b) + E(a',b')| \leq 2j. \quad (11)$$

If we define normalized observables $\frac{A(a,\lambda)}{j}$, the original CHSH inequality will need to be satisfied for local hidden variable theories, although stronger inequalities could be constructed.

In Peres' work on higher spin particles, the observable is defined by a mapping from the multistate operator to a two-state operator (Peres 1992). Under this mapping it was shown that Bell's inequality is violated for certain parameter settings of the detectors.

GHZ-type experiments. Changing the focus, we now first consider a three-particle version of GHZ-type experiments. All arguments known to us, in particular GHZ's

7.2 Hidden Variables in Quantum Mechanics

(Greenberger, Horne and Zeilinger 1989) own argument, the more extended one in Greenberger, Horne, Shimony and Zeilinger (1990) and Mermin's (1990a, 1993) proceed by assuming the existence of a deterministic hidden variable and then deriving a contradiction. It follows immediately from Theorem 1 that the nonexistence of a hidden variable is equivalent to the nonexistence of a joint probability distribution for the given observable random variables. The next theorem states this purely probabilistic GHZ result, and, more importantly, the proof is purely in terms of the observables, with no consideration of possible hidden variables.

THEOREM 7. *Let* $\mathbf{A}, \mathbf{B},$ *and* $\mathbf{C}$ *be* ± 1 *random variables having a joint probability distribution such that* $E(\mathbf{A}) = E(\mathbf{B}) = E(\mathbf{C}) = 1.$ *Then* $E(\mathbf{ABC}) = 1.$

Proof. Since $E(A) = 1$, $P(\overline{a}) = P(\overline{a}bc) = P(\overline{ab}c) = P(\overline{a}b\overline{c}) = P(\overline{abc}) = 0$, where $P(\overline{a}bc) = P(\mathbf{A} = -1, \mathbf{B} = 1, \mathbf{C} = 1)$, etc. By similar argument for $E(\mathbf{B})$ and $E(\mathbf{C})$, we are left with $P(abc) = 1$, which implies at once the desired result.

I now follow the quantum-mechanical argument given in Mermin (1990b). We start with the three-particle entangled state characteristic of GHZ-type experiments.

$$|\psi\rangle = \frac{1}{\sqrt{2}}(|+\rangle_1|+\rangle_2|+\rangle_3 + |-\rangle_1|-\rangle_2|-\rangle_3). \tag{12}$$

This state is an eigenstate of the following spin operators:

$$\mathbf{A} = \hat{\sigma}_{1x}\hat{\sigma}_{2y}\hat{\sigma}_{3y}, \quad \mathbf{B} = \hat{\sigma}_{1y}\hat{\sigma}_{2x}\hat{\sigma}_{3y}, \tag{13}$$

$$\mathbf{C} = \hat{\sigma}_{1y}\hat{\sigma}_{2y}\hat{\sigma}_{3x}, \quad \mathbf{D} = \hat{\sigma}_{1x}\hat{\sigma}_{2x}\hat{\sigma}_{3x}. \tag{14}$$

If we compute quantum mechanically the expected values for the correlations above, we obtain at once that $E_Q(\hat{\mathbf{A}}) = E_Q(\hat{\mathbf{B}}) = E_Q(\hat{\mathbf{C}}) = 1$ and $E_Q(\hat{\mathbf{D}}) = -1$. (To exhibit all the details of this setup is too lengthy to include here, but the argument is elementary and standard, in the context of quantum mechanics.) Moreover,

$$E_Q(\mathbf{ABC}) = (s_{1x}s_{2y}s_{3y})(s_{1y}s_{2x}s_{3y})(s_{1y}s_{2y}s_{3x}) \tag{15}$$

$$= s_{1x}s_{2x}s_{3x}(s_{1y}^2 s_{2y}^2 s_{3y}^2). \tag{16}$$

Since the s_{ij} can only be 1 or -1, $s_{1y}^2 = s_{2y}^2 = s_{3y}^2 = 1$, and we obtain from (16) that

$$E_Q(\mathbf{ABC}) = s_{1x}s_{2x}s_{3x} = E_Q(\mathbf{D}). \tag{17}$$

So we have a contradiction. Classically

$$E(\mathbf{ABC}) = 1$$

but quantum mechanically

$$E_Q(\mathbf{ABC}) = -1.$$

It is clear from the above derivation that one could avoid contradictions if we allowed the value of λ to depend on the experimental setup, i.e., if we allowed λ to be a contextual hidden variable. In other words, what the GHZ theorem proves is that noncontextual hidden variables cannot reproduce quantum mechanical predictions.

Detector inefficiencies. This striking characteristic of GHZ's predictions, however, has a major problem. How can one verify experimentally predictions based on correlation-one statements, since experimentally one cannot obtain events perfectly correlated?

Fortunately, the correlations present in the GHZ state are so strong that even if we allow for experimental errors, especially the inevitable detector inefficiencies, the nonexistence of a joint distribution can still be verified, as we show in the following theorem and its corollary.

THEOREM 8. (de Barros and Suppes 2000). *If **A**, **B**, and **C** are three ± 1 random variables, a joint probability distribution exists for the given expectations $E(\mathbf{A})$, $E(\mathbf{B})$, $E(\mathbf{C})$, and $E(\mathbf{ABC})$ if and only if the following inequalities are satisfied:*

$$-2 \leq E(\mathbf{A}) + E(\mathbf{B}) + E(\mathbf{C}) - E(\mathbf{ABC}) \leq 2, \tag{18}$$

$$-2 \leq E(\mathbf{A}) + E(\mathbf{B}) - E(\mathbf{C}) + E(\mathbf{ABC}) \leq 2, \tag{19}$$

$$-2 \leq E(\mathbf{A}) - E(\mathbf{B}) + E(\mathbf{C}) + E(\mathbf{ABC}) \leq 2, \tag{20}$$

$$-2 \leq -E(\mathbf{A}) + E(\mathbf{B}) + E(\mathbf{C}) + E(\mathbf{ABC}) \leq 2. \tag{21}$$

Proof. First we prove necessity. Let us assume that there is a joint probability distribution consisting of the eight atoms abc, $ab\bar{c}$, $a\bar{b}c$, ..., $\bar{a}\bar{b}\bar{c}$. Then,

$$E(\mathbf{A}) = P(a) - P(\bar{a}),$$

where

$$P(a) = P(abc) + P(a\bar{b}c) + P(ab\bar{c}) + P(a\bar{b}\bar{c}),$$

and

$$P(\bar{a}) = P(\bar{a}bc) + P(\bar{a}\bar{b}c) + P(\bar{a}b\bar{c}) + P(\bar{a}\bar{b}\bar{c}).$$

Similar equations hold for $E(\mathbf{B})$ and $E(\mathbf{C})$. For $E(\mathbf{ABC})$ we obtain

$$\begin{aligned} E(\mathbf{ABC}) &= P(\mathbf{ABC}=1) - P(\mathbf{ABC}=-1) \\ &= P(abc) + P(a\bar{b}\bar{c}) + P(\bar{a}b\bar{c}) + P(\bar{a}\bar{b}c) \\ &\quad - [P(a\bar{b}c) + P(ab\bar{c}) + P(\bar{a}bc) + P(\bar{a}\bar{b}\bar{c})]. \end{aligned}$$

Corresponding to the first inequality above, we now sum over the probability expressions for the expectations

$$F = E(\mathbf{A}) + E(\mathbf{B}) + E(\mathbf{C}) - E(\mathbf{ABC}),$$

and obtain the expression

$$\begin{aligned} F &= 2[P(abc) + P(\bar{a}bc) + P(a\bar{b}c) + P(ab\bar{c})] \\ &\quad - 2[P(\bar{a}\bar{b}\bar{c}) + P(\bar{a}\bar{b}c) + P(\bar{a}b\bar{c}) + P(a\bar{b}\bar{c})], \end{aligned}$$

and since all the probabilities are non-negative and sum to ≤ 1, we infer at once inequality (18). The derivation of the other three inequalities is very similar.

To prove the converse, i.e., that these inequalities imply the existence of a joint probability distribution, is slightly more complicated. We restrict ourselves to the symmetric case

$$P(a) = P(b) = P(c) = p,$$
$$P(\mathbf{ABC}=1) = q,$$

and thus

$$E(\mathbf{A}) = E(\mathbf{B}) = E(\mathbf{C}) = 2p - 1,$$
$$E(\mathbf{ABC}) = 2q - 1.$$

7.2 Hidden Variables in Quantum Mechanics

In this case, (18) can be written as
$$0 \leq 3p - q \leq 2,$$
while the other three inequalities yield just $0 \leq p + q \leq 2$. Let
$$x = P(\bar{a}bc) = P(a\bar{b}c) = P(ab\bar{c}),$$
$$y = P(\bar{a}\bar{b}c) = P(\bar{a}b\bar{c}) = P(a\bar{b}\bar{c}),$$
$$z = P(abc),$$
and
$$w = P(\bar{a}\bar{b}\bar{c}).$$

It is easy to show that on the boundary $3p = q$ defined by the inequalities the values $x = 0$, $y = q/3$, $z = 0$, $w = 1 - q$ define a possible joint probability distribution, since $3x + 3y + z + w = 1$. On the other boundary, $3p = q + 2$, so a possible joint distribution is $x = (1-q)/3$, $y = 0$, $z = q$, $w = 0$. Then, for any values of q and p within the boundaries of the inequality we can take a linear combination of these distributions with weights $(3p - q)/2$ and $1 - (3p - q)/2$, chosen such that the weighed probabilities add to one, and obtain the joint probability distribution:

$$x = \left(1 - \frac{3p-q}{2}\right)\frac{1-q}{3},$$
$$y = \left(\frac{3p-q}{2}\right)\frac{q}{3},$$
$$z = \left(1 - \frac{3p-q}{2}\right)q,$$
$$w = \left(\frac{3p-q}{2}\right)(1-q),$$

which proves that if the inequalities are satisfied, a joint probability distribution exists, and therefore a noncontextual hidden variable as well, thus completing the proof. The generalization to the asymmetric case is tedious but straightforward.

As a consequence of the inequalities above, one can show that the correlations present in the GHZ state can be so strong that even if we allow for experimental errors, the nonexistence of a joint distribution can still be verified (de Barros and Suppes, 2000).

COROLLARY 1. *Let* **A**, **B**, *and* **C** *be three* ± 1 *random variables such that*
(i) $E(\mathbf{A}) = E(\mathbf{B}) = E(\mathbf{C}) \geq 1 - \epsilon$,
(ii) $E(\mathbf{ABC}) \leq -1 + \epsilon$,

where ϵ represents a decrease of the observed GHZ correlations due to experimental errors. Then, there cannot exist a joint probability distribution of **A**, **B**, *and* **C** *if*
$$\epsilon < \frac{1}{2}. \tag{22}$$

Proof. To see this, let us compute the value of F defined above. We obtain at once that
$$F = 3(1 - \epsilon) - (-1 + \epsilon).$$
But the observed correlations are only compatible with a noncontextual hidden variable theory if $F \leq 2$, hence $\epsilon < \frac{1}{2}$. Then, there cannot exist a joint probability distribution

of **A**, **B**, and **C** satisfying (i) and (ii) if

$$\epsilon < \frac{1}{2}. \tag{23}$$

From the inequality obtained above, it is clear that any experiment that obtains GHZ-type correlations stronger than 0.5 cannot have a joint probability distribution. For example, the recent experiment made at Innsbruck (Bouwmeester et al., 1999) with three-photon entangled states supports the quantum mechanical result that no noncontextual hidden variable exists that explains their correlations. Thus, with this reformulation of the GHZ theorem it is possible to use strong, yet imperfect, experimental correlations to prove that a noncontextual hidden-variable theory is incompatible with the experimental results.

Second-order Gaussian theorems. A fundamental second-order representation theorem about finite sequences of continuous random variables is the following:

THEOREM 9. *Let n continuous random variables be given, let their means, variances and covariances all exist and be finite, with all the variances nonzero. Then a necessary and sufficient condition that a joint Gaussian probability distribution of the n random variables exists, compatible with the given means, variances and covariances, is that the eigenvalues of the correlation matrix be non-negative.*

A thorough discussion and proof of this theorem can be found in Loève (1978). It is important to note that the hypothesis of this theorem is that each pair of the random variables has enough postulated for there to exist a unique bivariate Gaussian distribution with the given pair of means and variances and the covariance of the pair. Moreover, if, as required for a joint distribution of all n variables, the eigenvalues of the correlation matrix are all non-negative, then there is a unique Gaussian joint distribution of the n random variables.

We formulate the next representation theorem to include cases like Bell's inequalities when not all the correlations or covariances are given.

THEOREM 10. *Let n continuous random variables be given such that they satisfy the locality hypothesis of Theorem 4, let their means and variances exist and be finite, with all the variances nonzero, and let $m \leq n(n-1)/2$ covariances be given and be finite. Then the following two conditions are equivalent.*

(i) There is a joint probability distribution of the n random variables compatible with the given means, variances and covariances.

(ii) Given the $m \leq n(n-1)/2$ covariances, there are real numbers that may be assigned to the missing correlations so that the completed correlation matrix has eigenvalues that are all non-negative.

Moreover, (i) or (ii) implies that there is a hidden variable λ satisfying Locality Condition II and the Second-Order Factorization Condition.

The proof of Theorem 10 follows directly from Theorem 9.

Using Theorem 9, we can also derive a nonlinear inequality necessary and sufficient for three Gaussian random variables to have a joint distribution. In the statement of the representation theorem $\rho(\mathbf{X}, \mathbf{Y})$ is the correlation of $\mathbf{X}$ and $\mathbf{Y}$.

THEOREM 11. *Let $\mathbf{X}, \mathbf{Y}$ and $\mathbf{Z}$ be three Gaussian random variables whose means, variances and correlations are given, and whose variances are nonzero. Then there exists a joint Gaussian distribution of $\mathbf{X}, \mathbf{Y}$ and $\mathbf{Z}$ (necessarily unique) compatible with the given means, variances and correlations if and only if*

$$\rho(\mathbf{XY})^2 + \rho(\mathbf{XZ})^2 + \rho(\mathbf{YZ})^2 \leq 2\rho(\mathbf{XY})\rho(\mathbf{YZ})\rho(\mathbf{XZ}) + 1.$$

The proof comes directly from the determinant of the correlation matrix. For a matrix to be non-negative definite the determinant of the entire matrix and all principal minors must be greater than or equal to zero,

$$Det \begin{pmatrix} 1 & \rho(\mathbf{XY}) & \rho(\mathbf{XZ}) \\ \rho(\mathbf{XY}) & 1 & \rho(\mathbf{YZ}) \\ \rho(\mathbf{XZ}) & \rho(\mathbf{YZ}) & 1 \end{pmatrix} \geq 0. \tag{24}$$

Including the conditions for the minors we have,

$$\begin{aligned} \rho(\mathbf{XY})^2 + \rho(\mathbf{XZ})^2 + \rho(\mathbf{YZ})^2 - 2\rho(\mathbf{XY})\rho(\mathbf{XZ})\rho(\mathbf{YZ}) &\leq 1 \\ \rho(\mathbf{XY})^2 &\leq 1 \\ \rho(\mathbf{YZ})^2 &\leq 1. \end{aligned} \tag{25}$$

The last two inequalities are automatically satisfied since the correlations are bounded by ± 1.

Simultaneous observations and joint distributions. When observations are simultaneous and the environment is stable and stationary, so that with repeated simultaneous observations satisfactory frequency data can be obtained, then there exists a joint distribution of all of the random variables representing the simultaneous observations. Note what we can then conclude from the above: in all such cases there must be, therefore, a factorizing hidden variable because of the existence of the joint probability distribution. From this consideration alone, it follows that any of the quantum mechanical examples that violate Bell's inequalities or other criteria for hidden variables must be such that not all the observations in question can be made simultaneously. The extension of this criterion of simultaneity to a satisfactory relativistic criterion is straightforward.

7.3 Weak and Strong Reversibility of Causal Processes[13]

It is widely known, and often commented upon, that the basic equations of classical physics remain valid under a transformation from time t to $-t$. It is often also remarked that this invariance under a change of direction of time is completely contrary to ordinary experience. So the invariance of well-established classical physics, as well as other parts of physics, for example, relativistic mechanics and the Schrödinger equation in quantum mechanics, creates a natural philosophical tension about the nature of causal processes. The purpose of the present section is not to resolve in any complete way this tension between physics and ordinary experience, but to at least reduce the tension by introducing two concepts of reversibility. The first is that of *weak reversibility*, which

---

[13] Most of the material in this section also appeared in Suppes (2001b).

is what is exemplified in the invariance of the equations of classical physics under time reversal. This is the *weak sense* of reversibility, because we can usually distinguish by observation whether a system of particle mechanics is running one way or the other. The appearance is not the same when we go to a reversal of time. Elementary examples would be measuring the velocity of a particle accelerating from rest, moving in a straight line and hitting an impenetrable wall. The picture of change of velocity, especially, will be very different under time reversal. Under such reversal, the particle has a very large velocity immediately, as motion begins, and will continue to decelerate until it finally reaches the state of rest, quite contrary to what would be observed in the usual direction of time. Of course, this difference is well accepted in discussions of the invariance of classical physics under time reversal.

The concept of *strong reversibility* introduced here, and used in various analyses of a variety of natural phenomena, including stochastic processes, has a much stronger condition for reversibility. A process, deterministic or stochastic, is said to be strongly reversible if we are unable to distinguish whether the process is running forward or backward. To put it in vivid terms often used, the backward movie, that is, running the film in reverse, of the observation of the process, is indistinguishable by observation from the forward movie. In what follows, these ideas are made more precise, especially for stochastic processes. I should mention that the concept of strong reversibility used here corresponds to what is often defined in the stochastic-process literature as simply *reversible*. So the condition given below for a Markov chain to be strongly reversible is the same as the standard concept to be found, for example, in Feller (1950, p. 342) for Markov chains.

Although many of the formal distinctions introduced in what follows are familiar in the literature of probability theory, especially of stochastic processes, it is my impression that the distinction introduced here has not been sufficiently emphasized, as a distinction, in the philosophical discussions of time reversibility. Its introduction is meant to reduce the tension I mentioned at the beginning, in the sense that there is a close alignment between ordinary experience not being strongly reversible, and, similarly, for many trajectories of idealized particles or other processes of importance in physics.

Weak reversibility. As already remarked, it is well known that classical particle mechanics and relativistic particle mechanics have the property of being weakly reversible, i.e., the transformation changing the direction of the time always carries systems of classical particle mechanics into systems of classical particle mechanics and, correspondingly, for the relativistic case. Detailed proofs of these results are to be found in McKinsey and Suppes (1953) for classical particle mechanics and in Rubin and Suppes (1954) for relativistic particle mechanics.

The proof that first-order Markov chains are carried into first-order Markov chains under time reversal is straightforward. Here is the elementary proof for first-order chains with a finite number of states. In the proof it is assumed that all the probabilities that occur in the denominators of expressions have probability greater than zero.

7.3 WEAK AND STRONG REVERSIBILITY

THEOREM 1. *All first-order Markov chains are weakly reversible.*

Proof. (for finite-state, discrete-time processes)

$$\begin{aligned}
P(i_{n-1} \mid j_n k_{n+1}) &= \frac{P(i_{n-1} j_n k_{n+1})}{P(k_{n+1} \mid j_n) P(j_n)} \\
&= \frac{P(k_{n+1} \mid j_n) P(j_n \mid i_{n-1}) P(i_{n-1})}{P(k_{n+1} \mid j_n) P(j_n)} \\
&= \frac{P(j_n i_{n-1})}{P(j_n)} \\
&= P(i_{n-1} \mid j_n).
\end{aligned}$$

The situation in quantum mechanics is somewhat more complicated concerning weak reversibility. Certainly, the Schrödinger equation is so reversible. It is weakly reversible under transformations changing the direction of time. On the other hand, in the standard accounts, this is not true of the measurement process (von Neumann 1932/1955, Ch. 5).

Finally, from a common-sense standpoint, ordinary experience is certainly not weakly reversible. Time has a fixed direction and the empirical evidence in terms of human experience of this direction is overwhelming.

$$\text{Forward} \neq \text{Backward}$$

People almost never walk up stairs backward. No races are run backward, etc. This is not to claim the 'backward movies' violate laws of mechanics, only inductive laws and facts of experience, most especially the universally accepted nature of human memory:

$$\text{unknown future} \neq \text{known past}.$$

Some definitions. Before going further, it is desirable to introduce some standard stochastic concepts, implicitly assumed in Theorem 1. First, let $\mathbf{X}(t)$ be a stochastic process such that for each time t, $\mathbf{X}(t)$ takes values in a finite set, a restriction imposed only for simplicity of exposition. Any finite family $\mathbf{X}(t_1), \mathbf{X}(t_2), \ldots, \mathbf{X}(t_n)$ has a joint probability distribution. Using an obvious simplification of notation, a process is (*first-order*) *Markov* if and only if for any times $t_1 < t_2, \ldots, t_n$,

$$P(x_n \mid x_{n-1}, \ldots, x_1) = P(x_n \mid x_{n-1}), \tag{1}$$

where the simplified notation is defined as follows:

$$P(x_n \mid x_{n-1}) = P(\mathbf{X}(t_n) = j_n \mid \mathbf{X}(t_{n-1}) = i_{n-1}).$$

A Markov process is *homogeneous* if the transition probabilities such as (1) do not depend on time, and it is *irreducible* if any state j can be reached from any other state in a finite number of steps. For a discrete-time homogeneous Markov process, I write the transition probabilities in several different, but useful notations

$$\begin{aligned}
P_{ij} = p_{i,j} &= P(j_n \mid i_{n-1}) = P(\mathbf{X}_n = j \mid \mathbf{X}_{n-1} = i) \\
&= P(\mathbf{X}(t_n) = j_n \mid \mathbf{X}(t_{n-1}) = i_{n-1}),
\end{aligned}$$

all of these often used in the probability literature.

Let $P_{ii}(n)$ be the probability of returning to state i in n transitions. The *period* $r(i)$ of a state i is the greatest common divisor of the set of integers n for which $P_{ii}(n) > 0$. A state is *aperiodic* if it has period 1. Moreover, it is easy to show that if a Markov process is irreducible every state has the same period. So an aperiodic process is one in which every state has period 1.

Hereafter, when I refer to a Markov *chain*, I am assuming a Markov process that is finite-state, discrete-time, homogeneous, irreducible and aperiodic.

The definitions of being Markov, homogeneous and irreducible all apply without change to continuous-time processes, but there are some further conditions to impose to avoid physically unlikely cases. First, we require such a process to remain in any state for a positive length of time, and, second, the process cannot pass through an infinite sequence of states in a finite time. Corresponding to the transition probability for discrete-time processes, the *transition rate* for continuous-time processes is defined as:

$$q_{ij} = \lim_{\tau \to 0} \frac{P(X(t+\tau) = j \mid X(t) = i)}{\tau}, \quad i \neq j,$$

and for definitional purposes, we set $q_{ii} = 0$. Hereafter, when I refer to a Markov *process*, I will mean a continuous-time one. Such a process remains in each state for a length of time exponentially distributed with the parameter $q(i)$ defined as:

$$q(i) = \sum_j q_{ij},$$

and when it leaves state i it moves to state j with the probability

$$p_{ij} = \frac{q_{ij}}{q_i}.$$

Note that in general

$$\sum_j q_{ij} \neq 1,$$

i.e., transition rates from i to another state need not add up to 1, as must be the case for transition probabilities from state i in the discrete-time processes. (A good discussion of such continuous-time processes may be found in Kelly (1979).)

Strong reversibility. The intuitive idea of strong reversibility is easily characterized in terms of movies. A movie is strongly reversible if an observer cannot tell whether the movie is being run forward or backward. There is no perceptual difference in the two directions. Now, as our ordinary movie experience tells us at once, this is a pretty exceptional situation. On the other hand, there are important physical processes, a few of which we will discuss, that do have strong reversibility. An interesting case to begin with is this. When is a Markov chain strongly reversible? The right setting for this is to restrict ourselves at once to stationary, ergodic Markov chains. A Markov chain is *stationary* if its mean distribution is the same for all times. It is *ergodic* if it has a unique asymptotic distribution independent of the initial distribution. Note that under these definitions, a Markov chain can be ergodic but not stationary if its initial distribution is different from its unique, asymptotic one. But here I assume ergodic chains are stationary.

An ergodic Markov chain is *strongly reversible* if

$$\pi_i p_{ij} = \pi_j p_{ji} \tag{2}$$

7.3 Weak and Strong Reversibility

for all states i and j, where π is the unique asymptotic density. The equations (2) are often called the *detailed balance* conditions for strong reversibility. Bernoulli processes, such as coin tossing, are strongly reversible processes, because there is no memory of past states: $p_{ij} = p_j$ and $p_{ji} = p_i$, so we satisfy (2). Moreover, any 2-state ergodic Markov chain is strongly reversible. Here is the proof. Since the process is ergodic

$$\begin{aligned} \pi_1 &= \pi_1 p_{11} + \pi_2 p_{21} \\ &= \pi_1(1 - p_{12}) + \pi_2 p_{21}, \end{aligned}$$

so

$$\pi_1 p_{12} = \pi_2 p_{21}.$$

On the other hand, many 3-state chains are not strongly reversible. Here is an example of a three-color spinner.

	B	R	Y
B	$\frac{1}{2}$	$\frac{1}{4}$	$\frac{1}{4}$
R	$\frac{3}{8}$	$\frac{1}{2}$	$\frac{1}{8}$
Y	$\frac{1}{8}$	$\frac{3}{8}$	$\frac{1}{2}$

We have at once for the mean density π of the three states:

$$\begin{aligned} \pi_b &= \tfrac{1}{2}\pi_b + \tfrac{3}{8}\pi_r + \tfrac{1}{8}(1 - \pi_b - \pi_r) \\ \pi_r &= \tfrac{1}{4}\pi_b + \tfrac{1}{2}\pi_r + \tfrac{3}{8}(1 - \pi_b - \pi_r); \end{aligned}$$

solving the two equations, we find:

$$\pi_b = \frac{13}{37}, \quad \pi_r = \frac{14}{37}, \quad \pi_y = \frac{10}{37}.$$

Here is the proof that it is not strongly reversible:

$$\pi_b p_{br} = \frac{13}{17} \cdot \frac{1}{4} \neq \frac{14}{37} \cdot \frac{3}{8} = \pi_r p_{rb}.$$

We can summarize what we have been saying about Markov chains in the following theorem, where in the notation of the theorem, $\stackrel{d}{=}$ means equal in distribution.

THEOREM 2. *If an ergodic Markov chain is strongly reversible, then it is impossible to distinguish whether a movie of it is running forward or backward, that is, any finite sequence of random variables of the process has the same distribution when the order of the random variables is reversed:*

$$(\mathbf{X}_0, \mathbf{X}_1, \ldots \mathbf{X}_n) \stackrel{d}{=} (\mathbf{X}_n, \mathbf{X}_{n-1}, \ldots \mathbf{X}_0).$$

I give the proof for $n = 2$:

$$\begin{aligned} P(\mathbf{X}_0 = i, \mathbf{X}_1 = j) &= P(\mathbf{X}_1 = j | \mathbf{X}_0 = i) P(\mathbf{X}_0 = i) \\ &= \pi_i p_{ij} \\ &= \pi_j p_{ji} \quad \text{by (2)} \\ &= P(\mathbf{X}_1 = i | \mathbf{X}_0 = j) P(\mathbf{X}_0 = j) \\ &= P(\mathbf{X}_1 = i, \mathbf{X}_0 = j). \end{aligned}$$

The concepts introduced hold for finite-state continuous-time changes by replacing transition probabilities by transition rates $q_{ij}, i \neq j$ with $q_{ij} \geq 0$, as defined above. The detailed balance conditions defining strong reversibility are the same in form as those of (2) with the $\pi_i \geq 0$ and summing to 1, as before:

$$\pi_i q_{ij} = \pi_j q_{ji}. \tag{3}$$

With a proof just like that of Theorem 2, we have at once:

THEOREM 3. *If an ergodic continuous-time Markov process is strongly reversible, then it is impossible to distinguish whether a movie of it is running forward or backward.*

In the present context, there is nothing special about first-order Markov chains. A second-order ergodic Markov chain is strongly reversible iff

$$\pi_i \pi_{ij} p_{ijk} = \pi_k \pi_{kj} p_{kji}, \tag{4}$$

where a chain is second-order Markov if

$$P(x_n \mid x_{n-1}, \ldots, x_1) = P(x_n \mid x_{n-1}, x_{n-2}),$$

corresponding to (1) for first-order.

THEOREM 4. *If a second-order Markov chain is ergodic and strongly reversible, then it is impossible to distinguish a movie of it running forward or backward, i.e.,*

$$(\mathbf{X}_0, \mathbf{X}_1, \ldots \mathbf{X}_n) \stackrel{d}{=} (\mathbf{X}_n, \mathbf{X}_{n-1}, \ldots \mathbf{X}_0).$$

Proof for $n = 3$:

$$P(\mathbf{X}_0 = i, \mathbf{X}_1 = j, \mathbf{X}_2 = k)$$
$$= P(\mathbf{X}_2 = k | \mathbf{X}_1 = j, \mathbf{X}_0 = i) P(\mathbf{X}_1 = j | \mathbf{X}_0 = i) P(\mathbf{X}_0 = i)$$
$$= p_{ijk} \pi_{ij} \pi_i$$
$$= \pi_k \pi_{kj} p_{kji} \quad \text{by (4)}$$
$$= P(\mathbf{X}_2 = i | \mathbf{X}_1 = j, \mathbf{X}_0 = k) P(\mathbf{X}_1 = j | \mathbf{X}_0 = k) P(\mathbf{X}_0 = k)$$
$$= P(\mathbf{X}_2 = i, \mathbf{X}_1 = j, \mathbf{X}_0 = k).$$

As is obvious, this proof also works for continuous-time second-order Markov processes. Without too much effort, this same sort of proof can be extended to chains of infinite order that are ergodic and stationary. The line of reasoning required is developed in Lamperti and Suppes (1959).

Turning again from Markov chains to continuous-time Markov processes, simple examples of strongly reversible processes are ergodic *birth and death* processes, defined by the transition rates being zero except for $q(i, i + 1) > 0$, representing a birth, and $q(i, i - 1) > 0$ representing a death. The detailed balance conditions (3) then assume the form

$$\pi_i q_{i,i+1} = \pi_{i+1} q_{i+1,i}, \tag{5}$$

where the π_i's form the stationary probability density of the finite set of states.

Ehrenfest model. A simplified model of statistical mechanics nicely exemplifies a birth and death process that is strongly reversible. In the spirit of Maxwell's demon, there are two containers of particles, thought of as ideal gas molecules. Let $\mathbf{X}(t)$ be the number of particles in container I, and, so, $k - \mathbf{X}(t)$ is the number of particles in

7.3 Weak and Strong Reversibility

container II. The "birth and death" process corresponds to increasing or decreasing the number of particles in container I, with the transition rates being:

$$q_{i,i-1} = i\lambda, \qquad i = 1, 2, \ldots, k,$$
$$q_{i,i+1} = (k-i)\lambda, \qquad i = 0, 1, \ldots, k-1,$$

where λ is the rate parameter. The stationary density may be shown to be

$$\pi_i = 2^{-k}\binom{k}{i} \qquad i = 0, 1, \ldots, k,$$

where

$$\binom{k}{i} = \frac{k!}{(k-i)!i!} \text{ and } 0! = 1.$$

That the process is strongly reversible may be seen by checking the detailed balance conditions (5).

$$\begin{aligned}
\pi_i q_{i,i+1} &= 2^{-k}\binom{k}{i}(k-i)\lambda \\
&= \frac{2^{-k}k!(k-i)\lambda}{(k-i)!i!} \\
&= \frac{2^{-k}k!\lambda}{(k-(i+1))!i!} && \text{since } \frac{k-i}{(k-i)!} = \frac{1}{(k-(i+1))!} \\
&= \frac{2^{-k}k!(i+1)\lambda}{(k-(i+1))!(i+1)!} && \text{since } \frac{1}{i!} = \frac{i+1}{(i+1)!} \\
&= 2^{-k}\binom{k}{i+1}(i+1)\lambda \\
&= \pi_{i+1}q_{i+1,i}.
\end{aligned}$$

Without attempting a detailed discussion of entropy for this process, I note that even though it is stationary and strongly reversible, the mean fluctuation from moment to moment is always stronger toward the equal distribution of particles in the two containers rather than away from this equal distribution, for

$$q_{i,i-1} = i\lambda < (k-i)\lambda = q_{i,i+1}$$

if and only if $i < \frac{k}{2}$, independent of the rate parameter λ. If we compute the entropy by the relative frequency $f(t)$ at time t of the number of particles in container I, then the 'momentary' entropy of the process at t is

$$H(t) = -(f(t)\log f(t) + (1-f(t))\log(1-f(t)))$$

and by the analysis just given the mean rate of change of entropy is always increasing, i.e., is positive, in spite of the strong reversibility of the process. But there is, obviously, no contradiction between these two features of the model, even at equilibrium, i.e., when stationary. An excellent detailed analysis of the Ehrenfest model and also of the Boltzmann equation, and of the associated "paradoxes" of Loschmidt and Zermelo, is to be found in Kac (1959).

Deterministic systems. I now turn to deterministic classical physical systems. It should be evident enough that only very special ones are strongly reversible. Any sort of dissipative system will, in general, not be. A real billiard ball on a real table is not strongly reversible. We start the motion and it comes to a halt somewhere, due to friction on the table. We can easily distinguish the forward from the backward picture. On the other hand, an idealized billiard ball on an idealized table with total conservation of energy and exact satisfaction of the equality of the angle of incidence and the angle of reflection can be put in periodic motion and, once in motion, will continue forever. Moreover, this idealized motion will, of course, be strongly reversible. By "observing" it, we cannot decide whether we are seeing the forward or the backward movie. (I put 'observing' in scare quotes, for there is, in fact, no such billiard case, but we can artificially simulate it well for a finite period of time.)

A simple, but important, example of a classical system that has strong reversibility is an undamped and undriven one-dimensional harmonic oscillator. The differential equation for such an oscillator is:

$$\frac{d^2x}{dt^2} + \omega^2 x = 0,$$

where ω is the natural frequency of the oscillator when undamped and undriven, and the initial conditions at time $t = 0$ are:

$$x_0 = \alpha,$$
$$\frac{dx_0}{dt} = 0.$$

(When the model is a pendulum, the initial conditions correspond to the pendulum being at rest at time $t = 0$ with displacement α.) The general solution of (1) is:

$$x(t) = A\cos\omega t + B\sin\omega t,$$

and

$$\frac{dx}{dt} = -A\omega\sin\omega t + B\omega\cos\omega t,$$

so at $t = 0$, $B = 0$, whence

$$A\cos\omega t = \alpha \text{ at } t = 0, \text{ and } A = \alpha.$$

Then

$$\frac{d^2x}{dt} = -A\omega^2\cos\omega t$$

and finally

$$x(t) = A\cos\omega t. \tag{6}$$

We have at once that such an oscillator is strongly reversible, since for all t

$$x(-t) = A\cos-\omega t = A\cos\omega t = x(t).$$

Notice that this solution holds for $-\infty < t < \infty$, and the conditions at $t = 0$ are not really initial conditions, but just conditions for some t that yield a solution simple in form.

On the other hand, the damped, undriven harmonic oscillator, whose differential equation is:

$$\frac{d^2x}{dt^2} + 2k\frac{dx}{dt} + \omega^2 x = 0,$$

where k is the damping coefficient, is easily shown not to be strongly reversible. This result is scarcely surprising, since a damped oscillator (in one dimension) is one of the simplest physical examples of a dissipative system. In the standard case where the damping is not too heavy, the oscillating function decreases in amplitude with time.

7.3 WEAK AND STRONG REVERSIBILITY

More specifically the envelope of the oscillations is a negative exponential of the form e^{-kt} on the positive side and $-e^{-kt}$ on the negative side.

Without question, if the aim is to find examples of systems in classical physics that are not strongly reversible, and, therefore, the causal analysis is not strongly reversible, the place to look is among the many kinds of dissipative systems. The great prevalence of such systems in nature reinforces our personal experience to support the common-sense view that, as the saying goes, "...of course time is not reversible. Who could ever think otherwise?"

8

Representations of Language

In this chapter various representations of languages, for example, context-free languages, are given. The first representations are in terms of automata of various strengths, and the representations go both ways, i.e., languages in terms of automata, and automata represented by languages. That the two are so closely entwined is perhaps surprising, but understandable once the problem of how a language is to be parsed or compiled is raised. These matters occupy the first two sections. The content is now part of the education of every young computer scientist. I taught it for many years in the 1970s as part of a course on the theory of automata.

The next two sections are more special and directly related to earlier work of my own. Section 3 is concerned to show how simple abstract ideas of conditioning and reinforcement are sufficient, when properly axiomatized, to show that stimulus-response structures of learning can represent any finite automaton, and thereby the regular languages the automata can represent. The last part of the section extends the representation to any computable function via the stimulus-response representation of unlimited register machines. It is probably true that most linguists are deeply skeptical that simple mechanisms of association or conditioning are adequate to account for complex and subtle linguistic phenomena. What is proved in this section will be too abstract mathematically to persuade many of them, but that is not the task undertaken. The objective is just to show in a rigorous way how "in principle" such a reduction can be made. Some of the kinds of further detailed work needed are to be found in Sections 5 and 6 of this chapter.

Section 4 focuses on the relation of stimulus-response models, as developed in the previous section, to learning models formulated in terms only of behavioral variables. The main theorem is about the representation of the latter models in terms of the former ones. The long proof is testimony to the difficulties of proving detailed representation theorems to reduce one theory to another, even for relatively simple cases. The representation theorem also has some interest in showing how models that have current events depending on the remote past can be represented by first-order Markov processes, once additional theoretical notions are introduced. Almost everyone believes, for example, that any lasting effects of childhood events on an adult must be embodied in the current physical and mental structure of the person, and not by direct, unmediated action of the events at a distance across the years. But describing in detail the current structural embodiment is not, for most cases, possible with the data and scientific concepts now available. The example of this section indicates the kind of considerations that must be entered into even in the most simplified cases of such structural analysis.

Section 5 is devoted to machine learning of robotic natural language. The phrase 'machine learning' refers, in fact, to the development of computer programs that are learning programs. Here the results of such learning are expressed as comprehension grammars that have an associated semantics. The relevant sense of representation is that of learning how simple robotic commands are represented in any one of ten languages, English, Chinese (Mandarin), etc. This case study, as it were, is meant to show how a systematic fragment of a natural language can be learned by using principles that, broadly speaking, apply to human language learning as well. However, the machine learning exemplified here is by no means meant to be a simulation of human language learning, as is evident enough from the details given in the section.

Section 6 turns away from learning to survey some of my own most recent work with younger colleagues on language and the brain. The focus is on brain-wave representations of words and sentences. The current work is aimed at trying to represent correctly how the brain represents language, at least immediately after hearing or seeing some single words or sentences. Some interesting invariance results have already been found. A more general aim of this section is to give a somewhat detailed sense of the methodology used in these studies, especially the probabilistic aspects.

In this chapter no invariance theorems are stated or proved. Some are, of course, possible, but they lie outside the main development of the concepts and theories considered. But some significant empirical invariances, found experimentally in the neural processing of language, are presented in Section 6.

Section 7, the final one, is an epilogue on representation and reduction in science. The remarks are mainly historical and are meant to provide a perspective on the detailed work earlier in this chapter and those that preceded.

8.1 Hierarchy of Formal Languages

We begin with the definitions of formal languages much used in contemporary linguistics and computer science. The basic concepts of this hierarchy are originally due to Chomsky (1959). We begin with the general concept of an arbitrary phrase-structure grammar. By adding restrictions we get grammars whose powers of expression are ever more restricted.

Let V be a finite, nonempty set. Then V^* is the set of all finite sequences of elements of V. Let $\emptyset$ be the empty sequence. Then $V^+ = V^* - \{\emptyset\}$. V is the *vocabulary*. V^* is the set of *strings*.

DEFINITION 1. *A structure $G = (V, N, P, S)$ is a* phrase-structure grammar *iff*

1. *V, N and P are nonempty finite sets,*
2. *$N \subseteq V$,*
3. *$P \subseteq V^+ \times V^*$,*
4. *$S \in N$.*

The standard terminology is this. The set N is the set of *nonterminal* symbols, or

8.1 HIERARCHY OF FORMAL LANGUAGES

variables. $V_T = V - N$ is the set of *terminal* symbols or words. P is the set of productions. If $(\alpha, \beta) \in P$ we ordinarily write: $\alpha \to \beta$, to indicate that from α we may "produce" β. Finally, S is the *start* symbol with which derivations begin. The weakness of the axioms matches the great generality of the definition of phrase-structure grammars. Most interest centers around various restricted classes defined below.

If γ and δ are in V^* then $\gamma\alpha\delta \Rightarrow_G \gamma\beta\delta$ iff $\alpha \to \beta$ is a production, i.e., $(\alpha, \beta) \in P$. We say in this case that the string $\gamma\beta\delta$ is *directly derivable* from the string $\gamma\alpha\delta$ in grammar G. We next define the notion of a derivation in a grammar in terms of direct derivations. Intuitively a derivation is the way we show that a particular string belongs to the language generated by the grammar.

If $\gamma_1, \gamma_m \in V^*$, then $\gamma_1 \Rightarrow_G^* \gamma_m$ iff $\exists \gamma_2, \ldots, \gamma_{m-1} \in V^*$ such that $\gamma_1 \Rightarrow_G \gamma_2, \ldots, \gamma_{m-1} \Rightarrow_G \gamma_m$. The *language generated by a grammar* G, denoted $L(G)$, is:

$$L(G) = \{w | w \in V_T^* \,\&\, S \Rightarrow_G^* w\}.$$

Note that for a string to be in the language it must consist only of terminal words and its derivation in G must begin with the start symbol S.

We say that two grammars G_1 and G_2 are *weakly equivalent* if and only if they generate the same language, i.e., $L(G_1) = L(G_2)$. Notice that this does not say very much about the structure of the utterances. It is just that they produce the same set of strings without attention to structure.

An important application of formal languages is to the analysis of the structure of natural languages. Indeed, it can be claimed that natural languages fit within the hierarchy that we shall be defining. The rather complicated question of whether this is true in any satisfactory way will not really be examined here. Certainly the evidence is substantial that important fragments of natural languages can be analyzed with the grammatical concepts introduced in this section. We now consider one simple example.

$$N = \{S, NP, VP, PN, AdjP, Adj, CN, Art, IV, TV\},$$

where S is the start symbol, NP is for noun phrases, VP for verb phrases, PN for proper nouns, $AdjP$ for adjectival phrases, Adj for adjectives, CN for common nouns, Art for articles, IV for intransitive verbs, and TV for transitive verbs.

$V_T = \{$you pick the terminal English words with their usual grammatical categories$\}$,

$P = $ the following set of production rules, with the lexical rules omitted for the terminal words—a lexical rule is one whose right-hand member is a terminal word:

1. $S \to NP + VP$
2. $NP \to PN$
3. $NP \to AdjP + CN$
4. $AdjP \to AdjP + Adj$
5. $AdjP \to Adj$
6. $AdjP \to Art$
7. $VP \to IV$
8. $VP \to TV + NP$.

These rules should seem familiar but of course incomplete. In stating the rules the plus sign is used just to show concatenation, in order to avoid confusion in concatenation of strings.

Here is a sample derivation.

1.	S	Start symbol
2.	$NP + VP$	By Rule 1
3.	$AdjP + CN + VP$	By Rule 3
4.	$AdjP + Adj + CN + VP$	By Rule 4
5.	$Art + Adj + CN + VP$	By Rule 6
6.	$Art + Adj + CN + TV + NP$	By Rule 8
7.	$Art + Adj + CN + TV + PN$	By Rule 2
8.	*The large ball hit John*	By Lexical Rules.

In many respects more natural than a linear derivation is the derivation tree. The example just given generates the following tree.

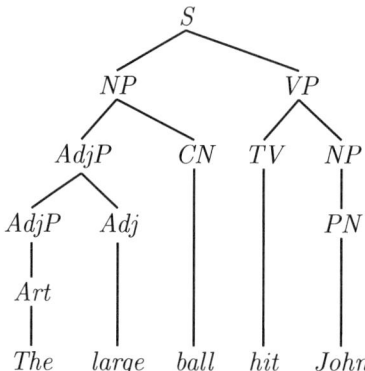

We shall not formally define derivation trees in spite of their importance for context-free grammars.

Types of grammars. We turn now to types of grammar. *Type-0 grammars* are characterized by Definition 1. To obtain a *type-1* or *context-sensitive grammar*, we add the restriction that for every production $\alpha \to \beta$, $|\alpha| \leq |\beta|$, where $|\alpha|$ is the length of α, or the number of symbols in α. This restriction can be shown equivalent to

$$\alpha_1 A \alpha_2 \to \alpha_1 \beta \alpha_2$$

with $\beta \neq \emptyset$ and $A \in N$.

To obtain *type-2* or *context-free grammars*, we add the stronger restriction: if $\alpha \to \beta \in P$ then

(i) α is a variable, i.e., $\alpha \in N$, so productions are of the form $A \to \beta$,

(ii) $\beta \neq \emptyset$.

To obtain *type-3* or *regular grammars*, we add the still stronger restriction: any production must be of the form

(i) $A \to aB$ or
(ii) $A \to a$,

with $A, B \in N$, $a \in V_T$. (We can have instead of (i), $(i') A \to Ba$, but the change from right-linear to left-linear is trivial, even though all rules of form (i) must be changed.)

8.1 Hierarchy of Formal Languages

Note that we could write (i) as $A \to a + B$, to show the familiar use of the plus sign to denote concatenation in writing individual rules. The definitions of context-sensitive, context-free, and regular grammars are such that the languages they generate exclude the empty string, which is not true of type-0 grammars as such. For some technical purposes it is desirable not to make this exclusion, but it simplifies the exposition to do so here.

For an example of a regular grammar, we may slightly change our previous example, by making V_T just the symbols of the grammatical categories of some words.

$$N = \{S, NP, VP, AdjP\}$$
$$V_T = \{PN, CN, Adj, Art, IV, TV\}.$$

If we drop Rule 8 and rewrite Rule 1 as $1'$, $S \to NP + IV$, we can immediately check that the grammar is regular. To rewrite Rule 8 is more troublesome and requires more variables, i.e., elements of N. I only sketch the attack. We add rules like these

$$S \to X + CN$$
$$X \to Y + Adj$$
$$X \to Y + Art$$
$$Y \to VP' + Adj$$
$$Y \to VP' + Art$$
$$VP' \to TV.$$

In general, it is not easy to see if a language has a grammar of a given type.

The abstractness of the definitions of various types of phrase-structure grammars is typical of what happens in every area of science when theories are axiomatically defined as certain set-theoretical predicates. Without further restrictions on the kinds of objects that may occur in the sets of vocabulary, etc., intuitively very odd grammars can be constructed—odd not because of their structural characteristics but because of their material features.

In this section a few important theorems about the types of grammars defined are considered. I first state some results on normal forms for grammars of a given language, then some theorems about set-theoretical operations on languages, followed by a sample of results on unsolvable problems and problems of ambiguity. An important potential natural-language application of context-sensitive languages as defined by lexical-functional grammars is the final topic touched on in this section.

It is not my intention to give anything like a thorough development of formal languages in this chapter, but only a sense of the kinds of questions that have been investigated for these theoretical structures.

Normal forms. I begin with a theorem about context-sensitive and type-0 languages. Note that a language is context-sensitive, for example, if there is at least one context-sensitive grammar that generates the language.

THEOREM 1. *Every context-sensitive language can be generated by a grammar in which all productions are of the form $a \to b$, where a and b are strings of variables only, or of the form $A \to b$, where A is a variable and b is a terminal word. Moreover,* *every type-0 language can be generated by a grammar whose productions are of the same two forms.*

Proof. Let $G = (V, N, P, S)$ be a context-sensitive grammar. For each a in V_T, let X_a be a new symbol not in N. Define now the grammar $G' = (V', N', P', S)$ where

$$N' = N \cup \{X_a | a \in V_T\},$$
$$V' = V \cup N',$$

and P' is defined as follows. First, every production of the form $X_a \to a$, for $a \in V_T$, is in P'. Second, if $a \to b$ is in P, then $\alpha_1 \to \beta_1$ is in P' where α_1 is obtained from a by replacing each occurrence of a in V_T by X_a, and β_1 is obtained from b in similar fashion. It is easy to see by this construction that $L(G') = L(G)$. It is also apparent that the proof is similar if G is a type-0 grammar.

The simplicity of this proof is characteristic of many of the interesting theorems about formal languages. The whole problem is to define the right new entity—in the present case the grammar G'. The proof that the entity defined has the appropriate properties is then usually straightforward, but often tedious if carried out in full detail. In almost all cases, the proof implicitly characterizes an algorithm for constructing the new grammar or other entity mentioned in the theorem.

THEOREM 2. (Chomsky Normal Form, 1959). *Any context-free language can be generated by a grammar in which all productions are of the form $A \to BC$ or $A \to a$, where A, B and C are variables and a is a terminal word.*

Proof. Let $G = (V, N, P, S)$ be a context-free grammar that generates the context-free language. From G we construct a new grammar G' satisfying the requirements of the theorem. We first characterize P':

(i) Any production in P of the form $A \to BC$, where A, B, C are in N, put in P',

(ii) Any production in P of the form $A \to a$, where A is in N and a is in V_T, put in P',

(iii) For each production $A \to X_1...X_k$ in P, where $k > 2$ and $A, X_1, ..., X_k$ are in V, put in P' the following productions:

$$A \to X'_1 Y_2$$
$$Y_2 \to X'_2 Y_3$$
$$\cdot$$
$$\cdot$$
$$\cdot$$
$$Y_{k-2} \to X'_{k-2} Y_{k-1}$$
$$Y_{k-1} \to X'_{k-1} X'_k,$$

where each Y_j is a new variable not in N, and each X'_i is X_i if X_i is in N and is a new variable not in N if X_i is in V_T.

(iv) For each production of the form $A \to X_1 X_2$, where either X_1 or X_2 or both are in V_T, put in P' the production $A \to X'_1 X'_2$,

(v) For any X_i of (iii) or (iv) with X_i in V_T, put in P' the production $X'_i \to X_i$.

8.1 HIERARCHY OF FORMAL LANGUAGES

To complete the construction of G', let N' be N and all the new variables introduced in the construction of P', let $V' = V \cup N'$, and $S' = S$. Explicit proof that $L(G') = L(G)$ is left to the reader. Q.E.D.

We state without proof another well-known normal form for context-free languages.

THEOREM 3. (Greibach Normal Form, 1965). *Every context-free language can be generated by a grammar in which every production is of the form* $A \to a\beta$, *where A is a variable, a is a terminal word, and β is a possibly empty string of variables.*

As in other cases of study concerned with formal systems, the normal forms we have considered are useful in simplifying and standardizing the notation used for context-free or other languages. As in the case of disjunctive and conjunctive normal forms in propositional logic, such forms also play a direct role in proving various properties of languages. For example, the Chomsky normal form for context-free grammars is used in a direct way to prove that there is an algorithm to determine if a given context-free grammar G generates a finite or infinite language, i.e., to determine whether $L(G)$ is a finite or infinite set.

Operations on languages. Let G and G' be two phrase-structure grammars. There are various natural operations on the languages $L(G)$ and $L(G')$, especially when considered as sets. The question, what classes of languages are closed under these operations, then arises. Union and intersection are obvious operations. Complementation creates problems because of the empty string $\emptyset$. As remarked earlier, this string can be in a type-0 language, but not as defined above in context-sensitive, context-free, or regular languages. For purposes of this subsection on operations, we shall extend the definitions to include as a possibility the empty sequence, and we shall not be concerned here to make explicit the production rule for generating it. Given this extension, we define complementation in the obvious way. Let V_T be a finite, nonempty set, and let L be a phrase-structure language such that $L \subseteq V_T^*$, i.e., L is a subset of all finite sequences whose elements belong to V. Then the complement of L, relative to V_T, is $V_T^* - L$.

We state without proof the following theorem.

THEOREM 4. *Let V_T be a finite, nonempty set. Then the family of all subsets L of V_T^* which are regular languages forms a Boolean algebra of sets.*

In other words, if $L, L' \subseteq V_T^*$, and L and L' are regular languages, then $L \cup L'$, $L \cap L'$ and $V_T^* - L$ are also regular languages.

This theorem does not hold for context-free languages, which are not closed under intersection or complementation. A counterexample for intersection is this. Let

$$V = V' = \{A, B, a, b, c\}$$
$$N = N' - \{A, B\}$$
$$P = \{A \to Ac, A \to B, B \to aBb, B \to ab\}$$
$$P' = \{A \to aA, A \to B, B \to bBc, B \to bc\}$$
$$S' = S.$$

Then

$$L = L(G) = \{a^n b^n c^i | n \geq 1 \ \& \ i \geq 0\}$$
$$L' = L(G') = \{a^j b^n c^n | n \geq 1 \ \& \ j \geq 0\}.$$

But
$$L(G) \cap L(G') = \{a^n b^n c^n | n \geq 1\},$$
and the language $L(G) \cap L(G')$ is a standard example of a language that is not context-free, a fact I shall not prove here (for details, see Aho and Ullman 1972, vol. 1, pp. 195–196). Context-free languages are closed under union, a fact which we shall return to. So it follows by the counterexample just given that they cannot be closed under complementation, for
$$L \cap L' = -(-L \cup -L').$$
We define the *product* or *concatenation* of two languages L and L' by
$$LL' = \{x | x = uv, u \text{ is in } L, v \text{ is in } L'\}.$$
We then have the following theorem for all four types of languages. The proof is rather long and is therefore omitted.

THEOREM 5. *Let V_T be a finite, nonempty set. Then the families of regular, context-free, context-sensitive and type-0 languages that are subsets of V_T^* are each closed under union and concatenation.*

Unsolvable problems. A problem about formal languages is said to be *unsolvable* iff there is no algorithm or mechanical decision procedure for deciding it. Problems of this kind are ordinarily said to be *undecidable* in logic. For example, it is undecidable whether any arbitrarily given sentence in elementary number theory is provable—one of Gödel's famous results. It is even undecidable whether an arbitrary formula of first-order logic is valid—a result due to Church and, independently, Turing. Here are some sample results for context-free languages–for a more detailed discussion, see Hopcroft and Ullman (1969, Ch. 14):

(i) It is unsolvable whether or not the intersection of two arbitrary context-free languages is empty;

(ii) It is unsolvable whether or not a context-free grammar generates the set of all strings over its terminal vocabulary;

(iii) It is unsolvable whether or not two context-free grammars generate the same language.

We turn now to some results about ambiguity. We first define the concept of a *leftmost derivation*. We say that a derivation is *leftmost* iff at every step (n) the variable being replaced by a production rule of the grammar has no variable to its left in the string of step (n). For example, if in a derivation, step (n) has string BA and the production rule $A \to a$ is used to derive as step $(n+1)$ the string Ba, the derivation is not leftmost. It can be shown that any derivation of a terminal string u in a context-free grammar can be replaced by a leftmost derivation of u. A context-free grammar G is said to be *ambiguous* iff there is a string in $L(G)$ that has at least two distinct leftmost derivations. A context-free language L is said to be *inherently ambiguous* iff every context-free grammar generating L is ambiguous. The following theorem about ambiguity can then be proved.

THEOREM 6. *It is unsolvable whether an arbitrary context-free grammar is ambiguous. Moreover, it is unsolvable whether a context-free grammar generates an inherently ambiguous context-free language.*

A consequence of this theorem is that there can be no automatic checking of the absence of ambiguity in a new programming language that is context-free. The absence of ambiguity will have to be established, or proved not to be the case, by nonalgorithmic methods of proof.

Natural-language applications. After Chomsky (1952) introduced transformations, it was felt for many years that no phrase-structure grammar could do justice to the complexities of a natural language like English. On the other hand, a great many problems arose in defining transformations with just the right power, for it was shown early that rather simple transformations, when added to a context-free language, could generate a recursively enumerable language that was not recursive, even though there was general agreement that a natural language should at least be recursive.

In the past several decades, the center of attention has turned back to phrase-structure grammars, e.g., in the form of the lexical-functional grammars proposed by Bresnan (1982) and Kaplan and Bresnan (1982). In this approach the constituent structure generated by a context-free grammar is augmented by a functional structure which imposes additional grammatical structure. From a formal standpoint the functional structure can be viewed as imposing a set of functional equations as additional constraints between adjacent nodes of a derivation tree of the context-free grammar. The functional equations must have a unique solution for a string to be accepted by the lexical-functional grammar. It can be shown that the imposition of the functional structure in general implies equivalence to a context-sensitive grammar.

In many ways a still more promising direction is the development of generalized phrase-structure grammars. An initial systematic exposition is given in Gazdar, Klein, Pullum and Sag (1985). The class of the grammars they consider are weakly equivalent to the context-free grammars defined earlier, but, of course, not structurally equivalent. For a recent overview of English grammar, see Huddleston and Pullum (2002).

8.2 Representation Theorems for Grammars

I begin with finite automata and the languages they accept in order to develop the machinery needed to prove the representation theorem for regular grammars.

Finite automata. I shall give in this section several definitions of finite automata and several examples. Because the general definition is so simple and straightforward I begin with the definitions, to be followed by an example. This first definition is close to that of Rabin and Scott (1959).

DEFINITION 1. *A structure* $\mathfrak{A} = (A, V, M, s_0, F)$ *is a* finite (deterministic) automaton *iff*

(i) A *is a finite, nonempty set* (*the set of states of* $\mathfrak{A}$),
(ii) V *is a finite, nonempty set* (*the alphabet or vocabulary*),

(iii) M is a function from the Cartesian product $A \times V$ to A (M defines the transition table of $\mathfrak{A}$),

(iv) s_0 is in A (s_0 is the initial state of $\mathfrak{A}$),

(v) F is a subset of A (F is the set of final states of $\mathfrak{A}$).

The generality of this definition is apparent. It is also indicative of the weakness in a certain sense of the general notion of a finite automaton. About the only restrictions are that the sets A and V be finite. On the other hand, these finite restrictions are critical.

It is evident that we can have some extremely trivial models of Definition 1. Here is the simplest: $A = \{s_0\}, V = \{0\}, M(s_0, 0) = s_0, F = \{\ \}$.

Perhaps the simplest nontrivial example of a finite automaton is the following. We have a two-letter alphabet, which we may as well take to be the symbols '1' and '0', and two internal states, which we may take to be s_0 and s_1. The transition function of the automaton is defined by the following table:

	s_0	s_1
$s_0 0$	√	
$s_0 1$		√
$s_1 0$		√
$s_1 1$	√	

Finally, we select the internal state s_1 as the only member of the set F of final states. I have said that this is the simplest nontrivial automaton. I mean by this that the transition table depends both on the internal state and the input letter. From a more general conceptual standpoint, it is clear that the device is itself pretty trivial.

If we examine our ordinary experience for instances of finite automata that are not quite so trivial, that do important jobs and yet that are simple enough to describe rather easily and quickly, it is pretty clear that one of the most evident areas of application is that of elementary arithmetic. Intuitively most of us recognize that there is something mechanical or algorithmic about the arithmetical operations we are taught to perform in elementary school. Without any formal definitions it is clear in a general way that the standard algorithms of arithmetic can be defined in a completely objective and mechanical fashion and the work can be turned over to a machine that is not very clever at all. In fact, in an intuitive sense the machine has no cleverness whatsoever because it always does exactly the same thing when presented with a similar situation and there is no consideration of strategy or contingencies whatsoever in its makeup. What I have said here is vague, but I intend to convey an intuitive idea we all have a fairly good grasp of.

There is one trouble about Definition 1 when we consider our ordinary way of doing arithmetic. We are expected to output an answer, and not simply to move into a final state. For theoretical work and for a good many applications Definition 1 has great simplicity. However, it is also natural to seek a second closely related definition. Later we shall study briefly the relation between these two definitions. In this new definition we eliminate the set of final states and introduce an output function as well as an output vocabulary or alphabet.

8.2 REPRESENTATION THEOREMS FOR GRAMMARS

DEFINITION 2. *A structure* $\mathfrak{A} = (A, V_I, V_O, M, Q, s_0)$ *is a finite (deterministic) automaton with output iff*

(i) *A is a finite, nonempty set,*
(ii) *V_I and V_O are finite nonempty sets (the input and output vocabularies respectively),*
(iii) *M is a function from the Cartesian product $A \times V_I$ to A (M defines the transition table),*
(iv) *Q is a function from the Cartesian product $A \times V_I$ to V_O (Q is the output function),*
(v) *s_0 is in A (s_0 is the initial state).*

As an example of a finite automaton with output, that is, a finite automaton in the sense of Definition 2, we may characterize an automaton that will perform column addition of two integers in standard base-10 representation.

$$A = \{0, 1\},$$
$$V_I = \{(m, n) : 0 \leq m, n \leq 9\},$$
$$V_O = \{0, 1, ..., 9\},$$
$$M(k, (m, n)) = \begin{cases} 0 & \text{if } m + n + k \leq 9, \\ 1 & \text{if } m + n + k > 9, \end{cases} \text{ for } k = 0, 1,$$
$$Q(k, (m, n)) = (k + m + n) \bmod 10,$$
$$s_0 = 0.$$

Thus the automaton operates by adding first the ones' column, storing as internal state 0 if there is no 'carry', 1 if there is a 'carry', outputting the sum of the ones' column modulus 10, and then moving on to the input of the two tens' column digits, etc. The initial internal state s_0 is 0 because at the beginning of the problem there is no 'carry'. The characterizations of similar automata for performing other arithmetical operations are left as exercises.

Because on the surface Definition 1 and Definition 2 seem to be defining somewhat different devices, it will be useful to make some study of the relations between finite automata in the sense of Definition 1 and finite automata with output (as defined in Definition 2).

For this purpose it will be desirable to introduce the standard notions of isomorphism for the two senses of finite automata.

DEFINITION 3. *Let $\mathfrak{A} = (A, V, M, s_0, F)$ and $\mathfrak{A}' = (A', V', M', s_0', F')$ be finite automata. Then $\mathfrak{A}$ and $\mathfrak{A}'$ are isomorphic iff there exists a function f such that*

(i) *f is one-one,*
(ii) *Domain of f is $A \cup V$ and range of f is $A' \cup V'$,*
(iii) *For every a in $A \cup V$*

$$a \in A \text{ iff } f(a) \in A',$$

(iv) *For every s in A and σ in V*

$$f(M(s, \sigma)) = M'(f(s), f(\sigma)),$$

(v) *$f(s_0) = s_0'$,*

(vi) For every s in A
$$s \in F \text{ iff } f(s) \in F'.$$
It is apparent that conditions (i)-(iii) of the definition imply that for every a in $A \cup V$
$$a \in V \text{ if and only if } f(a) \in V',$$
and consequently this condition on V need not be stated. From the standpoint of the general algebraic or set-theoretical concept of isomorphism, it would have been more natural to define an automaton in terms of a basic set $B = A \cup V$, and then require that A and V both be subsets of B. Rabin and Scott (1959) avoid the problem by not making V a part of the automaton. They define the concept of an automaton $\mathfrak{A} = (A, M, s_0, F)$ with respect to an alphabet V, but for our purposes it is also desirable to include the alphabet V in the definition of $\mathfrak{A}$ in order to make explicit the natural place of the alphabet in the general scheme of things, and above all, to provide a simple setup for going from one alphabet V to another V'. In any case, exactly how these matters are handled is not of central importance.

DEFINITION 4. Let $\mathfrak{A} = (A, V_I, V_O, M, Q, s_0)$ and $\mathfrak{A}' = (A', V_I', V_O', M', Q', s_0')$ be finite automata with output. Then $\mathfrak{A}$ and $\mathfrak{A}'$ are isomorphic iff there exists a function f such that

(i) f is one-one,
(ii) Domain of f is $A \cup V_I \cup V_O$ and range of f is $A' \cup V_I' \cup V_O'$,
(iii) For every s in $A \cup V_I \cup V_O$
$$s \in A \text{ iff } f(s) \in A',$$
(iv) For every σ in $A \cup V_I \cup V_O$
$$\sigma \in V_I \text{ iff } f(\sigma) \in V_I',$$
(v) For every s in A and σ in V_I
$$f(M(s, \sigma)) = M'(f(s), f(\sigma)),$$
(vi) For every s in A and σ in V_I
$$f(Q(s, \sigma)) = Q'(f(s), f(\sigma)),$$
(vii) $f(s_0) = s_0'$.

Definitions 3 and 4 are so similar it is somewhat redundant to write them both out explicitly. However, with both definitions in front of us it should be apparent that general set-theoretical methods for defining the isomorphism of like structures apply here as well. In the usual sense the two definitions of isomorphism are axiom-free, in the sense that they do not depend on the substantive assumptions in Definition 1 or 2. As has already been remarked, these assumptions are just the assumptions of finiteness. For this reason the axiom-free character of the definitions of isomorphism is not too evident in these two instances.

Languages accepted by finite automata. We want to describe the language that a given finite automaton can accept or recognize. The definitions given at this point are for a finite automaton $\mathfrak{A} = (A, V, M, s_0, F)$, in the sense of Definition 1.

8.2 REPRESENTATION THEOREMS FOR GRAMMARS

First, V^* is the set of all finite sequences of elements of V, including the empty sequence $\emptyset$. The elements of V^* are called *sentences, strings* or *tapes*. If $\sigma_1, \ldots, \sigma_n$ are in V, then $\sigma_1\sigma_2\ldots\sigma_n$ is in V^*; in other words, we shall usually show elements of V^* by juxtaposing names of elements of V.

Second, the function M can be extended to a function from $A \times V^*$ to A by the following recursive definition for s in A, x in V^* and σ in V

$$M(s, \emptyset) = s$$
$$M(s, x\sigma) = M(M(s, x), \sigma).$$

Thus, take the two-state, two-letter alphabet automaton

	s_0	s_1
$s_0 0$	√	
$s_0 1$		√
$s_1 0$		√
$s_1 1$	√	

Let us compute M for the string $x = 101$ by using this recursive definition.

$$\begin{aligned} M(s_0, 101) &= M(M(s_0, 10), 1) \\ &= M(M(M(s_0, 1), 0), 1) \\ &= M(M(s_1, 0), 1) \\ &= M(s_1, 1) \\ &= s_0, \end{aligned}$$

and so the string 101 is not accepted, because the final state is s_0, not s_1. More formally, we have:

DEFINITION 5. *A string x of V^* is accepted by $\mathfrak{A}$ iff $M(s_0, x)$ is in F.*

We shall usually call strings that are accepted by $\mathfrak{A}$, *sentences* of $\mathfrak{A}$. The language *accepted* by $\mathfrak{A}$ is the set $L(\mathfrak{A})$ of all sentences of $\mathfrak{A}$. In the literature, the language accepted by $\mathfrak{A}$ is often called the *set of tapes accepted by* $\mathfrak{A}$.

There is a natural notion of equivalence for automata which is weaker than the strict concept of isomorphism defined above.

DEFINITION 6. *Two automata are (weakly) equivalent iff they accept the same language.*

To say that the relation of isomorphism for automata is always stronger than that of equivalence is false, for two automata, it is apparent, can be isomorphic but not equivalent, because of a different vocabulary or alphabet. We can easily change Definition 6 to take account of this fact. We require not sameness of language but rather a one-one function from V onto V'.

A finite automaton is *connected* iff for every state s there is a string x such that $M(s_0, x) = s$. We have then the obvious theorem.

THEOREM 1. *Any finite automaton is equivalent to a connected finite automaton.*

The next theorem states a simple finitistic criterion for the language accepted by a finite automaton not to be empty.

THEOREM 2. *The set of sentences accepted by a finite automaton with n states is nonempty iff the finite automaton accepts a sentence of length less than n.*

Proof. If a word of length less than n is accepted, then the language is nonempty, and by way of contradiction, let the smallest sentence w accepted be of length $\geq n$. The first point to note is that $\mathfrak{A}$ must pass through some state at least twice in processing w—this follows at once from the definition of M for any string. Let this state be s. We can break up w then into three parts with

$$w = w_1 w_2 w_3$$

and $w_2 \neq \emptyset$, and by assumption on s

$$M(s_0, w_1) = M(s, w_2) = s,$$

and

$$M(s, w_3) \in F.$$

Now either w_1 or $w_3 \neq \emptyset$, for otherwise $w_2 = w$ and we have not broken up w properly in terms of repetitions of s. So we now have

$$M(s_0, w_1 w_3) = M(s_0, w_1 w_2 w_3) \in F,$$

and

$$|w_1 w_3| < |w_1 w_2 w_3|,$$

contrary to supposition. Q.E.D.

More surprising, there also exists a strict finitistic test of when the language accepted by a finite automaton is infinite.

THEOREM 3. *The language (i.e., set of sentences) accepted by a finite automaton with n states is infinite iff the automaton accepts a sentence w of length equal to or greater than n and less than $2n$, i.e., $n \leq |w| < 2n$.*

Proof. First, assume there is a sentence w accepted and $n \leq |w| < 2n$. Then as in the proof of Theorem 2, we can find a repeated state s such that

$$w = w_1 w_2 w_3$$
$$w_2 \neq \emptyset$$
$$M(s_0, w_1) = M(s, w_2) = s$$
$$M(s, w_3) = F.$$

It then follows that for all i,

$$w_1 w_2^i w_3$$

is accepted, where w^i is a string of w's of length i, and this set of sentences is clearly infinite.

On the other hand, let us now assume that $\mathfrak{A}$ accepts an infinite number of sentences but none is of length between n and $2n - 1$ (inclusive). Let w be a sentence of minimum length $\geq 2n$ that is accepted. Then by the argument given in the proof of Theorem 2 we can find a repeated state s and a sentence w_2 of length $1 \leq |w_2| < n$ such that

$$w = w_1 w_2 w_3$$

8.2 REPRESENTATION THEOREMS FOR GRAMMARS

and w_1w_3 is accepted. Then either $n \leq |w_1w_3| < 2n$, or the argument can be repeated if $|w_1w_3| \leq 2n$. In a finite number of steps we will obtain $w'_1w'_3$ such that $n \leq |w'_1w'_3| < 2n$, contrary to our supposition.

Q.E.D.

It follows from Theorems 2 and 3 that there exists the following algorithm.

THEOREM 4. *There is an algorithm for determining whether the language accepted by a finite automaton is empty, finite but nonempty, or infinite.*

Regular grammars and finite automata. To establish the equivalence between regular grammars and finite automata, it is in many ways natural to bring in probabilistic or nondeterministic automata. We have not previously considered nondeterministic automata. Roughly speaking, we can get from a probabilistic automaton a nondeterministic automaton by ignoring the actual probabilities of transitions and only permit the transitions with positive probability. The concept is not in my opinion a very intuitive one, and so we should use probabilistic rather than nondeterministic automata, but I note that the approach I take here differs on this point from much of the standard literature. The definition of probabilistic automata is a direct generalization of the one for deterministic automata.

DEFINITION 7. *A structure $\mathfrak{A} = (A, V, p, s_0, F)$ is a (finite) probabilistic automaton if and only if*

(i) *A is a finite, nonempty set,*

(ii) *V is a finite, nonempty set,*

(iii) *p is a function on $A \times V$ such that for each s in A and σ in V, $p_{s,\sigma}$ is a probability density over A, i.e.,*

    (a) *for each s' in A, $p_{s,\sigma}(s') \geq 0$,*

    (b) $\sum_{s' \in A} p_{s,\sigma}(s') = 1$,

(iv) *s_0 is in A,*

(v) *F is a subset of A.*

Let $\mathfrak{A} = (A, V, p, s_0, F)$ be a probabilistic automaton as just defined. Then we may extend p to V^*

$$p_{s,\emptyset}(s') = \begin{cases} 1 \text{ if } s' = s \\ 0 \text{ otherwise.} \end{cases}$$

$$p_{s,x\sigma}(s') = \sum_{s'' \in A} p_{s,x}(s'') \cdot p_{s'',\sigma}(s').$$

A sentence x is *accepted* by $\mathfrak{A}$ if $\sum_{s \in F} p_{s_0,x}(s) > 0$, i.e., if the probability of entering a final state, i.e., a state in F is positive. Note that all variables are bound in this definition of *accept* and so we may simply write $p(x)$ for the probability of accepting string x in V^*. Using this definition of acceptance we want to show that the language accepted by a probabilistic automaton is also accepted by a finite automaton. Later we shall briefly discuss a different definition of acceptance for which this is not true.

THEOREM 5. *If a language L is accepted by a probabilistic automaton, then it is accepted by a finite automaton.*

Proof. The most important step is to identify the states of the finite automaton as sets of states of the probabilistic automaton. The rest follows as the night the day.

Let $\mathfrak{A} = (A, V, p, s_0, F)$ be a probabilistic automaton that accepts language $L(\mathfrak{A})$. Define the deterministic automaton as follows:

$$A' = 2^A = \text{the power set of } A,$$
$$V' = V \quad \text{(the same vocabulary)},$$
$$s'_0 = \{s_0\},$$

$$M(B, \sigma) = \{s' : p_{s,\sigma}(s') > 0 \text{ for some } s \text{ in } B\},$$

For B in A', i.e., B a subset of A and σ in V

$$F' = \{B : B \in A' \ \& \ B \cap F \neq \emptyset\}.$$

We now need to show $\mathfrak{A}' = (A', V', M, s'_0, F')$ accepts the same language as $\mathfrak{A}$.

We show by induction on the length of x that

$$M(s'_0, x) = B \text{ iff}$$
$$p_{s_0, x}(s') > 0 \leftrightarrow s' \in B, \text{ i.e.}$$
$$M(s'_0, x) = \{s' : p_{s_0, x} f(s') > 0\}.$$

If $|x| = 0$

$$M(s'_0, \emptyset) = s'_0 = \{s_0\},$$

and

$$p_{s_0, \emptyset}(s_0) = 1, \text{ so there is nothing else to show.}$$

We now use strong induction on the length of x. Suppose the result holds for $|x| < n$. Consider then $x\sigma$. By the inductive hypothesis

$$M(s'_0, x) = \{s'' : p_{s_0, x}(s'') > 0\} = B$$

and by definition of M

$$M(B, \sigma) = \{s' : p_{s'', \sigma}(s') > 0 \text{ for some } s'' \text{ in } B\}.$$

So

$$M(s'_0, x\sigma) = \{s' : \sum_{s'' \in A} p_{s_0, x}(s'') p_{s'', \sigma}(s') > 0\},$$

i.e.,

$$M(s'_0, x\sigma) = \{s' : p_{s_0, x\sigma}(s') > 0\}.$$

Finally it is easy to verify that

$$M(s'_0, x) \in F \text{ iff } \sum_{s \in F} p_{s_0, x}(s) > 0,$$

and so

$$L(\mathfrak{A}) = L(\mathfrak{A}').$$

Q.E.D.

However, as has already been noted, we may strengthen the definition of acceptance for probabilistic automata and get a different result. Let $\mathfrak{A}$ be a probabilistic automaton

8.2 Representation Theorems for Grammars

and let $\lambda \in [0,1]$. Then $L(\mathfrak{A},\lambda)$ is the set of sentences accepted with *cutpoint* λ. Formally:
$$L(\mathfrak{A}, \lambda) = \{x \colon x \in V^* \ \& \ p(x) > \lambda\}.$$

An example may be found in Rabin (1963) for which $L(\mathfrak{A},\lambda)$ is *not* a regular language. If λ is rational, $L(\mathfrak{A},\lambda)$ is regular.

Now to the main business at hand. We first prove that any regular language is accepted by some finite automaton. The proof proceeds by first constructing an appropriate probabilistic automaton and then using the preceding theorem. The natural correspondence is to have the nonterminal vocabulary N map onto the states of the automaton and the terminal vocabulary V_T onto the vocabulary V of the automaton. Indeterministic or probabilistic transitions are expected in most grammars, e.g., we have several rewrite or production rules for noun phrases or verb phrases.

THEOREM 6. *If G is a regular grammar, there is a finite automaton that will accept $L(G)$.*

Proof. Let $G = (V, N, P, S)$ be a regular grammar. For the automaton
$$A = N \cup \{q\}, \ q \notin N$$
(in other words the states are N together with one other symbol, q).
$$V' = V_T$$
$$s_0 = S$$
$$F = \{q\}.$$

We now define p:

(i) if $\alpha \to a$ is a production, i.e., in P,
$$p_{\alpha,a}(q) > 0,$$

(ii) if $\alpha \to a\beta$ is in P $(\alpha, \beta \in N)$
$$p_{\alpha,a}(\beta) > 0,$$

(iii) $p_{q,a}(q) = 1$.

So $\mathfrak{A} = (A, V', p, s_0, F)$ is the probabilistic automaton constructed from the grammar G. We first show that $L(G) \subseteq L(\mathfrak{A})$. Let x be a sentence in L(G). Then there is a derivation of $x = a_1 \ldots a_n$ in G of the form
$$S \Rightarrow a_1 \alpha_1 \Rightarrow a_1 a_2 \alpha_2 \Rightarrow \ldots \Rightarrow a_1 \ldots a_{n-1} \alpha_{n-1} \Rightarrow a_1 \ldots a_n.$$

From the definition of the discrete probability distribution p as given by (i)-(iii), it is clear that each step of the derivation has positive probability, say, $\epsilon_i > 0$, and therefore
$$\prod_{i=1}^{n} \epsilon_i > 0,$$
whence by the definition of acceptance for probabilistic automata, x is accepted by $L(\mathfrak{A})$.

We next show by a similar sort of argument that $L(\mathfrak{A}) \subseteq L(G)$. Let $x \in L(\mathfrak{A})$. Then there exists a sequence of states $S, \alpha_1, \ldots, \alpha_{n-1}, q$ such that $p_{S,a_1}(\alpha_1) > 0$, $p_{\alpha_{i-1},a_i}(\alpha_i) > 0, p_{\alpha_{n-1},a_n}(q) > 0$, where, of course, $x = a_1 \ldots a_n$. But it then follows at once from the definition of p that there must be a derivation of x in G, and so $x \in L(G)$.

Given what has been proved, namely, that if G is a regular grammar then there is a probabilistic (finite) automaton that will accept $L(G)$, we use Theorem 5 to conclude that $L(G)$ is accepted by some finite automaton.

Q.E.D.

THEOREM 7. *Given a finite automaton $\mathfrak{A}$, there exists a regular grammar G such that $L(G) = L(\mathfrak{A})$.*

Proof. Let $\mathfrak{A} = (A, V, M, s_0, F)$ be a finite automaton. We then define a grammar $G = (V', N, P, S)$ as follows:

$$\begin{aligned} V' &= A \cup V \\ N &= A \qquad (\text{So } V'_T = V) \\ S &= s_0. \end{aligned}$$

The set P of production rules is the critical item, but the manner of this definition should be clear from the proof of the converse theorem (Theorem 6):

(i) $s \to as' \in P$ if $s, s' \in A, a \in V, M(s, a) = s'$,
(ii) $s \to a \in P$ if $s \in A, a \in V, M(s, a) = s'$ for some $s' \in F$,

and of course only productions of form (i) or (ii) can be in P.

Completion of the proof we leave as an exercise. As before, it is natural to break what remains to be shown into two parts:

Proof that $L(\mathfrak{A}) \subseteq L(G)$, and proof that $L(G) \subseteq L(\mathfrak{A})$. Q.E.D.

Theorems 6 and 7 together provide a mutual representation of regular grammars and finite automata in terms of each other. In principle a procedural account is thereby provided for regular grammars, but a perusal of the proofs of the two theorems shows that the representing structures are too close to the original ones to give a very rich impression of representation. It is doubtful that anyone would look at this representation and then say, "Ah, I now see how a regular grammar works". The one exception is that the representation theorem does show that a regular grammar can be processed by a strictly finite machine, and this is not necessarily obvious from the form of such grammars.

Many may feel that the following structural theorem about regular languages, due to Kleene (1956), provides a more interesting representation, even though it is not procedural in character. A more psychological analysis of processing is attempted in the next section.

THEOREM 8. (Kleene 1956). *Let V_T be a finite nonempty set. Then the class of subsets of V_T^* that are regular languages or, equivalently, languages accepted by finite automata, is the smallest class that contains all finite subsets of V_T^* and that is closed under union, concatenation and closure.*

The *closure* of a language L is the set of all finite strings that results from concatenating strings in L. For example, if $L = \{1, 01\}$ then the closure of L is $\{1, 01, 11, 011, 101, 111, \ldots\}$. Often the empty string is included in the definition of closure. On this point see the next paragraph. The definition of concatenation is given just

8.2 Representation Theorems for Grammars

before Theorem 5.

Remark on the empty sequence. In the definition of regular, context-free and context-sensitive grammars, we have not permitted a production rule of the form

$$\alpha \to \emptyset,$$

with $\alpha \in V^*$. It is possible to extend matters to include this special rule—or rather, rules of this special form. Tedious account then has to be taken of their presence or absence in the proof of theorems like the two for regular languages just considered. For a good discussion of the empty-sequence type production rule, see Hopcroft and Ullman (1969, pp. 15 ff.).

Pushdown automata and context-free languages. First, we define a probabilistic pushdown automaton. The probabilistic part is a natural generalization of the concept of a probabilistic finite automaton. The new idea, and the important one, is that of the pushdown store, which is a restricted form of memory. The automaton can store structured information in this store, but it can make state transitions only in terms of the top symbol on the store—the store operates like a stack of cafeteria trays on the principle of "first-in and last-out". A special store vocabulary is provided the machine. The transition function depends now, not just on the input symbol and the current state, but also on the top symbol on the store. Of course, in the probabilistic case we have a conditional probability distribution depending on these three things: s in A, α in V, and γ in Γ (the store vocabulary).

An indeterminate or *nondeterministic* pushdown automaton is obtained from a probabilistic one by ignoring the probability distribution. This remark is made more precise later.

DEFINITION 8. *A structure* $\mathfrak{A} = (A, V, \Gamma, p, s_0, \gamma_0, F)$ *is a probabilistic pushdown automaton iff*

(1) A *is a nonempty, finite set (set of states).*

(2) V *is a nonempty, finite set (input vocabulary).*

(3) Γ *is a nonempty, finite set (store vocabulary).*

(4) For each s *in* A, σ *in* $V \cup \{\emptyset\}$, γ *in* Γ, $p_{s,\sigma,\gamma}$ *is a probability density on a finite subset of* $A \times \Gamma^*$ *(probability transition table).*

(5) $s_0 \in A$ *(initial state).*

(6) $\gamma_0 \in \Gamma$ *(start store symbol).*

(7) $F \subseteq A$ *(set of final states).*

Several remarks about Definition 8 are in order. First, although Γ^*, not just Γ is involved in the definition of p, only a fixed finite number of different words (the elements of Γ) can be stored, although the sequence of them can be indefinitely long. Second, state transitions can take place without input, because σ can be $\emptyset$, as can be seen from clause (4) of the definition. Third, we get a nondeterministic automaton from a probabilistic pushdown automaton by replacing the conditional probability density p

with a relation from $A \times (V \cup \{\emptyset\}) \times \Gamma$ to $A \times \Gamma^*$:

$$s, \sigma, \gamma M s', \alpha \text{ iff } p_{s,\sigma,\gamma}(s', \alpha) > 0,$$

where $\alpha \in \Gamma^*$. (It follows from (4) that the range of M is finite.)

A pair (s, α), with s in A and α in Γ^* is a fixed *configuration*. We say that a pushdown automaton $\mathfrak{A}$ is in configuration (s, α) if $\mathfrak{A}$ is in state s and α is the string in the pushdown store, the leftmost symbol of α being the top symbol on the store. We use the notation

$$\sigma, (s, \gamma\alpha) \vdash (s', \beta\alpha)$$

to mean that given input σ in V and configuration $(s, \gamma\alpha)$ the automaton goes to configuration $(s', \beta\alpha)$, where $\beta \in \Gamma^*$. (Here 'goes to' with positive probability.)

We extend this definition to input sequences x in V^* in the same manner as we extended the transition function for finite automata. In particular we define by induction:

$$\emptyset, (s, \gamma) \vdash^* (s, \gamma)$$

and

$$x\sigma, (s, \gamma) \vdash^* (s', \alpha') \text{ iff } \exists s'' \text{ in } A, \alpha'' \text{ in } \Gamma^*$$

such that

$$x, (s, \gamma) \vdash^* (s'', \alpha'') \text{ and } \sigma, (s'', \alpha'') \vdash (s', \alpha').$$

One explicit point, technically, about this definition is this. We want to permit terms of x to be blanks or $\emptyset$. Thus really x is in $(V^* \cup \{\emptyset\})$. We leave the details to the reader.

For $\mathfrak{A}$ a pushdown automaton, we define the *language accepted by final state* as

$$L(\mathfrak{A}) = \{x : x, (s_0, \gamma_0) \vdash^* (s, \alpha) \text{ for } s \text{ in } F \text{ and } \alpha \text{ in } \Gamma^*\}.$$

An intuitively appealing, closely related definition is of the *language* accepted by empty store:

$$L_E(\mathfrak{A}) = \{x : x, (s_0, \gamma_0) \vdash^* (s, \emptyset) \text{ for any } s \text{ in } A\}.$$

We need one preliminary result.

THEOREM 9. *A language is accepted by final state by some pushdown automaton iff it is accepted by empty store by some pushdown automaton.*

In view of Theorem 9 we may speak without ambiguity of a language being accepted by a pushdown automaton without specifying by final state or by store.

THEOREM 10. *If a language is context-free, then there is a pushdown automaton that accepts the language.*

Proof. In view of Theorem 3, without loss of generality we may take the grammar $G = (V, N, P, S)$ of the language to be in Greibach normal form. The most enlightening aspect of the construction of the pushdown automaton $\mathfrak{A}$ that processes $L(G)$ is that $\mathfrak{A}$ need have only one state:
$A = \{s_0\}, V' = V_T, \Gamma = N, \gamma_0 = S, F = \emptyset$ (so we accept by empty store), and

$$p_{s_0,s,\gamma}(s_0, a) > 0 \text{ iff } \gamma \to sa \in P.$$

8.2 Representation Theorems for Grammars

The proof is completed by an inductive argument showing that

$$S \Rightarrow_G^* x \quad \text{iff} \quad x, (s_0, \gamma_0) \vdash^* (s_0, \emptyset).$$

We leave the argument as an exercise. Q.E.D.

As the proof shows, the pushdown store is really doing all the computational work, so that in fact context-free languages can be represented by a narrow class of pushdown automata, namely, those with only one state.

THEOREM 11. *If $\mathfrak{A}$ is a pushdown automaton, then $L(\mathfrak{A})$ is a context-free language.*

Proof. Let $\mathfrak{A} = (A, V, \Gamma, p, s_0, \gamma_0, \emptyset)$ and so $\mathfrak{A}$ accepts by empty store. We define the context-free grammar that accepts $L(\mathfrak{A})$ as follows:

$$V_T' = V$$
$$N = A \times \Gamma \times A \cup \{S\}, \ S \in A \times \Gamma \times A.$$
$$\text{Thus } V' = N \cup V_T'.$$

More important, note that the elements of N are triples, except for S. The production rules in P are of three forms:

(i) $S \to (s_0, \gamma_0, s)$ for every s in A,
(ii) $(s, \gamma, s') \to \sigma$ iff $p_{s,\sigma,\gamma}(s', \emptyset) > 0$,
(iii) $(s, \gamma, s') \to \sigma(s_1, \gamma_1, s_2)(s_2, \gamma_2, s_3) \ldots (s_m, \gamma_m, s')$ iff $p_{s,\sigma,\gamma}(s_1, \gamma_1 \ldots \gamma_m) > 0$ with $\sigma \in V \cup \{\emptyset\}$.

The proof is completed by an inductive argument showing that

$$S \Rightarrow_G^* x \quad \text{iff} \quad x, (s_0, \gamma_0) \vdash^* (s, \emptyset) \text{ for some } s \text{ in } A,$$

—very similar in form to the proof of Theorem 10, but somewhat more complex. We leave this step as an exercise. Q.E.D.

Turing machines and linear bounded automata. First, we need to introduce Turing machines.[1] A Turing machine has a single read-write head. Given an internal state and the scan of an input letter, the Turing machine executes a *move* by

(i) changing internal state,
(ii) printing a nonblank symbol on the cell scanned, thereby replacing the symbol scanned,
(iii) moving its head left or right one cell.

DEFINITION 9. *A structure $\mathfrak{A} = (A, V, \Gamma, M, s_0, F)$ is a Turing machine iff*

(1.) A *is nonempty and finite (set of states),*
(2.) V *is nonempty and finite (input alphabet),*
(3.) Γ *is nonempty and finite, and $V \subseteq \Gamma$ (set of tape symbols with blank symbol B in $\Gamma - V$),*
(4.) $M: A \times \Gamma \to A \times (\Gamma - \{B\}) \times \{L, R\}$, *where M is the next move function,*
(5.) $s_0 \in A$ *(start state),*
(6.) $F \subseteq A$ *(set of final states).*

---
[1] Turing machines were discussed briefly in Section 3.5 in connection with register machines and computability.

A *linear bounded automaton* is a Turing machine that stays within the squares of the tape on which the input is placed. The representation theorems for context-sensitive and type-0 languages are given in the next two theorems, which we do not prove. For a variety of reasons the representations for regular and context-free languages already discussed are of greater conceptual importance. On the other hand, the mutual representation between Turing machines and partial recursive functions is of great importance. It was proved in Chapter 3.

THEOREM 12. *A language is context sensitive iff it is accepted by some linear bounded automaton.*

THEOREM 13. *A language is type-0 iff it is accepted by some Turing machine.*

Finally, we state an important theorem about context-sensitive languages. We say that a grammar $G = (V, N, P, S)$ is *recursive* iff there is an algorithm to decide for each string x in V_T^* whether $x \in L(G)$ or not.

THEOREM 14. *Any context-sensitive grammar is recursive.*

Basic idea of proof: Since the vocabulary V is finite, the number of production rules is finite, and the production rules of a context-sensitive grammar never decrease the length of a string, for a string x of length n it is easy to compute an upper bound as a function of n on the number of strings of length n that can be derived. (If deletions are permitted as in type-0 grammars, this is not possible.)

8.3 Stimulus-response Representation of Finite Automata

The central idea of this section is quite simple—it is to show how, by applying accepted principles of conditioning, an organism may theoretically be taught by an appropriate reinforcement schedule to respond as a finite automaton. In order to show that an organism obeying general laws of stimulus conditioning and sampling can be conditioned to become an automaton, it is necessary first of all to interpret within the usual run of psychological concepts the notion of a letter of an alphabet and the notion of an internal state. In my own thinking about these matters, I was first misled by the perhaps natural attempt to identify the internal state of the automaton with the state of conditioning of the organism. This idea, however, turned out to be clearly wrong. In the first place, the various possible states of conditioning of the organism correspond to various possible automata that the organism can be conditioned to become. Roughly speaking, to each state of conditioning there corresponds a different automaton. Probably the next most natural idea is to look at a given conditioning state and use the conditioning of individual stimuli to represent the internal states of the automaton. In very restricted cases this correspondence works, but in general it does not, for reasons that become clear below. The correspondence that turns out to work is the following: the internal states of the automaton are identified with the responses of the organism.

The correspondence, to be made between letters of the alphabet that the automaton will accept and the appropriate objects within stimulus-response theory, is fairly obvious. The letters of the alphabet correspond in a natural way to sets of stimulus

8.3 STIMULUS-RESPONSE REPRESENTATION OF FINITE AUTOMATA

elements presented on a given trial to an organism. It may seem like a happy accident, but the correspondences between inputs to the automata and stimuli presented to the organism, and between internal states of the machine and responses of the organism, are conceptually very natural.

Because of the conceptual importance of the issues that have been raised by linguists for the future development of psychological theory, perhaps above all because language behavior is the most characteristically human aspect of our behavior patterns, it is important to be as clear as possible about the claims that can be made for a stimulus-response theory whose basic concepts seem so simple and to many so woefully inadequate to explain complex behavior, including language behavior. I mention two examples that present very useful analogies. The first is the reduction of standard mathematics to the concept of set and the simple relation of an element being a member of a set. From a naive standpoint, it seems unbelievable that the complexities of higher mathematics can be reduced to a relation as simple as that of set membership. But this is indubitably the case, and we know in detail how the reduction can be made. This is not to suggest, for instance, that in thinking about a mathematical problem or even in formulating and verifying it explicitly, a mathematician operates simply in terms of endlessly complicated statements about set membership. By appropriate explicit definition we introduce many additional concepts, the ones actually used in discourse. The fact remains, however, that the reduction to the single relationship of set membership can be made and in fact has been carried out in detail. The second example, which is close to our present inquiry, is the status of simple machine language for computers. Again, from the naive standpoint it seems incredible that modern computers can do the things they can in terms either of information processing or numerical computing when their basic language consists essentially just of finite sequences of 1's and 0's; but the more complex computer languages that have been introduced are not at all for the convenience of the machines but for the convenience of human users. It is perfectly clear how any more complex language, like Visual C, can be reduced by a compiler or other device to a simple machine language. The same attitude, it seems to me, is appropriate toward stimulus-response theory. We cannot hope to deal directly in stimulus-response connections with complex human behavior. We can hope, as in the two cases just mentioned, to construct a satisfactory systematic theory in terms of which a chain of explicit definitions of new and ever more complex concepts can be introduced. It is these new and explicitly defined concepts that will be related directly to the more complex forms of behavior.

Before turning to specific mathematical development, it will be useful to make explicit how the developments in this section may be used to show that many of the common conceptions of conditioning, and particularly the claims that conditioning refers only to simple reflexes like those of eye blinking, are mistaken. The mistake is to confuse particular restricted applications of the fundamental theory with the range of the theory itself. Experiments on classical conditioning do indeed represent a narrow range of experiments from a broader conceptual standpoint. It is important to realize, however, that *experiments* on classical conditioning do not define the range and limits of conditioning theory itself. The main aim of the present section is to show how any finite automaton, no matter how complicated, may be constructed purely within stimulus-response theory. But from the standpoint of automata, classical conditioning represents a particularly trivial example of an automaton. Classical conditioning may be represented by an automaton having a one-letter alphabet and a single internal state. The next simplest case corresponds to the structure of classical discrimination experiments. Here there is more than a single letter to the alphabet, but the transition table of the automaton depends in no way on the internal state of the automaton. In the case of discrimination, we may again think of the responses as corresponding to the internal states of the automaton. In this sense there is more than one internal state, contrary to the case of classical conditioning, but what is fundamental is that the transition table of the automaton does not depend on the internal states but only on the external stimuli presented according to a schedule fixed by the experimenter. It is of the utmost importance to realize that this restriction, as in the case of classical conditioning experiments, is not a restriction that is in any sense inherent in conditioning theory itself. It merely represents concentration on a certain restricted class of experiments.

Leaving the technical details for later, it is still possible to give a very clear example of conditioning that goes beyond the classical cases and yet represents perhaps the simplest nontrivial automaton. By nontrivial I mean: there is more than one letter in the alphabet; there is more than one internal state; and the transition table of the automaton is a function of both the external stimulus and the current internal state. As an example, we may take a rat being run in a maze. The reinforcement schedule for the rat is set up so as to make the rat become a two-state automaton. We will use as the external alphabet of the automaton a two-letter alphabet consisting of a black or a white card. Each choice point of the maze will consist of either a left turn or a right turn. At each choice point either a black card or a white card will be present. The following table describes both the reinforcement schedule and the transition table of the automaton.

	L	R
LB	1	0
LW	0	1
RB	0	1
RW	1	0

Thus the first row shows that when the previous response has been left (L) and a black stimulus card (B) is presented at the choice point, with probability one the animal is reinforced to turn left. The second row indicates that when the previous response is left and a white stimulus card is presented at the choice point, the animal is reinforced 100% of the time to turn right, and so forth, for the other two possibilities. From a formal standpoint this is a simple schedule of reinforcement, but already the double aspect of contingency on both the previous response and the displayed stimulus card makes the schedule more complicated in many respects than the schedules of reinforcement that are usually run with rats.

There is no pretense that this simple two-state automaton is in any sense adequate to serious learning. I am not proposing, for example, that there is much chance of teaching even a simple regular grammar to rats. I am proposing to psychologists, however, that already automata of a small number of states present immediate experimental challenges in terms of what can be done with animals of each species. For example, what is the most complicated automaton a monkey may be trained to imitate? In this case, there seems some possibility of approaching at least reasonably complex regular grammars (using Theorem 10 of the previous section that any regular grammar is representable by a finite-state automaton). In the case of the lower species, it will be necessary to exploit to the fullest the kind of stimuli to which the organisms are most sensitive and responsive in order to maximize the complexity of the automata they can imitate.

Stimulus-response theory. The formalization of stimulus-response theory given here follows closely that in Suppes (1969b). The theory is based on six primitive concepts, each of which has a direct psychological interpretation. The first one is the set S of stimuli, which we shall assume is not empty, but which we will not restrict to being either finite or infinite on all occasions. The second primitive concept is the set R of responses and the third primitive concept the set E of possible reinforcements. As in the case of the set of stimuli, we need not assume that either R or E is finite, but in the present applications to the theory of finite automata we shall make this restrictive assumption.

The fourth primitive concept is that of a measure μ on the set of stimuli. In case the set S is finite, this measure is often the number of elements in S. For the general theory we shall assume that the measure of S itself is always finite, i.e., $\mu(S) < \infty$.

The fifth primitive concept is the sample space Ω. Each element ω of the sample space represents a possible experiment, that is, an infinite sequence of trials. In the present theory, each trial may be described by an ordered quintuple (C, T, s, r, e), where C is the conditioning function, T is the subset of stimuli presented to the organism on the given trial, s is the sampled subset of T, r is the response made on the trial, and e is the reinforcement occurring on that trial. It is not possible to make all the comments here that are required for a full interpretation and understanding of the theory. For those wanting a more detailed description, the two references already given will prove useful. A very comprehensive set of papers on stimulus-sampling theory has been put together in the collection edited by Neimark and Estes (1967); see also Estes (1959a,b). The present version of stimulus-response theory should in many respects be called stimulus-sampling theory, but I have held to the more general stimulus-response terminology to emphasize the juxtaposition of the general ideas of behavioral psychology on the one hand and linguistic theory on the other. In addition, in the theoretical applications to be made here the specific sampling aspects of stimulus-response theory are not as central as in the analysis of experimental data.

Because of the importance to be attached later to the set T of stimuli presented on each trial, its interpretation in classical learning theory should be explicitly mentioned. In the case of simple learning, for example, in classical conditioning, the set T is the same on all trials and we would ordinarily identify the sets T and S. In the case of discrimination learning, the set T varies from trial to trial, and the application we are making to the theory of automata falls generally under the discrimination case. The conditioning function C is defined over the set R of responses and C_r is the subset of S conditioned or connected to response r on the given trial. How the conditioning function changes from trial to trial is made clear by the axioms. As an example of a conditioning function, let $T = S = \{\sigma_1, \sigma_2, \sigma_3\}$, $R = \{r_1, r_2\}$, and $C_{r_1} = C(r_1) = \{\sigma_1, \sigma_2\}$ and $C(r_2) = \emptyset$.[2] (Ordinarily it is convenient to use subscript notation rather than function notation for C.) In this example no stimuli are conditioned to r_2 and the stimulus σ_3 is conditioned to no response.

From the quintuple descriptions of a given trial it is clear that certain assumptions about the behavior that occurs on a trial have already been made. In particular it is apparent that we are assuming that only one sample of stimuli is drawn on a given trial, that exactly one response occurs on a trial and that exactly one reinforcement occurs on a trial. These assumptions have been built into the set-theoretical description of the sample space Ω and will not be an explicit part of our axioms.

Lying behind the formality of the ordered quintuples representing each trial is the intuitively conceived temporal ordering of events on any trial, which may be represented by the following diagram:

State of conditioning at beginning of trial		presentation of stimuli		sampling of stimuli		response		reinforcement		state of conditioning at beginning of new trial
C_n	$\to$	T_n	$\to$	s_n	$\to$	r_n	$\to$	e_n	$\to$	C_{n+1}

The sixth and final primitive concept is the probability measure P on the appropriate σ-algebra of cylinder sets of Ω. The exact description of this σ-algebra is rather complicated when the set of stimuli is not finite, but the construction is standard, and we shall assume the reader can fill in details familiar from general probability theory. It is emphasized that all probabilities must be defined in terms of the measure P.

We also need certain notation to take us back and forth between elements or subsets of the sets of stimuli, responses, and reinforcements to events of the sample space Ω. First, r_n is the event of response r on trial n, that is, the set of all possible experimental realizations or elements of Ω having r as a response on the nth trial. Similarly, $e_{r,n}$ is the event of response r's being reinforced on trial n. The event $e_{0,n}$ is the event of no reinforcement on trial n. In like fashion, C_n is the event of conditioning function C occurring on trial n, T_n is the event of presentation set T occurring on trial n, and so forth. Additional notation that does not follow these conventions will be explicitly noted.

We also need a notation for sets defined by events occurring up to a given trial. Reference to such sets is required in expressing that central aspects of stimulus conditioning and sampling are independent of the pattern of past events. If I say that Y_n

---

[2] Notice that to avoid ambiguity, I use the letter 's' throughout this section only for a set of sampled stimuli, and thus, as in the present example, 'σ' for stimuli, i.e., elements of S. But across sections ambiguity remains, for in Definition 1 of Section 2 of this chapter, the letter 's' is used for states of the basic set A of a finite automaton. Possible points of notational ambiguity arising from this double usage occur only a few times in this section and will be pointed out.

8.3 STIMULUS-RESPONSE REPRESENTATION OF FINITE AUTOMATA

is an n-cylinder set, I mean that the definition of Y_n does not depend on any event occurring after trial n. However, an even finer breakdown is required that takes account of the postulated sequence $C_n \to T_n \to s_n \to r_n \to e_n$ on a given trial, so in saying that Y_n is a C_n-cylinder set what is meant is that its definition does not depend on any event occurring after C_n on trial n, i.e., its definition could depend on T_{n-1} or C_n, for example, but not on T_n or s_n. As an abbreviated notation, I shall write $Y(C_n)$ for this set and similarly for other cylinder sets. The notation Y_n without additional qualification shall always refer to an n-cylinder set.

Moreover, to avoid an overly cumbersome notation, event notation of the sort already indicated will be used, e.g., $e_{r,n}$, for reinforcement of response r on trial n, but also the notation $\sigma \in C_{r,n}$ for the event of stimulus σ's being conditioned to response r on trial n. To simplify the formal statement of the axioms it is assumed without repeated explicit statement that any given events on which probabilities are conditioned have positive probability. Thus, for example, the tacit hypothesis of Axiom S2 is that $P(T_m) > 0$ and $P(T_n) > 0$.

The axioms naturally fall into three classes. Stimuli must be sampled in order to be conditioned, and they must be conditioned in order for systematic response patterns to develop. Thus, there are naturally three kinds of axioms: sampling axioms, conditioning axioms, and response axioms. A verbal formulation of each axiom is given together with its formal statement. From the standpoint of formulations of the theory already in the literature, perhaps the most unusual feature of the present axioms is not to require that the set S of stimuli be finite. It should also be emphasized that for any one specific kind of detailed application, additional specializing assumptions are needed. Some indication of these will be given in the particular application to the theory of automata, but it would take us too far afield to explore these specializing assumptions in any detail and with any faithfulness to the range of assumptions needed for different experimental applications.

DEFINITION 1. *A structure* $\mathfrak{S} = (S, R, E, \mu, \Omega, P)$ *is a* stimulus-response model *if and only if the following axioms are satisfied:*

Sampling Axioms

S1. $P(\mu(s_n) > 0) = 1$.
   (*On every trial a set of stimuli of positive measure is sampled with probability 1.*)

S2. $P(s_m|T_m) = P(s_n|T_n)$.
   (*If the same presentation set occurs on two different trials, then the probability of a given sample is independent of the trial number.*)

S3. If $s \cup s' \subseteq T$ and $\mu(s) = \mu(s')$, then $P(s_n|T_n) = P(s'_n|T_n)$.
   (*Samples of equal measure that are subsets of the presentation set have an equal probability of being sampled on a given trial.*)

S4. $P(s_n|T_n, Y_n(C_n)) = P(s_n|T_n)$.
   (*The probability of a particular sample on trial n, given the presentation set of stimuli, is independent of any preceding pattern $Y_n(C_n)$ of events.*)

Conditioning Axioms

C1. If $r, r' \in R, r \neq r'$ and $C_r \cap C_{r'} \neq \emptyset$, then $P(C_n) = 0$.
(*On every trial with probability 1 each stimulus element is conditioned to at most one response.*)

C2. There exists $c > 0$ such that for every $\sigma, C, r, n, s, e_r,$ and Y_n
$P(\sigma \in C_{r,n+1} | \sigma \notin C_{r,n}, \sigma \in s_n, e_{r,n}, Y_n) = c$.
(*The probability is c of any sampled stimulus element's becoming conditioned to the reinforced response if it is not already so conditioned, and this probability is independent of the particular response, trial number, or any preceding pattern Y_n of events.*)

C3. $P(C_{n+1} | C_n, e_{0,n}) = 1$.
(*With probability 1, the conditioning of all stimulus elements remains the same if no response is reinforced.*)

C4. $P(\sigma \in C_{r,n+1} | \sigma \in C_{r,n}, \sigma \notin s_n, Y_n) = 1$.
(*With probability 1, the conditioning of unsampled stimuli does not change.*)

Response Axioms

R1. If $\bigcup_{r \in R} C_r \cap s \neq \emptyset$ then

$$P(r_n | C_n, s_n, Y(s_n)) = \frac{\mu(s \cap C_r)}{\mu(s \cap \cup C_r)}.$$

(*If at least one sampled stimulus is conditioned to some response, then the probability of any response is the ratio of the measure of sampled stimuli conditioned to this response to the measure of all the sampled conditioned stimuli, and this probability is independent of any preceding pattern $Y(s_n)$ of events.*)

R2. If $\cup_{r \in R} C_r \cap s = \emptyset$ then there is a number ρ_r such that

$$P(r_n | C_n, s_n, Y(s_n)) = \rho_r.$$

(*If no sampled stimulus is conditioned to any response, then the probability of any response r is a constant guessing probability ρ_r that is independent of n and any preceding pattern $Y(s_n)$ of events.*)

Representation of finite automata. A useful beginning for the analysis of how we may represent finite automata by stimulus-response models is to examine what is wrong with the most direct approach possible. The difficulties that turn up may be illustated by the simple example of a two-letter alphabet (i.e., two stimuli σ_1 and σ_2, as well as the "start-up" stimulus σ_o) and a two-state automaton (i.e., two responses r_1 and r_2). Consideration of this example will be useful for several points of later discussion.

By virtue of Axiom S1, the single presented stimulus must be sampled on each trial, and we assume that for every n,

$$0 < P(\sigma_0 \in s_n), P(\sigma_1 \in s_n), P(\sigma_2 \in s_n) < 1.$$

8.3 STIMULUS-RESPONSE REPRESENTATION OF FINITE AUTOMATA

Suppose, further, the transition table of the machine is:

	r_1	r_2
$r_1\sigma_1$	1	0
$r_1\sigma_2$	0	1
$r_2\sigma_1$	0	1
$r_2\sigma_2$	1	0

which requires knowledge of both r_i and σ_j to predict what response should be next. The natural and obvious reinforcement schedule for imitating this machine is:

$$P(e_{1,n}|\sigma_{1,n}, r_{1,n-1}) = 1,$$
$$P(e_{2,n}|\sigma_{1,n}, r_{2,n-1}) = 1,$$
$$P(e_{2,n}|\sigma_{2,n}, r_{1,n-1}) = 1,$$
$$P(e_{1,n}|\sigma_{2,n}, r_{2,n-1}) = 1,$$

where $\sigma_{i,n}$ is the event of stimulus σ_i's being sampled on trial n. But for this reinforcement schedule the conditioning of each of the two stimuli continues to fluctuate from trial to trial, as may be illustrated by the following sequence. For simplification and without loss of generality, we may assume that the conditioning parameter c is 1, and we need indicate no sampling, because, as already mentioned, the single stimulus element in each presentation set will be sampled with probability 1. We may represent the states of conditioning (granted that each stimulus is conditioned to either r_1 or r_2) by subsets of $S = \{\sigma_0, \sigma_1, \sigma_2\}$. Thus, if $\{\sigma_1, \sigma_2\}$ represents the conditioning function, this means both elements σ_1 and σ_2 are conditioned to r_1; $\{\sigma_1\}$ means that only σ_1 is conditioned to r_1, and so forth. Consider then the following sequence from trial n to $n+2$:

$$(\{\sigma_2\}, \sigma_2 \in s, r_1, e_2) \to (\emptyset, \sigma_2 \in s, r_2, e_2) \to (\emptyset, \sigma_2 \in s, r_2, e_1).$$

The response on trial $n+1$ satisfies the machine table, but already on $n+2$ it does not, for $r_{2,n+1}\sigma_{2,n+2}$ should be followed by $r_{1,n+2}$. It is easy to show that this difficulty is fundamental and arises for any of the four possible conditioning states. (In working out these difficulties explicitly, the reader should assume that each stimulus is conditioned to either r_1 or r_2, which will be true for n much larger than 1 and $c = 1$.)

What is needed is a quite different definition of the states of the Markov chain of the stimulus-response model. (For proof of a general Markov-chain theorem for stimulus-response theory, see Estes and Suppes 1959b, 1974.) Naively, it is natural to take as the states of the Markov chain the possible states of conditioning of the stimuli in S, but this is wrong on two counts in the present situation. First, we must condition the *patterns* of responses and presentation sets, so we take as the set of stimuli for the model, $R \times S$, i.e., the Cartesian product of the set R of responses and the set S of stimuli. What the organism must be conditioned to respond to on trial n is the pattern consisting of the preceding response given on trial $n-1$ and the presentation set occurring on trial n.

It is still not sufficient to define the states of the Markov chain in terms of the states of conditioning of the elements in $R \times S$, because for reasons that are given explicitly and illustrated by many examples in Estes and Suppes (1959b) and Suppes and Atkinson (1960), it is also necessary to include in the definition of state the response r_1 that actually occurred on the preceding trial. The difficulty that arises if $r_{i,n-1}$ is not included in the definition of state may be brought out by attempting to draw the tree in the case of the two-state automaton already considered. Suppose just the pattern $r_1\sigma_1$ is conditioned and the other four patterns, $\sigma_0, r_1\sigma_2, r_2\sigma_2,$ and $r_2\sigma_2$, are not. Let us represent this conditioning state by C_1, and let τ_j be the noncontingent probability of $\sigma_j, 0 \leq j \leq 2$, on every trial with every $\tau_j > 0$. Then the tree looks like this.

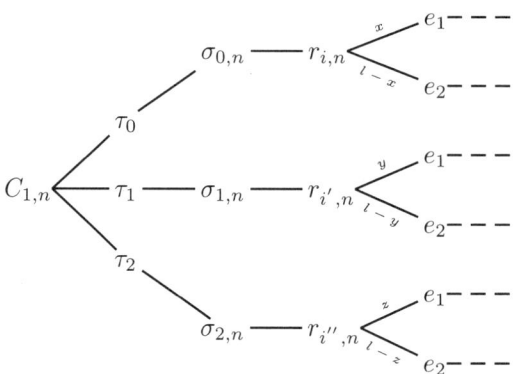

The tree is incomplete, because without knowing what response actually occurred on trial $n - 1$ we cannot complete the branches (e.g., specify the responses), and for a similar reason we cannot determine the probabilities x, y, and z. Moreover, we cannot remedy the situation by including among the branches the possible responses on trial $n - 1$, for to determine their probabilities we would need to look at trial $n - 2$, and this regression would not terminate until we reached trial 1.

So we include in the definition of state the responses on trial $n - 1$. On the other hand, it is not necessary in the case of deterministic finite automata to permit among the states all possible conditioning of the patterns in $R \times S$. We shall permit only two possibilities—the pattern is unconditioned or it is conditioned to the appropriate response because conditioning to the wrong response occurs with probability zero. Thus with p internal states or responses and m letters in $\sum$, there are $(m+1)p$ patterns, each of which is in one of two states, conditioned or unconditioned, and there are p possible preceding responses, so the number of states in the Markov chain is $p2^{(m+1)p}$. Actually, it is convenient to reduce this number further by treating σ_0 as a single pattern regardless of what preceding response it is paired with. The number of states is then $p2^{mp+1}$. Thus, for the simplest 2-state, 2-alphabet automaton, the number of states is 64. We may denote the states by ordered $mp + 2$-tuples

$$(r_j, i_0, i_{0,1}, \ldots, i_{0,m}, \ldots, i_{p-1,m}),$$

where $i_{k,l}$ is 0 or 1 depending on whether the pattern $r_k\sigma_l$ is unconditioned or conditioned with $0 \leq k \leq p - 1$ and $1 \leq l \leq m$; r_j is the response on the preceding trial, and i_0 is the state of conditioning of σ_0. What we want to prove is that starting in the purely unconditioned set of states $(r_j, 0, 0, \ldots, 0)$, with probability i the system will always ultimately be in a state that is a member of the set of fully conditioned states

8.3 STIMULUS-RESPONSE REPRESENTATION OF FINITE AUTOMATA

$(r_j, 1, 1, \ldots, 1)$. The proof of this is the main part of the proof of the basic representation theorem.

Before turning to the theorem, we need to define explicitly the concept of a stimulus-response model's asymptotically becoming an automaton. As has already been suggested, an important feature of the definition is this. The basic set S of stimuli corresponding to the alphabet Σ of the automaton is not the basic set of stimuli of the stimulus-response model, but rather, this basic set is the Cartesian product $R \times S$, where R is the set of responses.[3] Moreover, the definition has been framed in such a way as to permit only a single element of S to be presented and sampled on each trial; this, however, is an inessential restriction used here in the interest of conceptual and notational simplicity. Without this restriction the basic set would be not $R \times S$, but $R \times \mathcal{P}(S)$, where $\mathcal{P}(S)$ is the power set of S, i.e., the set of all subsets of S, and then each letter of the alphabet Σ would be a subset of S rather than a single element of S. What is essential is to have $R \times S$ rather than S as the basic set of stimuli.

For example, the pair (r_i, σ_j) must be sampled *and* conditioned as a pattern, and the axioms are formulated to require that what is sampled and conditioned be a subset of the presentation set T on a given trial. In this connection, to simplify notation I shall often write $T_n = (r_{i,n-1}, \sigma_{j,n})$ rather than

$$T_n = \{(r_i, \sigma_j)\},$$

but the meaning is clear. T_n is the presentation set consisting of the single pattern (or element) made up of response r_i on trial $n-1$ and stimulus element σ_j on trial n, and from Axiom S1 we know that the pattern is sampled because it is the only one presented.

The use of $R \times S$ is formally convenient, but is not at all necessary. The classical S-R tradition of analysis suggests a formally equivalent, but psychologically more realistic approach. Each response r produces a stimulus σ_r, or more generally, a set of stimuli. Assuming again, for formal simplicity just one stimulus element σ_r, rather than a set of stimuli, we may replace R by the set of stimuli S_R, with the purely contingent presentation schedule

$$P(\sigma_{r,n}|r_{n-1}) = 1,$$

and in the model we now consider the Cartesian product $(S_R \times S)$ rather than $R \times S$. Within this framework the important point about the presentation set on each trial is that one component is purely subject-controlled and the other purely experimenter-controlled—if we use familiar experimental distinctions. The explicit use of S_R rather than R promises to be important in training animals to perform like automata, because the external introduction of σ_r reduces directly and significantly the memory load on the animal. The importance of S_R for models of children's language learning is less clear.

Still another interpretation of the theory can be given by going from psychology to neuroscience in the spirit of the final section of this chapter. This move means replacing the response itself by the brain's representation of the response. Again, this biologically

---
[3] This distinction is important to keep in mind in what follows. Note that replacing R by S_R is discussed two paragraphs further on, to help emphasize it.

more realistic move would not change the formal structure of the basic argument of this section, but an empirically realistic account would entail many more details of the kind given in Section 6.

DEFINITION 2. *Let* $\mathfrak{S} = (R \times S, R, E, \mu, \Omega, P)$ *be a stimulus-response model where*

$$R = \{r_0, \ldots, r_{p-1}\},$$
$$S = \{\sigma_0, \ldots, \sigma_m\},$$
$$E = \{e_0, \ldots, e_{p-1}\},$$

and $\mu(S')$ *is the cardinality of* S' *for* $S' \subseteq S$. *Then* $\mathfrak{S}$ *asymptotically becomes the automaton* $\mathfrak{A}(\mathfrak{S}) = (R, S - \{\sigma_0\}, M, r_0, F)$ *if and only if*

(i) *as* $n \to \infty$ *the probability is 1 that the presentation set* T_n *is* $(r_{i,n-1}, \sigma_{j,n})$ *for some* i *and* j,

(ii) $M(r_i, \sigma_j) = r_k$ *if and only if* $\lim_{n \to \infty} P(r_{k,n} | T_n = (r_{i,n-1}, \sigma_{j,n})) = 1$ *for* $0 \leq i \leq p-1$ *and* $1 \leq j \leq m$,

(iii) $\lim_{n \to \infty} P(r_{0,n} | T_n = (r_{i,n-1}, \sigma_{0,n})) = 1$ *for* $0 \leq i \leq p-1$,

(iv) $F \subseteq R$.

A minor but clarifying point about this definition is that, as already remarked, the stimulus σ_0 is not part of the alphabet of the automaton $\mathfrak{A}(\mathfrak{S})$, because a stimulus is needed to put the automaton in the initial state r_0, and from the standpoint of the theory being worked out here, this requires a stimulus to which the organism will give response r_0. That stimulus is σ_0. The definition also requires that asymptotically the stimulus-response model $\mathfrak{S}$ is nothing but the automaton $\mathfrak{A}(\mathfrak{S})$. It should be clear that a much weaker and more general definition is possible. The automaton $\mathfrak{A}(\mathfrak{S})$ could merely be embedded asymptotically in $\mathfrak{S}$ and be only a part of the activities of $\mathfrak{S}$. The simplest way to achieve this generalization is to make the alphabet of the automaton only a proper subset of $S - \{\sigma_0\}$ and correspondingly for the responses that make up the internal states of the automaton; they need be only a proper subset of the full set R of responses. This generalization will not be pursued here, although something of the sort will be necessary to give an adequate account of the semantical aspects of language.

THEOREM 1. (Representation Theorem for Finite Automata). *Given any connected finite automaton, there is a stimulus-response model that asymptotically becomes isomorphic to it. Moreover, the stimulus-response model may have all responses initially unconditioned.*

Proof. Let $\mathfrak{A} = (A, \Sigma, M, s_0, F)$ be any connected finite automaton. As indicated already, we represent the set A of internal states by the set R of responses r_i, for $0 \leq i \leq p-1$, where p is the number of states. We represent the alphabet Σ by the set of stimuli $\sigma_1, \ldots, \sigma_m$, and, for reasons already made explicit, we augment this set of stimuli by σ_0, to obtain

$$S = \{\sigma_0, \sigma_1, \ldots, \sigma_m\}.$$

For subsequent reference let f be the function defined on $A \cup \Sigma$ that establishes the natural one-one correspondence between A and R, and between Σ and $S - \{\sigma_0\}$. (To avoid some trivial technical points, I shall assume that A and Σ are disjoint.)

We take as the set of reinforcements

$$E = \{e_0, e_1, \ldots, e_{p-1}\},$$

and the measure $\mu(S')$ is the cardinality of S' for $S' \subseteq S$, so that as in Definition 2, we are considering a stimulus-response model $\mathfrak{S} = (R \times S, R, E, \mu, \Omega, P)$. In order to show that $\mathfrak{S}$ asymptotically becomes an automaton, we impose five additional restrictions on $\mathfrak{S}$. They are these.

First, in the case of reinforcement e_0 the schedule is this:

$$P(e_{0,n}|\sigma_{0,n}) = 1, \tag{1}$$

i.e., if $\sigma_{0,n}$ is part of the presentation set on trial n, then with probability 1 response r_0 is reinforced—note that the probability of the reinforcing event $e_{0,n}$ is independent of the actual occurrence of the response $r_{0,n}$.

Second, the remaining reinforcement schedule is defined by the transition table M of the automaton $\mathfrak{A}$. Explicitly, for $j, k \neq 0$ and for all i and n

$$P(e_{k,n}|\sigma_{j,n}r_{i,n-1}) = 1 \text{ if and only if } M(f^{-1}(r_i), f^{-1}(\sigma_j)) = f^{-1}(r_k). \tag{2}$$

Third, essential to the proof is the additional assumption beyond (1) and (2) that the stimuli $\sigma_0, \ldots, \sigma_m$ each have a positive, noncontingent probability of occurrence on each trial (a model with a weaker assumption could be constructed but it is not significant to weaken this requirement). Explicitly, we then assume that for any cylinder set $Y(C_n)$ such that $P(Y(C_n)) > 0$

$$P(\sigma_{i,n}) = P(\sigma_{i,n}|Y(C_n)) \geq \tau_i > 0 \tag{3}$$

for $0 \leq i \leq m$ and for all trials n.

Fourth, we assume that the probability ρ_i of response r_i occurring when no conditioned stimuli is sampled is also strictly positive, i.e., for every response r_i

$$\rho_i > 0, \tag{4}$$

which strengthens Axiom R2.

Fifth, for each integer $k, 0 \leq k \leq mp+1$, we define the set Q_k as the set of states that have exactly k patterns conditioned, and $Q_{k,n}$ is the event of being in a state that is a member of Q_k on trial n. We assume that at the beginning of trial 1, no patterns are conditioned, i.e.,

$$P(Q_{0,1}) = 1. \tag{5}$$

It is easy to prove that given the sets R, S, E and the cardinality measure μ, there are many different stimulus-response models satisfying restrictions (1)-(5), but for the proof of the theorem it is not necessary to select some distinguished member of the class of models because the argument that follows shows that all the members of the class asymptotically become isomorphic to $\mathfrak{A}$.

The main thing we want to prove is that

$$\lim_{n \to \infty} P(Q_{mp+1,n}) = 1. \tag{6}$$

We first note that if $j < k$ the probability of a transition from Q_k to Q_j is zero, i.e.,

$$P(Q_{j,n}|Q_{k,n-1}) = 0, \tag{7}$$

moreover,
$$P(Q_{j,n}|Q_{k,n-1}) = 0, \tag{8}$$
even if $j > k$ holds, unless $j = k+1$. In other words, in a single trial, at most one pattern can become conditioned.

To show that asymptotically (6) holds it will suffice to show that there is an $\epsilon > 0$ such that on each trial n for $0 \le k \le mp < n$ if $P(Q_{k,n}) > 0$,
$$P(Q_{k+1,n+1}|Q_{k,n}) \ge \epsilon. \tag{9}$$

To establish (9) we need to show that there is a probability of at least ϵ of a stimulus pattern that is unconditioned at the beginning of trial n becoming conditioned on that trial. The argument given will be a uniform one that holds for any unconditioned pattern. Let $r^*\sigma^*$ be such a pattern on trial n.

Now it is well known that for a connected automaton, for every internal state s, there is a tape x such that[4]
$$M(s_0, x) = s \tag{10}$$
and the length of x is not greater than the number of internal states. In terms of stimulus-response theory, x is a finite sequence of length not greater than p of stimulus elements. Thus we may take $x = \sigma_{i_1}, \ldots, \sigma_{i_p}$ with $\sigma_{i_p} = \sigma^*$. We know by virtue of (3) that
$$\min_{0 \le i \le m} \tau_i = \tau > 0. \tag{11}$$

The required sequence of responses $r_{i_1}, \ldots, r_{i_p-1}$ will occur either from prior conditioning or if any response is not conditioned to the appropriate pattern, with guessing probability r_i. By virtue of (4)
$$\min_{0 \le i \le p-1} \rho_i = \rho > 0. \tag{12}$$

To show that the pattern $r^*\sigma^*$ has a positive probability ϵ of being conditioned on trial n, we need only take n large enough for the tape x to be 'run', say, $n > p+1$, and consider the joint probability
$$P^* = P(\sigma_n^*, r_{n-1}^* \sigma_{i_{p-1}, n-1}, \ldots, r_{0, n-i_p-1} \sigma_{0, n-i_p-1}). \tag{13}$$

The basic axioms and the assumptions (1)-(5) determine a lower bound on P^* independent of n. First we note that for each of the stimulus elements $\sigma_0, \sigma_1, \ldots, \sigma^*$, by virtue of (3) and (11)
$$P(\sigma_n^*|\ldots) \ge \tau, \ldots, P(\sigma_{0,n-i_p-1}) \ge \tau.$$

Similarly, from (4) and (12), as well as the response axioms, we know that for each of the responses $r_0, r_{i_1}, \ldots, r^*$
$$P(r_{n-1}^*|\ldots) \ge \rho, \ldots, P(r_{0,n-i_p-1}|\sigma_{0,n-i_p-1}) \ge \rho.$$

Thus we know that
$$P^* \ge \rho^p \tau^{p+1},$$

---

[4]In equation (10) we have one of the few cases in this section where 's' and 's_0' refer to states of the automaton being modeled, not sets of sampled stimuli.

and given the occurrence of the event $\sigma_n^* r_{n-1}^*$, the probability of conditioning is c, whence we may take
$$\epsilon = c\rho^p \tau^{p+1} > 0,$$
which establishes (9) and completes the proof.

Given the theorem just proved, there are several significant corollaries whose proofs are almost immediate. The first combines the representation theorem for regular languages with that for finite automata to yield:

THEOREM 2. *Any regular language is generated by some stimulus-response model at asymptote.*

Once probabilistic considerations are introduced, we can in several different ways go beyond the restriction of stimulus-response generated languages to regular languages. This is done later in this section.

I suspect that many psychologists or philosophers who are willing to accept the sense given here to the reduction of finite automata and regular languages to stimulus-response models will be less happy with the claim that one well-defined sense of the concepts of *intention, plan,* and *purpose* can be similarly reduced. However, without any substantial new analysis on my part this can be done by taking advantage of an analysis already made by Miller and Chomsky (1963). The story goes like this. In 1960 Miller, Galanter and Pribram published a provocative book entitled *Plans and the Structure of Behavior.* In this book they severely criticized stimulus-response theories for being able to account for so little of the significant behavior of humans and the higher animals. They especially objected to the conditioned reflex as a suitable concept for building up an adequate scientific psychology. It is my impression that a number of cognitively oriented psychologists felt that the critique of S-R theory in this book was devastating.

As I indicated earlier, I would agree that conditioned reflex *experiments* are indeed far too simple to form an adequate scientific basis for analyzing more complex behavior. This is as hopeless as would be the attempt to derive the theory of differential equations, let us say, from the elementary algebra of sets. Yet the more general theory of sets does encompass in a strict mathematical sense the theory of differential equations.

The same relation may be shown to hold between stimulus-response theory and the theory of plans, insofar as the latter theory has been systematically formulated by Miller and Chomsky. This point was developed informally by Millenson (1967). The theory of plans is formulated in terms of tote units ('tote' is an acronym for the cycle test-operate-test-exit). A plan is then defined as a tote hierarchy, which is just a form of oriented graph, and every finite oriented graph may be represented by a finite automaton. So we have the result:

THEOREM 3. *Any tote hierarchy in the sense of Miller and Chomsky is isomorphic to some stimulus-response model at asymptote.*

Response to criticisms.[5] Arbib's (1969) comments on the stimulus-response theory of finite automata raise issues that continue to divide cognitive and stimulus-response

---

[5] The remainder of this section draws on Suppes (1969c, 1989).

psychologists. I am not persuaded by any of his arguments that a viable alternative to stimulus-response theory has yet been defined, although I am willing to admit that a fully adequate stimulus-response theory of complex learning and behavior has yet to be developed. Arbib's comments and arguments can be analyzed under five headings. I deal with each in turn, and in several cases, amplify my own views as well.

Matters of proof. Arbib alleges that he has given a much simpler derivation of my main theorem: given any connected finite automaton, there is a stimulus-response model asymptotically isomorphic to it. He simplifies the proof by assuming that "for each stimulus-response table there is a stimulus-response model that asymptotically becomes isomorphic to it [p. 507]"; but as Bertrand Russell said long ago, such procedures of postulation instead of proof have all the virtues of theft over honest toil. In other words, Arbib has not shown how the fundamental result follows from simple general assumptions about stimulus-response connections—my main objective. To put the matter technically, he is faced with the problem of showing that his very powerful postulate, which corresponds to a theorem in my framework, is consistent with the simple assumptions about conditioning and sampling of stimuli that are an integral part of any standard stimulus-response theory.

Stimulus traces and mediating responses. Arbib seems to have a rather simple view of how an organism can remember its previous responses. He mentions the natural device used in experimentation with animals of having the previous response externalized by an appropriate stimulus. Exactly this approach has been used in the conditioning of pigeons to behave as simple automata, but it is not a general approach. Certainly it is not adequate to the problems of language learning to which my own research is primarily addressed, or even to the problems of children learning arithmetic.

Within classical stimulus-response theory, it is natural to talk about stimulus traces, and I mentioned this possibility originally. In terms of later stimulus-response theories, especially those associated with Osgood, Maltzman, Berlyne, the Kendlers, and others, it is natural to talk about mediating responses internal to the organism. Either of these concepts provides a natural framework to replace or to extend the very narrow externalization approach used by Arbib. He objects to having some representation of the previous response in the organism for fear "that the input channels would be completely overloaded if they must carry not only representations of external stimuli but also complete information about the current state of execution of the TOTE hierarchy [p. 509]". This comment, like most of Arbib's other comments, however, is not backed up by a detailed analysis. It is not at all clear that this is a problem. Put in this general way, one may as well express concern over the internal processing also and simply say that it is found to be an overweening mystery of how the organism can work at all.

Number of states. In many ways I think the most serious and important issue raised by Arbib is the query of whether it is possible for an organism ever to have adequate time to learn the apparently large number of stimulus-response connections needed for it to become a suitably large automaton. This is a central issue that has been raised continually, particularly by psycholinguists critical of stimulus-response theories of language learning. The negative claims that Arbib makes concerning a TOTE hierarchy involving eight components with four states each are familiar, and his arguments also assume a familiar theological tone of saying it simply cannot be done.

Because of the many arguments I have had about these matters over the years, I have come to make a distinction between negative dogma and negative proof. A wide range of psycholinguists and cognitive psychologists have asserted the negative dogma that stimulus-response ideas can never account for complex learning, because suitable conditioning connections could never be learned in sufficient time by organisms exhibiting complex behavior. This negative dogma seems to be an article of faith, and not an article of proof on the part of almost all who assert it, including Arbib. Certainly he does not give a proof that a device with 4^8 states is needed for any cognitive processing he cares to define. In view of his familiarity with the formal literature in the theory of automata on these problems, I suspect he is somewhat more wary than many of the psycholinguists in venturing to give what would appear to be a detailed argument. For an example of a presumed negative proof as opposed to a negative dogma, but what is in fact a mélange of confusions from end to end, see the proposed proof by Bever, Fodor, and Garrett (1968) that a formal limitation of stimulus-response theory or associationism can be established.

It is generally recognized in mathematics and associated disciplines that negative arguments must be formulated in a more formal and explicit fashion than positive arguments. There is also a long history in mathematics and philosophy of establishing explicit and carefully defined systematic standards for evaluating the validity of a negative argument. It seems to me that Arbib does not work at all within this classical tradition. He makes a few casual statements about number of states, but offers no serious negative argument about the size of automata required for any cognitive processes understood well enough to be characterized in relatively exact terms. I will return to this point subsequently.

I think that the problem of the number of states required for various tasks is indeed a fundamental one, and I am certainly not prepared to establish small upper bounds on the number of states needed for a very large number of tasks. On the other hand, I am skeptical of the large claims so often made by psycholinguists or automaton theorists. I mention just two examples. At the end of Suppes (1969b), I gave a detailed analysis of the kind of automaton required for column addition of two numbers, and I showed how it can be related in realistic fashion to the actual behavior of students.[6] The number of states required is just two. Casual thought without detailed analysis might lead to the suggestion that the number of states would need to be large. This is not the case.

My second example, which I will not give in detail, concerns efforts to write probabilistic grammars for the spoken speech of 2- and 3-year-old children. Although I do not want to claim that these efforts are wholly successful, it is clear from scanning the length of sentences uttered by young children that the number of states required for production of the grammar is relatively small, and probably a larger number of states is required just for the handling of the lexical vocabulary. I should add that I have yet to

---

[6]This example is also given earlier in this section.

meet anyone, linguist or automaton theorist, who would want to claim that vocabulary can be acquired by any way other than some rather direct empirical process of learning. I would conjecture that the number of states required for any cognitive task, including language comprehension or production, will turn out to be a lot smaller than any of us originally thought. I readily admit that this is simply a conjecture, but I have enough confidence in it to insist that those who wish to claim that stimulus-response theory is inadequate, because of the problem of the number of states that must be conditioned, should offer negative proofs and not simply negative dogmas. The central focus of my own research efforts is to convert my own positive dogmas into positive proofs. (On the question of number of states, it is perhaps worth noting that a universal Turing machine can be constructed with only seven states.)

A simple example will illustrate an important point in talking about the number of states. If we toss a coin 100 times, the number of possible outcomes is 2^{100}. Each of us can perform this experiment, but we could set the entire population of the world to tossing coins and not approach the number of possible outcomes. The number of *actual* states in a process, it is essential to note, is usually incredibly smaller, by orders and orders of magnitude, than the number of possible states, and so it is with the relation between the number of actual and the number of possible conditioning connections.

Real versus metaphorical learning. Arbib argues in several places that learning simply could not take place according to stimulus-response conceptions because the number of states required is too large and the amount of time needed for conditioning is too restricted. I have already expressed my central argument against his claims about the number of states. I am willing to agree that if the number of states that must be conditioned individually is very large, then there will not be time for the task. What is interesting is the vagueness and weakness of the alternatives considered by Arbib. He expresses clear preference for TOTE hierarchies, without considering these hierarchies as special cases of stimulus-response models, according to the first corollary of my main theorem. What he does not say or even sketch is how, within the framework of TOTE hierarchies, a theory can be given of the organism's acquiring cognitive skills. He does mention casually some references to neurons and biochemical changes in the brain, but no serious ideas are set forth in this physiological framework. The reader is left totally uninformed as to what serious alternative Arbib would propose, with the single possible exception of his final remark about the computer as a metaphorical brain. Presumably, metaphorical brains engage in metaphorical learning. If this is what he is suggesting as a theory of learning, he cannot mean for his remarks in this direction to be taken seriously. An account of how organisms learn must be stated with sufficient definiteness and in sufficiently nonmetaphorical terms for it to be tested experimentally. Certainly Arbib himself makes no positive or definite suggestions of how to deal with any of the main difficulties facing theories of machine or human learning.

Indeed, it is surprising that in view of his opinions about the adequacy of psychological theories of learning he does not have more to say about the virtues of computer theories of learning, i.e., machine learning. I think the reasons, however, are really clear enough. It does not take a very serious or systematic study of the literature of artificial

8.3 STIMULUS-RESPONSE REPRESENTATION OF FINITE AUTOMATA

intelligence to reach the conclusion that, although psychological theories of learning are certainly far from being sufficiently developed, so are theories of how computers should be programmed to learn.[7]

Problem of hierarchies. I have stated already that the issue about the number of states raised by Arbib is indeed a serious one. I think the other serious issue raised by him is the problem of hierarchies. It seems to be a matter of belief on the part of almost all cognitively oriented psychologists that conditioning theories must treat each simple conditioning connection as separate and equal. The concept of a hierarchy, it seems to be suggested, is contrary to the spirit of stimulus-response ideas. That a clear theoretical counterexample exists is shown at once by the corollary to the main theorem of my paper, namely, the theorem that every TOTE hierarchy in the sense of Miller and Chomsky is isomorphic to some stimulus-response model at asymptote. The abstraction of this result may be unsatisfactory to some readers, but its meaning is clear. Any intuitive hierarchy that may be represented formally by a TOTE hierarchy, a finite oriented graph, a finite tree, or a finite automaton may be represented just as well by a stimulus-response model; in fact, better, for the stimulus-response model also includes an account of how the hierarchy is learned if the stimulus-response connections are not already coded in the genes.

Criticism of the representation theorem. A technically detailed and interesting criticism of Theorem 1, the representation theorem for finite automata, has been given by Kieras (1976). The intuitive source of Kieras' confusion in his claim that the theorem as stated is too strong is easy to identify.[8] Because I identified the *internal states* of a given automaton with the *responses* of the representing stimulus-response model, Kieras inferred I had unwittingly restricted my analysis to automata that have a one-one correspondence between *their* internal states and responses. On the basis of this confusion on his part he asserts that the representation theorem is not correct as it stands.

My purpose now is to lay out this dispute in an explicit and formal way in order to show unequivocally that Kieras is mistaken and the representation theorem is correct as originally stated. From a mathematical standpoint, Kieras' mistake rests on a misunderstanding of representation theorems. The isomorphism of a representation theorem is a formal one. In the case of Theorem 1 above, the isomorphism is between the internal states of the automaton and the responses of the representing stimulus-response model. The Rabin-Scott definition of automata used in my 1969 article does not have an explicit response mechanism, but that this is a trivial addition to their definition is shown by Definition 2 of Section 2, for finite automata with output, as well as by the *general* definition of a sequential machine with output given by Harrison (1965, p. 294), who is referenced by Kieras and who acknowledges he is mainly following the terminology of

---

[7]What kind of details are needed in the theory of machine learning of language, even for a simple fragment of natural language, is well illustrated in Section 5 of this chapter. This is not to suggest the specific approach used there is the only one. It is just, at the present time anyway, hard to see how a host of explicitly worked out details can be avoided.

[8]Criticisms by Anderson (1976, pp. 81–89) are based on essentially the same confusion.

Rabin and Scott (1959). An automaton or sequential machine with output is for Harrison just an automaton in the sense of Rabin and Scott with the additional condition that the set F of final states are those 'giving a one output'. As Harrison and others have remarked, a restriction of output to 1's and 0's is no restriction on the generality of the sequential machine.

The addition of this output apparatus to the formal definitions I gave in the original article is trivial. We just pick two responses r_0 and r_1 not used to represent internal states, but one of them, say r_0, represents 0 and the other 1. Whenever the machine is in an internal state that is not a final state but a response is required, it outputs r_0. When it is in a final state it outputs r_1. To modify Definition 1 to take account of these output responses is easy. I note once again that the two output responses are in no way intended to correspond to internal states of the automata being represented. Other responses of the stimulus-response model represent the internal states. I emphasize, also, that this modification of adding output responses would not be correcting an error in Theorem 1 but would only be providing an additional closely related result.

Problem of determinate reinforcement. As has been noticed by several critics, and as I have remarked on several occasions, for example, Suppes (1975), in many kinds of learning environments strict determinate reinforcement of the response is not feasible, contrary to the demands of the proof of Theorem 1. For example, if I ask a child the sum of $7+5$, then a determinate reinforcement would be giving the correct answer, 12, when he gave an incorrect answer. An example of nondeterminate reinforcement would be simply to tell him that the answer was incorrect and to ask him to try again. When determinate reinforcement is used, it is clear that in some sense the responses have to be observable in order to correct each incorrect response. Theorem 1 used an assumption of determinate reinforcement. It seems to me that it is this assumption of determinate reinforcement rather than any of the informal remarks I made about responses being observable or internal states being identifiable with outputs that is the central limitation of the result.

From the standpoint of giving an account of complex learning, especially in natural settings as opposed to simple laboratory situations, it was clear to me before Theorem 1 was published that the most essential extension was to obtain similar results with nondeterminate reinforcement. This problem was tackled in conjunction with my former student, William Rottmayer, and detailed results are embodied in his 1970 dissertation. An informal and rather brief statement of the results appears in a survey article we published on automata (Suppes and Rottmayer 1974).

Because the formal statement of stimulus-response theory with nondeterminate reinforcement is rather complicated, I shall give only a brief informal statement similar to that in Suppes and Rottmayer (1974). Before doing so, let me formulate a canonical class of problems that can be used for intuitive reference in digesting the content of the individual axioms. The organism is presented with a potentially infinite class of stimulus displays, for example, line drawings. A subclass of the entire class is characterizable by a finite automaton. The problem for the learner is to learn the concept characterized by the finite automaton, given on each trial only the information of whether or not his classification of a given stimulus display is correct. I have mentioned line drawings here because I do not want to concentrate entirely on language, but it would also be possible to think of the learning in terms of recognizing grammatical strings of some regular language. Because I do not like to think of language learning wholly in terms of grammar, I prefer in the present context a geometrical example. Let us suppose that each line drawing consists of a finite number of line segments. A typical example might be a line drawing consisting of three segments forming a triangle but with one of the line segments, and only one, extending beyond the triangle. On each trial the learner is asked to say whether the particular line drawing shown is an instance of the display and after he has given this information he is told simply that his answer is correct or incorrect.

What the theory postulates is a sequence of implicit responses or, if you prefer, internal responses by the learner prior to giving the answer of 'yes' or 'no' to classify the display. Informally it is assumed that the implicit or internal responses are not available for observation and cannot be directly reinforced. The theory does not require this as an assumption, but it is implicit in any experimental application of the theory. Reinforcement takes place not after each internal response occurs, which is considered a subtrial, but only after a trial consisting of a sequence of subtrials.

In other words, putting the matter in standard experimental terms, a subtrial corresponds to what we usually think of as a trial, but no reinforcement or conditioning takes place and we cannot observe the response that was actually made. Conditioning occurs only after a sequence of subtrials, and the whole sequence of subtrials is called a trial. In automaton terms, a subtrial corresponds to an automaton making one transition, that is, from one internal state to another, and a trial to processing an entire tape or input string.

Conditioning occurs on trials that have a correct response, and deconditioning occurs on trials that have an incorrect response. Thus learning occurs on all trials, regardless of whether the response is correct or not. On the basis of these axioms, the following theorem, which represents a considerable improvement of the basic theorem in my 1969 article, can be proved. The improvement is due to the weakening of the methods of reinforcement.

THEOREM 4. *If $\mathcal{D}$ is any set of perceptual displays and G is a subset of $\mathcal{D}$ that can be recognized by a finite automaton, then there is a stimulus-response model that can also learn to recognize G, with performance at asymptote matching that of the automaton.*

One important point to note is that with nondeterminate reinforcement the theorem is, as one would expect, weaker. In the case of determinate reinforcement the stimulus-response model at asymptote becomes isomorphic to the given finite automaton. In the present case, the result can only be one of behavioral equivalence or, in the ordinary language of automaton theory, the result is one of weak equivalence.

It is clear that the nondeterminate reinforcement used in the theory I have just formulated is about the weakest version of reinforcement that is interesting, with the possible exception of giving only partial reinforcement, that is, reinforcement on certain trials. In actual learning, for example, in the learning of mathematics or in the learning of language, there are many situations in which much more decisive and informative reinforcement is given. It is not difficult to show that the more determinate the reinforcement, the faster learning will be in general for organisms of a given capacity. In the long and tangled history of the concept of reinforcement it has not been sufficiently emphasized that reinforcement is delivery of information, and a particular structure of information is implicit in any particular scheme of reinforcement. An exhausting but not exhaustive analysis of different structures of reinforcement is to be found in Jamison, Lhamon and Suppes (1970). (So many detailed theoretical computations were made in this article that it has scarcely been read by anyone; it does provide a good sense of how complex things rapidly become when reinforcement schemes that have even mildly complex information structures are used.)

It is important for the present discussion to consider one of the weakest structures of nondeterminate reinforcement and to emphasize the point that it is the nondeterminate character of the reinforcement that moves us out of the arena of classical observability of responses and permits the introduction of a repertoire of implicit or internal responses that are not in general observable. The reinforcement does not create the implicit responses, but when we have determinate reinforcement the theory is not applicable to situations in which implicit or internal responses occur.

It is also important to note that it is really a matter of terminology, and not of substantive theory, whether these implicit responses are called responses as such or are called internal states. It would be easy enough to reformulate the axioms given above and to replace responses with internal states except for the response that occurs at the end of a trial. This terminological change would not affect the axioms in any way.

It is worth mentioning that the implicit responses that we might want to baptize as internal states are often observed as taking place even when we do not know their exact form. A good example occurs in the case of subvocalized articulatory responses that are characteristic of most silent adult readers. Self-awareness of such subvocal responses is unusual, and I hasten to add that it is not usually possible to 'read off' from the subvocal responses the words being read.

Another misconception: restriction to finite automata. Another misconception in the literature is that stimulus-response theory can only deal with machines that have the power of finite automata. The purpose of this subsection is to show that this is not the case by giving the construction of register machines, which are equivalent to Turing machines. The development here extends and modifies substantially that in Suppes (1977b). To give the results formal definiteness, we shall develop a learning theory for any partial recursive function. Such functions can be defined explicitly in a fairly direct way, but we shall not do so here. I shall rely upon the fact that partial recursive functions are computable functions. We then use the basic theorem in the literature, whose technical framework we shall expand upon somewhat later, that any function is partially recursive if and only if it is computable by a register machine or, equivalently, by a Turing machine. The concept of a register machine used here was introduced in Section 3.5. The reason for using register machines rather than Turing

8.3 STIMULUS-RESPONSE REPRESENTATION OF FINITE AUTOMATA

machines is that their formal structure is simpler. For example, the proof of equivalence between a function being a partial recursive function and being computable by a register machine is much simpler than the corresponding proof for Turing machines. First, let me recall from Section 3.5 how simple a classical register machine for a finite vocabulary is. All we have is a potentially infinite list or sequence of registers, but any given program uses only a finite number. Exactly three simple kinds of instructions are required for each register. The first is to place any element of the finite vocabulary at the top of the content of register n; the second is to delete the bottommost letter of the content of register n if the register is nonempty; because any computation takes place in a finite number of steps, the content of any register must always be finite in length. The third instruction is a jump instruction to another line of the program, if the content of register n is such that the bottommost or beginning letter is a_i; in other words, this is a conditional jump instruction. Thus, if we think of the contents of registers as being strings reading from left to right, we can also describe the instructions as placing new symbols on the right, deleting old symbols on the left, and using a conditional jump instruction in the register when required.

It is straightforward to give a formal definition of programs for such an unlimited register machine, but I delay this for the moment. It is clear that a program is simply made up of lines of instructions of the sort just described. The potentially infinite memory of an unlimited register machine both in terms of the number of registers and the size of each register is a natural mathematical idealization. It is also possible to define a single-register machine with instructions of the kind just stated and to show that a single register is also adequate.

An important point about the revision of stimulus-response theory given here is that the internal language used for encoding stimulus displays is all that is dealt with. In other words, in the present formulation of the register-machine theory I shall not enter into the relation between the set of external stimuli and the encoding language, but deal only with the already encoded representation of the display. This level of abstraction seems appropriate for the present discussion, but of course is not appropriate for a fully worked out theory. It is a proper division of labor, however, with the proper modularity. I am assuming that the sensory system passes to the central nervous system such encoded information, with the first level of encoding taking place well outside the central nervous system. Thus, in one sense the concept of stimulus becomes nonfunctional as such, but only because the encoding is already assumed. It is obvious enough that no serious assumptions about the actual perceptual character of stimuli is a part of classical S-R theory. Secondly, the concept of a program internally constructed replaces the direct language of responses being conditioned to stimuli. A natural question would be why not try to give a more neural version of this construction. Given how little we know about the actual way in which information is transduced to the central nervous system and then used for encoding and programming, it seems premature. Certainly what does seem to be the case is that there is internal programming. I am not suggesting that the abstract simple theory of a register machine catches the details of that internal programming—it is only a way of representing it—, and it is a matter for detailed additional theory to modify the abstract representation to make it more realistic.

On the other hand, without giving anything like a detailed neural analysis, the register-machine programs can be replaced by computationally equivalent stimulus-response connections, but without further specification such postulated S-R conditioning connections are no more concrete, i.e., closer to empirical realization, than the register-machine programs. It seems to me that it is therefore better to think of the programs as being realized by neural 'hardware' we cannot presently specify. What is presented in the remainder of this section is formally adequate, but can surely be improved upon in many ways, either to more closely imitate the learning of different organisms or to make machine learning more efficient. Moreover, given some feature coding of presented stimuli, there is good reason to think that to any software program there is a corresponding neural network, and vice versa, for solving a particular class of problems with essentially the same rate of learning.

To make matters more explicit and formal but without attempting a complete formalization, I recall the following definitions from Section 3.5, in particular, the subsection on recursive functions over an arbitrary finite alphabet. First, $<n>$ is the content of register n before carrying out an instruction; $<n'>$ is the content of register n after carrying out an instruction. Second, a *register machine* has (1) a denumerable sequence of registers numbered 1,2,3, ..., each of which can store any finite sequence of symbols from the finite alphabet V, and (2) three basic kinds of instructions:

(a) $P_N^{(i)}(n)$: Place a_i on the right-hand end of $<n>$.

(b) $D_N(n)$: Delete the leftmost letter of $<n>$ if $<n> \neq \emptyset$.

(c) $J_N^{(i)}(n)[q]$: Jump to line q if $<n>$ begins with a_i.[9]

If the jump is to a nonexistent line, then the machine stops. The parameter N shown as a subscript in the instructions refers to the set of feature registers holding sensory data and not used as working computation registers. (This point is made more explicitly in the definition given below.)

A *line* of a program of a register machine is either an ordered couple consisting of a natural number $m \geq 1$ (the line number) and one of the instructions (a) or (b), or an ordered triple consisting of a natural number $m \geq 1$, one of the instructions (c), and a natural number $q \geq 1$. The intuitive interpretation of this definition is obvious and will not be given.

A *program* (of a register machine) is a finite sequence of k lines such that (1) the first number of the ith line is i, and (2) the numbers q that are third members of lines are such that $1 \leq q \leq k+1$. The parameter k is, of course, the number of lines of the program. I shall also refer to programs as *routines*. How a register machine *follows* a program or routine is intuitively obvious and will not be formally defined. *Subroutines* are defined like programs except (1) subroutines may have several exits, and (2) third members of triples may range over $q_1, \ldots, q_k$—these variables being assigned values in a given program.

The formal definition of a partial recursive function defined over the alphabet V is given at the end of Section 3.5. I repeat briefly the essentials. Such a function is

---

[9] A slight change in the notation of Section 3.5 is made here, using 'q' rather than 'E1' for the line to jump to. It is convenient for the proof given later.

8.3 STIMULUS-RESPONSE REPRESENTATION OF FINITE AUTOMATA

any intuitively computable function. Given V, the finite vocabulary, then, as usual in such matters, V^* is the set of finite sequences of elements of V; in the present context, I shall call the elements of V^* feature codings. Let f be a function of n arguments from $V^* \times \cdots \times V^*$ (n times) to V^*. The basic definition is that f is computable by a register machine if and only if for every register x_i, y and N with $y \neq x_i$ for $i = 1, \ldots, n$ and $x_1, \ldots, x_n, y \leq N$ there exists a routine $R_N(y = f(x_1 \ldots, x_n))$ such that if $< x_1 >, \ldots, < x_n >$ are the initial contents of registers $x_1, \ldots, x_n$ then

1. if $f(< x_1 >, \ldots, < x_n >)$ is undefined the machine will not stop;
2. if $f(< x_1 >, \ldots, < x_n >)$ is defined, the machine will stop with $< y >$, the final content of register y, equal to $f(< x_1 >, \ldots, < x_n >)$, and with the final contents of all registers $1, 2, \ldots, N$, except y, the same as initially.

Axioms for register learning models. I turn now to the axioms for register learning models that in a very general way parallel those given for stimulus-response models with nondeterminate reinforcement in Suppes and Rottmayer (1974). I axiomatize only the model, and not the full probability space that serves as a formal framework for the learning trials. Extension to the latter, possibly via random variables and leaving the probability space implicit, is straightforward but tedious.

The axioms are based on the following structural concepts:

(i) the set R of registers,

(ii) the vocabulary V of the model,

(iii) the subset F of feature registers,

(iv) the subset C of computation registers,

(v) the subset Rp of response registers,

(vi) the working memory WM,

(vii) the long-term memory LTM,

(viii) the responses r_0 and r_1,

(ix) the real parameters p and c.

It will perhaps be useful to say something briefly and informally about each of the primitive concepts. The feature registers in F just encode the features of the presented stimulus. This encoding and computation are done by using the finite vocabulary V. The computer registers in C are working registers available as needed for computation. The working memory WM stores programs being constructed. For simplicity here I shall assume there is only one such memory, but clearly this is too restrictive for general purposes. The long-term memory LTM is where programs that are found by repeated trials to be correct are stored.

One distinction is essential between the two memories and the registers. The memories store the program, so the feature vocabulary $v_1, \ldots, v_n$ in V is added notation for the three types of instruction: P for placing or adding on the right, D for deleting on the left, and J for a jump instruction. The vocabulary V must also include notation for referring to registers used and to program lines. For this purpose I add the single digit 1 (thus $2 = 11$, $3 = 111$, etc.), the most rudimentary counting notation.

The set Rp of response registers is also here, for simplicity, assumed to be a singleton set. This register corresponds, in the general register machine characterized earlier, to the register that holds the value of the partial recursive function being computed. An inessential simplifying assumption is that learning will be restricted to concept learning, which is in principle no restriction on the set of computable functions. In the present case, given that the program is completed, if the register is cleared, the response is r_0, which means that the stimulus displayed—whose features are encoded in F—is an instance of the concept being learned; and if the register is not empty the response is r_1, which means the stimulus presented is not an instance of the concept. Moreover, if the program at any step is halted before completion, the response is r_0 with guessing probability p, and r_1 with probability $1-p$.

The two real parameters p and c enter in the axioms in quite different ways. As just indicated, p is the response guessing probability, and c is the constant probability of stopping construction of a program. These parameters, and others introduced implicitly in the axioms, are surely context dependent, and will naturally vary from task to task.

As formulated here, each line of a program is run as it is selected for the program construction and placed in working memory (WM). A program is transferred to long-term memory (LTM) only when it is completed and is successful in correctly identifying an instance of the concept being learned. The mechanism of completely erasing a constructed program that is in error is too severe, but is a simplifying assumption that holds for some animal learning, e.g., the all-or-none elimination of habituation in aplysia by sensitiving stimuli (Kandel 1985).

The three types of register-machine instructions—adding on the right, deleting on the left, or conditional jump—mentioned earlier are modified in one respect. To jump to a nonexistent line and thereby halt the program, rather than jumping to $m+1$ where m is the number of lines, the jump is to 0, which is a possible number for no line. The reason for this change should be apparent. As the program is probabilistically constructed line by line by the learning model, there is no way of knowing in advance how long the program will be. So it is convenient to have in advance a fixed 'place' to jump to in order to halt the program.

DEFINITION 3. *A structure* $\mathfrak{R} = (R, V, F, C, Rp, WM, LTM, r_0, r_1, p, c)$ *is a* register learning model of concept formation *if and only if the following axioms are satisfied:*

Register Structure Axioms

R1. *The subsets F, C, and Rp of registers are nonempty and pairwise disjoint.*

R2. *Subsets F and Rp, and the set V are finite and nonempty.*

R3. *Each register in R can hold any word of V_1^*, i.e., any finite string of elements of* $V_1 = V - \{1, P, D, J\}$.

Stimulus Encoding Axiom

D1. *At the start of each trial, the stimulus presented is encoded as having features $<f>$ in the registers f of F.*

Program Construction Axioms

P1. *If at the start of the trial, the LTM is nonempty, no program construction occurs.*

8.3 STIMULUS-RESPONSE REPRESENTATION OF FINITE AUTOMATA

P2. *Given that LTM is empty:*

(i) *With probability $c, 0 < c < 1$, construction of the program in WM terminates after each line, independent of the trial number and any preceding subsequence of events;*

(ii) *Given that a line is to be added to the program, the probability of sampling an instruction of any type with any argument is positive, independent of the trial number and any preceding subsequence of events; in the case of the line number n to which a jump is to be made the probability is geometrically distributed.*

Program Execution Axioms

E1. *If LTM is nonempty, the contents are copied into WM, and then the program is executed.*

E2. *If LTM is empty, then a program is constructed probabilistically, line by line according to Construction Axioms P1 and P2, and is executed as each line is constructed.*

E3. *When a jump instruction is executed, there is a fixed positive probability the program is halted after one step, with this probability being independent of the trial number and any preceding subsequence of events.*

Response Axioms

Rp1. *If when the program is complete, register Rp is empty, the response is r_0.*

Rp2. *If when the program is complete, register Rp is nonempty, the response is r_1.*

Rp3. *If the program is halted by Axiom E3, response r_0 is made with guessing probability p, and response r_1 with probability $1 - p$; the probability p is independent of the trial number and any preceding subsequence of events.*

Reinforcement Axioms

Er1. *If positive reinforcement occurs at the end of a trial, the program in WM is copied in LTM if LTM is empty.*

Er2. *If negative reinforcement occurs at the end of a trial, the program in WM is erased and so is the program in LTM if it is nonempty.*

A few of the axioms require comments that were not made earlier in the informal discussion. The probabilistic program construction axiom P2 is similar to a stimulus sampling axiom which guarantees accessibility for conditioning of all relevant stimuli. Axiom P2 is obviously formulated in such a way as to bound sampling probabilities away from asymptotically approaching zero except in the case of the geometric distribution for sampling line numbers. The stopping probability required in program execution axiom E3 is required in order to prevent staying with programs that generate infinite loops. Finally, the informal concept of reinforcement used in the axioms has an obvious meaning and is easily formalized. Positive reinforcement here just means that the concept classification of a stimulus by the response r_0 or r_1 is correct, and negative reinforcement that it is incorrect. Obviously, more informative reinforcement methods can and are widely used in learning and without question facilitate the speed of learning. More is said on this point in the final remarks on hierarchical learning.

On the basis of the axioms stated above we may prove an asymptotic learning theorem corresponding in a general way to Theorem 1 for stimulus-response models.

THEOREM 5. *Let f be any partial function of n arguments over the finite alphabet V and having just two values in V. Then f is a partial recursive function if and only if f is asymptotically learnable with probability one by a register learning model $\mathfrak{R}$ of concept formation.*

Proof. Let $\mathcal{P}$ be a program for $\mathfrak{R}$ that computes f. We know there must be such a program by virtue of the fact that a function f over a finite alphabet is partial recursive if and only if it is computable by a register machine. Furthermore, given a definition of f we have a constructive method for producing $\mathcal{P}$. Our objective is to show that in the learning environment described by the axioms there is a positive probability of constructing $\mathcal{P}$ on each trial.

Let $C \subseteq V^* \times \cdots \times V^*$ (n times) be the set of encoded stimulus instances of the f-computable concept C—without loss of generality in this context I identify the concept with its set of instances, and let $\neg C$ be the complement of C. We take as a presentation distribution of stimuli, where $(<f_1>,\ldots,<f_n>)$ is the feature-encoding representation of a stimulus,

$$P((<f_1>,\ldots,<f_n>) \in C) = P((<f_1>,\ldots,<f_n>) \in \neg C) = \frac{1}{2}.$$

Moreover, we design the experiment to sample from C and $\neg C$ in the following geometric fashion. Let f_i be the coding in V^* of feature i of stimulus σ and let $|f_i|$ be the number of symbols in f_i. Then $\sum |f_i|$ is the total number of symbols used to encode σ. We use a geometric distribution for the total number of symbols, and a uniform distribution for selecting among those of the same total number of symbols. (In a completely formalized theory, these assumptions about probabilistic selection of presented stimuli would be part of the axioms, which I have restricted here just to the register learning model, and have not included axioms on stimulus presentation or reinforcement procedures in any detail.)

Suppose now that initially LTM is nonempty. If the program stored in LTM correctly computes f, we are done. If the program does not for some stimulus σ, then by the assumptions just stated there is a fixed positive probability that σ will be presented on every trial and hence with probability one asymptotically LTM will be cleared by virtue of Axiom Er2.

The probability of then constructing $\mathcal{P}$ is positive on every trial. The detailed calculation is this. First, let $\mathcal{P}$ have m lines. By Axiom P2(i), the probability of constructing a program of exactly m lines is equal to $c(1-c)^{m-1}$. If line i is not a jump instruction, then by Axiom P2(ii), the probability of line i being of the desired form is greater than some $\epsilon_1 > 0$. And if line i is a conditional jump instruction, where the jump is to line n_i, then also by Axiom P2(ii), the probability of line i being exactly line i of program $\mathcal{P}$ is equal to $\epsilon_2^2(1-\epsilon_2)^{n_i-1}$ for some $\epsilon_2 > 0$.

So, independent of trial number, the finite product of these probabilities is positive on every trial. Explicitly, let $i_1,\ldots,i_{m_1}$ be the lines that are not jump instructions and

8.3 STIMULUS-RESPONSE REPRESENTATION OF FINITE AUTOMATA

let $j_1, \ldots, j_{m_2}$ be the lines that are, with $m = m_1 + m_2$. Then

$$\text{Prob of } \mathcal{P} > \epsilon_1^{m_1} \Pi_{i=j_1}^{i=j_{m_2}} \epsilon_2^2 (1-\epsilon_2)^{n_i - 1} \cdot c(1-c)^{m-1} > 0. \qquad (I)$$

From this inequality, we infer at once that asymptotically $\mathcal{P}$ will be learned with probability one, which completes the proof, except to remark that to prove the constructed program characterizes a partial recursive function is straightforward.

Role of hierarchies and more determinate reinforcement. For the theory of register-model concept learning, as formulated in Definition 3, we cannot improve on inequality (I). Treating it as an equality it is evident that for programs $\mathcal{P}$ of any length learning will be very slow, much slower than we observe in most human learning and even much animal learning.

Within the framework of the present theory, the only practical hope for learning to occur in a reasonable time is to organize learning into a hierarchy of relatively small tasks to be mastered. It might be thought that this conclusion could be avoided by making the reinforcement more informative or determinate than what was assumed in Axioms Er1 and Er2 above. There is something correct and important about this view, and it can be supported by detailed computations on significant examples. On the other hand, there is also a question of interpretation. For the completely deterministic reinforcement used in the proof of Theorem 1, we could regard conditioning of each internal state of the finite automaton as a task—here task is defined by what gets reinforced, and in this view, the most fine-grained hierarchy is created by completely deterministic reinforcement.

It will be useful to end with application of the theory to a small, familiar task, to show that the theory can be brought down to earth and applied to data. Of course, in the present context I shall not try to be serious about actual parameter estimation. The task selected is that of 5-year-old children learning the concept of a triangle by recognizing triangles when presented with triangles, quadrilaterals and pentagons.

I make the following assumptions about the register model being used by the children. (It has the sort of simplifications necessary in such matters.)

(i) The language V_1 has a single element, α, which is used for counting.
(ii) There are two feature registers, #1 for number of segments and #2 for size, with α = small, $\alpha\alpha$ = medium and $\alpha\alpha\alpha$ = large.
(iii) The conditional jump is either to a previous line or to 0 (for a nonexistent line and stop).
(iv) To simplify formulation, computations are made directly on the feature registers rather than first copying their contents to a working register. (Characterizing copying from one register to another in terms of the three types of primitive instructions is straightforward.)
(v) Rp is the single response register.
(vi) Let a be the probability of selecting the delete instruction, b the probability for the jump instruction, and $1 - a - b$ the probability of the place or add instruction.
(vii) Let p be the probability of selecting feature register 1, and $1 - p$ that of selecting feature register 2 for reference in a line of program.

A simple correct program is:

1. D(1) Delete α from register 1.
2. D(1) Delete α from register 1.
3. D(1) Delete α from register 1.
4. Copy(1,Rp) Copy the contents of register 1 in the response register Rp.

All programs, in the short form used here, must end in copying the contents of a feature or working register to the response register. A response is then made. So the probability of lines 1–3 is: $p^3 a^3 c(1-c)^2$, where c is the parameter for the distribution of number of lines introduced in Axiom P2(i).

It is important to recognize that many different programs will produce the correct response, and so the probability of a correct response is considerably greater than $p^3 a^3 c(1-c)^2$. The complexity of a full analysis even for the simple experimental situation considered is much greater if the task is recognition of quadrilaterals rather than triangles. Still, under reasonable assumptions, the probabilities of the correct programs that are near the minimum length should dominate the theoretical computation of a correct response.

The learning setup defined axiomatically here is, in terms of its scope, comparable to the definition of partial recursive functions or the definition of register machines for computing such functions—namely, the definitions apply to each function considered individually. But for extended learning of a hierarchy of concepts, the structure must be enriched to draw upon concepts that have been previously learned in order to reach a practical rate of learning. Here is a very simple example to illustrate the point. Consider a disjunctive concept made up of n disjoint cases. Only one register is required, the alphabet V_1 is the set $\{\alpha, \beta\}$, and there is no jump instruction, but only the four instructions for deleting letters on the left or adding them on the right. Let the program be at most 10 lines for each case. Then assuming a uniform distribution on sampling of instructions and of the number of lines (1 to 10), the probability of each program of at most 10 lines can be directly computed. More importantly in the present instance, we can easily compute the possible number of programs: 4 of length 1, 16 of length 2, and in general 4^n of length n, with $1 \leq n \leq 10$, for a total of $(4^{11} - 4)/3$, which is approximately 4^{10}. If now at the second stage programs are put together using only original instructions and the n subroutines from individual cases, with programs of length at most $2n$ permitted, then there are $[(n+4)^{2n+1} - (n+4)]/(n+3)$ possible programs, which is approximately $(n+4)^{2n}$. On the other hand, if a single program is developed in one step with $10n$ lines, the number of possible programs is approximately 4^{10n}. Consider, for example, the case $n = 3$. Then 4^{30} is many orders of magnitude larger than $7^6 + 3(4^{10})$. The details of this example are not important, and I have not attempted to fix them sufficiently to determine in each of the two approaches the number of possible programs that are correct. Ordinarily in both the hierarchical and nonhierarchical approach this number would be a very small percentage of the total. The gain from the hierarchical approach is evident enough.

More generally, clever ways of dynamically changing the probability of using a previously defined concept, i.e., its recognition program, are critical to actual machine learning, for example, and sound hypotheses about such methods seem essential to any sophisticated study of human or animal learning of an extended hierarchy of concepts. Of equal importance is the introduction of forms of information feedback richer than the simple sort postulated in Definition 2, but the mathematical study of alternatives seems still to be in its infancy—only the extreme cases are relatively well understood. Much human learning depends upon verbal instruction and correction, but an approximately adequate theory of this process of feedback is as yet out of reach from a fundamental standpoint. Various gross simplifying assumptions, as used, for example, in protocol analysis, seem uneliminable at the present time. This is one measure of how much remains to be done.

8.4 Representation of Linear Models of Learning by Stimulus-sampling Models[10]

To show how intricate and technical even conceptually relatively simple representations can be of one theory in terms of another, I prove here a theorem about two closely related theories of learning. The theorem and its proof are taken from an unpublished technical report of a good many years ago (Estes and Suppes 1959b). A simpler but restricted version of the theorem and its proof was published by Norman (1972). The restriction of Norman's is severe. It is a restriction to the reinforcement schedule depending only on events that happen on the same trial that the reinforcement occurs.

The example considered here is simple and mathematically elementary compared to the intricacies of giving a really satisfactory proof of the reduction of classical thermodynamics to statistical mechanics. This latter reduction is one, of course, of great scientific importance, but it is probably not sufficiently realized how complicated it is to give an adequate theoretical development of the reduction and what restrictions have to be imposed to get a completely satisfactory mathematical argument.[11]

The present simple example, already intricate and technical in its own right, provides some sense of the difficulties of establishing significant and satisfactory reductions of one theory to another by means of appropriate representation theorems in any area of science. It is important to note that the reduction of one scientific theory to another is conceptually a very different thing from proving a representation theorem for a theory of measurement or a theory of geometry. In the latter kind of case the relationship between the structure of models of theory and the particular representing models is straightforward and apparent. Indeed, usually the theory of measurement or geometric theory has been constructed with a very definite numerical interpretation in mind. The axioms are guided to a large extent by this intended numerical interpretation. The reduction of one scientific theory to another is operating in quite a different conceptual framework ordinarily.

In this section we represent various linear models which have been much discussed in the literature (Bush and Mosteller 1955). Most applications of linear models have been to simple learning experiments, that is, experiments for which the presentation set

---

[10]This highly technical section may be omitted without loss of continuity. It does provide a real example of reduction of one theory to another via isomorphism of models.

[11]Compare remarks in Ruelle (1969) on the corresponding situation in statistical mechanics.

of stimuli is constant over trials. In Estes and Suppes (1959a) there is given a rather extensive analysis of a certain class of such models based on a single learning parameter θ. So each ω in Ω represents an infinite sequence of trials. Each trial of ω is represented by an ordered pair (r_i, e_k) where $r_i \in R$, the finite set of possible responses, and $e_k \epsilon E$, the finite set of reinforcements. Let ω_n be the sequence of responses and reinforcements of a given subject through trial n. The axioms given in Estes and Suppes (1959a) for linear models postulate linear relationships for the probabilities of responses given the finite sequence ω_n.

DEFINITION 1. *A structure $(R, E, \Omega, P, \theta)$ is a (single parameter) linear model for simple learning if, and only if, Ω is the sample space of sequences ω, P is a probability measure on $\mathfrak{B}(\Omega)$, the σ-algebra of cylinder sets, the learning parameter θ is in the half-open interval $(0 < \theta \leq 1)$, and the following three axioms are satisfied, for every n, every ω in Ω and every r_i in R and e_k in E:*

Axiom 1. If $P(e_{i,n} r_{i',n} \omega_{n-1}) > 0$ then
$$P(r_{i,n+1}|e_{i,n}r_{i',n}\omega_{n-1}) = (1-\theta)P(r_{i,n}|\omega_{n-1}) + \theta.$$

Axiom 2. If $P(e_{k,n} r_{i',n} \omega_{n-1}) > 0, k \neq i$, and $k \neq 0$, then
$$P(r_{i,n+1}|e_{k,n}r_{i',n}\omega_{n-1}) = (1-\theta)P(r_{i,n}|\omega_{n-1}).$$

Axiom 3. If $P(e_{0,n} r_{i',n} \omega_{n-1}) > 0$ then
$$P(r_{i,n+1}|e_{0,n}r_{i',n}\omega_{n-1}) = P(r_{i,n}|\omega_{n-1}).$$

Note that e_0 represents the event of no reinforcement.

The present section consists of four parts. In the first part the general axioms for stimulus-sampling theory, as stated in the previous section, are slightly modified. In the second part some preliminary theorems are established. In the third part theorems involving actual sequences of responses r_i and reinforcements e_j are formulated. Finally, in the fourth part special limit assumptions are introduced and the general representation theorem is proved.

Modification of general axioms. Turning back now to the stimulus-response models, it is desirable to have a notation for the *number* of elements conditioned to a given response on trial n, the number of elements sampled, etc. In general, if A is any set, we designate the cardinality of A by $N(A)$. Following usage in learning theory, we let $N = N(S)$, where S is the set of stimuli. Moreover, $N(s_n)$ is the set of all sequences ω in Ω that on the nth trial have exactly $N(s)$ sampled stimulus elements. In like fashion we use the notation $N(T), N(T_n), N(C_i), N(C_{i,n}), N(s_i), N(s_{i,n})$. The distinction between $N(s)$ and $N(s_n)$, and so forth, needs explicit emphasis. $N(s)$ is an integer standing for the finite number of elements in s, a subset of S. In contrast, $N(s_n)$ is an infinite set of sample-space points. We also continually use, as in the previous section,

$$T, T', T'', \ldots \text{ for presentation sets of stimuli},$$
$$C, C', C'', \ldots \text{ for conditioning functions},$$
$$s, s', s'', \ldots \text{ for sampled sets of stimuli}.$$

Using the explicit notation $N(s)$, we also strengthen the earlier axioms by always making the set s finite and always interpreting $\mu(s)$ for any subset s of S as the number of elements in s, and therefore write $N(s)$ instead of $\mu(s)$, and reserve $\mu(i)$ for the probabilistic mean of some quantity i.

We also require the notion of an *experimenter's partition,* already introduced in Estes and Suppes (1959a). We give a verbal definition of such a partition. An experimenter's partition $H(n)$ for trial n, with elements $\eta, \eta', \eta'', \ldots$, is a partition of the sample space Ω such that each element η of $H(n)$ is defined only in terms of events $T_{n'}$, $r_{i,n''}$, $e_{j,n'''}$, with $1 \leq n', n'', n''' \leq n$. And it should be clear that the elements of an experimenter's partition $H(n)$ are simply special sorts of n-dimensional cylinder sets.

Reinforcement axiom.

E1. The probability of a reinforcement depends only on previous observables, namely, preceding presentation sets of stimuli, preceding responses, and preceding outcomes.

There is an experimenter's partition $H(n)$ such that if W_{n-1} is any $n-1$ cylinder set, $Y \subseteq W_{n-1} \cap C_n \cap s_n, \eta \in H(n)$, and $P(Y \cap \eta) > 0$ then

$$P(e_{j,n}|Y \cap \eta) = P(e_{j,n}|\eta).$$

Presentation axiom.

P1. The probability of a stimulus presentation set depends only on previous observables.

There is an experimenter's partition $H(n-1)$ such that if W_{n-1} is any $n-1$ cylinder set, $Y \subseteq W_{n-1} \cap C_n, \eta \in H(n-1)$ and $P(Y \cap \eta) > 0$, then

$$P(T_n|Y \cap \eta) = P(T_n|\eta).$$

Preliminary theorems. We establish a number of theorems and corollaries which are preliminary to the main theorem. Because we are restricting ourselves to simple learning, all theorems of the section are based on the hypothesis

(I) *The set S of stimuli is the presentation set on every trial.*

By virtue of (I) we may omit all consideration of presentation sets T in the sequel. To obtain results independent of n, we also assume that the probability of response i given the empty sample s^o is ρ_i, i.e., for every i and n

(II) $\rho_{i,n} = \rho_i$.

Third, as a matter of notation, we explicitly define the mean of the number of sampled stimuli:

(III) $\displaystyle \overline{s} = \sum_{N(s)=1}^{N} N(s) P(N(s)).$

The corollaries of the first five theorems show more clearly than do the theorems themselves how the theorems provide a stepping stone for deriving the linear model. The first two theorems vary the response on trial n. The second two change the reinforcing event on trial n. The fifth theorem deals with the case of no reinforcement on trial n.

In the first theorem we need the fact that the second raw moment of the hypergeometric distribution is the following:

$$\sum_{a=0}^{A} a^2 \frac{\binom{A}{a}\binom{N-A}{s-a}}{\binom{N}{s}} = \frac{1}{N}\left\{As + \frac{A(A-1)s(s-1)}{N-1}\right\}.$$

THEOREM 1. *If* $P(e_{i,n}r_{i,n}N(C_{i,n})) > 0$ *then*

$$P(r_{i,n+1}|e_{i,n}r_{i,n}N(C_{i,n})) = \rho_i P(s^o) + (1 - P(s^o)) \cdot$$

$$\left[\frac{N(C_i)^2}{N^2} + \frac{\bar{s}N(C_i)}{N^2} - \frac{N(C_i)}{N^2} - \frac{N(C_i)(N(C_i)-1)(\bar{s}-1)}{N^2(N-1)}\right]$$

$$/ \left[(1 - P(s^o))\frac{N(C_i)}{N} + \rho_i P(s^o)\right]$$

Proof. By the usual methods of expansion we have:

$$P(r_{i,n+1}|e_{i,n}r_{i,n}N(C_{i,n})) = \qquad (1)$$

$$\sum_{s=s_i}^{N}\sum_{s_i=0}^{C_i} P(r_{i,n+1}|e_{i,n}r_{i,n}s_{i,n}s_n C_{i,n})P(e_{i,n}|r_{i,n}s_{i,n}s_n C_{i,n})$$

$$\cdot P(r_{i,n}|s_{i,n}s_n)P(s_i|s_n C_{i,n})P(s_n)/P(e_{i,n}|r_{i,n}C_{i,n})P(r_{i,n}|C_{i,n})P(C_{i,n}),$$

where on the right-hand side of (1), for economy of notation we have replaced $N(C_{i,n})$, $N(s_{i,n})$ and $N(s_n)$ by $C_{i,n}, s_{i,n}$ and s_n respectively, and similarly for $N(C_i)$, etc. This convention will be followed throughout this section in proofs, but not in the statement of theorems. We may simplify (1) by noting first that in simple learning, because of Axiom E1, the conditional probabilities for $e_{i,n}$ are the same in the numerator and denominator. Second, from the general learning theorems proved in Estes and Suppes (1959b) we know that

$$P(r_{i,n+1}|e_{i,n}r_{i,n}s_{i,n}s_n C_{i,n}) = P(r_{i,n+1}|C'_{i,n+1}) = \frac{C_i + s - s_i}{N},$$

where $C'_i = C_i \cup (s - s_i)$. We may then easily prove that

$$P(r_{i,n}|s_{i,n}s_n) = \frac{s_i}{s},$$

and

$$P(s_{i,n}|s_n C_{i,n}) = \frac{\binom{C_i}{s_i}\binom{n-C_i}{s-s_i}}{\binom{N}{s}}.$$

(The proof of the first equation uses Theorem 2.4.1 and proof of the second Theorem 2.3.4 in Estes and Suppes (1959b).)

Moreover, if the sample is empty, then

$$P(r_{i,n+1}|e_{2,n}r_{i,n}s_n^o C_{i,n}) = P(r_{i,n}|C_{i,n}),$$

8.4 Representation of Linear Models of Learning

and for the empty sample case the two terms cancel; also, of course, $P(r_{i,n}|s_n^o) = \rho_i$. Combining these various remarks and applying them to (1), we infer:

$$P(r_{i,n+1}|e_{i,n}r_{i,n}C_{i,n}) = \tag{2}$$

$$\rho_i P(s^o) + \left[\sum_{s=s_i}^{N} \sum_{s_i=1}^{C_i} \frac{(C_i + s - s_i)}{N} \cdot \frac{s_i}{s} \cdot \frac{\binom{C_i}{s_i}\binom{N - C_i}{s - s_i}}{\binom{N}{s}} P(s) \right] / P(r_{i,n}|C_{i,n}).$$

Summing out s_i in (2) and using the expression for the second raw moment of the hypergeometric distribution mentioned before the theorem, and the fact that the expected value of the distribution is $\frac{sC_i}{N}$, we have:

$$P(r_{i,n+1}|e_{i,n}r_{i,n}C_{i,n}) = \tag{3}$$

$$\rho_i P(s^o) + \sum_{s=1} \left[\frac{C_i^2}{N^2} + \frac{sC_i}{N^2} - \frac{1}{sN^2}\left(sC_i + \frac{C_i(C_i - 1)s(s - 1)}{N - 1}\right) \right] P(s) / P(r_{i,n}|C_{i,n}).$$

Summing now over s, we obtain:

$$P(r_{i,n+1}|e_{i,n}r_{i,n}C_{i,n}) = \tag{4}$$

$$\rho_i P(s^o) + (1 - P(s^o)) \left[\frac{C_i^2}{N^2} + \frac{\bar{s}C_i}{N^2} - \frac{C_i}{N^2} - \frac{C_i(C_i - 1)(\bar{s} - 1)}{N^2(N - 1)} \right] / P(r_{i,n}|C_{i,n}).$$

It is easy to show we may replace $P(r_{i,n}|C_{i,n})$ by $(1 - P(s^o))\frac{C_i}{N} + \rho_i P(s^o)$. We then obtain the desired result. Q.E.D.

Obviously the theorem has a much simpler form if we assume that the probability of the empty sample is zero.

COROLLARY 1. If $P(s^o) = 0$, then

$$P(r_{i,n+1}|e_{i,n}r_{i,n}N(C_{i,n})) = \frac{(N - \bar{s})}{N - 1}\frac{N(C_i)}{N} + \frac{\bar{s} - 1}{N - 1}.$$

Proof. On the basis of the hypothesis of the Corollary, we have at once from the theorem

$$\begin{aligned}
P(r_{i,n+1}|e_{i,n}r_{i,n}C_i) &= \tfrac{N}{C_i}\left[\tfrac{C_i^2}{N^2} + \tfrac{\bar{s}C_i}{N^2} - \tfrac{C_i}{N^2} - \tfrac{C_i(C_i-1)(\bar{s}-1)}{N^2(N-1)}\right] \\
&= (1 + \tfrac{1-\bar{s}}{N-1})\tfrac{C_i}{N} + \tfrac{\bar{s}-1}{N} + \tfrac{\bar{s}-1}{N(N-1)} \\
&= \tfrac{N-\bar{s}}{N-1}\tfrac{C_i}{N} + (\bar{s} - 1)(\tfrac{1}{N-1}) \quad\text{Q.E.D.}
\end{aligned}$$

The second corollary states a limiting result as $N \to \infty$. The limiting process we use will be discussed later in more detail, but in essence for a fixed experimenter's schedule we choose a sequence of stimulus-sampling models $\mathfrak{S}(N)$ such that for each i, $\lim_{N \to \infty} \frac{N(C_{i,1})}{N}$ exists, $\lim_{N \to \infty} \frac{\bar{s}(N)}{N}$ exists, and the variance $\frac{1}{N^2}\sum_{N(s)=0}^{N}(N(s) - \bar{s})^2 P(s)$ has a limit which is zero. (That such a sequence can be selected is made clear later.) For purposes of making connections with the linear model intuitively clear, we define the learning parameter θ of the linear model as follows:

$$\theta = \lim_{N \to \infty} \frac{\bar{s}(N)}{N}.$$

We then have:

COROLLARY 2.
$$\lim_{N\to\infty} P(r_{i,n+1}|e_{i,n}r_{i,n}N(C_{i,n})) = (1-\theta)\lim_{N\to\infty}\frac{C_i}{N} + \theta.$$

The next four theorems and their corollaries are similar in structure to the first group, so routine details will be omitted in proofs. In the second theorem we need the fact that a certain cross-moment of a multivariate hypergeometric distribution is given by the following expression.

$$\sum_{a=0}^{r}\sum_{b=0}^{B} ab\frac{\binom{A}{a}\binom{B}{b}\binom{N-A-B}{s-a-b}}{\binom{N}{s}} = \frac{ABs(s-1)}{N(N-1)}.$$

THEOREM 2. If $i \neq j$ and $P(e_{i,n}r_{j,n}N(C_{i,n})N(C_{j,n})) > 0$ then

$$P(r_{i,n+1}|e_{i,n}r_{j,n}N(C_{i,n})N(C_{j,n})) = \frac{\rho_j P(s^o)[(1-P(s^o))\frac{N(C_i)}{N} + \rho_i P(s^o)]}{(1-P(s^o))\frac{N(C_j)}{N} + \rho_j P(s^o)}$$

$$+ (1-P(s^o))[\frac{N(C_i)N(C_j)}{N^2} + \frac{\bar{s}N(C_j)}{N^2} - \frac{N(C_i)N(C_j)(\bar{s}-1)}{N^2(N-1)}]$$

$$/ [(1-P(s^o))\frac{N(C_j)}{N} + \rho_j P(s^o)].$$

Proof. To begin with we have:

$$P(r_{i,n+1}|e_{i,n}r_{j,n}C_{i,n}C_{j,n}) \qquad (5)$$

$$= \sum_{s=s_i+s_j}^{N}\sum_{s_i=0}^{C_i}\sum_{s_j=0}^{C_j} P(r_{i,n+1}|e_{i,n}r_{j,n}s_{i,n}s_{j,n}s_n C_{i,n})P(r_{j,n}|s_{j,n}s_n)$$

$$\cdot P(s_{i,n}s_{j,n}|s_n C_{i,n}C_{j,n})P(s_n) / P(r_{j,n}|C_{j,n}).$$

On the right-hand side of (5) we have canceled the conditional probability of $e_{i,n}$ from the numerator and denominator. To obtain $P(s_{i,n}s_{j,n}|s_n C_{i,n}C_{j,n})$, we apply Theorem 2.3.5 of Estes and Suppes (1959b).[12] Thus by the same lines of argument as used for Theorem 1, we infer from (5):

$$P(r_{i,n+1}|e_{i,n}r_{j,n}C_{i,n}C_{j,n}) = \rho_j P(s^o)P(r_{i,n}|C_{i,n}) / P(r_{j,n}|C_{j,n}) \qquad (6)$$

$$+ \left[\sum_{s=s_i+s_j}^{N}\sum_{s_i=0}^{C_i}\sum_{s_j=1}^{C_j} \cdot \frac{\binom{C_i}{s_i}\binom{C_j}{s_j}\binom{N-C_i-C_j}{s-s_i-s_j}}{\binom{N}{s}} P(s)\right] / P(r_{j,n}|C_{j,n}).$$

---

[12]This theorem is a straightforward sampling theorem:

Theorem 2.3.5 If $P(N(s_n)C_n) > 0$ then $P(\cap_{i=1}^{r} N(s_{i,n})|N(s_n)C_n) = \pi_{i=1}^{r}\binom{C_i}{s_i}/\binom{N}{s}$, where r is the number of responses.

8.4 REPRESENTATION OF LINEAR MODELS OF LEARNING

Summing out s_i and s_j in (6) and using the cross-moment of the hypergeometric distribution stated prior to the theorem, we obtain:

$$P(r_{i,n+1}|e_{i,n}r_{j,n}C_{i,n}C_{j,n}) = \rho_j P(s^o) P(r_{i,n}|C_{i,n}) \ / \ P(r_{j,n}|C_{j,n}) \qquad (7)$$
$$+ \sum_{s=1}^{N} [\frac{C_i C_j}{N^2} + \frac{sC_j}{N^2} - \frac{1}{sN}(\frac{C_i C_j s(s-1)}{N(N-1)})] P(s) \ / \ P(r_{j,n}|C_{j,n}).$$

Summing now over s we obtain:

$$P(r_{i,n+1}|e_{i,n}r_{j,n}C_{i,n}C_{j,n}) = \rho_j P(s^o) P(r_{i,n}|C_{i,n}) \ / \ P(r_{j,n}|C_{j,n}) \qquad (8)$$
$$+(1-P(s^o))[\frac{C_i C_j}{N^2} + \frac{\overline{s}C_j}{N^2} - \frac{C_i C_j(\overline{s}-1)}{N^2(N-1)}] \ / \ P(r_{j,n}|C_{j,n}).$$

Using then Theorem 2.4.5 of Estes and Suppes (1959b) to replace $P(r_{i,n}|C_{i,n})$ and $P(r_{j,n}|C_{j,n})$ in (8), we obtain the desired result.[13]   Q.E.D.

We state without proof the two corollaries.

COROLLARY 1. If $P(s^o) = 0$, then

$$P(r_{i,n+1}|e_{i,n}r_{j,n}N(C_{i,n})N(C_{j,n})) = \left(\frac{N-\overline{s}}{N-1}\right)\frac{N(C_i)}{N} + \frac{\overline{s}}{N}.$$

COROLLARY 2.

$$\lim_{N\to\infty} P(r_{i,n+1}|e_{i,n}r_{j,n}N(C_{i,n})N(C_{j,n})) = (1-\theta)\lim_{N\to\infty}\frac{N(C_i)}{N} + \theta.$$

We now consider the two cases which arise when the reinforcement on trial n is different from e_i.

THEOREM 3. If $k \neq i, k \neq 0$ and $P(e_{k,n}r_{i,n}N(C_{i,n})) > 0$, then

$$P(r_{i,n+1}|e_{k,n}r_{i,n}N(C_{i,n}))$$
$$= \rho_i P(s^o) + (1-P(s^o))\left[\frac{N(C_i)^2}{N^2} - \frac{N(C_i)}{N^2} - \frac{N(C_i)(N(C_i)-1)(\overline{s}-1)}{N^2(N-1)}\right]$$
$$\Big/ \left[(1-P(s^o))\frac{N(C_i)}{N} + \rho_i P(s^o)\right].$$

Proof. Proceeding in the same manner as in the proof of Theorem 1, but accommodating in the first term after the summation for the different reinforcement, we obtain:

$$P(r_{i,n+1}|e_{k,n}r_{i,n}N(C_{i,n})) = \rho_i P(s^o) \qquad (9)$$
$$+ \left[\sum_{s=i}^{N}\sum_{s_i=1}^{C_i}\frac{(C_i-s_i)}{N}\cdot\frac{s_i}{s}\frac{\binom{C_i}{s_i}\binom{N-C_i}{S-s_i}}{\binom{N}{s}}P(s)\right] \ / \ P(r_{i,n}|C_{i,n}).$$

---

[13]This theorem expresses a direct conditional response probability given the number of elements conditioned to that response on a given trial:
Theorem 2.4.5 If $P(C_{i,n}) > 0$ then $P(r_{i,n}|C_{i,n}) = (1-P(s_n^o))\frac{N(C_i)}{N} + P(s_n^o)\rho_{i,n}$.

Summing out s_i and then s, we have:

$$P(r_{i,n+1}|e_{k,n}r_{i,n}N(C_{i,n})) = \rho_i P(s^o) \qquad (10)$$
$$+ (1 - P(s^o))\left[\frac{C_i^2}{N^2} - \frac{C_i}{N^2} - \frac{C_i(C_i-1)(\bar{s}-1)}{N^2(N-1)}\right] / P(r_{i,n}|C_{i,n}),$$

and the desired result is immediately obtained from (10) by substituting for $P(r_{i,n}|C_{i,n})$.
Q.E.D.

Again we have two corollaries whose obvious proofs are omitted.

COROLLARY 1. *If* $P(s^o) = 0$ *then*

$$P(r_{i,n+1}|e_{k,n}r_{i,n}N(C_{i,n})) = \left(\frac{N-\bar{s}}{N-1}\right)\frac{N(C_i)}{N} - \frac{N-\bar{s}}{N(N-1)}.$$

COROLLARY 2.

$$\lim_{N\to\infty} P(r_{i,n+1}|e_{k,n}r_{i,n}N(C_{i,n})) = (1-\theta)\lim_{N\to\infty}\frac{N(C_i)}{N}.$$

The next theorem is for different responses on trials n and $n+1$.

THEOREM 4. *If* $k \neq i$, $k \neq 0, j \neq i$ *and* $P(e_{k,n}r_{j,n}N(C_{i,n})N(C_{j,n})) > 0$ *then*

$$P(r_{i,n+1}|e_{k,n}r_{j,n}N(C_{i,n})N(C_{j,n})) = \frac{\rho_j P(s^o)[(1 - P(s^o))\frac{N(C_i)}{N} + \rho_i P(s^o)]}{(1 - P(s^o))\frac{N(C_j)}{N} + \rho_j P(s^o)}$$

$$+(1-P(s^o))\left[\frac{N(C_i)N(C_j)}{N^2} - \frac{N(C_i)N(C_j)(\bar{s}-1)}{N^2(N-1)}\right] / \left[(1-P(s^o))\frac{N(C_j)}{N} + \rho_j P(s^o)\right].$$

Proof. Proceeding as in the case of Theorem 2, but making changes for the different reinforcement, we have, corresponding to (7) of the proof of Theorem 2,

$$P(r_{i,n+1}|e_{i,n}r_{j,n}C_{i,n}C_{j,n}) = \rho_j P(s^o)P(r_{i,n}|C_{i,n}) / P(r_{j,n}|C_{j,n})$$
$$+(1-P(s^o))\left[\frac{C_iC_j}{N^2} - \frac{C_iC_j(\bar{s}-1)}{N^2(N-1)}\right] / P(r_{j,n}|C_{j,n}).$$

And by applying, as before, Theorem 2.4.5 of Estes and Suppes (1959b), the desired result is easily obtained. Q.E.D.

COROLLARY 1. *If* $P(s^o) = 0$ *then*

$$P(r_{i,n+1}|e_{k,n}r_{j,n}N(C_{i,n})N(C_{j,n})) = \left(\frac{N-\bar{s}}{N-1}\right)\frac{N(C_i)}{N}.$$

COROLLARY 2.

$$\lim_{N\to\infty} P(r_{i,n+1}|e_{k,n}r_{j,n}N(C_{i,n})N(C_{j,n})) = (1-\theta)\lim_{N\to\infty}\frac{N(C_i)}{N}.$$

Finally, we have the theorem for nonreinforcement on trial n. The result, it will be noticed, is the same, whether $j = i$ or $j \neq i$.

THEOREM 5. *If* $P(e_{0,n}r_{j,n}N(C_{i,n})N(C_{j,n})) > 0$ *then*

$$P(r_{i,n+1}|e_{0,n}r_{j,n}N(C_{i,n})N(C_{j,n})) = P(r_{i,n}|N(C_{i,n})).$$

8.4 REPRESENTATION OF LINEAR MODELS OF LEARNING

Proof. By virtue of Theorem 2.5.6 of Estes and Suppes (1959b) and the obvious extension to the present case[14]

$$P(r_{i,n+1}|e_{0,n}r_{j,n}N(C_{i,n})N(C_{j,n})) = P(r_{i,n+1}|N(C_{i,n+1}))$$
$$= P(r_{i,n}|N(C_{i,n})). \qquad \text{Q.E.D.}$$

Theorems involving the sequence ω_n. Corresponding to the theorems of the previous subsection, we now have a series involving conditionalization on the sequence ω_{n-1}.[15] For greater simplicity of formulation we shall assume in all of these theorems that $P(s^o) = 0$. We shall not be concerned with corollaries dealing with $N \to \infty$, for we shall want to let N approach infinity at a somewhat different stage.

THEOREM 6. *If $P(s^o) = 0$ and $P(e_{i,n}r_{i,n}\omega_{n-1}) > 0$ then*

$$P(r_{i,n+1}|e_{i,n}r_{i,n}\omega_{n-1}) = \frac{\frac{N-\bar{s}}{N-1}\sum_{N(C_i)=0}^{N} P(r_{i,n}|N(C_{i,n})^2 P(N(C_{i,n})|\omega_{n-1})}{P(r_{i,n}|\omega_{n-1})} + \frac{\bar{s}-1}{N-1}.$$

Proof. We expand in terms of $N(C_i)$. (As in the case of proofs in the preceding section, we write C_i instead of $N(C_i)$, etc.)

$$P(r_{i,n+1}|e_{i,n}r_{i,n}\omega_{n-1}) = \sum_{C_i=0}^{N} P(r_{i,n+1}|e_{i,n}r_{i,n}C_{i,n})P(r_{i,n}|C_{i,n})$$
$$\cdot P(C_{i,n}|\omega_{n-1}) \,/\, P(r_{i,n}|\omega_{n-1})$$
$$= \sum_{C_i}\left[\frac{(N-\bar{s})}{(N-1)}\frac{C_i}{N} + \frac{\bar{s}-1}{N-1}\right]P(r_{i,n}|C_{i,n})P(C_{i,n}|\omega_{n-1})$$
$$/P(r_{i,n}|\omega_{n-1})$$
$$= \left[\frac{N-\bar{s}}{N-1}\sum_{C_i}P(r_{i,n}|C_{i,n})^2 P(C_{i,n}|\omega_{n-1})\right.$$
$$\left. + \frac{\bar{s}-1}{N-1}P(r_{i,n}|\omega_{n-1})\right]/P(r_{i,n}|\omega_{n-1}),$$

and we now obtain the conclusion by cancelling $P(r_{i,n}|\omega_{n-1})$ in the numerator and denominator of the second term on the right. Q.E.D.

Note that in the first line the conditional probability of $e_{i,n}$ cancels in the numerator and denominator on the right because of the relevant reinforcement axioms and the fact that the full sequence ω_{n-1} is given.

Proofs of the next four theorems are similar, with each relying on the corresponding theorem and corollary of the preceding subsection, and so we omit most of them.

---

[14]The theorem referred to is the following:

Theorem 2.5.6 *If $P(e_{o,n}C_{i,n}) > 0$ then $P(C'_{i,n+1}|e_{o,n}C_{i,n}) = \begin{cases} 1 & \text{if } C'_i = C_i \\ 0 & \text{otherwise.} \end{cases}$*

[15]In a more explicit, but cumbersome equivalence-class notation, it would be appropriate to write $[\omega]_n$ rather than ω_n, where then ω_n refers only to trial n. But to avoid this additional notation, I write always ω_n rather than $[\omega]_n$. Note that the formal notation is defined as follows:

$$[\omega]_n = \{\omega' : \omega'_i = \omega_i, \text{ for } 1 \leq i \leq n\}.$$

This definition was already stated informally just prior to Definition 1.

THEOREM 7. If $P(s^o) = 0$, $i \neq j$ and $P(e_{i,n}r_{j,n}\omega_{n-1}) > 0$, then

$$P(r_{i,n+1}|e_{i,n}r_{j,n}\omega_{n-1}) = \frac{N-\bar{s}}{N-1} \sum_{N(C_i)=0}^{N} \sum_{N(C_j)=0}^{N-N(C_i)} P(r_{i,n}|N(C_{i,n}))$$

$$\cdot P(r_{j,n}|N(C_{j,n}))P(N(C_{i,n})N(C_{j,n})|\omega_{n-1})/P(r_{j,n}|\omega_{n-1}) + \frac{\bar{s}}{N}.$$

Proof. We expand in terms of $N(C_{i,n})N(C_{j,n})$ and use Theorem 2.

$$P(r_{i,n+1}|e_{i,n}r_{j,n}\omega_{n-1}) = \sum_{C_i=0}^{N} \sum_{C_j=0}^{N-C_i} P(r_{i,n+1}|e_{i,n}r_{j,n}C_{i,n})P(r_{j,n}|C_{j,n})$$

$$\cdot P(C_{i,n}C_{j,n}|\omega_{n-1})/P(r_{j,n}|\omega_{n-1})$$

$$= \sum_{C_i} \sum_{C_j} \left[\frac{(N-\bar{s})}{(N-1)}\frac{C_i}{N} + \frac{\bar{s}}{N}\right]\frac{C_j}{N} P(C_{i,n}C_{j,n}|\omega_{n-1})$$

$$/ P(r_{j,n}|\omega_{n-1})$$

$$= \frac{N-\bar{s}}{N-1} \sum_{C_i} \sum_{C_j} P(r_{i,n}|C_{i,n})P(r_{j,n}|C_{j,n})$$

$$\cdot P(C_{i,n}C_{j,n}|\omega_{n-1}) / P(r_{j,n}|\omega_{n-1}) + \frac{\bar{s}}{N},$$

because

$$\frac{\bar{s}}{N} \sum_{C_i} \sum_{C_j} \frac{C_j}{N} P(C_{i,n}C_{j,n}|\omega_{n-1}) / P(r_{j,n}|\omega_{n-1}) = \frac{\bar{s}}{N} \sum_{C_j} P(r_{j,n}|C_{j,n})$$

$$\cdot P(C_{j,n}|\omega_{n-1}) / P(r_{j,n}|\omega_{n-1})$$

$$= \frac{\bar{s}}{N} \frac{P(r_{j,n}|\omega_{n-1})}{P(r_{j,n}|\omega_{n-1})} = \frac{\bar{s}}{N}. \qquad \text{Q.E.D.}$$

THEOREM 8. If $P(s^o) = 0, k \neq i, k \neq 0$ and $P(e_{k,n}r_{i,n}\omega_{n-1}) > 0$, then

$$P(r_{i,n+1}|e_{k,n}r_{i,n}\omega_{n-1}) = \frac{\frac{N-\bar{s}}{N-1}\sum_{N(C_i)=0}^{N} P(r_{i,n}|N(C_{i,n}))^2 P(N(C_{i,n})|\omega_{n-1})}{P(r_{i,n}|\omega_{n-1})} + \frac{N-\bar{s}}{N(N-1)}.$$

THEOREM 9. If $P(s^o) = 0, k \neq i, k \neq 0, j \neq i$ and $P(e_{k,n}r_{j,n}\omega_{n-1}) > 0$ then

$$P(r_{i,n+1}|e_{k,n}r_{j,n}\omega_{n-1}) = \frac{N-\bar{s}}{N-1} \sum_{N(C_i)=0}^{N} \sum_{N(C_j)=0}^{N-N(C_i)} P(r_{i,n}|N(C_{i,n}))P(r_{j,n}|N(C_{j,n}))$$

$$\cdot P(N(C_{i,n})N(C_{j,n})|\omega_{n-1})/P(r_{j,n}|\omega_{n-1}).$$

Corresponding to Theorem 5, we now need two theorems depending on whether $j = i$ or $j \neq i$.

THEOREM 10. If $P(s^o) = 0$ and $P(e_{0,n}r_{i,n}\omega_{n-1}) > 0$ then

$$P(r_{i,n+1}|e_{0,n}r_{i,n}\omega_{n-1}) = \sum_{N(C_i)=0}^{N} P(r_{i,n}|N(C_{i,n}))^2 P(N(C_{i,n})|\omega_{n-1})/P(r_{i,n}|\omega_{n-1}).$$

8.4 REPRESENTATION OF LINEAR MODELS OF LEARNING

THEOREM 11. If $P(s^o) = 0, j \neq i$ and $P(e_{0,n}r_{j,n}\omega_{n-1}) > 0$ then

$$P(r_{i,n+1}|e_{0,n}r_{j,n}\omega_{n-1}) = \sum_{N(C_i)=0}^{N} \sum_{N(C_j)=0}^{N-N(C_i)} P(r_{i,n}|N(C_{i,n}))P(r_{j,n}|N(C_{j,n}))$$
$$\cdot P(N(C_i)N(C_j)|\omega_{n-1})/P(r_{j,n}|\omega_{n-1}).$$

As the form of the six theorems just stated suggests, the next step is to prove that as $N \to \infty$

$$\sum_{N(C_i)=0}^{N} P(r_{i,n}|N(C_{i,n}))^2 P(N(C_{i,n})|\omega_{n-1}) = P(r_{i,n}|\omega_{n-1})^2 \qquad (11)$$

and

$$\sum_{N(C_i)=0}^{N} \sum_{N(C_j)=0}^{N-N(C_i)} P(r_{i,n}|N(C_{i,n}))P(r_{j,n}|N(C_{j,n}))P(N(C_{i,n})N(C_{j,n})|\omega_{n-1}) \qquad (12)$$
$$= P(r_{i,n}|\omega_{n-1})P(r_{j,n}|\omega_{n-1})$$

for $i \neq j$. For this purpose it is convenient to define certain variances and covariances.

$$\sigma_{i\omega}^2(N, n) = \sum_{N(C_i)=0}^{N} [P(r_{i,n}|N(C_{i,n})\omega_{n-1}) - P(r_{i,n}|\omega_{n-1})]^2 \qquad (13)$$
$$\cdot P(N(C_{i,n})|\omega_{n-1}),$$

$$\sigma_{ij\omega}(N, n) = \sum_{N(C_i)=0}^{N} \sum_{N(C_j)=0}^{N-N(C_i)} [P(r_{i,n}|N(C_{i,n})\omega_{n-1}) - P(r_{i,n}|\omega_{n-1})] \qquad (14)$$
$$\cdot [P(r_{j,n}|N(C_{j,n})\omega_{n-1}) - P(r_{j,n}|\omega_{n-1})]P(N(C_{i,n})N(C_{j,n})|\omega_{n-1}).$$

It is easy to show that

$$\sigma_{i\omega}^2(N, n) = \sum_{N(C_i)=0}^{N} P(r_{i,n}|N(C_{i,n}))^2 P(N(C_{i,n})|\omega_{n-1}) - P(r_{i,n}|\omega_{n-1})^2, \qquad (15)$$

and

$$\sigma_{ij\omega}(N, n) = \sum_{N(C_i)=0}^{N} \sum_{N(C_j)=0}^{N-N(C_i)} P(r_{i,n}|N(C_{i,n}))P(r_{j,n}|N(C_{j,n})) \qquad (16)$$
$$\cdot P(N(C_{i,n})N(C_{j,n})|\omega_{n-1}) - P(r_{i,n}|\omega_{n-1})P(r_{j,n}|\omega_{n-1}).$$

From (15) and (16) it is clear that if

$$\lim_{N \to \infty} \sigma_{i\omega}^2(N, n) = 0 \qquad (17)$$

and

$$\lim_{N \to \infty} \sigma_{ij\omega}(N, n) = 0, \qquad (18)$$

then (11) and (12) will be established, provided the limits exist of the individual terms on the right-hand sides of (15) and (16). And this existence does follow from our other results. Before we proceed to establish (17) and (18), we review our limit assumptions as $N \to \infty$.

Limit assumptions. For a fixed experimenter's schedule, which since the presentation set is constant is just a fixed reinforcement schedule, and for fixed probabilities of reinforcement $e_{k,n}$ given $r_{i,n}$, we select a sequence $(\mathfrak{S}(1), \mathfrak{S}(2),\ldots, \mathfrak{S}(N),\ldots)$ of stimulus-sampling models with the Nth member of the sequence having N stimulus elements and satisfying the following three conditions:

I. *For each N there are numbers $\gamma_i^{(N)} \geq 0$ with $\Sigma \gamma_i^{(N)} = 1$ such that for any conditioning function $C_i^{(N)}$ on trial 1*

$$P_N(C_i^{(N)}) = 0$$

unless $\frac{N(C_i^{(N)})}{N} = \gamma_i^{(N)}$, *and there are numbers $\gamma_i \geq 0$ with $\Sigma \gamma_i = 1$ such that*

$$\lim_{N \to \infty} \gamma_i^{(N)} = \gamma_i.$$

II. *Let $\bar{s}(N)$ be the expected sample size (on every trial) for $\mathfrak{S}(N)$, then $\lim_{N \to \infty} \frac{\bar{s}(N)}{N}$ exists, and we define θ as this limit.*

III. *As $N \to \infty$ the variance of the distribution of $\frac{N(s)}{N}$ approaches zero, that is,*

$$\lim_{N \to \infty} \frac{1}{N^2} \sum_{N(s)=0}^{N} (N(s) - \bar{s})^2 P_N(s) = 0.$$

Because these three stipulations place restrictions only on the conditioning functions on trial 1 of the models and on the sampling distributions, which are independent of n, it is clear that it is possible to select such a sequence for any numbers $\gamma_i \geq 0$ with $\Sigma \gamma_i = 1$, and θ, with $0 < \theta \leq 1$. The intuitive content of (II) and (III) seems transparent, but (I) is somewhat surprising. To represent a given linear model as the limit of a sequence of stimulus-sampling models, it is necessary to fix the limiting proportion of stimuli connected to each response on trial 1. In making this last statement we have in mind the characterization of linear models given earlier. We shall return to this point later. It should be clear that (I) may be weakened in several inessential ways.

It is necessary to establish (17) and (18) together, for we proceed by induction and for certain cases the argument for $n+1$ for either (17) or (18) depends on being able to assume the other holds for n. The desired representation theorem is an immediate consequence of this theorem.

THEOREM 12. *If $P(\omega_n) > 0$, then*

(i) $$\lim_{N \to \infty} \sigma_{i\omega}^2(N, n) = 0$$

and

(ii) $$\lim_{N \to \infty} \sigma_{ij\omega}(N, n) = 0.$$

Proof. As already mentioned, we proceed by induction on both parts of the theorem simultaneously.

For n = 1 the results are immediate from Selection Rule (I). For example, if $\frac{N(C_{i,1})}{N} = \gamma_i^{(N)}$, then $P(N(C_{i,1})) = 1$ and

$$P(r_{i,1}|N(C_{i,1}))^2 P(N(C_{i,1})) = P(r_{i,1})^2.$$

8.4 Representation of Linear Models of Learning

Assuming now that the theorem holds for n, we must consider several cases. (As usual, in proofs we write C_i rather than $N(C_i)$, etc., in what follows.)

Case 1. $\omega_n = e_{k,n} r_{i,n} \omega_{n-1}$.

For this case, the argument for (i) does not depend on (ii), but this will not be true for Case 2. We first establish (i).

$$\sum_{C_i} P(r_{i,n+1}|C_{i,n+1})^2 P(C_{i,n+1}|\omega_n) \tag{19}$$

$$= \sum_{C_i} \sum_{C_i'} \sum_{s=s_i}^{N} \sum_{s_i=0}^{C_i'} P(r_{i,n+1}|C_{i,n+1})^2 P(C_{i,n+1}|e_{k,n}r_{i,n}s_{i,n}s_n C_{i,n}')$$

$$\cdot P(r_{i,n}|s_{i,n}s_n) P(s_{i,n}|s_n C_{i,n}) P(s_n) P(C_{i,n}'|\omega_{n-1}) \;/\; P(r_{i,n}|\omega_{n-1}).$$

Now from earlier results we know that the conditional probability of $C_{i,n+1}$ is zero unless $C_{i,n+1} = C_i' - s_i$, whence we make this simplification, eliminating the summation over C_i, and substitute familiar expressions for the other terms on the right-hand side of (19) to obtain:

$$\sum_{C_i} P(r_{i,n+1}|C_{i,n+1})^2 P(C_{i,n+1}|\omega_n) \tag{20}$$

$$= \sum_{C_i'} \sum_{s} \sum_{s_i} \left(\frac{C_i' - s_i}{N}\right)^2 \frac{s_i}{s} P(s_{i,n}|s_n C_{i,n}) P(s_n) P(C_{i,n}'|\omega_{n-1})$$

$$/ \; P(r_{i,n}|\omega_{n-1}).$$

As $N \to \infty$, $\frac{s_i}{s} \to \frac{C_i}{N}$ and $\frac{\bar{s}}{N} \to \theta$, and the variance of the sampling distribution goes to zero.[16] Thus in the limit (20) reduces to:

$$\lim_{N \to \infty} \sum_{C_i} P(r_{i,n+1}|C_{i,n+1})^2 P(C_{i,n+1}|\omega_n) \tag{21}$$

$$= \lim \sum_{C_i} (1-\theta)^2 \left(\frac{C_i}{N}\right)^3 \cdot P(C_{i,n}|\omega_{n-1})/P(r_{i,n}|\omega_{n-1})$$

$$= \lim (1-\theta)^2 \sum_{C_i} P(r_{i,n}|C_{i,n})^3 \cdot P(C_{i,n}|\omega_{n-1})/P(r_{i,n}|\omega_{n-1}).$$

We now make use of our inductive hypothesis. We first note that the right-hand side of (21) is expressed in terms of the third raw moment α_3 on trial n of $P(r_{i,n}|C_{i,n})$, which may be expressed in terms of the third central moment μ_3, the second raw moment α_2 and the mean μ by:

$$\alpha_3 = \mu_3 + 3\alpha_2 \mu - 2\mu^3,$$

but by our inductive hypothesis all the central moments are zero, thus $\mu_3 = 0$, and $\alpha_2 = \mu^2$, whence

$$\alpha_3 = \mu^3 = \lim P(r_{i,n}|\omega_{n-1})^3, \tag{22}$$

---

[16] It may be noted that at this point we make crucial use of the assumption that the sampling variance $\frac{1}{N^2} \sum_{N(s)=0}^{N} (N(s) - \bar{s})^2 P(s)$ approaches zero as a limit, for if this limit were other than zero it would be necessary to calculate explicitly in the limit various raw moments of the hypergeometric distribution of s_i and also of the sampling distribution of s.

and combining (21) and (22), we have:

$$\lim \sum_{C_i} P(r_{i,n+1}|C_{i,n+1})^2 P(C_{i,n+1}|\omega_n) = (1-\theta)^2 \lim P(r_{i,n}|\omega_{n-1})^2. \qquad (23)$$

We now apply the fact that by virtue of our inductive hypothesis $\alpha_2 = \mu^2$ to Theorem 8 as $N \to \infty$ and obtain:

$$\lim P(r_{i,n+1}|\omega_n) = (1-\theta) \lim P(r_{i,n}|\omega_{n-1}). \qquad (24)$$

Equation (i) follows for Case 1 immediately from (23) and (24).

We now turn to (ii), and use the same arguments which took us from (19) to (20).

$$\alpha(i,j,\omega,n+1) = \sum_{C_i}\sum_{C_j} P(r_{i,n+1}|C_{i,n+1}) P(r_{j,n+1}|C_{j,n+1}) \qquad (25)$$
$$\cdot P(C_{i,n+1}C_{j,n+1}|\omega_n)$$
$$= \sum_{C_i}\sum_{C_j}\sum_{s}\sum_{s_i}\sum_{s_j} \frac{(C_i - s_i)}{N}\frac{(C_j - s_j)}{N}\frac{s_i}{s} P(s_{i,n}s_{j,n}|s_n C_{i,n} C_{j,n})$$
$$\cdot P(s_n) P(C_{i,n}C_{j,n}|\omega_{n-1}) \,/\, P(r_{i,n}|\omega_{n-1}).$$

Now again as $N \to \infty$, $\frac{s_i}{s} \to \frac{C_i}{N}$, $\frac{s_j}{s} \to \frac{C_j}{N}$ and $\frac{\bar{s}}{N} \to \theta$, and $\frac{1}{N^2}\sum_s (s - \bar{s})^2 P(s) \to 0$, whence in the limit (25) reduces to:

$$\lim \alpha(i,j,\omega,n+1) = (1-\theta)^2 \lim \sum_{C_i}\sum_{C_j} (\frac{C_i}{N})^2 (\frac{C_j}{N}) P(C_{i,n}C_{j,n}|\omega_{n-1}) \qquad (26)$$
$$/\, P(r_{i,n}|\omega_{n-1})$$
$$= (1-\theta)^2 \lim \sum_{C_i}\sum_{C_j} P(r_{i,n}|C_{i,n})^2 P(r_{j,n}|C_{j,n})$$
$$\cdot P(C_{i,n}C_{j,n}|\omega_{n-1}) \,/\, P(r_{i,n}|\omega_{n-1}).$$

We now use our inductive hypothesis to compute the raw cross-moment on trial n, which is denoted by the expression on the right-hand side of (26). Thus, dropping notation for ω and n, let

$$\alpha(i^2, j) = \sum_{C_i}\sum_{C_j} P(r_{i,n}|C_{i,n})^2 P(r_{j,n}|C_{j,n}) P(C_{i,n}C_{j,n}|\omega_{n-1}), \qquad (27)$$

$$\mu(i^2, j) = \sum_{C_i}\sum_{C_j} [P(r_{i,n}|C_{i,n}) - \mu(i)]^2 [P(r_{j,n}|C_{j,n}) - \mu(j)] \qquad (28)$$
$$\cdot P(C_{i,n}C_{j,n}|\omega_{n-1})$$

where $\mu(i)$ is the mean of $P(r_{i,n}|C_{i,n})$, and similarly for $\mu(j)$. And the definition of $\alpha(i,j)$ should be obvious. Now by explicit computation

$$\mu(i^2, j) = \alpha(i^2, j) - \alpha(i^2)\mu(j) - 2\alpha(i,j)\mu(i) + 2\mu(i)^2\mu(j) \qquad (29)$$
$$+ \mu(i)^2\mu(j) - \mu(i)^2\mu(j).$$

It easily follows from our induction that the central cross-moment $\mu(i^2, j) = 0$. Also by our inductive hypothesis $\mu(i^2) = \sigma_i^2 = 0$, whence

$$\alpha(i^2) = \mu(i)^2.$$

8.4 REPRESENTATION OF LINEAR MODELS OF LEARNING 417

And also
$$\mu(i,j) = \sigma_{ij} = \alpha(i,j) - \mu(i)\mu(j),$$
but again by our inductive hypothesis $\sigma_{ij} = 0$, whence
$$\alpha(i,j) = \mu(i)\mu(j). \tag{30}$$
Substituting these results into (29), we conclude
$$\alpha(i^2, j) = \mu(i)^2 \mu(j). \tag{31}$$
Applying (31) to (26) we obtain at once that
$$\lim \alpha(i,j,\omega,n+1) = (1-\theta)^2 \lim P(r_{i,n}|\omega_{n-1})P(r_{j,n}|\omega_{n-1}). \tag{32}$$

Now as in the argument for (i) by our inductive hypothesis $\alpha(i) = \mu(i)^2$ and (24) holds. Moreover, by virtue of (30) and Theorem 4 for $\omega_n = e_{k,n} r_{i,n} \omega_{n-1}$
$$\lim P(r_{j,n+1}|\omega_n) = (1-\theta) P(r_{j,n}|\omega_{n-1}). \tag{33}$$
Equation (ii) follows for Case 1 immediately from (24), (32) and (33).

We shall not consider all the other cases in detail, but we shall sketch the main argument for several more. The interest of Case 2 is that it makes clear why the induction must be simultaneous on both σ_i^2 and σ_{ij}.

Case 2. $\omega_n = e_{k,n} r_{j,n} \omega_{n-1}$.

As before, we first consider (i). By the methods used for Case 1 but accommodated to $j \neq i$, we have:
$$\alpha(i^2, n+1) = \sum_{C_i} P(r_{i,n+1}|C_{i,n+1})^2 P(C_{i,n+1}|\omega_n) \tag{34}$$
$$= \sum_{C_i} \sum_{C_j} \sum_{s} \sum_{s_i} \sum_{s_j} \left(\frac{C_i - s_i}{N}\right)^2 \frac{s_j}{s} P(s_i s_j | s_n C_{i,n} C_{j,n}) P(s_n) P(C_{i,n} C_{j,n}|\omega_{n-1})$$
$$/ \; P(r_{j,n}|\omega_{n-1}).$$

Thus as $N \to \infty$
$$\lim \alpha(i^2, n+1) = (1-\theta)^2 \lim \sum_{C_i} \sum_{C_j} \left(\frac{C_i}{N}\right)^2 \frac{C_j}{N} P(C_{i,n} C_{j,n}|\omega_{n-1}) \tag{35}$$
$$/ \; P(r_{j,n}|\omega_{n-1})$$
$$= (1-\theta)^2 \lim \sum_{C_i} \sum_{C_j} P(r_{i,n}|C_{i,n})^2 P(r_{j,n}|C_{j,n})$$
$$\cdot P(C_{i,n} C_{j,n}|\omega_{n-1}) \; / \; P(r_{j,n}|\omega_{n-1}).$$

The expression on the right of (35) is the same as that on the right of (26) and we apply exactly the same methods, using in the process the inductive hypothesis on σ_{ij}, to obtain:
$$\lim \alpha(i^2, n+1) = (1-\theta)^2 \lim P(r_{i,n}|\omega_{n-2})^2, \tag{36}$$
and equation (i) is easily obtained as before from (36) and Theorem 4.

Turning now to (ii), we have by the usual methods:

$$\alpha(i,j,n+1) = \sum_{C_i}\sum_{C_j}\sum_{s}\sum_{s_i}\sum_{s_j} \frac{(C_i - s_i)}{N} \frac{(C_j - s_j)}{N} \frac{s_j}{s} \qquad (37)$$
$$\cdot P(s_{i,n}s_{j,n}|s_n C_{i,n}C_{j,n})P(s_n)P(C_{i,n}C_{j,n}|\omega_{n-1}) \,/\, P(r_{i,n}|\omega_{n-1})$$

but the right-hand side of (37) is just like the right-hand side of (25) with i and j interchanged, and the argument proceeds in identical fashion.

Case 3. $\omega_n = e_{k,n}r_{j',n}\omega_{n-1}$.

The proof of (i) is the same as for Case 2 with j' replacing j. The situation is different with respect to the proof of (ii), for a new kind of raw cross-moment must be considered, namely $\alpha(i,j,j')$ which, however, may be treated in the same fashion as $\alpha(i^2,j)$.

$$\alpha(i,j,j',n+1) = \sum_{C_i}\sum_{C_j}\sum_{s}\sum_{s_i}\sum_{s_j}\sum_{s_{j'}} \frac{(C_i - s_i)}{N} \frac{(C_j - s_j)}{N} \frac{s_{j'}}{s} \qquad (38)$$
$$\cdot P(s_{i,n}s_{j,n}s_{j',n}|s_n C_{i,n}C_{j,n}C_{j',n})P(s_n)P(C_{i,n}C_{j,n}C_{j',n}|\omega_{n-1})$$
$$/\, P(r_{j',n}|\omega_{n-1}).$$

Thus as $N \to \infty$

$$\lim \alpha(i,j,j',n+1) = (1-\theta)^2 \lim \alpha(i,j,j',n) \,/\, \mu(j',n). \qquad (39)$$

But the central cross-moment (on trial n) is:

$$\mu(i,j,j') = \alpha(i,j,j') - \alpha(i,j)\mu(j') - \alpha(i,j')\mu(j) - \alpha(j,j')\mu(i) \qquad (40)$$
$$- 2\mu(i)\mu(j)\mu(j').$$

By our inductive hypothesis $\mu(i,j,j') = 0$, and $\alpha(i,j) = \mu(i)\mu(j)$, etc., whence

$$\alpha(i,j,j') = \mu(i)\mu(j)\mu(j'), \qquad (41)$$

as would be expected from earlier results. From (39) and (41) we have

$$\lim \alpha(i,j,j',n+1) = (1-\theta)^2 \lim \mu(i,n)\mu(j,n), \qquad (42)$$

and the remainder of the argument moves along standard lines.

The other six cases are proved in a similar way and consequently will be omitted. It should be clear that the argument always hinges on establishing a result like (30), (31) or (41). Q.E.D.

We now have as a consequence of the theorems in this section the following representation theorem for linear models.

THEOREM 13. (Representation Theorem). *Given any single parameter linear model* $(R, E, \Omega, P, \theta)$ *there exists a sequence* $(\mathfrak{S}(1),\ldots,\mathfrak{S}(N),\ldots)$ *of stimulus-sampling models such that for every* ω *and* n

$$P(\omega_n) = \lim_{N \to \infty} P_N(\omega_n),$$

where P_N *is the probability measure of* $\mathfrak{S}(N)$.

Proof. We use selection rules (I), (II), and (III), the preceding theorems of this section, and proceed by induction on n. For $n = 1$, we use selection rule (I). Assuming

8.5 Robotic Machine Learning of Comprehension Grammars for Ten Languages[17]

the theorem holds for n, we then use the fact that in the limit the axioms hold for the sequence of stimulus-sampling models. The remainder of the simple inductive proof is exactly analogous to that of Theorem 4.1 in Estes and Suppes (1959a) on linear models, and will not be repeated here.

As can be seen from the dates of the principal articles referred to in these two sections, the main results were obtained more than thirty years ago. I still like them as examples of representation theorems in mathematical psychology. The proof of Theorem 13 can, with some work, probably be simplified, but even if so, it will remain as testimony to the technical problems of proving exact reduction theorems even for relatively simple scientific theories. A new area of application, with possibly some changes in the theorem and its proof, may be to the reinforcement learning models now being used in experimental economics. At least in the recent past, economists have been more interested than psychologists in representation theorems.

Without going into all the details, I want to convey a rather clear intuitive sense of the process of learning of natural language in terms of the various events that happen when an utterance is given to a robot. (But it should be understood that the basic program of machine learning would apply without serious modification to other applications, for example, machine learning of physics word problems, which I briefly consider later.) Also, following standard learning usage, I shall often speak of trials where, of course, I mean that the trial begins with a command in the form of an utterance to be executed by the robot.

The most important way to describe conceptually the learning process the program embodies is in the description of the state of memory of a robot at the beginning of each trial. There are four aspects of this memory that are changed due to learning. The first is the association relation between words of a given language and internal symbols that have as denotations actions, objects, properties and relations in the robot's world. A central problem is to learn in each language what word is properly associated with a given internal symbol. A second aspect of the memory that changes is the denotational value of a given word, which will affect its probability of being associated. The third part that changes is the short-term memory that holds a given verbal command for the period of the trial on which it is effective. This memory content decays and is not available for access after the trial on which a particular command is given. This means that at the beginning of the trial, before a command is given, this short-term buffer is empty. What we have said thus far, could, with some stretching, fit into classical theories of association, but for language learning it is quite evident that this semantically restricted association relation and some simple features of short-term memory are certainly not enough.

The fourth aspect is the important one of learning grammatical forms. Consider

---

[17] This section is based on Suppes, Böttner and Liang (1996), and Suppes and Liang (1996). Two prior papers describing the approach, first presented at two conferences in 1991, are Suppes, Liang and Böttner (1992) and Suppes, Böttner and Liang (1995).

the verbal command *Get the nut*. This would be an instance of the grammatical form *A the O*, where *A* is the category of actions and *O* is the category of objects. This form actually represents a mild oversimplification, because we do not have just a single category of actions. There are several subcategories, depending upon the number of arguments required, and certain other natural semantical requirements as well. The example will illustrate how things work, however. The grammatical forms are derived by generalization only from actual instances of verbal commands given to the robot. No prior knowledge of any sort of the grammar of the natural language to be learned is available to the robot. Also important is the fact that associated with each grammatical form as it arises from generalization are the associations of the words which have been the basis for the generalization, along with their internal representations. For example, if *Get the nut* were the occurrence in which the grammatical form just stated was generated, then also stored with that grammatical form would be the associations $get \sim \$g$ and $nut \sim \$n$, where $\$g$ and $\$n$ are internal symbols whose denotations are known to the robot. When incorrect associations are deleted by further learning, the grammatical forms based on such associations are also deleted.

Comprehension grammar. Most linguistic analysis is concerned with grammars detailed enough to produce natural utterances of the language being studied. A comprehension grammar, in contrast, as we characterize it here, can generate a superset of utterances. The rules are required only to lead to the correct semantic interpretation of an utterance of the language. Robots, like very young children, can easily have the capacity to understand language before they can produce it. Although it is difficult and subtle to collect accurate and anything like complete data on the comprehension grammar generated by very young children, the evidence is overwhelming that they comprehend much more than they can produce.

Problem of denotation. In the probabilistic theory of machine learning of natural language which we have been developing, we have encountered in a new form a standard problem in the analysis of the semantics of natural language, namely, how to handle words that are nondenoting. We do not mean nondenoting in some absolute sense, but relative to a fixed set of semantic categories. These categories in the robotic case are, roughly speaking, the categories of actions, objects, properties and relations. It may well be that in some elaborate set-theoretical semantics of natural language, nondenoting words like the definite article *the* denote a complicated set-theoretical function, but the relevance of such an elaborate semantics to language learning is doubtful. In the robotic context, we have something simpler and closer to the common man's view of what denotations are. We take, as denoting words, color and object words, common nouns, familiar concrete action words, etc. We take ordinary prepositions in English, and sometimes other devices in other languages, to denote relations in most cases. On the other hand, our computational semantics centered on physics word problems, with an internal equational language of physical quantities, is further removed from common-sense ideas.

When a child learning a first language, or an older person learning a second language,

8.5 LEARNING COMPREHENSION GRAMMARS

first encounters utterances in that new language, there is no uniform way in which nondenoting words are marked. There is some evidence that various prosodic features are used in English and other languages to help the child. For example, in many utterances addressed to very young children, the definite or indefinite article is not stressed but rather the common noun it modifies, as in the expression *Hand me the cup*. But such devices do not seem uniform, and in any case are not naturally available to us in our machine-learning research, where we use written input of words without additional prosodic notation.

As has already been made clear, a central feature of our approach to machine learning is the probabilistic association between words of the natural language being learned and denoting symbols of the internal language. It is appropriate that at the beginning all words are treated equally, and so the associations are formed from sampling based on a uniform distribution. On the other hand, after many words have been learned and a good deal of language has been acquired by the robot, it is very unnatural, and also inefficient, if the robot is now given, for example, the esoteric command *Get the astrolabe*, to have the internal symbol *$ast* be associated with equal probability with the definite article *the* and *astrolabe*—we assume here that the association of *get* is already correctly fixed. After much linguistic experience, there should be very little chance of the robot's associating the definite article *the* with any denoting symbol.

To incorporate such learning many variant models are easily formulated. We have restricted ourselves to one which brings out the most salient distinctions to consider.

Background cognitive and perceptual assumptions. Before explicitly formulating the learning principles of association and denotation used, I first state informally assumptions we make about the cognitive and perceptual capacities of the class of robots, albeit as yet quite limited, we work with.

Internal language. The robot has a fully developed internal language, which it does not learn. It is technically important, but not conceptually fundamental, that in our case this language is Lisp. When we speak here of the internal language we refer only to the language of the internal representation, which is itself a language at a higher level of abstraction, relative to the concrete movements and perceptions of the robot. It is the language of the internal representations held in memory that provides the direct interface to the natural-language learning. In fact, most of the machine learning of a given natural language can take place through simulation of the robot's behavior by using just the language of the internal representation. The first associations learned are between the internal representation in memory of a coerced action and a contiguous verbal utterance in the natural language being learned.

The fundamental importance of this internal representation of a coerced action can be recognized by considering a parallel case of animal learning. When a dog is trained to *Get the paper* or *Get the ball* by being led through the desired action or by some related technique, the residue in memory of what we term the coerced action is surely drastically abstracted from the perceptually rich context of the demonstrated action desired, and it is that abstracted internal representation in memory that must be associated to the verbal stimulus in order for the dog later to perform the desired action upon hearing the verbal command. We are a long way from knowing even the general structure of the dog's internal representation in memory of the action. In this limited sense, life with a robot is much easier, for we ourselves create the form of its internal representation.

Objects, relations and properties. We furthermore assume the robot begins its natural-language learning with all the basic cognitive and perceptual concepts it will have. In other words, our first-language learning experiments are pure language learning. Any learning of new concepts is delayed to another phase. For example, we have assumed that the spatial relations frequently referred to in all, or at least all the languages we consider in detail, are already known to the robot. This is quite contrary to human language learning. For example, probably in no widely used natural language at least, do children at the age of thirty-six months use or fully understand the relations of left and right. To avoid misunderstanding, we emphasize that we consider it an important future task to have the robot also learn the familiar spatial and temporal relations.

Actions. What was just said about objects and relations applies also to actions, represented in English by such verbs as *pick up, get, place*, etc. The English, of course, must be learned, but not the underlying actions.

Associations and grammatical forms. Before stating any formal principles of learning, I describe as informally and intuitively as possible the learning setup used. Consider the English command *Pick up the screw*, no part of which has as yet been learned by the robot. The learning steps may be roughly schematized as follows:

(i) By coercion, or simulation of coercion, the robot creates in memory an internal representation of the coerced action of picking up the screw. For statement of learning principles we show this internal representation, not as a Lisp expression, but just as a schematic function $I(...)$ of the denoting terms in the Lisp expression. Here, by *denoting terms* we mean the names in the internal language of the actions, objects, properties and relations mentioned. The internal representation of *Pick up the screw* is then $I(\$p,\$u,\$s)$, where $\$p$ = the action of picking, $\$u$ = the direction up and $\$s$ = screw.

(ii) By contiguity the robot associates the verbal utterance and the internal representation

$$\text{Pick up the screw} \sim I(\$p,\$u,\$s),$$

where '$\sim$' is the symbol used for association.

(iii) By probabilistic association, the robot associates the internal denotations with the English words, with one possibility the following incorrect result:

$$\text{pick} \sim \$s, \ up \sim \$p, \ screw \sim \$u.$$

We need to observe the following:

(a) We assume from the beginning the robot knows word boundaries, as delineated by the typed input. This is an example of an assumption that is natural

8.5 Learning Comprehension Grammars

for robots, but clearly false for very young children.

(b) For our simple example, there are twenty-four possible ways of associating the three internal symbols to the four denoting words in the English utterance. We initially assign to each of these twenty-four possibilities equal probability, but as trials continue, modify the probability by dynamic changes in denotational values, as is explained later in detail.

(iv) After the associations are made, by the principle of generalization, which we call category generalization, each word is assigned the category of its associated internal symbol. In the present case $pick \in O$—the category of objects, $up \in A$—the category of actions and $screw \in R$—the category of relations. A grammatical form is then also generalized from the verbal command:

$$O\ A\ the\ R$$

which, like the assigned categories is wrong for English, but remember that this is just the starting point of learning. With this grammatical form is associated its internal representation $I(A, R, O)$ which characterizes its meaning.

(v) A new command is presented as the next step, say *Pick up the nut*. By coercion the internal representation $I(\$p, \$u, \$n)$ is created (see (i) above). The robot then first searches its memory to see if any of the words uttered are associated to one of the internal denotations. Here the result is $up \sim \$p$, and also the classification of *the* as a nondenoting word is found. There are then six possibilities of probabilistic association for *pick*, *the* and *nut*. Note that the earlier incorrect association of *pick* with $\$s$ does not appear here, which means that at this stage of learning it will be changed. So, let us suppose the new associations are

$$pick \sim \$u,\ nut \sim \$n.$$

We also have as a new grammatical form

$$R\ A\ the\ O$$

which though incorrect, now has only the confusion of the associations of *pick* and *up* as its source. To correct these associations we must separate the constant pairing of *pick* and *up*, which is what we do. In any case, we form at once the association to the internal representation:

$$R\ A\ the\ O \sim I(A, R, O).$$

(vi) Learning stops whenever the following steps of interpretation can be successfully completed upon giving the robot a verbal command:

(a) An association to an internal denotation or a nondenoting classification is found in memory for each word.

(b) The category of each word is found in memory.

(c) The grammatical form resulting from (b) is found with an associated internal representation in memory.

(d) The command is correctly executed on the basis of the internal representation.

I turn now to the detailed development of these ideas, beginning with the internal language and then the learning axioms.

Internal language. We use Lisp for the internal language of the study reported here. The internal language is stored in memory prior to learning and does not undergo any change during learning.

The set of expressions of the internal language is specified by the grammar in Table 1 with lexical categories A_1, A_2, A_3, A_5 (= action), REL (= relation), PROP (= property), OBJ (= object property) and phrasal categories A (= action), S (= set of objects), O (= object), G (= region), and DIR (= direction). The lexicon of our internal language is

TABLE 1   Grammar of internal language

I	A	$\to (fa_1\ r_1\ O)$
II	A	$\to (fa_2\ r_2\ G)$
III	A	$\to (fa_3\ r_3\ O\ G)$
IV	A	$\to (fa_5\ r_5\ DIR\ O)$
V	A	$\to (fa_5\ r_5\ O)$
VI	G	$\to (fr\ REL\ O)$
VII	DIR	$\to (fd\ REL)$
VIII	O	$\to (io\ S)$
IX	O	$\to (so\ S)$
X	S	$\to (fo\ PROP\ S)$
XI	S	$\to (fo\ OBJ\ *)$

given in Table 2. We refer to the elements of the lexical categories as *internal symbols*. The operations, such as fa_1 and fa_2, read *form action*, all have a straightforward procedural interpretation in a given robotic environment.

The English words used in Table 2 reflect an English lexicon, but the syntax of the internal language is Lisp, not English. Our categories match closely conventional linguistic categories with A corresponding to the category of a (imperative) sentence, A_1

TABLE 2   Lexicon of internal language

Categories							Semantic Operations
OBJ	PROP	REL	A_1	A_2	A_3	A_5	
$screw	$large	$up	$get	$go	$put	$pick	fa_1 (form-action)
$nut	$medium	$on			$place		fa_2
$washer	$small	$into					fa_3
$hole	$square	$above					fa_5
$plate	$hexagonal	$to					fr (form-region)
$sleeve	$round	$behind					$fdir$ (form-direction)
	$black						io (identify-object)
	$red						so (select-object)
	$gray						fo (form-object)

8.5 LEARNING COMPREHENSION GRAMMARS

corresponding to the category of transitive verbs, REL corresponding to the category of prepositions, PROP to the category of adjectives, OBJ corresponding to the category of a common noun, DIR to the category of adverbs, G to the category of prepositional phrases, O to the category of (determined) noun phrases, and S to the category of a nominal group. We chose, however, not to refer to these categories by their usual linguistic labels, since we think of them as semantic categories.

The grammar of the internal language would derive the following Lisp structure for the internal representation of the action corresponding to the English command *Get a screw*, where the asterisk * refers to the set of objects that are present in a certain visual environment:

$$(fa1 \ \$get \ (so \ (fo \ \$screw \ *))). \tag{1}$$

Let $\gamma = (fo \ \$screw \ *)$. Then γ itself is the *minimal* Lisp expression in (1) containing only the internal symbol $\$screw$, and $(so \ (fo \ \$screw \ *))$ is the *maximal* Lisp expression in (1) containing only the internal symbol $\$screw$. We use this distinction later.

General learning axioms. We now turn to our learning axioms, which naturally fall into two groups, those for computations using working memory and those for changes in the state of long-term memory. We use, as is obvious, a distinction about kinds of memory that is standard in psychological studies of human memory, but the details of our machine-learning process are not necessarily faithful to human learning of language, and we make no claim that they are. On the other hand, our basic processes of association, generalization, specification and rule-generation almost certainly have analogues in human learning, some better understood than others at the present time. In the general axioms formulated in this section we assume rather little about the specific language of the internal representation, although the examples that illustrate the axioms use the internal language just described.

Notation. Concerning notation used in the axioms, we generally use Latin letters for sentences or their parts, whatever the natural language, and we use Greek letters to refer to internal representations of sentences or their parts. Turning now to specific notation, the letters $a, b, \ldots$ refer to words in a sentence, and the Greek letters $\alpha, \beta, \ldots$ refer to internal symbols. The symbol s refers to an entire sentence, and correspondingly σ to an entire internal representation. Grammatical forms—either sentential or term forms—are denoted by g or also $g(X)$ to show a category argument of a form; correspondingly the internal representations of a grammatical form are denoted by γ or $\gamma(X)$. We violate our Greek-Latin letter convention in the case of semantic categories or category variables X, X', Y, etc. We use the same category symbols in both grammatical forms and their internal representations.

To insure the proper semantic meaning is carried from a natural-language sentence to its internal representation, or vice versa, we index, as needed, multiple occurrences of the same category in a given sentence and the corresponding occurrences in its internal representation. An example of this indexing is given later.

Computations Using Working Memory.[18]

W1. Probabilistic Association. On any trial, let s be associated to σ, let a be in the set of words of s not associated to any internal symbol of σ, and let α be in the set of internal symbols not currently associated with any word of s. Then pairs (a, α) are sampled, possibly using the current denotational value, and associated, i.e., $a \sim \alpha$.

The probabilistic sampling in the case *Get the screw* could lead to the incorrect associations *get* $\sim$ \$*screw*, *the* $\sim$ \$*get* and no association for *screw*, for there are only two symbols to be associated to in the internal representation.

W2. Form Generalization.[19] If $g(g_i') \sim \gamma(\gamma_i')$, $g_i' \sim \gamma_i'$, and γ' is derivable from X, then $g(X_i) \sim \gamma(X_i)$, where i is the index of occurrence.

From the associations given after Axiom W1 we would derive the incorrect generalization

$$OBJ\ A_1\ screw \sim (fa_1\ A_1\ (io\ (fo\ OBJ\ *))). \tag{2}$$

The correct one is

$$A_1\ the\ OBJ \sim (fa_1\ A_1\ (io\ (fo\ OBJ\ *))). \tag{3}$$

W3. Grammar-Rule Generation. If $g \sim \gamma$ and γ is derivable from X, then $X \to g$.

Corresponding to W3, we now get the incorrect rule

$$A \to OBJ\ A_1\ screw. \tag{4}$$

The correct one is

$$A \to A_1\ the\ OBJ. \tag{5}$$

W4. Form Association. If $g(g') \sim \gamma(\gamma')$ and g' and γ' have the corresponding indexed categories, then $g' \sim \gamma'$.

We get from (2) the incorrect form association

$$OBJ \sim (io\ (fo\ OBJ\ *)). \tag{6}$$

The correct one—to be learned from more trials—is derived from (3)

$$the\ OBJ \sim (io\ (fo\ OBJ\ *)). \tag{7}$$

W5. Form Specification. If $g(X_i) \sim \gamma(X_i)$, $g' \sim \gamma'$, and γ is derivable from X, then $g(g_i') \sim \gamma(\gamma_i')$.

As can be easily seen, Axiom W5 on form specification is essentially the converse of Axiom W2 on form generalization.

W6. Content Deletion. The content of working memory is deleted at the end of each trial.

---

[18] The learning program has a short-term working memory for processing the command it is presented. The memory holds its content for the time period of a single trial. The first group of learning axioms describe the association computations that take place in working memory during the course of a trial.

[19] A distinct principle of generalization generates grammatical forms and their associated semantic forms. The methods used combine those of context-free grammars and model-theoretic semantics for formal languages. The concept of generalization is widely used in psychological theories of learning. The kind of generalization used here is restricted to the generation of grammatical forms and grammatical rules from concrete utterances. For example, the phrase *the nut* generalizes to the grammatical form *the O*, where O is the category of objects.

8.5 LEARNING COMPREHENSION GRAMMARS

We now turn to the axioms governing long-term memory.

Changes in State of Long-term Memory[20]

L1. Denotational Value Computation. *If at the end of trial n a word a in the presented verbal stimulus is associated with some internal symbol α, then $d(a)$, the denotational value of a, increases and if a is not so associated $d(a)$ decreases. Moreover, if a word a does not occur on a trial, then $d(a)$ stays the same unless the association of a to an internal symbol α is broken on the trial, in which case $d(a)$ decreases.*

Because this axiom is conceptually less familiar, we give a more detailed example later.

L2. Form Factorization. *If $g \sim \gamma$ and g' is a substring of g that is already in long-term memory and g' and γ' are derivable from X, then g and γ are reduced to $g(X)$ and $\gamma(X)$. Also $g(X) \sim \gamma(X)$ is stored in long-term memory, as is the corresponding grammatical rule generated by Axiom W4.*

We illustrate this axiom by an example, which is really simple. It seems complex, because the premises take three lines, and we have two conclusions, an association and the corresponding grammatical rule. Let

$$
\begin{array}{rlcl}
g \sim \gamma: & A_1 \ the \ OBJ & \sim & (fa_1 \ A_1 \ (io \ (fo \ OBJ \ *))) \\
g' \sim \gamma': & the \ OBJ & \sim & (io \ (fo \ OBJ \ *)) \\
\hline
X: & O & \to & the \ OBJ
\end{array}
$$

Then

$$
\begin{array}{rcl}
A_1 \ O & \sim & (fa_1 \ A_1 \ O) \\
A & \to & A_1 \ O.
\end{array}
$$

L3. Form Filtering. *Associations and grammatical rules are removed from long-term memory at any time if they can be generated.*

In the previous example, $g \sim \gamma$ can now be removed from long-term memory, and so can $A \to A_1 \ the \ OBJ$, learned as an example of Axiom W3.

L4. Congruence Computation.[21] *If w is a substring of g, w' is a substring of g' and they are such that*

(a) *$g \sim \gamma$ and $g' \sim \gamma$,*

(b) *g' differs from g only in the occurrence of w' in place of w,*

(c) *w and w' contain no words of high denotational value,*

then $w' \approx w$ and the congruence is stored in long-term memory.

---

[20] This memory can also change from trial to trial, but it stores associations and grammatical forms which remain unchanged when they are correct for the application being considered. The way the state of long-term memory changes from trial to trial is described in this second part of the learning axioms.

[21] By using a concept of semantic congruence restricted to nondenoting words, we simplify the grammars and at the same time permit direct comparisons across languages. The intuitive idea of such congruence is simple. Two strings of a natural language are congruent when they have identical representations in the internal language.

Various strong and weak concepts of congruence can be used to get varying degrees of closeness of meaning. The idea is not to be caught in the search for a single concept of synonymy, just as in modern geometry we are not caught in a single concept of congruence. We have in affine geometry, for example, a weaker sense of congruence than in Euclidean geometry but it is also easy to get a commonsense notion of congruence that is stronger than the Euclidean one, namely congruence that requires sameness of orientation.

Using Axiom L4, reduction of the number of grammatical rules for a given natural language is further achieved by using congruence of meaning (Suppes 1973b, 1991). Consider the following associations of grammatical forms:

$$die\ Schraube \sim (io\ (fo\ \$screw\ *)) \tag{8}$$

$$der\ Schraube \sim (io\ (fo\ \$screw\ *)). \tag{9}$$

Association (8) and (9) differ only with respect to the article. The article in (8) is in the nominative and accusative case, the article in (9) is in the genitive and dative case. What is important here is that there is no difference in the respective internal representations. We therefore call (8) *congruent* with (9) and collect the differing elements into a congruence class $[DA] = \{die, der\}$ where DA = definite article. This allows us to reduce the two grammatical forms (8) and (9) into one:

$$[DA]\ Schraube \sim (io\ (fo\ \$screw\ *)). \tag{10}$$

Notice that reduction by virtue of congruence is risky in the following way. We may lose information about the language to be learned. For instance, collapsing the gender distinction exhibited by the difference between (8) and (9) will make us incapable of distinguishing between the following sentences:

$$Steck\ die\ Schraube\ in\ das\ Loch. \tag{11}$$

$$Steck\ die\ Schraube\ in\ die\ Loch. \tag{12}$$

Whereas (11) is grammatical, (12) is not. As long as our focus is on comprehension grammar, a command like (12) will probably not occur, but for purposes of production, congruence in the present form should not be used.

> L5. *Formation of Memory Trace.* The first time a form generalization, grammatical rule or congruence is formed, the word associations on which the generalization, grammatical rule or congruence is based are stored with it in long-term memory.

Using our original example after W3, the incorrect associations would be stored in long-term memory, but with more learning, later deleted (L6 (a)).

> L6. *Deletion of Associations.*[22]
>
> (a) *When a word in a sentence is given a new association, any prior association of that word is deleted from long-term memory.*
>
> (b) *If $a \sim \alpha$ at the beginning of a trial, a appears in the utterance s given on that trial but α does not appear in the internal representation σ of s, then the association $a \sim \alpha$ is deleted from long-term memory.*
>
> (c) *If no internal representation is generated from the occurrence of a sentence s, σ is then given as the correct internal representation, and there are several words in s associated to an internal symbol α of σ such that the number of*

---

[22] For wider applications, this axiom is too strong. In many familiar uses of natural language, a variety of associations are significant. Here we drastically simplify the associations to strict semantic ones. This highly restrictive simplification is in contrast to the general use of associations for many kinds of computations in the earlier sections of this chapter. The only justification of the restricted simplification is to make, at this early stage of development, the work in machine learning easier.

8.5 LEARNING COMPREHENSION GRAMMARS

occurrences of these words is greater than the number of occurrences of α in σ, then these associations are deleted.

L7. *Deletion of Form Association or Grammatical Rule. If $a \sim \alpha$ is deleted, then any form generalization, grammatical rule or congruence for which $a \sim \alpha$ is a memory trace is also deleted from long-term memory.*

Comments on Axioms. Of the thirteen axioms, only three need to be more specific to the study reported here. These three are (W1) Probabilistic association, (W4) Form association, and (L1) Denotational value computations, which are given a more specific technical formulation in the next subsection. Axiom W2, especially, is given a much more detailed formulation. Examples of these more specific axioms are also given.

Specialization of certain axioms and initial conditions.

Probabilistic Association. (Axiom W1′) On any trial n, let s be associated to σ, let A be the set of words of s not associated to any internal symbol of σ, let $d_n(a)$ be the current denotational value of each such a in A and let $\mathbb{A}$ be the set of internal symbols not currently associated with any word of s. Then

1. *An element α is uniformly sampled without replacement from $\mathbb{A}$.*

2. *At the same time an element a is sampled without replacement from A with the sampling probability*
$$p(a) = \frac{d_n(a)}{\sum_A d_n(a)}.$$

3. *The sampled pairs are associated, i.e., $a \sim \alpha$.*

4. *Sampling continues until either the set A or the set $\mathbb{A}$ of internal symbols is empty.*

Due to the probabilistic nature of this procedure (Axiom W1), there are several possible outcomes. Consider, for example, *Get the screw*, which has the internal representation $(fa_1\ \$get\ (io\ (fo\ \$screw\ *)))$. The sampling process might generate any one of six different possible pairs, like, for instance, $get \sim \$screw$ and $screw \sim \$get$. Since there are three words occurring in the verbal command, there are in principle six ways to associate the three words of the command to the two symbols of the internal expression.

Form Association. (Axiom W4′) Let $g \sim \gamma$ at any step of an association computation on any trial.

1. *If X occurs in g and $(fo\ X\ *)$ occurs in γ, then $X \sim (fo\ X\ *)$.*

2. *If (i) wX is a substring of g with $g \sim \gamma$ such that $w = a$, which is a word with low denotational value, or if X is preceded by a variable, or is the first symbol of g, $w = \varepsilon$, the empty symbol, and (ii) $\gamma'(X)$ is the maximal Lisp form of γ containing the occurrence of X and no other occurrence of categories, then*
$$wX \sim \gamma'(X).$$

3. If (i) $X_1 w_1 \cdots w_{m-1} X_m$ is a substring of g, where the $X_i, i = 1, \ldots, m$ are not necessarily distinct category names and w_i are substrings, possibly empty, or words that have no association to internal symbols on the given trial, and (ii) $\gamma'(X_{\pi(1)}, \ldots, X_{\pi(m)})$ is the minimal Lisp form of γ containing $X_{\pi(1)}, \ldots, X_{\pi(m)}$, then

$$X_1 w_1 \cdots w_{m-1} X_m \sim \tau(X_{\pi(1)}, \ldots, X_{\pi(m)}),$$

where π is a permutation of the numbers $1, \ldots, m$.

To show how Axiom $W4'$ works, assume we have arrived at the following association of grammatical forms:

$$A_1 \text{ the } PROP \text{ } OBJ \sim (fa_1 \text{ } A_1 \text{ } (io \text{ } (fo \text{ } PROP \text{ } (fo \text{ } OBJ \text{ }*)))) \tag{13}$$

which could be obtained as a generalization, for instance, from the command *Get the red screw* with the words correctly associated. To analyze this example, we make reference to Table 1.

From Axiom $W4'.1$, we may infer

$$OBJ \sim (fo \text{ } OBJ \text{ }*). \tag{14}$$

From Axiom $W4'.2$, we infer

$$PROP \text{ } OBJ \sim (fo \text{ } PROP \text{ } (fo \text{ } OBJ \text{ }*)). \tag{15}$$

From Axiom $W4'.3$, we infer

$$\text{the } PROP \text{ } OBJ \sim (io \text{ } (fo \text{ } PROP \text{ } (fo \text{ } OBJ \text{ }*))). \tag{16}$$

Using grammar-rule generation (Axiom W3), and the grammar of the internal language (Table 1), we infer from (14) and Rule *XI* of Table 1

$$S \to OBJ. \tag{17}$$

From (15), Rule *X* of Table 1 and form generalization (Axiom W2)

$$PROP \text{ } S \sim (fo \text{ } PROP \text{ } S), \tag{18}$$

and finally from grammar-rule generation (Axiom W3)

$$S \to PROP \text{ } S \tag{19}$$

as a rule of English grammar. We also derive from (16), (15) and the internal grammar using Axiom W2

$$\text{the } S \sim (io \text{ } S) \tag{20}$$

and then again by grammar-rule generation

$$O \to \text{the } S \tag{21}$$

as a rule for our English grammar.

Before the introduction of the axioms I promised to give an example of indexing of categories to preserve meaning. Such indexing can be avoided for the restricted corpora here, but is needed for more general purposes. Here is an example from the corpus showing how it works. Consider the sentence

Put the nut on the screw.

8.5 LEARNING COMPREHENSION GRAMMARS

The correct grammatical form and associated internal representation would, with indexing, look like this

A_3 the $OBJ1$ REL the $OBJ2 \sim (fa_3\ A_3\ (io\ (fo\ OBJ1\ *))(fr\ REL\ (io\ (fo\ OBJ2\ *))))$

where postscript numerals are used for indexing OBJ.

Denotational Value Computations. (*Axiom L1'*) *If at the end of trial n a word a in the presented verbal stimulus is associated with some internal symbol α of the internal representation σ of s, then*[23]

$$d_{n+1}(a) = (1-\theta)d_n(a) + \theta,$$

and if a is not associated with some denoting internal symbol α of the internal representation

$$d_{n+1}(a) = (1-\theta)d_n(a).$$

Moreover, if a word a does not occur on trial n, then

$$d_{n+1}(a) = d_n(a),$$

unless the association of a to an internal symbol α is broken on trial n, in which case

$$d_{n+1}(a) = (1-\theta)d_n(a).$$

To show how the computation of denotational value (Axiom L1) works, let us assume the associations given are $get \sim \$screw$, $the \sim \$get$. Let us further assume that at the end of this trial

$$\begin{aligned} d(get) &= 0.900 \\ d(screw) &= 0.950 \\ d(the) &= 0.700. \end{aligned}$$

On the next trial the verbal command is

Get the nut.

As a result, we end this trial with

$$get \sim \$get,\ nut \sim \$nut$$

and with the association of *the* deleted (Axiom L6 (a)). Using $\theta = 0.03$, as we usually do, we now have

$$d(get) = 0.903$$
$$d(the) = 0.679.$$

After, let us say, three more occurrences of *the* without any association being formed the denotational value would be further reduced to 0.620. If the command *Get the sleeve* is given and *sleeve* has not previously occurred and $get \sim \$get$, then we may see how *sleeve* has a higher probability of being associated to $\$sleeve$ than *the*. For under the hypotheses given, the sampling probabilities for *the* and *sleeve* would be:

$$p(the) = \frac{d(the)}{d(the) + d(sleeve)} = \frac{0.620}{0.620 + 1} = 0.383$$

---

[23] The learning model used here is just the linear learning model introduced at the beginning of the preceding section (4).

and
$$p(sleeve) = \frac{d(the)}{d(the) + d(sleeve)} = \frac{1}{1.620} = 0.617.$$

The dynamical computation of denotation value continues after initial learning even when no mistakes are being made. As a consequence high-frequency words like *a* and *the* in English and *ba* in Chinese have their denotational values approach zero rather quickly, as can be seen from the learning curves in Figures 1 and 2 given later. (From a formal point, it is useful to define a word as *nondenoting* if its asymptotic denotational value is zero, or, more realistically, below a certain threshold.)

This particular linear learning model with two parameters, $d_1(a)$, and θ, could easily be replaced by more elaborate alternatives.

Initial conditions. At the beginning of trial 1, the association relation $\sim$, the congruence relation $\approx$ and the set of grammatical rules is empty. Moreover, the initial denotational value $d_1(a)$ is the same for all words a.

The Corpora. In order to test our system we applied it to ten corpora, each of a different language. These languages were: English, Dutch, German, French, Spanish, Catalan, Russian, Chinese, Korean, Japanese.[24] The size of our corpora varied from 400 to 440 sentences. The corpora in the 10 languages cover an almost identical set of internal structures. They could not be made completely identical for the following reason: an internal language word that was translated by one word in one language, say $\mathcal{L}$, might require two or more words (depending on context) in another language $\mathcal{L}'$. As a consequence, $\mathcal{L}'$ might not be learnable from the corpus of 400. To arrive at a complete learning of all the words, we therefore either added sentences as, e.g., in Spanish, or removed sentences, as in Japanese.

The most important variation requirement on the various corpora was that two words of a given language, which were intended to correspond to two internal symbols, must not always co-occur if the correct meaning were to be learned. For example, if *nimm* and *Schraube* only occurred in the single command *Nimm die Schraube!* there would be no assurance that the intended associations *Schraube* $\sim$ $screw and *nimm* $\sim$ $get would ever be learned, no matter how many learning trials there were.

We also note that in the Japanese case we deleted all the sentences translating $above as an internal symbol, because in Japanese *above* and *on* are expressed by the same word *ue*. So this means that we deleted ten sentences from the basic corpus in the case of Japanese.

Despite careful instruction of our translators we are not sure whether the translations sound natural in all cases and would really get used in a robotic working environment. To check this would however go much beyond what can be done by standard translation methods and requires field studies of language use in a working environment. In cases of a

---

[24]This ordering reflects qualitative judgments about the nearness of the languages, as may be seen explicitly in Table 5 below showing the comprehension-grammar rules for each language. For example, Chinese, Korean and Japanese are grouped together, because of the placement of an imperative verb of action in final position.

8.5 LEARNING COMPREHENSION GRAMMARS

lexical gap we used very simple devices: e.g., French has no single idiomatic word for the nonspatial meaning of the English adjective *medium*, so we used the circumlocutionary phrase *de taille moyenne*. In some cases we avoided technical language where a single word consists of two morphemes and expresses two internal denoting symbols. This occurs, as might be expected, often in German. For example, *Rundschraube* expresses the idea of *round screw* that is the property *round* and the object *screw*.

In the case of Catalan, we set in advance the denotational value for a few words as different from initial value of one. The reason for this is that there were so many nondenoting words in our sense, and these nondenoting words uniformly co-occurred with certain action verbs, so that within our limited corpus the intuitively correct denotational value could be learned only with a probability less than one. So we set the initial denotational value for the words *d* and *de* at 0.05.

In order to obtain a successful learning performance on a corpus, it is often useful to make some modifications in the corpus. In the final tests of a theory such changes are, of course, undesirable. The changes we did make are described below.

The introduction of *pick up* created many problems because of the special nature of this verb and relation in English. In many languages the notion of picking something up is expressed by a single action verb with no additional preposition required. The problems created by this particular action, as expressed in various languages, led to an important problem. Individual words in the corpus of a given natural language sometimes denoted more than one internal symbol. Our solution, which was artificial, but the only really artificial solution we had to adopt, was to split such words into two parts:

French: *ramasse* into *ra* and *masse*
Spanish: *recoge* into *rec* and *oge*
Catalan: *recull* into *re* and *cull*
Russian: *podnimi* into *pod* and *nimi*
Korean: *cip&ela* into *cip* and *&ela*
Japanese: *toriagero* into *tori* and *agero*.

This absence of isomorphism of the semantic categories of words across languages is not surprising, as has been observed in Bowerman (1996) and Choi and Bowerman (1991). What is identified as the same action and different relations in one language may be identified as different actions and the same relation in another language.

Empirical results. In summarizing our results in this subsection, we first present some learning curves, followed by the congruence classes of nondenoting words in our sense. We then use these abstract classes to simplify and facilitate the summary table of semantically based grammatical rules of comprehension generated from the ten languages.

Lexicon. In Table 3 we specify for each natural language the number of words learned. We count as different words different inflected forms of what according to dictionary usage is the same word. As might be expected, Russian has the largest number of internal symbols with more than one association, just because of its rich inflectional patterns.

TABLE 3   Comprehension Vocabulary for Ten Comparable Corpora

	words	nondenoting words	symbols with more than 1 assoc.
English	28	2	0
Dutch	38	3	2
German	49	10	9
French	42	7	5
Spanish	40	7	6
Catalan	42	8	8
Russian	75	0	16
Chinese	33	8	0
Korean	31	4	1
Japanese	31	4	1

Learning. A natural question is whether or not the order of presentation of the sentences in the corpora was fixed in advance to facilitate learning. The answer is that the order was not fixed. For each language the sentences presented were chosen randomly without replacement from the corpus of approximately 400 sentences.

Compared to most neural-net rates of learning, the learning was rapid. In each of the ten languages one cycle through the entire corpus was sufficient to produce a comprehension grammar that was intuitively correct. In contrast, and somewhat paradoxically, the standard mean learning curves, theoretically based on random sampling of sentences are, even for a corpus of only 400 sentences, computationally not feasible, as we proved in Suppes, Liang and Böttner (1992). In this same earlier paper, based on some very large runs—in fact, mean learning curves computed from up to 10,000 sample paths—, we conjectured that the mean learning rate for the kind of corpora we have studied is polynomially bound. The learning of the ten languages studied quite clearly supports the conjecture for our theoretical framework of learning.

In Figures 1-3 we show mean learning curves for denotational value of words for English, Chinese and German. The averaging in this case is over the total number of denoting or nondenoting words in a given corpus. The number of nondenoting words in the three languages are 2, 8 and 10, respectively, as also shown in Table 3. As would be expected, the rate of learning the denotational value of nondenoting words is for the three languages inversely proportional to their number, an argument from pure frequency of occurrence in the corpora. This is not the whole story, as can be seen by comparing the Chinese and German mean curves, even though the number of nondenoting words is very close.

In Figure 4, we show the corresponding mean curves for 60 physics word problems in English, with again the only two nondenoting words being *the* and *a* (Suppes and Liang 1996). The internal language is completely different from the robotic case, but we will not attempt to describe it here, except to say that it is essentially a pure equational language for physical quantities. We emphasize, however, that the learning axioms were the same for the robotic commands and the physics word problems.

8.5 LEARNING COMPREHENSION GRAMMARS

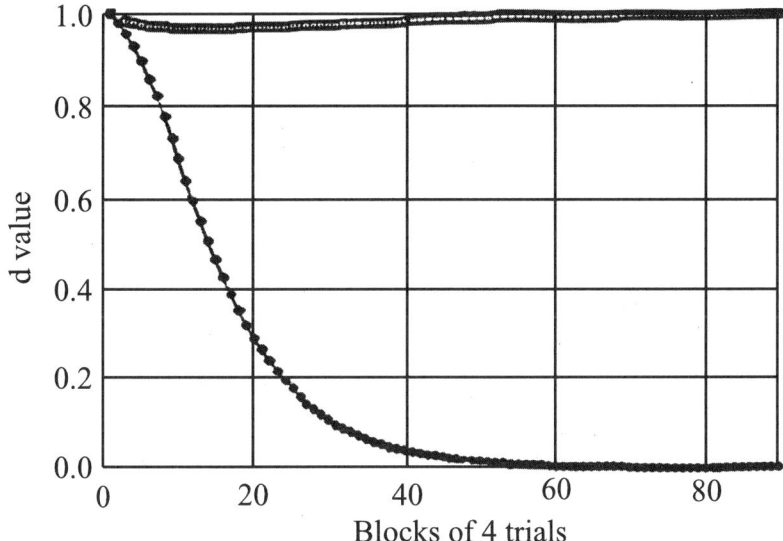

FIGURE 1 Mean denotational learning curves for English. The upper curve is for denoting words, with an asymptote of 1, and the lower curve is for nondenoting words with an asymptote of 0.

Congruence classes. In general, in order to compare the rules being used by different languages, we consolidated across languages as much as possible. The most important extension of congruence classes across languages was to introduce the empty word ϵ, so that when no nondenoting word appeared in a particular place, a given language could still be included in the group of languages using that rule. This has as a consequence that the grammatical rules were differentiated only by the order of occurrence of the semantic categories in the rule and not by the appearance of nondenoting words. In other words, two rules that have exactly the same semantic categories appearing in exactly the same order, independent of the appearance of nondenoting words, are treated as the same rule in Table 5 below. This kind of congruence reduction is desirable in order to get a real semantic comparison of the comprehension grammars generated for different languages. The occurrence of ϵ in Table 4 indicates that the language used the same grammatical rule as another language, but with no nondenoting word occurring.

A consequence of what has just been said above is that a particular grammatical rule with congruence notation may permit an instantiation in a given language of a grammatical form not instantiated in the corpus itself. For the purposes of comprehension, as opposed to production, this does not lead to difficulties.

There are a number of descriptive remarks to be made about Table 4. Congruence class γ_1 is made up of various forms of definite articles, including in the case of Russian and Japanese, the null word ϵ, because these two languages do not standardly use a definite article. The same remark applies to class γ_2 for indefinite articles. In introducing classes γ_1 and γ_2 across languages we are admittedly introducing a coarse-grained but useful clustering of nondenoting words that function rather similarly in different

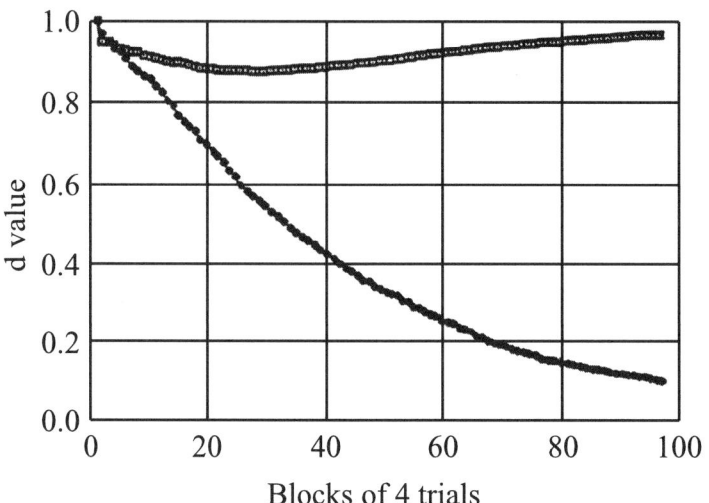

FIGURE 2   Mean denotational learning curves for Chinese.

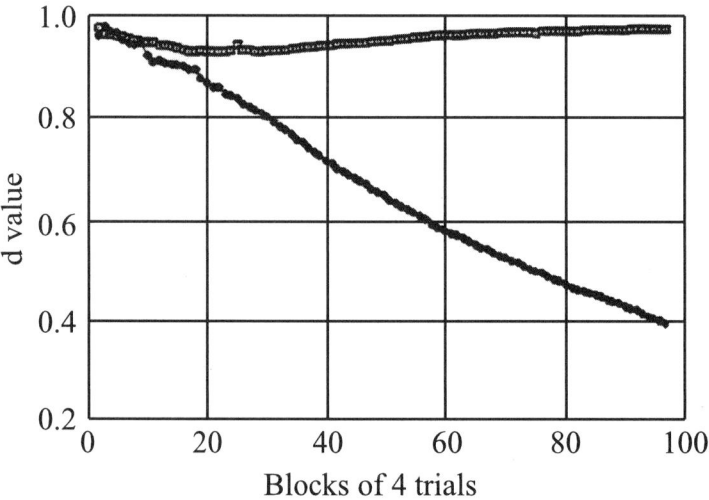

FIGURE 3   Mean denotational learning curves for German.

8.5 Learning Comprehension Grammars

TABLE 4 Congruence Classes

	E	D	G	F	S	C	R	Ch	K	J
γ_1	the	het de	die das der dem den	le l la	el l la	el l la	ϵ	nage zhege	ku ϵ	ϵ ϵ
γ_2	a	een	ein eine einem einen einer	un une	un una	ϵ	ϵ	yige	han ϵ	—
γ_3	—	—	—	de taille ϵ	ϵ	ϵ	—	—	—	—
γ_4	ϵ	ϵ	ϵ	ϵ	—	—	ϵ	ϵ	ϵ	no ϵ
γ_5	—	—	—	—	—	—	—	ϵ	ϵ	no tokoro no
γ_6	ϵ	ϵ	ϵ	de ϵ	de ϵ	d de ϵ	ϵ	ϵ	—	—
γ_7	—	—	—	—	—	—	—	—	—	o
γ_8	—	—	—	—	—	—	—	ba	ϵ	ϵ
γ_9	—	—	—	—	—	—	—	—	ul lul	o
γ_{10}	—	—	—	—	—	—	—	—	ϵ	ni
γ_{11}	—	—	—	—	—	—	—	zai ϵ	—	—
γ_{12}	—	—	—	—	—	—	—	chao ϵ	ϵ	ϵ
γ_{13}	—	—	—	—	—	—	—	na4 na4li ϵ	ϵ	ni
γ_{14}	ϵ	ϵ	ϵ	ϵ	junto ϵ	a l ϵ	ϵ	—	—	—

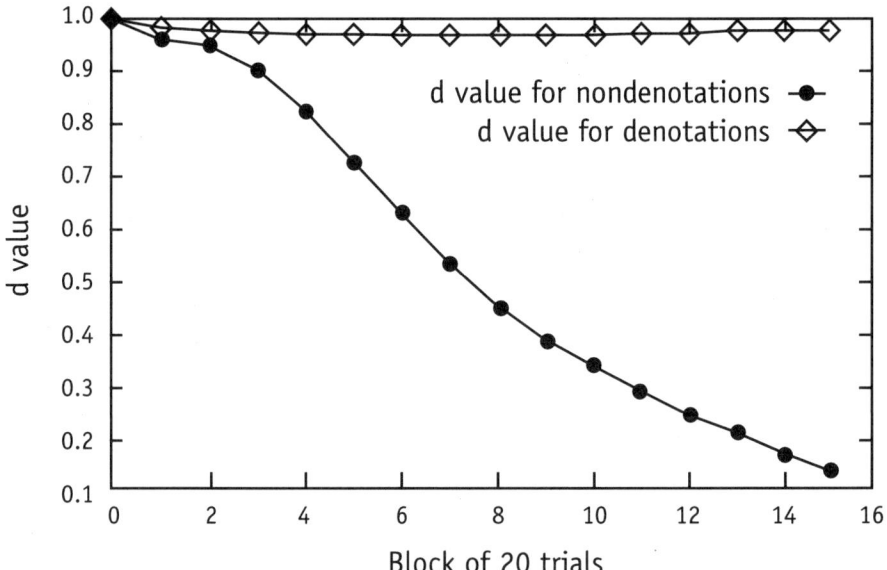

FIGURE 4   Mean denotational learning curves for 60 physics word problems in English.

languages. More refinements of congruence of meaning and use could lead to divisions of γ_1 and also γ_2 into several smaller classes.

The class γ_3 is special for French because several words are required to express the idea of *medium* as in *medium screw*, namely the phrase *de taille moyenne*. We recognize the resulting analysis is a distortion of the natural semantics of the French. The distortion arises from the English lexical bias, but not grammatical bias, built into our fixed set of denoting words in the internal language.

Given that Chinese is not an inflected language, as German is, the number of nondenoting words in Table 4 is large. We cannot discuss in detail all of these particles. The particle *ba* occurs before noun phrases describing direct objects of actions. The particle *zai* occurs before noun phrases describing objects of relations. We note that the grammar for Chinese given in Table 5 has the following two rules:

$$A \rightarrow [\gamma_8] \; O \; A_3 \; [\gamma_{11}] \; G$$

and

$$G \rightarrow REL \; [\gamma_6] \; O$$

But to put *zai* before a relation word is not correct Chinese. This superset causes no problem for comprehension, and the correct instances are when ϵ is used rather than *zai*. Remarks very similar to those for *zai* apply to *chao*. The difference in their use reflects a semantic distinction we do not need for comprehension but would need for production. The particle *zai* is used mainly in connection with relations of position such as *on* and *under*. In contrast *chao* is used in connection with the direction of motion.

Grammatical rules. An overview of the resulting grammars for the ten languages is given in Table 5. The table has eleven columns. In the first column we list the grammat-

8.5 Learning Comprehension Grammars

TABLE 5 Comprehension Grammars for Ten Comparable Corpora

		E	D	G	F	S	C	R	Ch	K	J
I	$A \to A_1 + O$	+	+	+	+	+	+	+			
	$A \to [\gamma_8] + O + [\gamma_9] + A_1$								+	+	+
II	$A \to A_2 + [\gamma_{14}] + G$	+	+	+	+	+	+	+			
	$A \to [\gamma_{12}] + G + [\gamma_{13}] + A_2$								+	+	+
III	$A \to A_3 + O + G$	+	+	+	+	+	+	+			
	$A \to [\gamma_8] + O + A_3 + [\gamma_{11}] + G$								+		
	$A \to O + [\gamma_9] + G + [\gamma_{10}] + A_3$									+	+
IV	$A \to A_5 + DIR + O$	+									
	$A \to A_5 + O + DIR$		+								
	$A \to DIR + O + A_5$			+							
	$A \to [\gamma_8] + O + [\gamma_9] + A_5 + DIR$								+	+	
	$A \to DIR + A_5 + O$				+	+	+	+			
	$A \to O + [\gamma_7] + DIR + A_5$										+
V	$A \to A_5 + O$	+	+	+	+			+	+		
	$A \to O + [\gamma_7] + A_5$										+
VI	$G \to REL + [\gamma_6] + O$	+	+	+	+	+	+	+	+		
	$G \to O + [\gamma_5] + REL$								+	+	+
VII	$DIR \to REL$	+	+	+	+	+	+	+	+	+	+
VIII	$O \to [\gamma_2] + S$	+	+	+	+	+	+	+	+	+	+
	$O \to [\gamma_7] + S$										+
IX	$O \to [\gamma_1] + S$	+	+	+	+	+	+	+	+	+	+
	$O \to [\gamma_7] + S$										+
X	$S \to PROP + [\gamma_4] + S$	+	+	+	+			+	+	+	+
	$S \to S + [\gamma_3] + PROP$					+	+	+			
XI	$S \to OBJ$	+	+	+	+	+	+	+	+	+	+

ical rules in the order of internal-language grammatical rules, cf. Table 1. In the next columns we list the languages in the following order: English, Dutch, German, French, Spanish, Catalan, Russian, Chinese, Korean, Japanese. To each internal grammatical rule corresponds in general a set of language specific rules. In the case of Rules *VII* and *XI* the set is a unit set for the obvious reason that no variation can arise in our limited internal language. Clearly in a more general setting, in the case of Rule *VII*, e.g., relations could have modifying properties.

The most important contrast between the Indo-European and Asian languages is that in the Indo-European languages the imperative verb expressing action is usually at the beginning of an utterance, but in the Asian languages it is usually in final position. See, for example, the two rules derived from Rule *II*. The first is for the seven Indo-European languages and the second for the three Asian languages. Similar remarks hold for the three rules derived from Rule *III*.

This well-known contrast between the two groups of languages leads to a more systematic question about the rules given in Table 5. Does the set of rules corresponding to each rule of the internal language exhaust the possible permutations of the order of the semantic categories? Surprisingly the answer is affirmative except for the set generated by Rule *III*, which has only three members rather than six.

However, reflecting only on the three languages controlled as native speakers by my two collaborators and me, we can give simple examples within the vocabulary and conceptual framework of our various corpora exhibiting two of the missing permutations:

	A_3		G			O	
E	Put	near	the	washer	a	screw.	
G	Leg	neben	die	Scheibe	eine	Schraube.	

	G			A_3		O	
E	Near	the	washer	put	a	screw.	
G	Neben	die	Scheibe	leg	eine	Schraube.	
Ch	Zai	nage	dianquan	fujin fang	yige	luosiding.	

For the third missing permutation, our Korean informant gave us the following example:

	G			O		A_3	
K	Ku	nasapati	yephey	(han)	nasa-lul	nohala.	
	The	washer	near	(one)	screw	put.	

The close correspondence between Tables 1 and 5 is, as a matter of principle, misleading. Although the grammatical rules of Table 5 are derived via Axiom W3 directly from the primitive grammatical rules of the internal language, as given in Table 1, this need not be the case. The larger corpus that is a superset of the one studied here for English, Chinese and German has examples requiring derived internal grammatical rules in applying Axiom W3. A German example of this phenomenon, taken from Suppes, Böttner and Liang (1995), is

$$A \to A_4 \ [einen] \ S \ D \ [der] \ PROP \ ist.$$

An instance of this rule is:

Heb eine Mutter hoch, die nicht groß ist.

Related work and unsolved problems. By far the most extensive research on language learning has been on children's learning their first language. Rather extensive theoretical treatments are to be found in Wexler and Culicover (1980) and in Pinker (1984, 1989). Some of this work is related to ours, but not in a central way in terms of the details of theoretical ideas. For example, Wexler and Culicover assume the learner already has a deep context-free grammar of the language in question, and learning is centered on the learning of transformations. In contrast, we begin with no grammar of the language to be learned. We make no detailed claims of the relevance of our work to children's learning, but connections undoubtedly exist at a certain level.

In the past decade or so there has been a relatively small number of articles or books on machine learning of natural language. Langley and Carbonell (1987) provide an excellent overview as of the date of their publication. Within the framework of this work, ours is more semantically than syntactically driven. This semantic commitment is also shared by the work of Feldman et al. (1990, 1996) and Siskind (1992, 1994), which is also the work closest to our own. Feldman et al. (1990) describe in direct and simple terms their original idea. First, the learning system is presented pairs of pictures and true natural language statements about the pictures. Second, the system is to learn the language well enough to determine whether or not a new sentence is true of the accompanying picture. Feldman et al. (1996)'s approach to language learning separates the learning of the grammar from the learning of the lexical concepts. The grammar is learned by use of Bayesian inference over a set of possible grammars and model merging. Siskind's original work (1992), his dissertation, was in the context of naive physics, but focused also on the algorithms children may use in learning language. This work is continued in Siskind (1994), but with any assumption of prior language knowledge eliminated. The concentration is on lexical acquisition via possible internal representations of meaning. Although Siskind (1994) concentrates entirely on lexical meaning, his 7-step procedure, which constitutes a learning algorithm, bears resemblance at the top level but not in detail to our procedure. Siskind (1991) has a concept that is certainly different but similar in certain respects to our concept of denotational value equal to 0 in the limit for nondenoting words. His ideas are, however, not probabilistic, and he does not present any learning curves. He does offer in his concept of temperature (1994) a treatment of homonymy, which we do not.

In spite of our considering ten languages, the present test of our theory must be regarded as very preliminary in character. The theory needs to be extended to solve a variety of pressing unsolved problems. We restrict ourselves to four problems, but it is a mark of the still primitive character of theoretical developments in this area, ours and others, that any informed reader can quickly double or triple this list. In our view, large-scale experimentation is premature for the kind of theory we are developing until more conceptual problems are solved:

- *Learning of anaphora.* Even very restricted and sometimes rather artificial uses of natural language are usually saturated with uses of anaphora. Physics word problems are a good example.
- *Learning temporal features.* The ordinary use of language marks temporal sequences of events in a rich variety of ways. Much systematic discourse in science and technology

requires continual time and tense distinctions that must be learned.
- *Learning multiple meanings.* Troublesome examples already exist in robotic use, e.g., *screw* as both a noun and a verb in English. Much more exotic are the many meanings of *washer* in English: ring of metal (our robotic case), a machine for washing, a raccoon, and so forth.
- *Learning concepts as well as words.* It can well be claimed that concepts should be learned, not just words that stand for them. At present our theory offers nothing in this respect, although we have begun some systematic work involving multivariate network models (Suppes and Liang 1998).

Final remarks on learning. The importance of the general idea of association or conditioning, the mechanism at the heart of learning in Sections 3-5, is undiminished. Now in the context of current neuroscience a main challenge is to find its physical embodiment. For this purpose many classical conditioning experiments with animals from aplysia up the evolutionary tree are essential for identifying when and where conditioning has occurred. Philosophers caught up in the rhetoric about behaviorism often do not realize that in developed theories of conditioning as axiomatized in Sections 3 and 5 the concept of association, conditioning state or even that of stimulus has a mainly theoretical status, not observable directly and only possible to estimate indirectly from behavioral response data.

On the other hand, it seems clear that highly specific mechanisms are used for different kinds of learning. It is almost certainly the case, for example, that a new word must be learned by a rather large population of neurons and never simply by a single neuron. This learning, it is also likely, takes the form of learning frequencies of given amplitudes and phases to represent words in the cortex. For other brain activity different representations can also be envisaged for learning by single neurons, as has now been demonstrated in a number of important cases, but not for human language, the subject to which I now turn.

8.6 Language and the Brain.

Some historical background. Aristotle said that the distinguishing feature of man as an animal is that he is a rational animal, but, in more biological and psychological terms, it is that of being a talking animal. Language is, in ways that we have not yet fully explored, the most distinguishing mark of man as an animal. Its processing is centered, above all, in the brain, not just for the production of speech, but for the intentional formation of what is to be said, or for the comprehension of what has been heard or read. So it is the brain's processing of language that is the focus of this section. I begin with a historical sketch of the discovery of electrical activity in the brain.

An early reference to electricity being generated by muscles or nerves of animals comes from a study by Francesco Redi (1671), who describes in this way an experiment he conducted in 1666: "It appeared to me as if the painful action of the *torpedine* (electric ray) was located in these two sickle-shaped bodies, or muscles, more than in any other part." Redi's work was done in Florence under the Medici's. These electrical observations were fragmentary and undeveloped. But the idea of electrical activity in the

8.6 LANGUAGE AND THE BRAIN.

muscles or nerves of various animals became current throughout the eighteenth century (Whittaker 1910). Yet it was more than 100 years after Redi before the decisive step was taken in Bologna by Luigi Galvani. He describes his first steps in the following manner:

> The course of the work has progressed in the following way. I dissected a frog and prepared it... Having in mind other things, I placed the frog on the same table as an electric machine. When one of my assistants by chance lightly applied the point of a scalpel to the inner crural nerves of the frog, suddenly all the muscles of the limbs were seen so to contract that they appeared to have fallen into violent tonic convulsions. Another assistant who was present when we were performing electrical experiments thought he observed that this phenomenon occurred when a spark was discharged from the conductor of the electrical machine. Marvelling at this, he immediately brought the unusual phenomenon to my attention when I was completely engrossed and contemplating other things. Hereupon I became extremely enthusiastic and eager to repeat the experiment so as to clarify the obscure phenomenon and make it known. I myself, therefore, applied the point of the scalpel first to one then to the other crural nerve, while at the same time some one of the assistants produced a spark; the phenomenon repeated itself in precisely the same manner as before. (Galvani 1791/1953, pp. 45–46)

Galvani's work of 1791 was vigorously criticized by the well-known Italian physicist Alessandro Volta (1745–1827), who was born in Como and was a professor of physics at the University of Pavia. Here are his words of criticism, excerpted from a letter by Volta to Tiberius Cavallo, read at the Royal Society of London:

> The name of animal electricity is by no means proper, in the sense intended by Galvani, and by others; namely, that the electric fluid becomes unbalanced in the animal organs, and by their own proper force, by some particular action of the vital powers. No, this is a mere artificial electricity, induced by an external cause, that is, excited originally in a manner hitherto unknown, by the connexion of metals with any kind of wet substance. And the animal organs, the nerves and the muscles, are merely passive, though easily thrown into action whenever, by being in the circuit of the electric current, produced in the manner already mentioned, they are attacked and stimulated by it, particularly the nerves. (Volta 1793/1918, pp. 203–208)

Galvani was able to meet these criticisms directly and in 1794 published anonymously a response containing the detailed account of an experiment on muscular contraction without the use of metals (Galvani 1794). The original and important nature of Galvani's work came to be recognized throughout Europe. The prominent German physicist Emil Du Bois-Reymond summarized in the following way Galvani's contribution:

1. Animals have an electricity peculiar to themselves, which is called Animal Electricity.
2. The organs to which this animal electricity has the greatest affinity, and in which it is distributed, are the nerves, and the most important organ of its secretion is the brain.
3. The inner substance of the nerve is specialized for conducting electricity, while the outer oily layer prevents its dispersal, and permits its accumulation.
4. The receivers of the animal electricity are the muscles, and they are like a Leyden jar, negative on the outside and positive on the inside.

5. The mechanism of motion consists in the discharge of the muscular fluid from the inside of the muscle via the nerve to the outside, and this discharge of the muscular Leyden jar furnishes an electrical stimulus to the irritable muscle fibres, which therefore contract. (Du Bois-Reymond 1848/1936, p. 159)

A next event of importance was the demonstration by Carlo Matteucci (1844) that electrical currents originate in muscle tissue. It was, however, about 100 years after Galvani, that Richard Caton (1875) of Liverpool detected electrical activity in an exposed rabbit brain, using the Thomson (Lord Kelvin) reflecting telegraphic galvanometer. In 1890, Adolf Beck of Poland detected regular electrical patterns in the cerebral cortex of dogs and rabbits. Beginning at the end of the nineteenth century Villem Einthoven, a Dutch physician and physiologist, developed a new electrocardiograph machine, based on his previous invention of what is called the string galvanometer, which was similar to the device developed to measure telegraphic signals coming across transatlantic cables. Using Einthoven's string galvanometer, significant because of its sensitivity, in 1914, Napoleon Cybulsky and S. Jelenska Macieszyna, of the University of Cracow in Poland, recorded a dog's epileptic seizures. Beginning about 1910, Hans Berger in Jena, Germany began an extensive series of studies that detected electrical activity through intact skulls. This had the great significance of being applicable to humans. His observations were published in 1929, but little recognized. Recognition came, however, when his findings were confirmed by Edward Douglas Adrian and B. H. C. Matthews of the University of Cambridge, who demonstrated Berger's findings at the Physiological Society in Cambridge in 1934, and the International Congress of Psychology in 1937. In the late 1930s and the early 1940s, research on electrical activity in brains, or what we now call electroencephalography (EEG), moved primarily to North America—W. G. Lennox and Erna and F. A. Gibbs at the Harvard Medical School, H. H. Jasper and Donald Linsley at Brown University, and Wilder Penfield at McGill University. One of the first English-language reports to verify Berger's work was by Jasper and Carmichael (1935). Nearly at the same time, Gibbs et al. (1935) began using the first ink-writing telegraphic recorder for EEG in the United States, built by Garceau and Davis (1935).[25] By the 1950s, EEG was widely used clinically, especially for the study of epilepsy, and for a variety of research on the nature of the electrical activity in the brain. This is not the place to summarize in any serious detail the work by a wide variety of scientists from 1950 to the present, but an excellent review of EEG, that is, of electrical activity, pertinent especially to cognition, is to be found in Rugg and Coles (1995).

Observing the brain's activity. The four main current methods of observing the brain are easy to describe. The first is the classical electroencephalographic (EEG) observations already mentioned, which,—and this is important—, have a time resolution of at least one millisecond. The second is the modern observation of the magnetic field rather than the electric field, which goes under the title of magnetoencephalography (MEG). This also has the same time resolution of approximately one millisecond. The third is positron emission tomography (PET), which has been widely used in the last several decades and is good for observing location, in some cases, of brain activity, but

---

[25] I obtained these last references from Geddes (2001).

8.6 Language and the Brain.

has a time resolution of only one second. Finally, the most popular current method is functional magnetic resonance imaging (fMRI), which does an excellent job of observing absorption of energy in well-localized places in the brain, but unfortunately, also has a time resolution of no less than a second.

Although many excellent things can be learned from PET and fMRI, they are not really useful if one wants to identify brain waves representing words or sentences, for the processing, although slow by modern computer standards, is much too fast to be able to accomplish anything with the time resolution of observation no better than one second. The typical word, for example, whether listened to or read, will be processed in not more than 4 or 5 hundred milliseconds, and often faster. My own serious interest, focused on the way the brain processes language, began from the stimulus I received by hearing a brilliant lecture in 1996 on MEG, by Sam Williamson, a physicist who has been prominent from the beginning in the development of MEG. I was skeptical about what he said, but the more I thought about it, the more I realized it would be interesting and important to try using MEG to recognize the processing of individual words. This idea suggested a program of brain-wave recognition, as recorded by MEG, similar in spirit to speech recognition. I was familiar with the long history of speech recognition from the 1940s to the present, and I thought maybe the same intense analytical effort could yield something like corresponding results. So, in 1996, assisted especially by Zhong-Lin Lu, who had just taken a Ph.D. with Sam Williamson and Lloyd Kaufman at New York University, we conducted an MEG experiment at the Scripps Institute of Research in San Diego, California. When we proceeded to analyze the results of the first experiment, the problem of recognizing which one of seven words was being processed on the basis of either having heard the word or having read it on a computer screen, we were not able to get very good recognition results from the MEG recordings. Fortunately, it was a practice at the Scripps MEG facility, which is technically very much more expensive and complicated to run than standard EEG equipment, to also record the standard 20 EEG sensors used for many years. We proceeded to analyze the EEG data as well, and here we had much better recognition results (Suppes, Lu and Han 1997).

In the standard EEG system, widely used throughout the world for observing electrical activity in the brain, sensors to record the electrical activity are arranged in what are commonly called the 10-20 system, as shown in Figure 5, with the location on the surface of the skull of the head shown in the approximate form of a circle, with ears left and right and eyes at the top in the figure. The first letters of the initials used in the locations correspond to references to the location in the part of the brain, for example, F for frontal, C for center, T for temporal, P for parietal, O for occipital. Second, you will note that the odd-numbered sensors are located on the skull on top of the left hemisphere and the even-numbered sensors are located over the right hemisphere, with three sensors located approximately along the center line. There are more opinions than deeply articulated and established facts about what takes place in the left hemisphere or in the right hemisphere, possibly both, in the processing of language. My own view is that there is probably more duality than has been generally recognized, but I will not try to make an empirical defense of that view in the present context, although I have published data supporting duality (Suppes, Han and Lu 1998, Table 2).

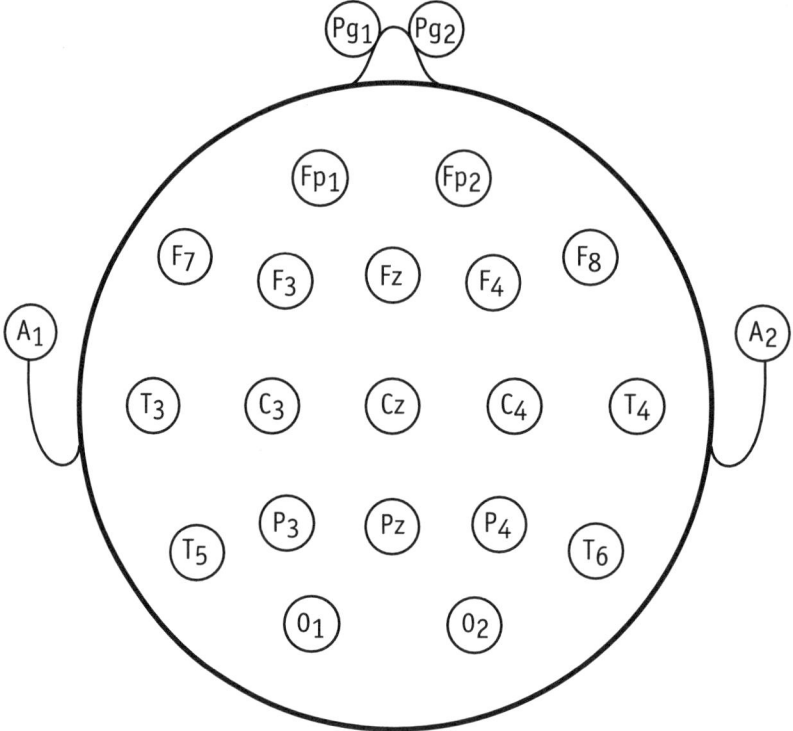

FIGURE 5   The 10-20 system of EEG sensors.

Figure 6 shows a typical trial, in which the subject was given a visual, i.e., printed, sentence, one word at a time, on a computer screen. The trial lasted for over 3000 milliseconds, with an observation of an amplitude of the observed wave plotted on the y-coordinate in microvolts every millisecond. Given that so much data are observed in a little over three seconds, just from one sensor, out of 20, it is easy to see that EEG recordings of language activity are rich in data and, in fact, we might almost say, swamped by data. It is not difficult to run an experiment with several subjects, each with a number of hours of recording, and have at the end between five and ten gigabytes of data of the kind to be seen in Figure 6. This means that the problem of trying to find waves corresponding to particular words and sentences is not going to be a simple matter. It is very different from behavioral experiments dealing with language, in which the observed responses of subjects are easily recorded in a few megabytes and then analyzed without anything like the same amount of computation.

Methods of data analysis. There is no royal road to finding words and sentences in the kind of data just described, so I will outline here the approach that I and my colleagues have used with some success in the past few years. The basic approach is very much taken from digital signal processing, but the application is a very different one from what is ordinarily the focus of electrical engineers or others using the extensive

8.6 LANGUAGE AND THE BRAIN.

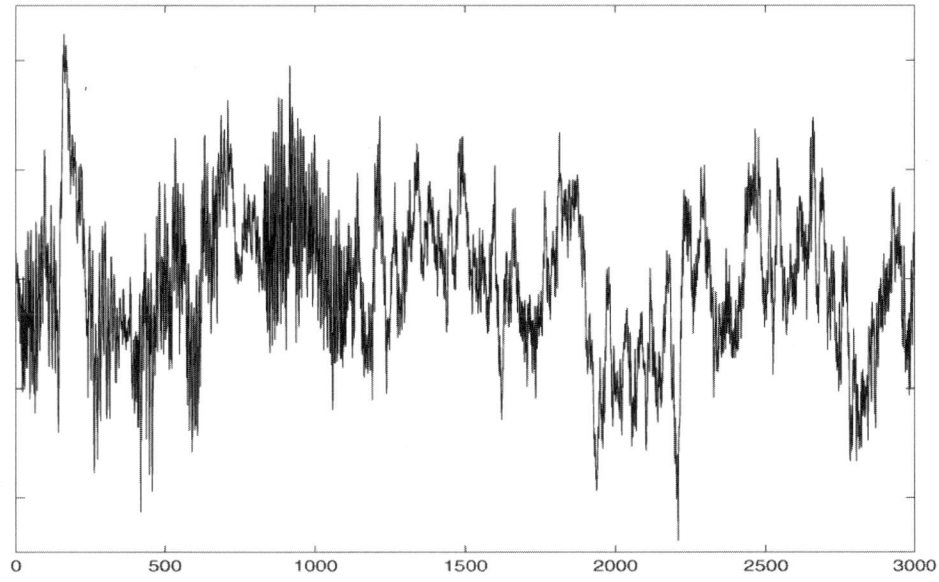

FIGURE 6   Unanalyzed EEG data from one trial and one sensor.

mathematical, quantitative and statistical techniques that have been developed in the last 50 years in digital signal processing. An excellent general reference is Oppenheim and Schafer (1975).

The approach is easily sketched, although the technical details are more complicated and will only be outlined. Generally, but with important exceptions to be noted, the first thing to do is to average the data, for example, for a given condition, a typical case being the brain response to a particular verbal stimulus, given either auditorily or visually. The purpose of this averaging is to eliminate noise, especially high-frequency noise, on the assumption that the signal is of much lower frequency. The next step is then to perform a Fourier transform on the averaged data, passing from the time domain to the frequency domain, perhaps the most characteristic feature of signal processing analysis. The third step, which can be done simultaneously with the second step, is to filter in the frequency domain, to reduce still further the bandwidth of the signal used for identifying the word or sentence being processed. An alternative, which we have explored rather thoroughly in some work, is to select in the frequency domain the frequencies with the highest energies, as measured by the absolute value of their amplitudes, and then to superpose the sine functions in the time domain.[26] In either case, by filtering or superposition, we get a much simpler signal as we pass by an inverse Fourier transform back to the time domain.

Speaking now of filtering, and ignoring for the moment the superposition approach, we go back to the time domain with a bandpass filter fixed by two parameters, the low frequency L and the high frequency H of the filter. We now select two more parameters

---

[26]Actually, in 'selecting a frequency ω_i with amplitude A_i', we are in fact selecting a sine wave $A_i \sin(\omega_i t + \varphi_i)$, where φ_i is its phase. More on this on the next page.

that are of importance. Namely, what should we take to be the beginning s and the ending e of the signals for the words or sentences in a given experimental condition. As in other brain matters, it is not at a glance obvious from the recorded brain wave when the representation of a particular word or sentence begins or ends in the continually active electrical waves that are observed. When a quadruple (L, H, s, e) is selected, we then use that quadruple of parameters to make a classification of the brain waves of the words or sentences that are the stimuli in a given experimental condition.

Our aim is to optimize or maximize the number of brain waves correctly classified. We keep varying the quadruple of parameters until we get what seems to be the best result that can be found. We are doing this in a four-dimensional grid of parameters, but I show in Figure 7 the optimal surface for two parameters of the filter, although here what is used as a measure on the ordinate or y-axis is the difference between the high-frequency and low-frequency filter rather than the high filter itself. So we have on the abscissa the low filter measured in Hz and on the ordinate the difference W, which is the width in Hz of the bandpass filter. The smoothness of the surface shown in Figure 7 is characteristic of what we always observe, and would be expected of observations, of an electrical field outside the skull, the place where we are observing them. (We are of course very fortunate that the electrical fields are strong enough to be observed without complete contamination by noise.) This isocontour map is for the first experiment with 48 geography sentences discussed in more detail below, but the map shows clearly that the best recognition rate was 43 of the 48 sentences (approximately 90%). As the parameters of the bandpass filter are changed, the contour map shows definitely lower recognition rates.

Fourier analysis of EEG data. In the background of the Fourier analysis is the standard theory of the Fourier integral, but in practice our data are finite. The finite-impulse data that we are interested in observing usually last for no more than a few seconds. For example, the sentences studied will ordinarily last not more than three or four seconds when spoken at a natural rate. To analyze frequencies with given amplitudes and phases, we use the discrete Fourier transform. As indicated, our goal is to find the frequencies that contain the signal and eliminate the noise. (The artifacts generated by eye blinks or other such events are discussed at the end of this section.)

Let N be the number of observations, equally spaced in time, usually one millisecond apart. We then represent the finite sequence of observations $x(n), 0 \leq n \leq N-1$ by Fourier series coefficients $\tilde{X}(k)$ as a periodic sequence of period N, so we have the dual pair

$$\tilde{X}(k) = \sum_{n=0}^{N-1} \tilde{x}(n) e^{-i(\frac{2\pi}{N})kn} \qquad (1)$$

$$\tilde{x}(n) = \frac{1}{N} \sum_{k=0}^{N-1} \tilde{X}(k) e^{i(\frac{2\pi}{N})kn}. \qquad (2)$$

8.6 LANGUAGE AND THE BRAIN.

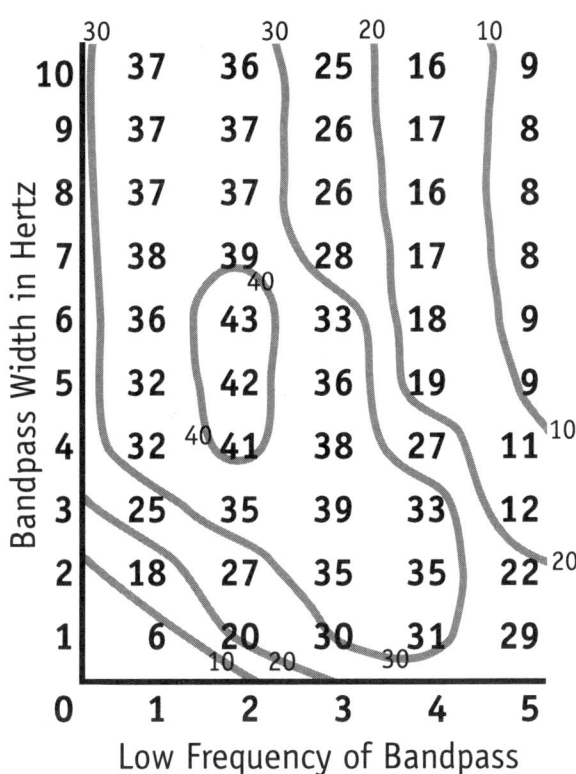

FIGURE 7 Typical contour map of recognition rate surface for bandpass-filter parameters L and W.

I first note the following:

1. The periodic sines and cosines are represented by the standard exponential terms.
2. $\tilde{x}(n) = x(n) = \tilde{x}(n + kN)$, the tilde shows periodicity of length N and is for duality of time and frequency.
3. The kn part of the exponent gives us distinct exponentials, and thus sine and cosine terms for integer submultiples of the period N. This way we get in the representation frequencies that are an integer multiple of $\frac{2\pi}{N}$.
4. Using the periodicity N gets us duality between the time and frequency domains.

The properties of the two equations (1) and (2) are:

1. Linearity: If $\tilde{x}_1(n)$ and $\tilde{x}_2(n)$ have period N, so does
$$\tilde{x}_3(n) = a\tilde{x}_1(n) + b\tilde{x}_2(n)$$

and
$$\tilde{X}_3(k) = a\tilde{X}_1(k) + b\tilde{X}_2(k).$$

2. Invariance under shift of a sequence
$$n \to n + m.$$

3. Various symmetry properties, e.g., $|\tilde{X}(k)| = |\tilde{X}(-k)|$.

4. Convolution of $\tilde{x}_1$ and $\tilde{x}_2$ of period N has period N:
$$\tilde{x}_3(n) = \sum_{m=0}^{N-1} \tilde{x}_1(m)\tilde{x}_2(n-m).$$

Of importance is the efficient fast discrete Fourier transform, an algorithm due to Cooley and Tukey (1965) and others, a variant of which was used in the computations reported below.

Filters. The principle of filter construction is simple. Details are not. A bandpass filter, e.g., 1-20 Hz simply "filters all the frequencies below 1 Hz and above 20 Hz". There are many developments in the electrical engineering literature on the theory and art of designing filters, which it is not possible to survey here. The important point is always to design a filter with some criterion of optimality.

If the signal is known, then the engineering objective is to optimize its transmission. Our problem, as already mentioned, is that in our experiments, the signal carrying the word or sentence in question is unknown. So our solution is to optimize the filter to predict the correct classification. The parameters we used have been discussed above. In addition we often make a smoothing correction around the edges of the filter by using a 4th-order Butterworth filter, although in the work reported here, something simpler would serve the purpose just about as well.[27]

Three experimental results. I turn now to three of the most important results we have obtained so far.

Invariance between subjects. In the first experiment, we presented 48 sentences about the geography of Europe to 9 subjects. The subjects were asked to judge the truth or falsity of the sentences, and while they were either listening to or reading the sentences displayed one word at a time on a computer screen, we made the typical EEG recordings. The semantic task was simple, but because the sentences were separated by only four seconds, the task of judging their truth or falsity was not trivial. Typical sentences were of the form *The capital of Italy is not Paris,* and *Warsaw is the largest city in Austria.* Taking now the data from five subjects to form prototypes of the 48 sentences, by averaging the data from the five subjects, and taking the other four subjects to form corresponding averaged test samples of each sentence, we applied the Fourier methods described above and found an optimal bandpass filter from a predictive standpoint.[28] We were able to recognize correctly 90% of the test samples, using as a criterion for

---

[27] Butterworth filters are well described in Oppenheim and Schafer (1975, pp. 211–218).

[28] The data are for the visual condition of reading the sentences, one displayed word at a time.

8.6 LANGUAGE AND THE BRAIN.

selection a classical least-squares fit between a test sample and each of the 48 prototypes, after filtering (Suppes, Han, Epelboim and Lu 1999a).[29]

The surprising invariance result is that the data for prototypes and for test samples came from different subjects. There was no overlap in the two groups. Theoretically this is an efficient aspect of any much used communication system. My brain-wave representation of words and sentences is much like yours, so it is easy to understand you. But it is a theoretical point that needs strong empirical support to have it accepted. Another angle of comment is that the electric activity in the cortex is more invariant across subjects performing the same task than is the detailed anatomical geometry of their brains. I return to this invariance between subjects a little later, when I respond to some skeptical comments.

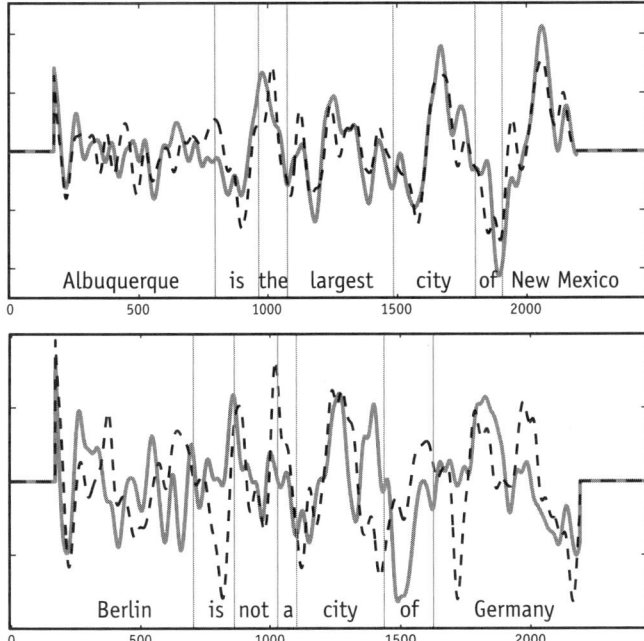

FIGURE 8 Prototypes (grey lines) and test samples (dashed black lines) generated by the best fitting sentence (upper panel) and worst fitting sentence (lower panel) correctly classified for subject S32. Time measurements after the onset of the sentence are shown in milliseconds on the abscissa.

---

[29]Let $x_i(n), 0 \leq n \leq N-1$, be the ith prototype (in the time domain), and $y_j(n), 0 \leq n \leq N-1$, the jth test sample. Then the sum of squared differences is S_{ij}, where

$$S_{ij} = \sum_{n=0}^{N-1} (x_i(n) - y_j(n))^2.$$

The test sample $y_j(n)$ is correctly classified if

$$S_{jj} = \min_i S_{ij},$$

with the minimum being unique.

One hundred sentences. I now turn to a second more recent experiment in which subjects were visually presented 100 different geography sentences (Suppes, Wong et al., to appear). I concentrate here only on the remarkable result of correct recognition of 93 of the 100 sentences for one subject (S32). Using the methods described, the best recognition rate achieved for a single subject (S32) was 93%, i.e., 93 of the 100 test samples. These results were achieved with $L = 1.25$ Hz, $W = 21.25$ Hz, $s = 180$ ms after onset of the visual presentation of the first word of each sentence, and $e = 2200$ ms, marking the ending of the recordings used for the least-squares criterion of fit. The best bipolar sensor was C4-T6. In Figure 8 we show at the top the best and at the bottom the worst fit, as measured by the least-squares criterion, for the 93 sentences correctly recognized. The sum of squares for the worst was more than three times that for the best.

Invariance between visual images and their names. The third experiment showed that the visual images generated on a computer screen, of a familiar shape, such as a circle or triangle, were very similar to the brain images generated by the corresponding word (Suppes, Han, Epelboim and Lu 1999b). This surprising result very much reinforced a classical solution of how the mind has general concepts. It is a famous episode in the history of philosophy in the eighteenth century that Berkeley and Hume strongly criticized Locke's conception of abstract or general ideas. Berkeley has this to say in *A New Theory of Vision* (1709/1901):

> It is indeed a tenet, as well of the modern as the ancient philosophers, that all general truths are concerning universal abstract ideas; without which, we are told, there could be no science, no demonstration of any general proposition in geometry. But it were no hard matter, did I think it necessary to my present purpose, to shew that propositions and demonstrations in geometry might be universal, though they who make them never think of abstract general ideas of triangles or circles.
>
> After reiterated efforts and pangs of thought to apprehend the general idea of a triangle, I have found it altogether incomprehensible. And surely, if any one were able to let that idea into my mind, it must be the author of the *Essay Concerning Human Understanding*: he, who has so far distinguished himself from the generality of writers, by the clearness and significancy of what he says. Let us therefore see how this celebrated author describes the general or which is the same thing, the abstract idea of a triangle. 'It must be,' says he, 'neither oblique nor rectangle, neither equilateral, equicrural, nor scalenum; but all and none of these at once. In effect it is somewhat imperfect that cannot exist; an idea, wherein some parts of several different and inconsistent ideas are put together.' (*Essay on Human Understanding*, B. iv. ch. 7. s. 9.) This is the idea which he thinks needful for the enlargement of knowledge, which is the subject of mathematical demonstration, and without which we could never come to know any general proposition concerning triangles. Sure I am, if this be the case, it is impossible for me to attain to know even the first elements of geometry: since I have not the faculty to frame in my mind such an idea as is here described. (Berkeley, pp. 188–189.)

Hume, in a brilliant exposition and extension of Berkeley's ideas, in the early pages of *A Treatise of Human Nature*, (1739/1951) phrased the matter beautifully in the opening paragraph of Section VII, entitled *Of Abstract Ideas*:

8.6 LANGUAGE AND THE BRAIN. 453

> A very material question has been started concerning *abstract* or *general* ideas, *whether they be general or particular in the mind's conception of them.* A great philosopher has disputed the receiv'd opinion in this particular, and has asserted, that all general ideas are nothing but particular ones, annexed to a certain term, which gives them a more extensive signification, and makes them recall upon occasion other individuals, which are similar to them. As I look upon this to be one of the greatest and most valuable discoveries that has been made of late years in the republic of letters, I shall here endeavour to confirm it by some arguments, which I hope will put it beyond all doubt and controversy.
>
> (Hume, *Treatise*, p. 17.)

Although not discussed by Berkeley and Hume, we also confirmed that the same is true of simple patches of color. In other words, a patch of red and the word 'red' generate similar brain images in the auditory part of the cortex.

The specific significant results were these. By averaging over subjects as well as trials, we created prototypes from brain waves evoked by stimuli consisting of simple visual images and test samples from brain waves evoked by auditory or visual words naming the visual images. We correctly recognized from 60% to 75% of the test-sample brain waves. Our general conclusion was that simple shapes and simple patches of color generate brain waves surprisingly similar to those generated by their verbal names. This conclusion, taken together with extensive psychological studies of auditory and visual memory, support the solution conjectured by Berkeley and Hume. The brain, or, if you prefer, the mind, associates individual visual images of triangles, e.g., to the word *triangle*. It is such an associative network that is the likely procedural replacement for the mistaken attempt by Locke to introduce abstract ideas.

Comparisons of averaged and filtered brain waves generated by visual images and spoken names of the images are shown in Figure 9. Time after the onset of the stimulus (visual image or word) is shown in milliseconds on the abscissa. In the upper panel the solid curved line is the prototype brain wave generated by the color blue displayed as a blank computer screen with a blue background. The dotted curved line is the test-sample brain wave generated by the spoken word *blue*. In the lower panel are the prototype brain wave (solid line), generated by display of a triangle on the screen, and the test-sample brain wave (dotted line), generated by the spoken word *triangle*. In neither case is the match perfect, for even when the same stimulus is repeated, the filtered brain waves do not match exactly, since the brain's electric activity continually changes from moment to moment in numerous ways. But, all the same, there are many invariances necessary for human communication, and even at this early stage we can identify some of them.

Criticisms of results and response. I first sketch the general nature of the criticisms. In many brain imaging experiments the data are very rich and complex. Consequently, a complicated procedure may also be used to find, for given conditions, an optimal value. The search for this optimal value, which here is the best correct recognition rate, is analogous to computing an extreme statistic for a given probability distribution. The basis of the analogy is that the search corresponds to possibly many repetitions of a null-hypothesis experiment. These repetitions require computation of the appropriate extreme statistic. Moreover, if several parameters are estimated in finding such

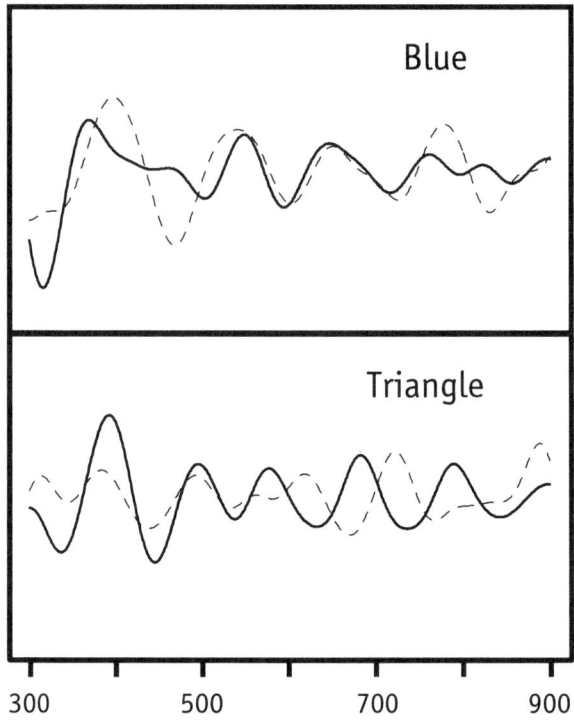

FIGURE 9 Comparison of filtered brain waves generated by visual images (solid curves) with those generated by the spoken names (dotted curves) of the images.

an optimal value, the significance of the value found may be challenged. The basis of such a challenge is the claim that for rich data and several estimated parameters, even a random assignment of the meaningful labels in the experiment may still produce a pretty good predictive result with a large enough number of chance repetitions.

Extreme statistics. We meet this criticism in two ways. The first is to derive the extreme-statistic distribution under the null hypothesis of a binomial distribution arising from such random assignments of labels. We compute how many standard deviations the scientifically interesting predictive result is from the mean of the extreme-statistic distribution. Physicists usually accept at least three or four standard deviations as a significant result. Other scientists and statisticians usually prefer a p value expressing the probability of the observed result under the null hypothesis. Here we report both measures.

The second approach is meant to answer those who are skeptical that the null hypothesis of a binomial distribution with a single parameter for the chance probability of a correct classification will adequately characterize the structure of the data even after a random permutation of the labels. To respond to such possible skeptics, we also compute a recognition rate for a sample of 50 random permutations of labels. We then fit a beta distribution to each such sample for a given experimental condition to compare

8.6 Language and the Brain.

with the corresponding extreme statistic distribution arising from the null hypothesis.

We first derive the extreme statistic under the null hypothesis.

Let p = probability of a success, a correct classification in our case, on a single trial, and $q = 1 - p$. Let $\mathbf{X}$ be the random variable whose value is the number k of successes in n independent trials. The probability of at least k successes is:

$$P(\mathbf{X} \geq k) = \sum_{j=k}^{n} P(\mathbf{X} = j) = \sum_{j=k}^{n} \binom{n}{j} p^j q^{n-j}. \tag{3}$$

Now we repeat the experiment governed by a binomial distribution. So we have r independent repetitions of the n independent trials. For r repetitions ($r = 21,000$ in the 100-sentences experiment), the random variable representing the extreme statistic is

$$\mathbf{Y} = \max(\mathbf{X}_1, \ldots, \mathbf{X}_r). \tag{4}$$

Let $P(\mathbf{Y} \geq k)$ be the probability that $\mathbf{Y}$ is at least k in at least one of the r repetitions, the extreme statistic of interest. Then clearly

$$P(\mathbf{Y} \geq k) = 1 - P(\mathbf{X} < k)^r, \tag{5}$$

$$= 1 - \left[\sum_{j=0}^{k-1} \binom{n}{j} p^j q^{n-j}\right]^r.$$

We also need the theoretical density distribution of $\mathbf{Y}$, to compare to various empirical results later. This is easy to compute from (3).

$$P(\mathbf{Y} = k) = P(\mathbf{Y} \geq k) - P(\mathbf{Y} \geq k+1), \tag{6}$$

$$= P(\mathbf{X} < k+1)^r - P(\mathbf{X} < k)^r,$$

$$= \left[\sum_{j=1}^{k} \binom{n}{j} p^j q^{n-j}\right]^r - \left[\sum_{j=1}^{k-1} \binom{n}{j} p^j q^{n-j}\right]^r.$$

From (4) we can compute the mean and standard deviation of the extreme statistic $\mathbf{Y}$.

Second, we report results for the beta distribution on $(0, 1)$ fitted to the empirical sample of extreme statistics. The density $f(x)$ of the beta distribution is:

$$f(x) = \begin{cases} \frac{\Gamma(a+b)}{\Gamma(a)\Gamma(b)} x^{a-1}(1-x)^{b-1}, & a, b > 0, \quad 0 < x < 1, \\ 0 & \text{otherwise,} \end{cases} \tag{7}$$

where $\Gamma(a)$ is the gamma function. If $\mathbf{Z}$ is a random variable with a beta distribution, then its mean and variance are given as simple functions of the parameters a and b.

$$\mu_{\mathbf{Z}} = E(\mathbf{Z}) = \frac{a}{a+b}, \tag{8}$$

$$\sigma_{\mathbf{Z}}^2 = \text{Var}(\mathbf{Z}) = \frac{ab}{(a+b)^2(a+b+1)}. \tag{9}$$

The probability that the random variable $\mathbf{Z}$ has a value equal to or greater than $\frac{k}{n}$ is:

$$P(\mathbf{Z} \geq \frac{k}{n}) = \frac{\Gamma(a+b)}{\Gamma(a)\Gamma(b)} \int_{\frac{k}{n}}^{1} x^{a-1}(1-x)^{b-1} dx. \tag{10}$$

The computation of $P(\mathbf{Z} \geq \frac{k}{n})$ is difficult for the extreme tail of the distribution. In some cases we use a mathematically rigorous upper bound that is not the best possible,

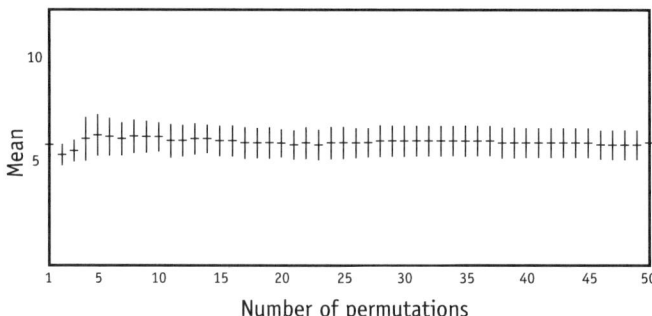

FIGURE 10 Cumulative mean and standard deviation of the recognition rate of the sample of random permutations.

but easy to compute, namely, just the area of the rectangle with height $f(\frac{k}{n})$ containing the tail of the distribution to the right of $f(\frac{k}{n})$:

$$P(\mathbf{Z} \geq \frac{k}{n}) \leq f(\frac{k}{n})(1 - \frac{k}{n}), \tag{11}$$

where $f(\frac{k}{n})$ is defined by (5).

Computation of extreme statistics. I begin with the second experiment using 100 sentences. As a check on the null hypothesis, we constructed an empirical distribution of the extreme statistic by sampling 50 random permutations. Several points are to be noted.

1. A permutation of the 100 sentence "labels" is randomly drawn from the population of 100! possible permutations, and the sentence test samples are relabeled using this permutation.
2. Exactly the same grid of parameters (L, W, s, e) is now run for each bipolar pair of sensors, as for the correct labeling on the data of subject S32, to obtain, by Fourier analysis, filtering and selection of temporal intervals (s, e), a best rate of recognition or classification for the random label assignment. For the 100-sentences experiment, the number of points on the grid tested for each random permutation is $7 \times 10 = 70$ for $L \times W$, $5 \times 4 = 20$ for $s \times e$ and 15 for the number of sensors, so the number of repetitions r, from the standpoint of the null hypothesis, is $70 \times 20 \times 15 = 21,000$.
3. This random sampling of label permutations is repeated, and the recognition results computed, until a sample of 50 permutations has been drawn.

In Figure 10, I show the cumulative computation of the mean m and standard deviation s for the sample of 50 label permutations for the data of subject S32. For the full sample of 50 the mean $m = 6.04$ and the standard deviation $s = 0.77$. In Figure 11, I show: (i) the frequency distribution of the null-hypothesis extreme statistic $\mathbf{Y}$ with $n = 100$, $p = 0.01$ and $r = 21,000$, (ii) the empirical histogram of the maximum number of successes obtained for the 50 sample points with $r = 21,000$, and (iii) the fitted beta distribution as well. From Figure 11 it is visually obvious that the correct

8.6 Language and the Brain.

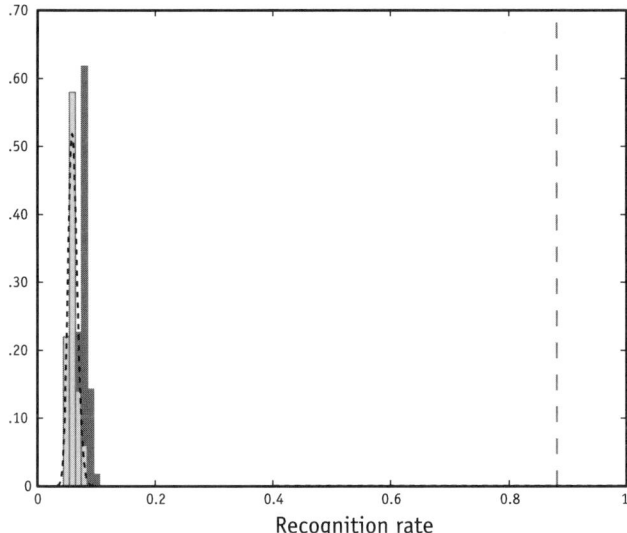

FIGURE 11 The frequency distribution (dark area) of the null-hypothesis extreme statistic **Y**, the histogram of the sample of 50 random permutations (lightly shaded areas), the fitted beta distribution (dotted line), and on the right (dashed vertical line) the recognition rate for S32.

classification of more than 80 of the 100 sentences for S32 is not compatible with either the distribution of the extreme statistic **Y** or the estimated beta distribution for the sample grid computations based on 50 random permutations of the labels. The fact that the beta distribution fits slightly better than the distribution of **Y** is not surprising, since no free parameters were estimated for the latter. A finer search, with much larger r, yielding the higher result of 93 out of 100, is discussed in the next paragraph.

What is perhaps surprising is that the mean $\mu = 6.95$ of the null-hypothesis extreme statistic **Y** is slightly larger than the mean $m = 6.04$ of the empirical sample distribution. Three points are worth noting. First, the standard deviation $s = 0.77$ of the empirical sample is larger than the standard deviation $s = 0.67$ of the extreme statistic **Y**. I comment on this difference below. Second, I show in Figure 12 the rate of growth of the recognition rate for the null-hypothesis extreme statistic **Y**, for $n = 100$ and $p = 0.01$, and some other values of p and r used later, as r is increased by one or more orders of magnitude. As can be seen, under the null hypothesis the correct-recognition growth rate is slow. As an important example, we refined by extensive search the grid for the data of S32. We did not use a complete grid, but refined and extended only in promising directions. Extended comparably in all directions, the order of magnitude of r would be 10^7, i.e., 10,000,000 repetitions. So we computed the null-hypothesis distribution of **Y** for this large value of r, which is much larger than any actual computation we made. Even for this large grid, the mean of the extreme statistic **Y** for $r = 10^7$ only moved to $\mu = 9.60$. With the standard deviation now reduced to 0.62, the number of standard

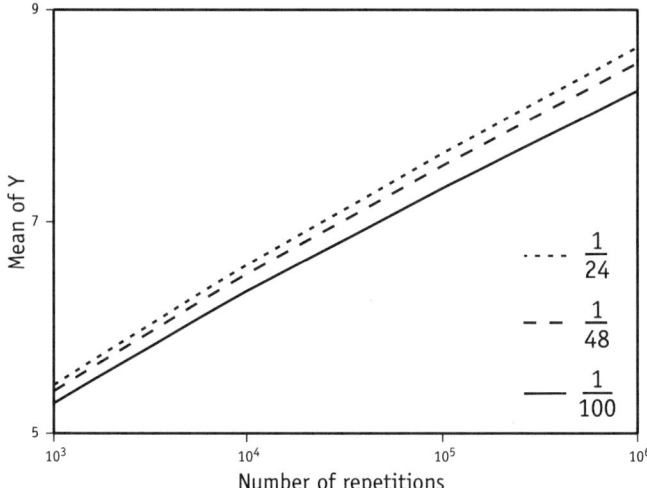

FIGURE 12   Mean of the null-hypothesis extreme statistic **Y** as a function of the number r of repetitions.

deviation units of the distance between 93 and 9.60 is 134.5, larger than before.[30]

The third point concerns the level of significance, or p value, we report for rejecting the null hypothesis. The p value of the result of $k = 93$, which is 134.5 standard deviations from the mean of the null-hypothesis extreme statistic **Y** is extravagantly low, at the very least $p < 10^{-100}$. Every other aspect of the experiment would have had to be perfect to support such a p value. (We did check the computation of 93 on two different computers running different programs.) So here, and in other cases later, we report only the inequality $p < 10^{-10}$ for such very small values, but the actual number of standard deviations from the mean is reported.

We also checked that using the rigorous upper bound of inequality (9), $P(\mathbf{X} \geq 93)$, computed for the fitted beta distribution, is also on the order of $p < 10^{-100}$. This is further support for the view that the p value inequality used later, namely, $p < 10^{-10}$, is highly conservative.

Analysis of earlier studies. I have emphasized the gain in predictive results from averaging across subjects as well as trials. The best result of the second experiment of 93% for one individual subject prompted us to review the best individual results in earlier experiments. In each experiment we have performed, the analysis of at least one individual subject's brain waves yielded a correct classification greater than 90%, with the exception of the 48-sentence experiment, mentioned already, which was 77%. (In

---

[30] In reflecting on these results, it is important to keep in mind that physicists are usually very happy with a separation of 6 or 7 standard deviations between the classical or null-hypothesis prediction and new observed results, e.g., in quantum entanglement experiments of the sort discussed in Section 7.2. The significance levels obtained in our brain experiments, and the very large distance in standard deviations of the observed results from the expected null-hypothesis results, are as yet seldom found in psychological or neuroscience experiments.

8.6 Language and the Brain.

Suppes, Han, Epelboim and Lu (1999a), this 77% was reported as 79%, because a finer grid was used.) Results for the best subject in the various experiments are summarized in Table 6.

With one exception, the p values shown in Table 6 are highly significant, by most standards of experimental work, extravagantly so. The exception is for the visual-image experiment in which 8 simple visual images were presented as stimuli. For four of the subjects, as shown in Table 6, we were able to classify all 8 brain waves correctly, but this perfect result of 100 percent was significant only at the level of $p < 0.01$ for the null hypothesis, because with enough repetitions the best guesses under the null hypothesis do pretty well also, with $\mu = 6.28$.

The lesson for experimental design of this last point is obvious. If the data are massive and complex, as in the brain experiments described, and extensive search for optimal parameters is required, then the probability p of a correct response under the null hypothesis should be small. Figure 12 graphically makes the point. When p is small the number of repetitions can be very large, without affecting very much the mean of $\mathbf{Y}$, the extreme statistic of r repetitions. As can be seen also, from Table 6, when $p = 0.01$, the binomial parameter of the 100-sentences experiment, even 10,000,000 repetitions under the null hypothesis of 100 trials, increases $E(\mathbf{Y})$ only slightly to 9.60. To put the argument dramatically, at the rate of 1 second per trial, it would take more time than the present estimated age of the universe to have enough repetitions to obtain $E(\mathbf{Y}) \geq 93$.

I say in the preceding paragraph that p should be small, but that is too simple. The other way out, used in the first two experimental conditions of Table 6, 7 visual words and 7 auditory words, is to increase the number of test samples. In those two conditions, $p = \frac{1}{7}$, but the number of test samples was 35, and, as can be seen from the table, the null hypothesis was rejected at a level better than 10^{-10}. Reanalysis of the data from the visual-image experiment with $p = \frac{1}{8}$, in a similar approach by increasing the number of test samples from 8 to 24 yielded some better levels of rejection of the null hypothesis. The details are reported below.

More skeptical questions. As in all regimes of detailed experimentation, there is no sharp point after which further experiments need not be conducted, because all relevant questions have been answered. Galison (1987) made a detailed study of several important research programs of experimentation in physics. It seems likely that the main aspects of his analysis apply to many other areas of science. He says:

> Amidst the varied tests and arguments of any experimental enterprise, experimentalists must decide, implicitly or explicitly, that their conclusions stand *ceteris paribus*: all other factors being equal. And they must do so despite the fact that the end of an experimental demonstration is not, and cannot be, based purely on a closed set of procedures. ...Certain manipulations of apparatus, calculations, assumptions, and arguments give confidence to the experimentalist: what are they? ...When do experimentalists stake their claim on the reality of an effect? When do they assert that the counter's pulse or the spike in a graph is more than an artifact of the apparatus or environment? In short: How do experiments end? (Galison, pp. 3–4.)

TABLE 6 Exceptional Recognition Rates[a]

Experiment	Subj.	Number of Successes	Chance Prob.	% Cor.	Repet. r	Statistic Y μ	m	σ	s	Significance # σ	# s	p value
7 visual words[1]	S1	32 of 35	1 of 7	91	2925	13.40		0.92		20.2		$< 10^{-10}$
7 auditory words[1]	S3	34 of 35	1 of 7	97	3600	13.55		0.91		22.5		$< 10^{-10}$
12 sentences[2]	S8	56 of 60	1 of 12	93	60,480	16.10		0.90		44.3		$< 10^{-10}$
24 visual sent.[3]	S18	24 of 24	1 of 24	100	30,800	6.83	5.64	0.63	0.79	27.3	23.2	$< 10^{-10}$
48 visual sent.[3]	S26	38 of 48	1 of 48	79	30,800	7.04	6.20	0.63	0.94	49.1	33.8	$< 10^{-10}$
8 visual images[4]	4 Ss	8 of 8	1 of 8	100	95,550	6.28	4.92	0.46	0.69	3.7	4.5	$< .01$
100 visual sent.	S32	88 of 100	1 of 100	88	21,000	6.95	6.04	0.67	0.77	121.0	106.4	$< 10^{-10}$
100 visual sent.	S32	93 of 100	1 of 100	93	10^7	9.60		0.62		134.5		$< 10^{-10}$

1. Suppes, Lu, and Han (1997); 2. Suppes, Han and Lu (1998)
3. Suppes, Han, Epelboim and Lu (1999a); 4. Suppes, Han, Epelboim and Lu (1999b)

[a] The first column lists the experiment, with the last two entries being for the 100-sentence one. The subjects, listed in the second column, are numbered continuously from the experiments first reported in Suppes, Lu and Han (1997). The third column shows the maximum number of test samples successfully recognized out of the total presented. The fourth column shows the chance probability of a correct classification, which is simply 1 divided by the number of prototypes. The fifth column records the percent correct, as computed from the third column. The sixth columns shows the number r of repetitions used in the particular experiment to compute the extreme statistic. The number r is also the number of repetitions originally used in the grid for the initial search with correct labels. The seventh column records the mean μ of the null-hypothesis extreme statistic Y. The eighth column records the mean of the empirical samples of extreme statistics for the experiments for which we made this computation. The ninth column records the standard deviation σ of the extreme statistic Y, and the tenth column the corresponding standard deviation s of the empirical samples. The eleventh column records the number $\frac{k-\mu}{\sigma}$, which is the number of standard deviations that the number k of successes recorded in column three is from the mean μ of the null-hypothesis distribution of the extreme statistic Y, and the twelfth column the corresponding number for the empirical sample. The thirteenth column shows a conservative bound for the p value of the observed number k of successes, with respect to the distribution of the extreme statistic Y, as given by equation (6). In the case of the four subjects in the visual-image experiment, m and s are the average for the four. The superscript on the description of each experiment is the reference to the published study. (The EEG sensor, or bipolar pair of sensors, and the optimal filter for each subject, except the four subjects of Suppes, Han, Epelboim and Lu (1999b), were as follows: S1:T6, 1-10 Hz; S3:T3, 3-11 Hz; S8:C4-T6, 2.5-9 Hz; S18:P4-T6, 0.5-10 Hz; S26:C4-C6, 1-15 Hz; S32:C4-T6, 1.25-22.5 Hz. The optimal parameters were often not unique.)

8.6 LANGUAGE AND THE BRAIN.

In the context of the present brain experiments, the question is not really when do they end, but when do the computations on the experimental data come to an end? I examine three more different, but typical skeptical questions that are asked about new research with strong statistical support for its validity.

Other pairs in the first experiment with 48 sentences. Some skeptics commented to us that we were just lucky in the particular 2-element partition of the subjects we analyzed. So, we ran all 510 2-element partitions of the 9 subjects, with the same optimal values as in Table 6, without trying all points on the grid (Suppes, Wong et al., to appear). In Figure 13 we show the histogram of these 510 partitions. The level of significance of the results is $p < 10^{-10}$ for all but 4 of the 510 possibilities, and one of these 4 has $p < 10^{-7}$. The best result is 46 out of 48, which holds for several partitions. So the brain-wave invariance between subjects argued for in the earlier study is robustly supported by the present more thorough statistical analysis. Another view of the same data is shown in Figure 14, where the number of subjects in the prototype of each 2-element partition is plotted on the abscissa and on the ordinate is shown the mean number of correct classifications of the 48 sentences for each type of prototype. Surprisingly, the mean results are good ($p < 10^{-10}$) when the prototype has only 1 subject or all but 1 subject, i.e., 8 subjects. The evidence is pretty convincing that our original choice of a partition was not just some happy accident.

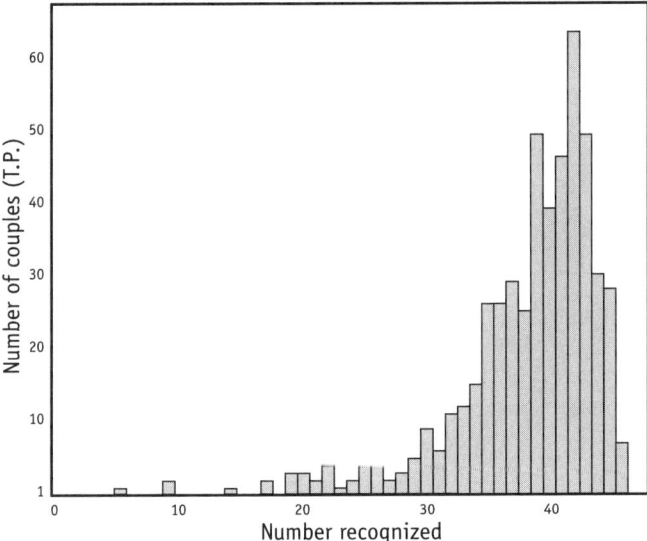

FIGURE 13 Histogram of the correct recognition rates for the 510 2-element partitions of the 9 subjects in the 48-sentences experiment.

Test of a timing hypothesis for the experiment with 100 sentences. In discussing with colleagues the high recognition rate of 93% obtained in the second exper-

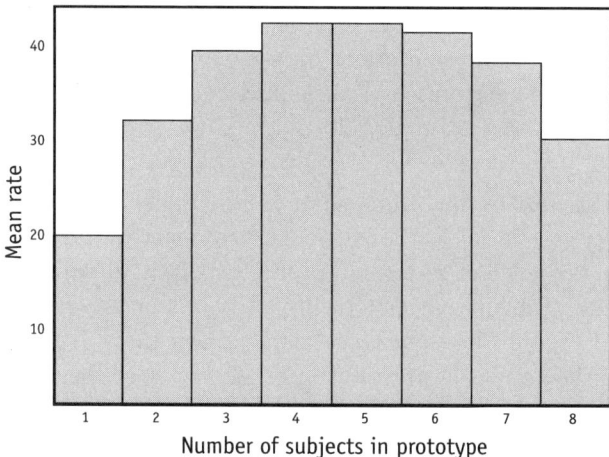

FIGURE 14 Histogram of mean number of the correct recognition rate for the 510 2-element partitions, indexed from 1-8 for the number of subjects in the prototype. So, e.g., 1 on the abscissa corresponds to all the 2-element partitions having exactly one subject used for the prototype.

iment reported above, and the earlier results summarized in Table 6, several persons skeptically suggested that perhaps our recognition rates are just coming from the different timing of the visual presentation of words in different sentences. Sentences were presented one word at a time in the center of the computer screen, with the onset time of each visual word the same as the onset time of the corresponding auditory presentation of the sentence. Visually displaying one word at a time avoids many troublesome eye movements that can disrupt the brain waves, and it has also been shown to be an effective fast way to read for detailed content (Rubin and Turano 1992). The duration of each visual word of a sentence also matched the auditory duration within a few milliseconds.

To test this timing idea, which is supported by the presence of an evoked response potential at the onset of most visual words, we used a recognition model that depended only on an initial segment of the brain-wave response to each word in a sentence (Suppes, Wong et al., to appear). The model replaces the two parameters s and e for the temporal interval by two different parameters. The first is α, which is the estimated time lag between the onset of each word in every sentence and the beginning of the corresponding brain wave in the cortex. The second is β, which is the proportion of the displayed length of each word, starting from its onset, used in the prototype for recognition after the delay time α for the signal to reach the cortex. Because of the variable length of words and sentences, we normalized the least squares computation by dividing by the number of observations used. If only timing, and not the full representation of the word, matters in recognition, then only a small portion of the initial segment of a word is needed, essentially the initial segment containing the onset-evoked response potential. On the other hand, if the full representation of the word is used in successful recognition, in terms of our least squares criterion, then the larger β is, the better for recognition. To

8.6 LANGUAGE AND THE BRAIN.

adjust β to the temporal length of each word displayed, we expressed β as a decimal multiple of the temporal display length of word i of each sentence. The best predictive result was for $\alpha = 200$ ms and $\beta = 1.25$, with a recognition rate of 92%. The recognition rate as a function of $0.125 \leq \beta \leq 2.00$ is shown in Figure 15. The rate of correct recognition increases monotonically with β up to $\beta = 1.25$ and then declines slowly after $\beta = 1.50$. These results support two conclusions. First, timing is important. The recognition rate of 45% for $\beta = 0.125$ is much greater than a chance outcome. But, second, the more complete the brain-wave representation of the words in a sentence, the better for recognition purposes.

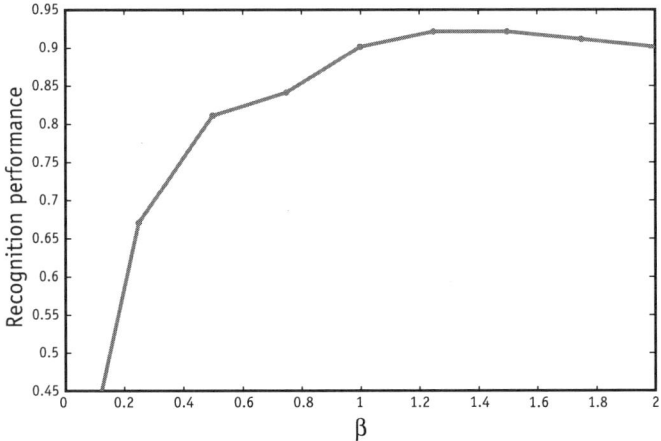

FIGURE 15 Classification results using different initial segments of brain-wave data after onset of each word in a sentence. Segment i begins after α ms following onset of word i in a sentence and segment i ends after $\beta \cdot l_i$ ms, where l_i is the length in ms of the visual display of word i.

Censoring data in the visual-image experiment. One kind of skeptical question that keeps the computations going is about artifacts. Perhaps the remarkable levels of statistical significance are due to some artifacts in the data.[31] Now there is a long history of the problems of artifacts in EEG research. A main source is eye blinks and saccadic or pursuit eye movements, another is ambient current in the environment, mainly due to the 60 Hz oscillation of the standard alternating current in any building in which experiments are ordinarily conducted, and still another source is in the instrumentation for observing and recording the electric brain waves. This list is by no means exhaustive. There is a large literature on the subject from several different angles, but it would be too much to survey it here.

---

[31] Given that the extreme statistics of the random permutations had a mean close to the low mean of the null hypothesis for the second experiment, it is extremely unlikely that any artifacts could account for the correct classification of 93% in the second experiment, or, in fact, in any of the others with $p < 10^{-10}$. But artifact removal remains an important topic, not so much to meet ill-informed skeptical questions, but to improve the classification results, as is the case for an example that follows.

I restrict myself to describing how we used a rather familiar statistical approach, rather than any visual inspection of the recorded data for eye blinks or other artifacts.[32] The approach was to censor the data, but to introduce a free parameter to optimize the censoring—optimize in the sense already described of maximizing the correct recognition rate.

Let $\mathbf{X}_{ik}$ = observation i on trial k, and let ω be the number of trials averaged to create a prototype or test sample. Then

$$m_i = \overline{\mathbf{X}}_i = \frac{1}{\omega} \sum_{k=1}^{\omega} \mathbf{X}_{ik}.$$

In similar fashion, we compute the variance s_i^2. Let α be the free parameter for censoring such that if

$$|\mathbf{X}_{ik} - m_i| > \alpha s_i$$

eliminate observation i of trial k from the data being averaged. The computational task is to find the $\hat{\alpha}$ that optimizes classification. In the example reported here, we ran a one-dimensional grid of 20 values of s_i to approximate the best value $\hat{\alpha}$.

The experiment for which the extreme statistics were not highly significant was the visual-image one already described, the one that was relevant to the eighteenth-century controversy about abstract ideas. The data, as I said earlier, supported in a direct way the skeptical views of Berkeley and Hume, but the statistical support was not very strong. So we reanalyzed the data, creating 24 rather than 8 test samples, and we also ran the experiment with four monolingual Chinese (Mandarin) speakers to confirm the verbal part, with auditory or visual words, in Chinese as well as English. The details are reported in Suppes, Wong et al. (to appear).

Table 7 shows the significant results for cross-modal classification. The first two conditions are for the original experiment using English. For the feminine-auditory-voice representing brain waves (AWF) as prototypes and the visual-image brain waves as test samples, 15 of the 24 test samples were correctly classified after censoring, an improvement from 11 of 24 without censoring, for a resulting significance level of $p < 0.016$. When the roles of prototype and test sample were reversed the results of censoring were better, 16 of the 24 test samples correctly classified after censoring, with $p < 0.001$. Note that the significance levels here are conservative, based on the complete grid search equal to $r = 1,470,000$, the number of repetitions under the extreme-statistic null hypothesis.

In the case of the Chinese, the best results were in the comparison of the auditory and visual presentation of the eight words, with the best result being for the visual Chinese words (VW) as prototypes and the auditory Chinese words (AW) as test samples, in the censored case, 17 of 24 correctly classified, with p approximately 0.0001, and with, as before, the number of repetitions r, under the null hypothesis, greater than a million. So we end by strengthening the case for Berkeley and Hume, without claiming the evidence is as yet completely decisive.

An appropriate stopping point for this analysis is to emphasize that censoring does

---

[32] In our larger experiments with more than 40,000 trials it is impractical to try to use traditional observational methods to detect artifacts.

TABLE 7   Cross-modal results in visual-image experiment with censored data

Exper.	R	Y μ	σ	Significance $\#\sigma$	p-value
English					
AWF/VI	11	11.57	.692	0.8	$\approx .500$
censor	15	13.01	.642	3.1	$< .016$
VI/AWF	11	11.57	.692	0.8	$\approx .500$
censor	16	13.01	.642	4.7	$< .001$
Chinese					
AW/VW	9	11.42	.690		
censor	16	12.86	.662	4.7	$< .001$
VW/AW	13	11.42	.690	2.3	$\approx .05$
censor	17	12.86	.662	6.3	$\approx .0001$

not guarantee improvement in classification. Most of the other results in Table 6 showed little improvement from censoring, but then for all of the experiments reported there, except for the visual-image one, the results were highly significant without censoring.

As in many areas of science, so with EEG recordings, statistical and experimental methods for removing artifacts and other anomalies in data constitute a large subject with a complicated literature. I have only reported a common statistical approach here, but I am happy to end with this one example of a typical method of "cleaning up" data. It is such censored data that should be used to form a representation suitable for serving as a test of some theory or, as is often the case, some congerie of theoretical ideas.

8.7 Epilogue: Representation and Reduction in Science[33]

The theme of reduction is as old as philosophy. Already in the 5th century BC, Leucippus and especially his student, Democritus, stated in clear general terms the atomic ideas for which they are justly famous. Matter is nothing but solid atoms moving in empty space. The weight, density and hardness of a physical object depends only on the arrangements of its atoms in the void, as the Greeks described empty space. But also, according to Democritus, nothing happens randomly, but according to law and by necessity. Although much criticized by Aristotle and his followers over the centuries, this reductionist mechanical view of nature was once again brought to the fore with the development of physics, or, as it was known then, natural philosophy, in the seventeenth and eighteenth centuries.

But the two most famous ancient atomists to come after Democritus added to this mechanical picture the effects of chance. Here is a well-known quotation from Epicurus, taken from his letter to Herodotus, written probably in the 4th century BC (Oates 1940, p. 5):

> And the atoms move continuously for all time, some of them falling straight down, others swerving, and others recoiling from their collisions. And of the latter, some are borne

---
[33]The remarks in this section supplement those on reductionism in Suppes (1984, pp. 120-125).

on, separating to a long distance from one another, while others again recoil and recoil, whenever they chance to be checked by the interlacing with others, or else shut in by atoms interlaced around them. For on the one hand the nature of the void which separates each atom by itself brings this about, as it is not able to afford resistance, and on the other hand the hardness which belongs to the atoms makes them recoil after collision to as great a distance as the interlacing permits separation after the collision. And these motions have no beginning, since the atoms and the void are the cause.

And this is was Lucretius had to say in *De Rerum Natura,* written in the first century BC (Oates 1940, p. 95):

> This point too herein we wish you to apprehend: when bodies are borne downwards sheer through void by their own weights, at quite uncertain times and uncertain spots they push themselves a little from their course: you just and only just can call it a change of inclination. If they were not used to swerve, they would all fall down, like drops of rain, through the deep void, and no clashing would have been begotten nor blow produced among the first-beginnings: thus nature never would have produced aught.

So Epicurus and Lucretius represent reduction of nature to probabilistic mechanics, a discipline not really well developed scientifically until the twentieth century. Of course, these atomists did not get a good press from Aristotle to Newton, with a few notable exceptions. A justly quotable passage from Newton is to be found in his well-known letter of February 25, 1692/3, written to Richard Bentley.

> It is inconceivable, that inanimate brute matter should, without the mediation of something else, which is not material, operate upon and affect other matter without mutual contact, as it must be, if gravitation, in the sense of Epicurus, be essential and inherent in it. And this is one reason why I desired you would not ascribe innate gravity to me. That gravity should be innate, inherent, and essential to matter, so that one body may act upon another at a distance through a vacuum, without the mediation of anything else, by and through which their action and force may be conveyed from one to another, is to me so great an absurdity, that I believe no man, who has in philosophical matters a competent faculty of thinking, can ever fall into it. Gravity must be caused by an agent acting constantly according to certain laws; but whether this agent be material or immaterial, I have left to the consideration of my readers.

From Newton onward the great contributors to the development of physics were, almost without exception, quite clear on the difference between establishing a thesis of reduction scientifically, as opposed to speculative ideas about it. About two hundred years after Newton, Maxwell states this view very clearly in his little book *Matter and Motion.*

> Of course if we begin by assuming that the real bodies are systems composed of matter which agrees in all respects with the definitions we have laid down, we may go on to assert that all phenomena are changes of configuration and motion, though we are not prepared to define the kind of configuration and motion by which the particular phenomena are to be explained. But in accurate science such asserted explanations must be estimated, not by their promises, but by their performances. The configuration and motion of a system are facts capable of being described in an accurate manner, and therefore, in order that the explanation of a phenomenon by the configuration and motion of a material system may be admitted as an addition to our scientific knowledge, the configurations, motions,

8.7 Epilogue: Representation and Reduction in Science

and forces must be specified, and shown to be consistent with known facts, as well as capable of accounting for the phenomenon."

Maxwell, *Matter and Motion,* 1877/1920, pp. 70-71

Maxwell makes clear enough the need to distinguish between reductions that are firmly established from those that are not. This does not mean that all speculation is a bad thing. We might even maintain that in most cases in science, the "next big thing" is preceded by speculation that it is out there to be discovered or invented. What is important to the modern temperament is that scientific speculations be taken straight as conjectures, not mixed and served in metaphysical cocktails of suspect ingredients. So one of the puritanical themes of this book is that scientific reduction is hard work. To finish it off with detailed and explicit representation theorems is even harder. In spite of the many pages in this book and the continual focus on representation theorems, not a single really significant representation theorem providing the stiff backbone for some important claim about reduction of one part of science to another has been proved or even stated. The reader who has missed some of my earlier remarks on this topic may ask, Why not?

The answer is obvious. The problem was too difficult for me. Perhaps the most famous case to consider is the reduction of thermodynamics to statistical mechanics. At a general conceptual level this is one of the great triumphs of what used to be called the mechanical philosophy. And at a general level of analysis a famous article on the topic was published by Ernest Nagel in 1949. But Nagel fully realized he was not able to give the full details or to begin to offer a rigorous mathematical proof of an appropriate representation theorem. Ruelle (1978) is one of the best examples of a rigorous mathematical approach, but his idealization requires that the physical systems to which it directly applied be actually infinite in all directions, i.e., completely space filling, in order to exhibit sharp phase transitions. The detailed explanation of phase transitions in macroscopic thermodynamic systems is one of the main objectives of classical equilibrium statistical mechanics. I also want to emphasize that phase transitions are among the most obvious systematic changes in natural phenomena that we would naturally want to explain. For example, we would like to derive from fundamental physical principles the result that under ordinary atmospheric pressure liquid water turns into steam at about 100 degrees Celsius. To derive this result rigorously from first principles for the kitchen tea kettle is still not possible. The same goes for the other dramatic phase transition, the freezing of a glass of water.

The absence of rigorous representation theorems for these everyday dramatic phenomena does not mean physicists are doing a bad job of analysis. It is just that we have come to realize in a very explicit way how difficult it is to give rigorous mathematical results for all kinds of ordinary natural phenomena. And physicists, clever problem solvers that they are, have gone ahead and learned much about such phenomena, including how to calculate, without rigorous justification, a large number of quantitative results.

Given how much this case generalizes to other important phenomena, with examples ranging from the weather to the stock market, it is, as in other matters, philosophically important to have a pluralistic approach to what can be accomplished. Weather predictions are much better now than they were forty years ago, even though we now realize that detailed predictions a month in advance are hopeless within our current theories and computing regimes for analyzing weather data. Physicists are adept at deriving useful and highly accurate equations explaining many phenomena that cannot be analyzed rigorously from a general mathematical standpoint. From such work we derive insights and intuitions about what representations and reductions are probably approximately correct, even though new surprises may be lurking in as yet undiscovered byways. And it is important for philosophers of science to have an understanding and overview of what is currently believed by physicists to be true of many physical phenomena, or, of other phenomena such as market behavior or language use by other scientists. In fact, in many areas of science, necessarily mainly empirical results on representation may be the only results that can be currently obtained. The brain-wave representations for language reported in Section 6 are an example.

But it is also part of such a pluralistic view to keep a place for rigorous results about representation and reduction. I have tried to give some sufficiently simple manageable examples in various areas of science in the last four chapters. I include in this list of examples the extended analysis of representations or interpretations of probability in Chapter 5. It is also evident that not all kinds of representation immediately suggest reduction. For example, the many representation theorems provable in the theory of measurement, of which I gave elementary examples in Chapter 3, do not conceptually suggest we are substantively reducing physical or psychological properties to numbers, but rather are using numbers to refine our analysis of the properties. In other words, this kind of representation is not like the physical reduction of the concept of temperature to the mean motion of particles.

On the other hand, the representation of computable functions by register machines or Turing machines, as stated in Chapter 3, or the representation of finite automata or register machines by stimulus-response models in Chapter 8, do suggest rather direct physical realizations of computational processes, and these realizations have at bottom much simpler mechanisms than were thought possible a hundred years earlier. The existence of such simple mechanisms for many complex phenomena—and thus their reduction in one sense—is a point much emphasized and exemplified in the recent lengthy tome *A New Kind of Science* (2002) by the theoretical physicist Stephen Wolfram. But Wolfram properly stresses, as he has in earlier publications, that many such processes are computationally irreducible, meaning that there is no simpler way to study their behavior than to observe the processes themselves.

Without entering into new technical complexities, it is evident that a concept like computational irreducibility is useful in helping to explain why accepted cases of reduction often do not result in any drastic change of concepts in our descriptions of ordinary phenomena. There are, of course, other reasons for there being little change in the face of what seems like a conceptual revolution. We will, I am sure, still talk in a hundred years about tables and chairs as we now do. And this holds for countless other familiar objects and processes.

On the other hand, to hold the thesis that such reductions have no impact on ordinary thought and talk is also a mistake. Numerical representations of time and temperature are now used by ordinary folk in ways that would have seemed strange and out of place

8.7 Epilogue: Representation and Reduction in Science

in the not too distant past. Some similar changes seem likely as we come to understand in ever more physical detail just how our minds and bodies work, but forecasting which representations of the future and their invariants will become the basis of new natural idioms is, even if not computationally irreducible, too difficult to attempt.

Summary Table of Representation and Invariance Theorems by Chapter[1]

Chapter 3. Theory of Isomorphic Representation

Representation theorems

3.1	Finite weak orderings	4.3	Bisection measurement
3.2	Infinite weak orderings	4.4	Conjoint measurement
4.1	Extensive measurement	5.1	Partial recursive function by unlimited register machine
4.2	Difference measurement		

Invariance theorems

None

Chapter 4. Invariance

Representation theorems

None

Invariance theorems

3.1	Ordinal measurement	3.4	Conjoint measurement
3.2	Extensive measurement	5.4	Entropy of ergodic Markov processes
3.3	Bisection measurement		

---

[1] Theorems are numbered by section and then consecutively within a section. So 3.1 is Theorem 1 of Section 3. There are no theorems in the first two chapters.

Chapter 5. Representations of Probability

REPRESENTATION THEOREMS

2.1	Finite Laplacean probability space	6.3	Propensity of response-strength
3.1	Finite sequences of outcomes	6.4	Propensity for heads in coin tossing
3.3	Infinite sequences of outcomes	6.5	Propensity for randomness in three-body problem
3.6	Random infinite sequence in Church's sense	7.3	Scott's
4.4	Complex finite sequence in Kolmogorov's sense	7.4	Qualitative expectations
		7.4	Qualitative probabilities
5.1	Carnap's regular measure	7.5	Qualitative conditional expectations
5.3	Chuaqui's Laplacean probability space for stochastic processes	7.5	Qualitative conditional probabilities
		7.6	Exchangeable sequences (de Finetti)
6.1	Standard sequences	7.7	Qualitative approximate belief measurement
6.2	Propensity to decay		

INVARIANCE THEOREMS

7.4	Qualitative expectations	7.5	Qualitative conditional probabilities
7.4	Qualitative probabilities	7.7	Qualitative approximate belief measurement
7.5	Qualitative conditional expectations		

Chapter 6. Representations of Space and Time

REPRESENTATION THEOREMS

1.1	Affine space	6.1	Visual space with restricted congruence
1.3	Euclidean space	8.5	Finite affine constructions

INVARIANCE THEOREMS

1.2	Affine space	6.1	Visual space with restricted congruence
1.4	Euclidean space	8.6	Finite affine constructions
2.1	Classical space-time		
3.1	Relativistic space-time		

Chapter 7. Representations in Mechanics

Representation theorems

1.4	Center of mass of system	2.5	Inequalities for three random variables
1.5	Rate of change of angular momentum of system	2.6	Bell's inequalities
1.7	Embedding system of particles	2.7	GHZ for perfect measurements
1.9	Potential energy function	2.8	Inequalities for GHZ experiments
2.1	Local hidden variable	2.9	Joint Gaussian distribution
2.2	Local hidden variable with symmetry	2.10	Joint Guassian distribution with missing covariances
2.3	Local hidden variable satisfying second-order factorization	2.11	Inequality for three Gaussian random variables
2.4	Hidden variable satisfying Locality Condition II		

Invariance theorems

1.6	Conservation of momentum and angular momentum	3.2	Temporal invariance of strongly reversible Markov chain
1.8	Conservation of energy	3.3	Temporal invariance of strongly reversible continuous-time Markov process
3.1	Temporal invariance of Markov property	3.4	Temporal invariance of second-order Markov chain

Chapter 8. Representations of Language

Representation theorems

1.2	Normal form (Chomsky) of context-free grammar	3.1	Finite automata by stimulus-response models
1.3	Normal form (Greibach) of context-free grammar	3.2	Regular languages by stimulus-response models
2.6	Regular grammars	3.3	Tote hierarchy in sense of Miller and Chomsky by stimulus-response model
2.7	Finite automata		
2.8	Regular languages by being closed under certain properties (Kleene)	3.4	Perceptual displays by finite automaton and by stimulus-response model
2.10	Context-free grammars		
2.11	Pushdown automata	3.5	Partial recursive functions by register learning models
2.12	Context-sensitive grammars		
2.12	Linear bounded automata	4.13	Linear learning models by stimulus-response models
2.13	Phrase-structure grammars		

Invariance theorems

None

References

Abelson, R. M. and R. A. Bradley (1954). A 2x2 factorial with paired comparisons. *Biometrics* 10(4):487–502.

Aczél, J. (1961). Über die Begründung der Additions- und Multiplikationsformeln von bedingten Wahrscheinlichkeiten. *Magyar Tudományos Akadémia Matematikai Kutató Int. Közleményei* 6:110–122.

Aczél, J. (1966). *Lectures on functional equations and their applications*. New York: Academic Press.

Aho, A. V. and J. D. Ullman (1972). *The theory of parsing, translation, and compiling*, vol. 1. Englewood Cliffs, NJ: Prentice Hall.

Alekseev, V. M. (1968–1969a). Quasirandom dynamical systems I, II, III. *Math. USSR Sbornik* 5,6,7:505–560 and 1–43.

Alekseev, V. M. (1969b). Quasirandom dynamical systems. (Russian). *Mat. Zametki* 6(4):489–498. English translation in Math. Notes 6 (1969), 749–753.

Alexandrov, A. D. (1950). On Lorentz transformations. (Russian). *Uspekhi Matematichekikh Nauk* 5, 3(37):187.

Alexandrov, A. D. (1975). Mapping of spaces with families of cones and space-time transformations. *Annali di Matematica Pura ed Applicata* 103:229–257.

Ames Jr., A. (1946). Binocular vision as affected by relations between uniocular stimulus patterns in common place environments. *American Journal of Psychology* 59:333.

Anderson, J. R. (1976). *Language, memory, and thought*. Hillsdale, NJ: Lawrence Erlbaum.

Angell, R. B. (1974). The geometry of visibles. *Noûs* 8:87–117.

Appell, P. (1941). *Traité de mécanique rationelle. Tome I: Statique, dynamique du point*. Paris: Gauthier-Villars, sixth edn.

Aquinas, T. (1944). Summa theologica. In A. C. Regis, ed., *Basic writings of Saint Thomas Aquinas*, vol. 1. New York: Random House.

Arbib, M. A. (1969). Memory limitations of stimulus-response models. *Psychological Review* 76:507–510.

Archimedes (1897). On the equilibrium of planes. In T. L. Heath, ed., *The works of Archimedes, edited in modern notation with introductory chapters*, pages 189–220. Cambridge, England: Cambridge University Press. Translation by T. L. Heath.

Aristotle (1960). *De Caelo (On the heavens)*. Cambridge, MA: Harvard University Press. English translation by W. K. C. Guthrie. First published in 1939.

Aristotle (1975). *De Anima (On the soul)*. Cambridge, MA: Harvard University Press, 4th edn. English translation by W. S. Hett. First published in 1936.

Arnold, V. I. (1978). *Mathematical methods of classical mechanics*. New York: Springer.

Arrow, K. J. (1951). *Social choice and individual values*. New York: Wiley.

Attias, H. (1999). Independent factor analysis. *Neural Computation* 11:803–852.

Balzer, W. (1978). *Empirische Geometrie und Raum-Zeit-Theorie in Mengentheoretischer Darstellung*. Königstein/Ts.: Scriptor Verlag.

Bayes, T. (1763). An essay towards solving a problem in the doctrine of chances. *Philosophical Transactions of the Royal Society* 53:370–418. Reprinted in *Facsimiles of two papers by Bayes* with a commentary by W. Edwards Deming, New York, Hafner Pub. Co., 1940.

Beck, A. (1890). Die Ströme der Nervencentren. *Centralblatt für Physiologie* 4:572–573.

Bell, J. S. (1964). On the Einstein-Podolsky-Rosen paradox. *Physics* 1:195–200.

Bentham, J. (1789/1907). *The principles of morals and legislation*. Oxford: Clarendon Press. First published in 1789.

Berger, H. (1929). Uber das Elektroenkephalogram des Menschen. *Archiv für Psychiatrie Nervenkrankheiten* 87(4):527–570. Translated by P. Gloor in *Electroencephalography Clinic Neurophysiology*, Supplement **28**, 37–73, 1969.

Berkeley, G. (1709/1901). An essay towards a new theory of vision. In A. C. Fraser, ed., *Berkeley's complete works*, vol. 1, pages 93–210. New York: Oxford University Press. First published in 1709.

Bernoulli, D. (1738). Specimen theoriae novae de mensura sortis. *Comentarii academiae scientiarum imperiales petropolitanae* 5:175–192. Translation by L. Sommer in *Econometrica* 1954, **22**, 23–26.

Bernoulli, J. (1713). *Ars conjectandi*. Basileae: Impensis Thurnisiorum fratrum. Reprinted on pp. 106-286 in Vol. 3 of *Die Werke von Jakob Bernoulli*, Birkhäuser Verlag, Basel, 1975.

Bertrand, J. L. (1888). *Calcul des probabiliti*és. Paris: Gauthier-Villars.

Bever, T. G., J. Fodor, and M. Garrett (1968). A formal limitation of associationism. In T. R. Dixon and D. L. Horton, eds., *Verbal behavior and general behavior theory*. Englewood Cliffs: Prentice-Hall.

Blank, A. A. (1953). The Luneburg theory of binocular visual space. *Journal of the Optical Society of America* 43:717–727.

Blank, A. A. (1957). The geometry of vision. *British Journal of Physiological Optics* 14:154–169.

Blank, A. A. (1958a). Analysis of experiments in binocular space perception. *Journal of the Optical Society of America* 48:911–925.

Blank, A. A. (1958b). Axiomatics of binocular vision: The foundations of metric geometry in relation to space perception. *Journal of the Optical Society of America* 48:328–334.

Blank, A. A. (1961). Curvature of binocular visual space: An experiment. *Journal of the Optical Society of America* 51:335–339.

Blumenfeld, W. (1913). Untersuchungen über die scheinbare Grösse in Sehraume. *Zeitschrift für Psychologie und Physiologie der Sinnesorgane* 65:241–404.

Bohm, D. (1965). *The special theory of relativity*. New York: W. A. Benjamin.

Borsuk, K. and W. Szmielew (1960). *Foundations of geometry*. Amsterdam: North Holland.

Bourbaki, N. (1950). The architecture of mathematics. *American Mathematical Monthly* 57:231–232.

Bouwmeester, D., J. W. Pan, M. Daniell, H. Weinfurter, and A. Zeilinger (1999). Observation of three-photon Greenberger-Horne-Zeilinger entanglement. *Physical Review Letters* 82(7):1345–1349.

Bowerman, M. (1989). Learning a semantic system. What role do cognitive predispositions play? In M. L. Rice and R. L. Schiefelbusch, eds., *The teachability of language*, pages 133–169. Baltimore: Paul H. Brookes.

Bowerman, M. (1996). The origin of children's spatial semantic categories: Cognitive vs. linguistic determinants. In J. J. Gumperz and S. C. Levinson, eds., *Rethinking linguistic relativity*, pages 145–176. Cambridge, England: Cambridge University Press.

Bradley, R. A. (1954a). Rank analysis of incomplete block designs, II. Additional tables for the method of paired comparisons. *Biometrika* 41:502–537.

Bradley, R. A. (1954b). Incomplete block rank analysis: On the appropriateness of the model for a method of paired comparisons. *Biometrics* 10:375–390.

Bradley, R. A. (1955). Rank analysis of incomplete block designs, III. Some large-sample results on estimation and power for a method of paired comparisons. *Biometrika* 42:450–470.

Bradley, R. A. and M. E. Terry (1952). Rank analysis of incomplete block designs, I. The method of paired comparisons. *Biometrika* 39:324–345.

Breiman, L. (1999). Prediction games and arcing algorithms. *Neural Computation* 11:1493–1518.

Bresnan, J. (1982). The passive in lexical theory. In J. Bresnan, ed., *The mental representation of grammatical relations*, pages 3–86. Cambridge, MA: MIT Press.

Brontë, C. (1847/1971). *Jane Eyre, an authoritative text, backgrounds, criticism, edited by Richard J. Dunn*. New York: W. W. Norton & Company. First published in 1847.

Bruns, H. (1887). Über die Integrale des Vielkörper-Problems. *Acta Mathematica* 11:25–96.

Brush, S. G. (1976). *The kind of motion we call heat: A history of the kinetic theory of gases in the 19th century*. Amsterdam: North-Holland Publishing Co.

Buffon, G. L. (1733). Communication. *Actes de l'académie des sciences - Paris* pages 43–45.

Buffon, G. L. (1777). *Essai d'arithmétique morale*.

Busemann, H. (1955). *The geometry of geodesics*. New York: Academic Press.

Bush, R. R. and W. K. Estes, eds. (1959). *Studies in mathematical learning theory*. Stanford, CA: Stanford University Press.

Bush, R. R. and F. Mosteller (1955). *Stochastic models for learning*. New York: Wiley.

Campbell, N. R. (1920). *Physics, the elements*. Cambridge University Press.

Campbell, N. R. (1928). *An account of the principles of measurement and calculation*. New York: Longmans, Green and Co.

Cantor, G. (1895). Beiträge zur Begründung der transfiniten Mengenlehre. *Mathematische Annalen* 46:481–512. Translation by P. E. B. Jourdain in *Contributions to the founding of the theory of transfinite numbers*, Dover, New York 1952, pp. 85–136 and pp. 137–201, with original pagination indicated.

Carnap, R. (1945). On inductive logic. *Philos. Sci.* 12:72–97.

Carnap, R. (1950). *The logical foundations of probability*. Chicago, IL: University of Chicago Press.

Carnap, R. (1952). *The continuum of inductive methods*. Chicago, IL: The University of Chicago Press.

Cartwright, N. (1999). *The dappled world: A study of the boundaries of Science*. Cambridge, England: Cambridge University Press.

Caton, R. (1875). The electric currents of the brain. *British Medical Journal* 2:278. Abstract only.

Chaitin, G. J. (1966). On the length of programs for computing finite binary sequences. *Journal of the ACM* 13(4):547–569.

Chaitin, G. J. (1969). On the length of programs for computing finite binary sequences. *Journal of the ACM* 16(1):145–159.

Choi, S. and M. Bowerman (1991). Learning to express motion events in English and Korean: The influence of language-specific lexicalization patterns. *Cognition* 41(1):83–121.

Chomsky, N. (1957). *Syntactic structures*. The Hague: Mouton.

Chomsky, N. (1959). On certain formal properties of grammars. *Information and Control* 2:137–167.

Chuaqui, R. (1977). A semantical definition of probability. In A. I. Arruda and N. C. A. da Costa, eds., *Non classical logic, model theory and computability*, pages 135–168. Amsterdam: North Holland.

Chuaqui, R. (1991). *Truth, possibility, and probability*. Amsterdam: North Holland.

Chuaqui, R. and P. Suppes (1990). An equational deductive system for the differential and integral calculus. In P. Martin-Löf and G. Mints, eds., *Lecture notes in computer science, Proceedings of COLOG-88 International Conference on Computer Logic*, pages 25–49. Berlin: Springer.

Chuaqui, R. and P. Suppes (1995). Free-variable axiomatic foundations of infinitesimal analysis: A fragment with finitary consistency proof. *The Journal of Symbolic Logic* 60:122–159.

Church, A. (1936). An unsolvable problem of elementary number theory. *American Journal of Mathematics* 58:345–363.

Church, A. (1940). On the concept of a random sequence. *Bulletin of American Mathematical Society* 46(2):130–135.

Clagett, M. (1959). *The science of mechanics in the middle ages*. Madison, WI: University of Wisconsin Press.

Compton, A. H. (1923). A quantum theory of the scattering of X-rays by light elements. *Physical Review* 21:483–502.

Cooley, J. W. and J. W. Tukey (1965). An algorithm for the machine computation of complex Fourier series. *Math. Computation* 19:297–301.

Copeland, A. H. (1928). Admissible numbers in the theory of probability. *Amer. Jour. Math.* 50(4):535–552.

Coren, S. and J. S. Girgus (1978). *Seeing is deceiving : the psychology of visual illusions*. Hillsdale, NJ: Lawrence Erlbaum.

Cover, T. M., P. Gacs, and R. M. Gray (1989). Kolmogorov's contribution to information theory and algorithmic complexity. *Annals of Probability* 17:840–863.

Cover, T. M. and J. A. Thomas (1991). *Elements of information theory*. New York: Wiley.

Crangle, C. and P. Suppes (1989). Geometrical semantics for spatial prepositions. In P. A. French, T. E. Uehling, Jr., and H. K. Wettstein, eds., *Midwest studies in philosophy*, vol. 14, pages 399–422. Notre Dame, IN: University of Notre Dame Press.

da Costa, N. C. A. and R. Chuaqui (1988). On Suppes' set theoretical predicates. *Erkenntnis* 29:529–589.

Daniels, N. (1972). Thomas Reid's discovery of a non-Euclidean geometry. *Philosophy of Science* 39:219–234.

David, F. N. (1962). *Games, gods and gambling: The origins and history of probability and statistical ideas from the earliest times to the Newtonian era*. London: Charles Griffin.

Davidson, D. and P. Suppes (1956). A finitistic axiomatization of subjective probability and utility. *Econometrica* 24:264–275.

Davidson, D., P. Suppes, and S. Siegel (1957). *Decision making: An experimental approach*. Stanford, CA: Stanford University Press. Midwahy Reprint, Chicago, IL: University of Chicago, Press, 1977.

Davis, M. (1958). *Computability and unsolvability*. New York: McGraw–Hill.

Davis, R. L. (1953). The number of structures of finite relations. *Proceedings of the American Mathematical Society* 4:486–495.

Dawid, A. P. (1986). Probability forecasting. In S. Kotz, N. L. Johnson, and C. B. Read, eds., *Encyclopedia of statistical sciences*, vol. 7, pages 210–218. New York: Wiley.

de Barros, A. J. and P. Suppes (2000). Inequalities for dealing with detector inefficiencies in Greenberger-Horne-Zeilinger-type experiments. *Physical Review Letters* 84:793–797.

de Finetti, B. (1931). Sul significato della probabilità. *Fundamenta Mathematicae* 17:298–329.

de Finetti, B. (1937/1964). La prévision: ses lois logiques, ses sources subjectives. *Annales de l'Institut Henri Poincaré* 7:1–68. Translation in Kyburg and Smokler (1964), 93–158.

de Finetti, B. (1951). Recent suggestions for the reconciliation of theories of probability. In J. Neyman, ed., *Proceedings of the second Berkeley symposium on mathematical statistics and probability*. Berkeley: University of California Press.

de Finetti, B. (1974). *Theory of probability*, vol. 1. New York: Wiley. Translated by A. Machi and A. Smith.

de Finetti, B. (1975). *Theory of probability*, vol. 2. New York: Wiley. Translated by A. Machi and A. Smith.

de Moivre, A. (1718/1756). *The doctrine of chances*. London: Chelsea, 3rd edn. First published in 1718.

Dedekind, R. (1872). *Stetigkeit und Irrationale Zahlen*. Vieweg: Braunschweig. Translated by W. W. Beman, Continuity and irrational numbers, in 'Essays on the theory of numbers', 1-27, New York: Dover, 1963.

Dembowski, P. (1968). *Finite geometries*. New York: Springer.

Dempster, A. P. (1967). Upper and lower probabilities induced by a multivalued mapping. *Ann. Math. Statist.* 38:325–340.

Descartes, R. (1637/1954). *La Géométrie*. New York: Dover. Translated from the French and Latin by D. E. Smith and M. L. Latham. First published in 1637.

Descartes, R. (1644/1983). *Principia philosophiae*. Amsterdam: Ludovicum Elzevirium. Translation by V. R. Miller and R. P. Miller, *Principles of philosophy*, Reidel, Dordrecht, 1983. First published 1644.

Descartes, R. (1649/1927). Passions of the soul. In R. M. Eaton, ed., *Descartes selections*, pages 361–403. New York: Charles Scribner's sons. First published in 1649.

Diaconis, P. (1977). Finite forms of de Finetti's theorem on exchangeability. *Synthese* 36:271–281.

Diaconis, P. and D. Freedman (1980). Finite exchangeable sequences. *The Annals of Probability* 8(4):745–764.

Dijksterhius, E. J. (1957). *Archimedes*. New York: Humanities Press.

Dirac, P. A. M. (1947). *The principles of quantum mechanics*. Oxford: Clarendon Press, 3rd edn. First published in 1930.

Domotor, Z. (1969). Probabilistic relational structures and their applications. Tech. Rep. 144, Stanford University, Institute for Mathematical Studies in the Social Sciences, Stanford, CA.

Domotor, Z. (1972). Causal models and space-time geometries. *Synthese* 24:5–57.

Doob, J. L. (1941). Probability as measure. *Ann. Math. Stat.* 12:206–214, 216–217.

Doob, J. L. (1960). Some problems concerning the consistency of mathematical models. In R. E. Machol, ed., *Information and Decision Processes*, pages 27–33. New York: McGraw–Hill.

Drösler, J. (1966). Das beidaeugige Raumsehen. In W. Metzger, ed., *Handbuch der Psychologie*, pages 590–615. Goettingen: Hogrefe.

Drösler, J. (1979a). Foundations of multi-dimensional metric scaling in Cayley-Klein geometries. *British Journal of Mathematical and Statistical Psychology* 32:185–211.

Drösler, J. (1979b). Grundlegung einer mehrdimensionalen metrischen Skalierung. *Zeitschrift fuer experimentelle und angewandte Psychologie* 26:1–36.

Drösler, J. (1979c). Relativistic effects in visual perception of real and apparent motion. *Arch. Psychol.* 131:249–266.

Drösler, J. (1987). Experimentelle untersuchung der Geometrie des monokularen Sehraumes. *Zeitschrift fuer experimentelle und angewandte Psychologie* 3:351–369.

Drösler, J. (1988). The psychophysical function of binocular space perception. *Journal of Mathematical Psychology* 32:285–297.

Drösler, J. (1992). Eine untersuchung des perspektivischen Sehens. *Zeitschrift fuer experimentelle und angewandte Psychologie* 34:515–532.

Du Bois-Reymond, E. (1848). *Untersuchungen über thierische Elektricität*. Berlin: Verlang von G. Reiner. Passage quoted translated by Hebbel E. Hoff, Galvani and the pre-Galvanic electrophysiologists, *Annals of Science*, 1 (1936), 157–172.

Einstein, A. (1905/1923). Zur electrodynamik bewegter Körper. *Annalen der Physik* 17. Translation in Lorentz et al. (1923).

Einstein, A. (1934). On the method of theoretical physics. *Philosophy of Science* 1:163–169.

Einstein, A. (1949). Remarks to the essays appearing in this collective volume. In P. A. Schilpp, ed., *Albert Einstein: Philosopher-scientist*, Library of Living Philosophers, Vol. 7, pages 663–688. Chicago: Open Court Publishing.

Epelboim, J. and P. Suppes (2001). A model of eye movements and visual working memory during problem solving in geometry. *Vision Research* 41:1561–1574.

Epicurus (1940). Epicurus to Herodotus. In Oates (1940), pages 3–15.

Estes, W. K. (1959a). Component and pattern models with Markovian interpretations. In Bush and Estes (1959).

Estes, W. K. (1959b). The statistical approach to learning theory. In S. Koch, ed., *Psychology: A Study of a Science*. New York: McGraw–Hill.

Estes, W. K. and P. Suppes (1959a). Foundations of linear models. In Bush and Estes (1959), pages 137–179.

Estes, W. K. and P. Suppes (1959b). Foundations of statistical learning theory, II. The stimulus sampling model. Tech. Rep. 26, Institute for Mathematical Studies in the Social Sciences, Stanford University.

Estes, W. K. and P. Suppes (1974). Foundations of stimulus sampling theory. In D. H. Krantz, R. C. Atkinson, R. D. Luce, and P. Suppes, eds., *Contemporary developments in mathematical psychology, Vol. 1: Learning, memory and thinking*, pages 163–183. San Francisco, CA: Freeman.

Euclid (1926). *Elements. The thirteen books of Euclid's Elements*. New York: Dover, 2nd edn. Translated from the text of Heiberg, with introduction and commentary, by Sir Thomas L. Heath.

Euclid (1945). Optics. *Journal of the Optical Society of America* 35. Translated by H. E. Burton.

Euclid (1996). *Euclid's Phaenomena: A translation and study of a Hellenistic treatise in spherical astronomy*. New York: Garland Publishing. Translated by J. L. Berggren and R. S. D. Thomas.

Euler, L. (1842). *Lettres de Euler à une Princesse d'Allemagne, sur divers sujets de physique et de philosophie*. Paris: L. Hachette.

Fagin, R. (1976). Probabilities on finite models. *Journal of Symbolic Logic* 41:50–58.

Feldman, J. A., G. Lakoff, D. Bailey, S. Narayanan, T. Regier, and A. Stolcke (1996). L0: The first five years. *Artificial Intelligence Review* 10:103–129.

Feldman, J. A., G. Lakoff, A. Stolcke, and S. Hollbach Weber (1990). Miniature language acquisition: A touchstone for cognitive science. In *Proceedings of the 12th Annual Conference of the Cognitive Science Society*, pages 686–693. Cambridge, MA: MIT Press.

Feller, W. (1950). *An introduction to probability theory and its applications*, vol. 1. New York: Wiley.

Feller, W. (1966). *An introduction to probability theory and its applications*, vol. 2. New York: Wiley.

Fine, A. (1982). Hidden variables, joint probability, and the Bell inequalities. *Physical Review Letters* 48:291–295.

Fine, T. L. (1973). *Theories of probability*. New York: Academic Press.

Finke, R. A. (1989). *Principles of mental imagery*. Cambridge, MA: MIT Press.

Fishburn, P. C. (1970). *Utility theory for decision making*. New York: Wiley.

Fitelson, B. (2001). *Studies in Bayesian Confirmation Theory*. Ph.D. thesis, University of Wisconsin-Madison, Ann Arbor, MI.

FitzGerald, G. F. (1889). The ether and the Earth's atmosphere. *Science* 13(328):390.

Fock, V. A. (1931/1978). *Fundamentals of quantum mechanics*. Moscow: Mir Publishers. Translated from the Russian by E. Yankovsky.

Fodor, J. and Z. Pylyshyn (1988). Connectionism and cognitive architecture: A critical analysis. *Cognition* 28:3–71.

Foley, J. M. (1964). Desarguesian property in visual space. *Journal of the Optical Society of America* 54:5:684–692.

Foley, J. M. (1965). Visual space: A scale of perceived relative direction. *Proceedings of the 73rd Annual Covention of the American Psychological Association* 1:49–50.

Foley, J. M. (1966). Locus of perceived equidistance as a function of viewing distance. *Journal of the Optical Society of America* 56:822–827.

Foley, J. M. (1969). Distance in stereoscopic vision: The three-point problem. *Vision Research* 9:1505–1521.

Foley, J. M. (1972). The size-distance relation and intrinsic geometry of visual space: Implications for processing. *Vision Research* 12:323–332.

Foley, J. M. (1978). Distance perception. In R. Held, H. Leibowitz, and H. L. Teuber, eds., *Handbook of sensory physiology, Vol. 8: Perception*, pages 181–213. New York: Springer.

Ford, L. R., Jr. (1957). Solution of a ranking problem from binary comparisons. *American Mathematical Monthly* pages 28–33.

Frank, P. (1909). Die Stellung des Relativitätsprinzips in System der Mechanik und der Elektrodynamik. *Sitzungsberichte der Akademie der Wissenschaften in Wien. Mathemat. Naturw. Klasse* 118(2a):373–446.

Frege, G. (1899). Letter to Hilbert. In Frege (1980), pages 35–36.

Frege, G. (1900). Letter to Hilbert. In Frege (1980), page 43.

Frege, G. (1903*a*). Über die Grundlagen der Geometrie. *Jahresbericht der Deutschen Mathematiker-Vereinigung* 12:319–324. Translation in Frege (1971).

Frege, G. (1903*b*). Über die Grundlagen der Geometrie II. *Jahresbericht der Deutschen Mathematiker-Vereinigung* 12:368–375. Translation in Frege (1971).

Frege, G. (1906*a*). Über die Grundlagen der Geometrie I. *Jahresbericht der Deutschen Mathematiker-Vereinigung* 15:293–309. Translation in Frege (1971).

Frege, G. (1906*b*). Über die Grundlagen der Geometrie (Fortsetzung) II. *Jahresbericht der Deutschen Mathematiker-Vereinigung* 15:377–403. Translation in Frege (1971).

Frege, G. (1906*c*). Über die Grundlagen der Geometrie (Schluss) III. *Jahresbericht der Deutschen Mathematiker-Vereinigung* 15:423–430. Translation in Frege (1971).

Frege, G. (1971). *On the foundations of geometry and formal theories of arithmetic*. London and New Haven: Yale University Press. Translation by E. H. Kluge.

Frege, G. (1980). *Philosophical and mathematical correspondence*. Chicago, IL: University of Chicago Press. Edited by G. Gabriel.

Freudenthal, H. (1965). Lie groups in the foundations of geometry. *Advances in Mathematics* 1:145–190.

Friedman, J., T. Hastie, and R. Tibshirani (2000). Additive logistic regression: A statistical view of boosting. *Annals of Statistics* 28:337–407.

Fu, K. S. (1974). *Syntactic methods in pattern recognition*. New York: Academic Press.

Gaifman, H. (1964). Concerning measures in first order calculi. *Israel J. Math.* 2:1–18.

Gaifman, H. and M. Snir (1982). Probabilities of rich languages, testing and randomness. *Journal of Symbolic Logic* 47:495–548.

Galavotti, M. C. (1987). Comments on Patrick Suppes "Propensity interpretations of probability". *Erkenntnis* 26:359–368.

Galavotti, M. C. (In press). Between logicism and subjectivism: Harold Jeffreys's probabilistic epistemology. *British Journal for the Philosophy of Science*.

Galison, P. (1987). *How experiments end*. Chicago: University of Chicago Press.

Galvani, L. (1791/1953). De viribus electricitatis in motu musculari. *De Bononiensi Scientiarum et Artium Instituto atque Academia, Comm.* 7:363–418. Translated by Margaret Glover Foley, in *Luigi Galvani, Commentary on the Effects of Electricity on Muscular Motion*. Notes and introduction by I. Bernard Cohen. Norwalk, CT: Burndy Library, 1953.

Galvani, L. (1794). *Dell'uso, e dell'attivitá, dell'arco conduttore nelle contrazioni dei muscoli*. Bologna: A. S. Tommaso d'Aquino. Published anonymously. Portion translated in: B. Dibner, *Galvani-Volta: A controversy that led to the discovery of useful electricity*. Norwalk, CT: Burndy Library, 1952.

Garceau, L. and H. Davis (1935). An ink-writing electroencephalogram. *Archives of Neurology and Psychiatry* 34:1292–1294.

Gazdar, G., E. Klein, G. Pullum, and I. Sag (1985). *Generalized phrase structure grammar*. Cambridge, MA: Harvard University Press.

Geddes, L. A. (2001). Contributions of the vacuum tube to early electrophysiological research. *IEEE Engineering in Medicine and Biology* pages 118–126.

Gibbs, F. A., H. Davis, and W. G. Lennox (1935). The electroencephalogram in epilepsy as a condition of impaired consciousness. *Archives of Neurology and Psychiatry* 34(6):1135–1148.

Giere, R. N. (1973). Objective single-case probabilities and the foundations of statistics. In P. Suppes, L. Henkin, G. C. Moisil, and A. Joja, eds., *Logic, methodology and philosophy of science*, vol. 4, pages 467–483. Amsterdam: North-Holland Publishing Co.

Gödel, K. (1949). A remark about the relationship between relativity theory and idealistic philosophy. In P. A. Schilpp, ed., *Albert Einstein: Philosopher-scientist*, Library of Living Philosophers, Vol. 7, pages 555–562. Chicago: Open Court Publishing.

Gogel, W. C. (1956a). Relative visual direction as a factor in relative distance perceptions. *Psychological Monographs* 70(11). Whole No. 418.

Gogel, W. C. (1956b). The tendency to see objects as equidistant and its inverse relation to lateral separation. *Psychological Monographs* 70(4). Whole No. 411.

Gogel, W. C. (1963). The visual perception of size and distance. *Vision Research* 3:101–120.

Gogel, W. C. (1964a). The perception of depth from binocular disparity. *Journal of Experimental Psychology* 67:379–386.

Gogel, W. C. (1964b). Size cue to visually perceived distance. *Psychological Bulletin* 62:217–235.

Gogel, W. C. (1965). Equidistance tendency and its consequences. *Psychological Bulletin* 64:153–163.

Goldblatt, R. (1987). *Orthogonality and space-time geometry*. New York: Springer.

Goldreich, O., S. Goldwasser, and S. Micali (1986). How to construct random functions. *Journal of the ACM* 33(4):792–807.

Good, I. J. (1962). Subjective probability as the measure of a non-measurable set. In E. Nagel, P. Suppes, and A. Tarski, eds., *Logic, methodology and philosophy of science: Proceedings of the 1960 International Congress*, pages 319–329. Stanford: Stanford University Press.

Good, I. J. (1983). *Good thinking: The foundations of probability and its applications*. Minneapolis: University of Minnesota Press.

Green, G. (1958). *An essay on the application of mathematical analysis to the theories of electricity and magnetism*. Göteborg, Sweden: Wezäta-Melins Aktiebolag. Facsimile reprint of the 1828 ed., printed for the author, by T. Wheelhouse, Nottingham.

Greenberger, D. M., M. A. Horne, A. Shimony, and A. Zeilinger (1990). Bell's theorem without inequalities. *Amer. J. Phys.* 58:1131–1143.

Greenberger, D. M., M. A. Horne, and A. Zeilinger (1989). Going beyond Bell's theorem. In M. Kafatos, ed., *Bell's theorem, quantum theory, and conceptions of the universe*. Dordrecht: Kluwer Academic Press.

Greenspan, D. G. (1973). *Discrete models*. Reading, MA: Addison–Wesley.

Greibach, F. A. (1965). A new normal form theorem for context-free phrase-structure grammars. *Journal of ACM* 12:42–52.

Griffith, R. M. (1949). Odds adjustment by American horse-race bettors. *American Journal of Psychology* 62:290–94.

Grünbaum, A. (1963). *Philosophical problems of space and time*. New York: Knopf.

Hacking, I. (1975). *The emergence of probability: A philosophical study of early ideas about probability, induction and statistical inference*. Cambridge, England: Cambridge University Press.

Hahn, R. (1967). Laplace's first formulation of scientific determinism in 1773. *Actes du XIe Congrès International d'Histoire des Sciences 1965* 2:167–171.

Hamel, G. (1958). Die Axiome der Mechanik. In *Handbuch der Physik*, vol. 5, pages 1–42. Berlin: Springer.

Hardy, L. H., G. Rand, M. C. Rittler, A. A. Blank, and P. Boeder (1953). *The geometry of binocular space perception*. Elizabeth, NJ: Schiller.

Harré, R. (1999). Models and type-hierarchies: Cognitive foundations for iconic thinking. In R. C. Paton and I. Neilson, eds., *Visual representations and interpretations*, pages 97–111. London: Springer.

Harrison, M. A. (1965). *Introduction to switching and automata theory*. New York: McGraw–Hill.

Hastie, T., R. Tibshirani, and J. Friedman (2001). *The elements of statistical learning*. New York: Springer.

Helmholtz, H. (1868). Über die Tatsachen, die der Geometrie zugrunde liegen. *Wissenschaftliche Abhandlungen* 2:618–637.

Henkin, L., P. Suppes, and A. Tarski, eds. (1959). *The axiomatic method with special reference to geometry and physics*. Amsterdam: North-Holland Publishing Co. Proceedings of an international symposium held at the University of California, Berkeley, December 16, 1957–January 4, 1958.

Hermes, H. (1938). Eine Axiomatisierung des allgemeinen Mechanik. *Forschungen zur Logik und zur Grundlegung der exakten Wissenschaften, Neue Folge* 3.

Hilbert, D. (1897/1930). *Grundlagen der Geometrie*. Leipzig and Berlin: B. G. Teubner, seventh edn. First published in 1897.

Hilbert, D. (1899). Letter to Frege. In Frege (1980), pages 39–40.

Hintikka, J. (1966). A two-dimensional continuum of inductive logic. In Hintikka and Suppes (1966), pages 113–132.

Hintikka, J. and P. Suppes, eds. (1966). *Aspects of inductive logic*. Amsterdam: North-Holland Publishing Co.

Holland, P. W. and P. R. Rosenbaum (1986). Conditional association and unidimensionality in monotone latent trait models. *Annals of Statistics* 14:1523–1543.

Hopcroft, J. E. and J. D. Ullman (1969). *Formal languages and their relation to automata*. Reading, MA: Addison–Wesley.

Hopf, E. (1934). On causality, statistics and probability. *Journal of Mathematics and Physics* 17:51–102.

Huddleston, R. and G. Pullum (2002). *The Cambridge grammar of the English language*. Cambridge, England: Cambridge University Press.

Hudgin, R. H. (1973). Coordinate-free relativity. In P. Suppes, ed., *Space, time and geometry*, pages 366–382. Dordrecht: Reidel.

Hume, D. (1739/1951). *A treatise of human nature*. London: John Noon. Quotations taken from L. A. Selby-Bigge's edition, Oxford University Press, London, 1951.

Humphreys, P. (1985). Why propensities cannot be probabilities. *Philosophical Review* 94:557–570.

Huygens, C. (1657). *De ratiociniis in ludo aleae*. Ludg. Batav., Ex officinia J. Elsevirii.

Huygens, C. (1673/1986). *The pendulum clock, or geometrical demonstration of pendula as applied to clocks*. Ames, IA: Iowa State University Press. First published in 1673. Translated by R. J. Rockwell.

Indow, T. (1967). Two interpretations of binocular visual space: Hyperbolic and Euclidean. *Annals of the Japan Association for Philosophy of Science* 3:51–64.

Indow, T. (1968). Multidimensional mapping of visual space with real and simulated stars. *Perception and Psychophysics* 3:45–53.

Indow, T. (1974a). Applications of multidimensional scaling in perception. In E. C. Carterette, ed., *Handbook of perception: Vol. 2. Psychophysical judgment and measurement*, pages 493–531. New York: Academic Press.

Indow, T. (1974b). On geometry of frameless binocular perceptual space. *Psychologia* 17:50–63.

Indow, T. (1975). An application of MDS to study of binocular visual space. In *Theory, methods and applications of multidimensional scaling and related techniques*. University of California, San Diego. U.S.-Japan Seminar (sponsored by the National Science Foundation and Japan Society for the Promotion of Science).

Indow, T. (1979). Alleys in visual space. *Journal of Mathematical Psychology* 19:221–258.

Indow, T. (1982). An approach to geometry of visual space with no a priori mapping functions: Multidimensional mapping according to Riemannian metrics. *Journal of Mathematical Psychology* 26:204–236.

Indow, T., E. Inoue, and K. Matsushima (1962a). An experimental study of the Luneburg theory of binocular space perception (1): The 3- and 4-point experiments. *Japanese Psychological Research* 4:6–16.

Indow, T., E. Inoue, and K. Matsushima (1962b). An experimental study of the Luneburg theory of binocular space perception (2): The alley experiments. *Japanese Psychological Research* 4:17–24.

Indow, T., E. Inoue, and K. Matsushima (1963). An experimental study of the Luneburg theory of binocular space perception (3): The experiments in a spacious field. *Japanese Psychological Research* 5:10–27.

James, W. (1890/1950). *Principles of psychology*. New York: Dover. First published in 1890.

Jamison, D., D. Lhamon, and P. Suppes (1970). Learning and the structure of information. In J. Hintikka and P. Suppes, eds., *Information and inference*, pages 197–259. Dordrecht: Reidel.

Jammer, M. (1961). *Concepts of mass in classical and modern physics*. Cambridge, MA: Harvard University Press.

Jasper, H. H. and L. Carmichael (1935). Electrical potentials from the intact human brain (special article). *Science* 81:51–53.

Jaynes, E. T. (1978). Where do we stand on maximum entropy? In Rosenkrantz (1983), pages 211–314.

Jaynes, E. T. (1979). Concentration of distributions at entropy maxima. In Rosenkrantz (1983), pages 315–336.

Jeffrey, R. C. (1965). *The logic of decision*. New York: McGraw-Hill.

Jeffreys, H. (1948). *Theory of probability*. Oxford: Clarendon Press, 2nd edn. First published in 1939.

Jeffreys, H. (1957). *Scientific inference*. Cambridge, England: Cambridge University Press, 2nd edn.

Jeffreys, H. and D. Wrinch (1919). On certain aspects of the theory of probability. *Philosophical Magazine* 38:715–731.

Jones, A. (2000). Pappus' notes to Euclid's Optics. In P. Suppes, J. M. Moravcsik, and H. Mendell, eds., *Ancient and mediaeval traditions in the exact sciences*, pages 49–58. Stanford, CA: CSLI Publications.

Jones, O. (1867). *Examples of Chinese ornament selected from objects in the South Kensington Museum and other collections*. London: S. & T. Gilbert.

Joyce, J. (1934). *Ulysses*. New York: Random House.

Kac, M. (1959). *Probability and related topics in the physical sciences*. New York: Interscience.

Kandel, E. R. (1985). Cellular mechanisms of learning and the biological basis of individuality. In E. R. Kandel and J. H. Schwartz, eds., *Principles of neural science*. Amsterdam: Elsevier, 2nd edn.

Kant, I. (1781/1997). *Critique of pure reason*. New York: Cambridge University Press. First published in 1781. Translated by P. Guyer and A. W. Wood.

Kant, I. (1786/1970). *Metaphysical foundations of natural science*. New York: Bobbs-Merrill Company, Inc. First published in 1786. Translated, with introduction and essay, by J. Ellington.

Kaplan, R. M. and J. Bresnan (1982). Lexical-functional grammar: A formal system for grammatical representation. In J. Bresnan, ed., *The Mental representation of grammatical relations*, pages 173–281. Cambridge, MA: MIT Press.

Kaufmann, S. A. (1982). The crystallization and selection of dynamical order in the evolution of metazoan gene regulation. In H. Haken, ed., *Evolution of order and chaos in physics, chemistry, and biology*, page 28. New York: Springer.

Keller, J. B. (1986). The probability of heads. *American Mathematical Monthly* 93:191–197.

Kelly, F. P. (1979). *Reversibility and stochastic networks*. New York: Wiley.

Kelvin (1846). On the mathematical theory of electricity in equilibrium. *Cambridge and Dublin Mathematical Journal* 1:75–96. (Lord Kelvin, also W. Thomson).

Keynes, J. M. (1921). *A Treatise on probability*. London: Macmillan.

Khinchin, A. I. (1949). *Statistical mechanics*. New York: Dover.

Kieras, D. E. (1976). Finite automata and S-R models. *Journal of Mathemathical Psychology* 13:127–147.

Kleene, S. C. (1936). Lambda-definability and recursiveness. *Duke Mathematical Journal* 2:340–353.

Kleene, S. C. (1956). Representation of events in nerve nets and finite automata. In C. E. Shannon and J. McCarthy, eds., *Automata studies*, pages 3–41. Princeton, NJ: Princeton University Press.

Klein, F. (1893). A comparative review of recent researches in geometry. *Bulletin of the New York Mathematical Society* 2:215–249.

Knorr, W. R. (1975). *The evolution of the Euclidean elements*. Dordrecht: Reidel.

Knorr, W. R. (1986). *The ancient tradition of geometric problems*. Boston, MA: Birkhäuser.

Kolmogorov, A. N. (1933/1950). *Foundations of the theory of probability*. New York: Chelsea Publishing Company. Translated by N. Morrison. First published in German in 1933. The transcription of Russian into English for the last syllable of Kolmogorov's name was generally changed from *-off* to *-ov* around 1970.

Kolmogorov, A. N. (1959). Entropy per unit time as a metric invariant of automorphisms (Russian). *Doklady Akademii Nauk SSSR* 124:754–755.

Kolmogorov, A. N. (1963). On tables of random numbers. *Sankhyā: The Indian Journal of Statistics: Series A* 25:369–376.

Kolmogorov, A. N. (1965). Three approaches to the quantitative definition of information. *Problemy Peredači Informacii* 1:4–7.

Kolmogorov, A. N. (1968). Logical basis for information theory and probability theory. *IEEE Trans. Information Theory* IT-14:662–664.

Koopman, B. O. (1940a). The axioms and algebra of intuitive probability. *Annals of Mathematics* 41(2):269–292.

Koopman, B. O. (1940b). The bases of probability. *Bulletin of the American Mathematical Society* 46:763–774. Reprinted in Kyburg and Smokler (1964), 159–172.

Koopman, B. O. (1941). Intuitive probabilities and sequences. *Annals of Mathematics* 42(1):169–187.

Kosslyn, S. M. (1973). Scanning visual images: Some structural implications. *Perception & Psychophysics* 14:90–94.

Kosslyn, S. M. (1975). Information representation in visual images. *Cognitive Psychology* 7:341–370.

Kosslyn, S. M. (1976). Can imagery be distinguished from other forms of internal representation? Evidence from studies of information retrieval times. *Memory & Cognition* 14:93–111.

Kosslyn, S. M. (1978). Measuring the visual angle of the mind's eye. *Cognitive Psychology* 10:356–389.

Kosslyn, S. M. (1980). *Image and mind*. Cambridge, MA: Harvard University Press.

Kosslyn, S. M., B. J. Reiser, M. J. Farah, and F. S. L. (1983). Generating visual images: Units and relation. *Journal of Experimental Psychology, general* 112:278–303.

Kraft, C. H., J. W. Pratt, and A. Seidenberg (1959). Intuitive probability on finite sets. *The Annals of Mathematical Statistics* 30:408–419.

Krantz, D. H., R. D. Luce, P. Suppes, and A. Tversky (1971). *Foundations of measurement Vol. I: Additive and polynomial representations*. New York: Academic Press.

Krauss, P. H. (1969). Representation of symmetric probability models. *J. Symbolic Logic* 34:183–193.

Kuratowski, C. (1921). Sur la notion de l'ordre dans la theorie des ensembles. *Fundamenta Mathematicae* 2:161–171.

Kyburg Jr., H. E. (1974). *The logical foundations of statistical inference*. Dordrecht: Reidel.

Kyburg Jr., H. E. and H. E. Smokler, eds. (1964). *Studies in subjective probability*. New York: Wiley.

Lagrange, J. L. (1873). *Oeuvres de Lagrange*, vol. 6. Paris: Gauthier-Villars. Edited by J. A. Serret.

Lamb, H. (1919). The kinematics of the eye. *Philosophical Magazine* 38:685–695.

Lamperti, J. and P. Suppes (1959). Chains of infinite order and their application to learning theory. *Pacific Journal of Mathematics* 9:739–754.

Langley, P. and J. G. Carbonell (1987). Language acquisition and machine learning. In B. MacWhinney, ed., *Mechanisms of language acquisition*, pages 115–155. Hillsdale, NJ: Lawrence Erlbaum.

Laplace, P. S. (1773/1776). Recherches, 1 sur l'intégration des équations différentielles aux différences finies, et sur leur usage dans la théorie des hasards. 2, sur le principe de la gravitation universelle, et sur les inégalités séculaires des planètes qui en dépendent. *Mémoires de mathématique et de physique présentés á l'Académie Royale des Sciences par divers Sçavans, année 1773 (often cited as Savants étrangers)* 7:37–232.

Laplace, P. S. (1774a). Memoire sur les suites récurro-récurrentes et sur leurs usages dans la théorie des hasards. *Mémoires de mathématique et de physique présentés á l'Académie Royale des Sciences par divers Sçavans (often cited as Savants étrangers)* 6:353–371.

Laplace, P. S. (1774b). Memoire des la probabilité des causes par les événements. *Mémoires de mathématique et de physique présentés á l'Académie Royale des Sciences par divers Sçavans (often cited as Savants étrangers)* 6:621–656.

Laplace, P. S. (1799/1966). *Mécanique céleste*. Bronx, N.Y.: Chelsea Pub. Co. Vols. 1-4 is a reprint of the English translation by Nathaniel Bowditch of the edition published in Boston, 1829-1839.

Laplace, P. S. (1812). *Théorie analytique des probabilités*. Paris.

Laplace, P. S. (1951). *A philosophical essay on probabilities*. New York: Dover, sixth edn. Introduction of his *Théorie analytique des probabilités* (Laplace 1812), translated by F. W. Truscott and F. L. Emory.

Latzer, R. W. (1972). Nondirected light signals and the structure of time. *Synthese* 24:236–280.

Leibniz, G. W. (1692/1966). Essay de dynamique. In G. H. Pertz, ed., *Gesammelte Werke*. Hildesheim, Germany: George Olms. First published in 1692.

Leibniz, G. W. (1695). Specimen dynamicum. *Acta Eruditorum* 14:145–157. Translation by R. Francks and R.S. Woolhouse in G.W. Leibniz, *Philosophical texts*, Oxford University Press, New York, 1998.

Levelt, W. J. M. (1982). Cognitive styles in the use of spatial direction terms. In R. J. Jarvella and W. Klein, eds., *Speech, place, and action*, pages 251–268. New York: Wiley.

Levelt, W. J. M. (1984). Some perceptual limitations on talking about space. In A. J. Van Doorn, W. A. van de Grand, and K. J., eds., *Limits in perception*, pages 323–358. Utrecht: VNU Press.

Levi, I. (1980). *The enterprise of knowledge*. Cambridge, MA: MIT Press.

Lewes, G. H. (1847). The reality of Jane Eyre. *Fraser's Magazine* pages 690–693. Reprinted in "Jane Eyre", by Charlotte Brönte (1971), Richard J. Dunn's edition, p. 448.

Li, M. and P. B. M. Vitanyi (1993). *An introduction to Kolmogorov complexity and its applications*. New York: Springer.

Lindberg, D. C., ed. (1970). *John Pecham and the science of optics: Perspectiva communis*. Madison, WI: University of Wisconsin Press.

Lindenbaum, A. and A. Tarski (1934-35/1983). On the limitations of the means of expression of deductive theories. In A. Tarski, ed., *Logic, semantics, metamathematics: Papers from 1923 to 1938*, pages 384–392. Indianapolis, IN: Hackett Pub. Co. First published in 1934-35 as Uber die Beschränktheit der Ausdrucksmittel deduktiven Theorien, in *Ergebnisse eines mathematischen Kolloquiums*, **7**, pp. 15-22.

Lindley, D. V. (1971). *Bayesian statistics: A review*. Philadelphia: Soc. For Industrial and Applied Mathematics.

Lindsay, R. B. and H. Margenau (1936). *Foundations of physics*. New York: Wiley.

Loève, M. (1978). *Probability theory II*. New York: Springer, 4th edn.

Lorentz, H. A. (1892). The relative motion of the earth and the ether. *Versl. K. Ak. W. Amsterdam* 1:74. Also in H. A. Lorentz *Collected papers*, Vol. 4 p. 219–223, Nijhoff, the Hague, 1937.

Lorentz, H. A. (1895). *Versuch einer Theorie der elektrischen und optischen Erscheinungen in bewegten Körpern*. Leiden: E. J. Brill. Also in H.A. Lorentz *Collected papers*, Vol. 5 p. 1–138, Nijhoff, the Hague, 1937.

Lorentz, H. A., A. Einstein, H. Minkowski, and H. Weyl (1923). *The principle of relativity. A collection of original memoirs on the special and general theory of relativity*. New York: Dover. With notes by A. Sommerfeld. Translated by W. Perrett and G. B. Jeffery.

Luce, R. D. (1959). *Individual choice behavior*. New York: Wiley.

Luce, R. D. (1968). On the numerical representation of qualitative conditional probability. *Annals of Mathematical Statistics* 39:481–491.

Luce, R. D. (2000). *Utility of gains and losses: Measurement, theoretical and experimental approaches*. Mahwah, NJ: Erlbaum Associates.

Luce, R. D., D. H. Krantz, P. Suppes, and A. Tversky (1990). *Foundations of measurement Vol. III: Representation, axiomatization, and invariance*. San Diego, CA: Academic Press.

Luce, R. D. and P. Suppes (1965). Preference, utility and subjective probability. In R. D. Luce, R. R. Bush, and E. H. Galanter, eds., *Handbook of mathematical psychology*, vol. 3, pages 249–410. New York: Wiley.

Lucretius (1940). De rerum natura. In Oates (1940), pages 69–219.

Luneburg, R. K. (1947). *Mathematical analysis of binocular vision*. Princeton, NJ: Princeton University Press.

Luneburg, R. K. (1948). Metric methods in binocular visual perception. In *Studies and essays presented to R. Courant on his 60th birthday*, pages 215–240. New York: Wiley(Interscience).

Luneburg, R. K. (1950). The metric of binocular visual space. *Journal of the Optical Society of America* 40:627–642.

Mach, E. (1883/1942). *The science of mechanics, a critical and historical account of its development*. Chicago: Open Court Publishing, 5th edn. First published in 1883. Translated from the German by Thomas J. McCormack, containing additions and alterations up to the ninth (final) edition, with two hundred and fifty cuts and illustrations.

MacLane, S. and G. D. Birkhoff (1967). *Algebra*. New York: Macmillan.

Makridakis, S., A. Andersen, R. Carbone, R. Fildes, M. Hilbon, R. Lewandowski, J. Newton, E. Parzen, and R. Winkler (1984). *The forecasting accuracy of major time series methods*. Chichester: Wiley.

Markov, A. A. (1951). The theory of algorithms (Russian). *Trudy Matematicheskogo Instituta imeni V. A. Steklova* 38:176–189.

Martin-Löf, P. (1966). The definition of random sequences. *Information and Control* 9:606–619.

Martin-Löf, P. (1969). Literature on Von Mises' kollektivs revisited. *Theoria* 35:12–37.

Matsushima, K. and H. Noguchi (1967). Multidimensional representation of binocular visual space. *Japanese Psychological Research* 9:85–94.

Matteucci, C. (1844). *Traité des phenomènes electrophysiologiques des animaux*. Paris: Fortin Masson.

Maxwell, J. C. (1860). Illustrations of the dynamical theory of gases. *Philosophical Magazine* Reprinted in Maxwell (1890), Vol. 1, 377–409.

Maxwell, J. C. (1861-1862). On physical lines of force. *Philosophical Magazine, Series 4* 21, 23:161–175, 281–291 and 338–348, 12–24, 85–95. Reprinted in Maxwell (1890), Vol. 1, 451–513.

Maxwell, J. C. (1867). On the dynamical theory of gases. *Philosophical Transactions* 157:49–88. Reprinted in Maxwell (1890), Vol. 2, 26–78.

Maxwell, J. C. (1877). *Matter and motion*. London. Reprinted by Macmillian, London, in 1920.

Maxwell, J. C. (1890). *The scientific papers of James Clerk Maxwell*. Cambridge, England: Cambridge University Press. Edited by W. D. Niven. Reprinted by Dover, New York, in 1965.

Mayo, D. G. (1996). *Error and the growth of experimental knowledge*. Chicago: University of Chicago Press.

McGlothlin, W. H. (1956). Stability of choices among uncertain alternatives. *American Journal of Psychology* 69:604–615.

McKinsey, J. C. C., A. C. Sugar, and P. Suppes (1953). Axiomatic foundations of classical particle mechanics. *Journal of Rational Mechanics and Analysis* 2:253–272.

McKinsey, J. C. C. and P. Suppes (1953). Transformations of systems of classical particle mechanics. *Journal of Rational Mechanics and Analysis* 2:273–289.

Mehlberg, H. (1935). Essai sur la théorie causale du temps. I. *Studia Philosophica* 1:119–260.

Mehlberg, H. (1937). Essai sur la théorie causale du temps. II. *Studia Philosophica* 2:111–231.

Mellor, D. H. (1971). *The matter of chance*. Cambridge, England: Cambridge University Press.

Mermin, N. D. (1990a). Quantum mysteries revisited. *American Journal of Physics* 58(8):731–734.

Mermin, N. D. (1990b). Extreme quantum entanglement in a superposition of macroscopically distinct states. *Physical Review Letters* 65(15):1838–1840.

Mermin, N. D. (1993). Hidden variables and the two theorems of John Bell. *Reviews of Modern Physics* 65(3):803–815.

Michelson, A. A. and E. W. Morley (1887). On the relative motion of the earth and the luminiferous ether. *American Journal of Science* 34:333–336.

Millenson, J. R. (1967). An isomorphism between stimulus-response notation and information processing flow diagrams. *The Psychological Record* 17:305–319.

Miller, G. A. and N. Chomsky (1963). Finitary models of language users. In R. D. Luce, R. R. Bush, and E. Galanter, eds., *Handbook of Mathematical Psychology*, vol. 2. New York: Wiley.

Miller, G. A., E. Galanter, and K. H. Pribram (1960). *Plans and the structure of behavior*. New York: Holt.

Milne, E. A. (1948). *Kinematic relativity*. London: Oxford University Press.

Montgomery, R. (2001). A new solution to the three-body problem. *Notices of the American Mathematical Society* 48:471–481.

Moody, E. A. and M. Clagett, eds. (1952). *The medieval science of weights (Scientia de ponderibus)*. Madison, WI: The University of Wisconsin Press.

Moser, J. (1973). *Stable and random motions in dynamical systems with special emphasis on celestial mechanics*. Herman Weyl Lectures, the Institute for Advanced Study. Princeton, NJ: Princeton University Press.

Mosteller, F. and D. L. Wallace (1964/1984). *Applied Bayesian and classical inference: The case of the federalist papers*. Springer Series in Statistics. New York: Springer.

Moulines, C. U. (1975). A logical reconstruction of classical equilibrium thermodynamics. *Erkenntnis* 9:101–130.

Moulines, C. U. (1976). Approximate application of empirical theories: A general explication. *Erkenntnis* 10:201–227.

Moulines, C. U. and J. D. Sneed (1979). Suppes' philosophy of physics. In R. J. Bogdan, ed., *Patrick Suppes*, pages 59–91. Dordrecht: Reidel.

Mundy, B. (1986a). Optical axiomatization of Minkowski space-time geometry. *Philosophy of Science* 37:1–30.

Mundy, B. (1986b). The physical content of Minkowski geometry. *British Journal for the Philosophy of Science* 37:25–54.

Murray, N. and M. Holman (2001). The role of chaotic resonances in the solar system. *Nature* 410:773–779.

Nagel, E. (1939a). The formation of modern conceptions of formal logic in the development of geometry. *Osiris* 7:142–224. Reprinted in E. Nagel *Teleology revisited*, New York, Columbia University Press, 1979, pp. 195–259.

Nagel, E. (1939b). Principles of the theory of probability. In *International Encyclopedia of Unified Science*, vol. 1 (6). Chicago, IL: University of Chicago Press.

Nagel, E. (1949). The meaning of reduction in the natural sciences. In R. C. Stauffer, ed., *Science and civilization*, pages 99–135. Madison, WI: University of Wisconsin Press.

Neimark, E. D. and W. K. Estes, eds. (1967). *Stimulus sampling theory*. Holden-Day series in Psychology. San Francisco: Holden-Day.

Nelson, E. (1987). *Radically elementary probability theory*. Princeton, NJ: Princeton University Press.

Neugebauer, O. (1957). *The exact sciences in antiquity*. Providence, RI: Brown University Press, 2nd edn.

Newton, I. (1687/1946). *Principia*. Berkeley, CA: University of California Press. First published in 1687. Translation by F. Cajori.

Newton, I. (1704/1931). *Opticks*. London: Bell. Reprinted from the 4th edition. First published in 1704.

Neyman, J. (1971). Foundations of behavioristic statistics. In V. Godambe and D. Sprott, eds., *Foundations of statistical inference*, pages 1–19. Toronto: Holt, Rinehart and Winston of Canada.

Nishikawa, Y. (1967). Euclidean interpretation of binocular visual space. *Japanese Psychological Research* 9:191–198.

Noether, E. (1918). Invariante Variationsprobleme. *Nachr. Konig. Gesell. Wissen. Göttingen, Math.-Phys. Klasse 2* pages 235–257. Reprinted in Noether *Collected papers*, Springer, New York 1983.

Noll, W. (1955). Die Herleitung der Grundgleichungen der Thermomechanik der Kontinua aus der statistischen Mechanik. *Journal of Rational Mechanics and Analysis* 4:627–646.

Noll, W. (1974). *The foundations of mechanics and thermodynamics: Selected papers by W. Noll*. New York: Springer. With a preface by C. Truesdell.

Norman, M. F. (1972). *Markov processes and learning models*. New York: Academic Press.

Oates, W. J., ed. (1940). *The Stoic and Epicurean philosophers: The complete extant writings of Epicurus, Epictetus, Lucretius, Marcus Aurelius*. New York: Random House, by arrangement with Oxford University Press.

Oppenheim, A. V. and R. W. Schafer (1975). *Digital signal processing*. Englewood Cliffs, NJ: Prentice-Hall.

Ornstein, D. S. (1970). Bernoulli shifts with the same entropy are isomorphic. *Advances in Mathematics* 4:337–352.

Ornstein, D. S. and B. Weiss (1991). Statistical properties of chaotic systems. *Bull. Am. Math. Soc.(New Series)* 24:11–116.

Padoa, A. (1902). Un nouveau système irréductible de postulats pour l'algèbre. *C. R. Deuxième Congrès International des Mathématiciens, Paris* pages 249–256.

Pais, A. (1982). *'Subtle is the Lord ...' The science and life of Albert Einstein*. New York: Oxford University Press.

Palladio, A. (1570/1965). *The four books of architecture*. New York: Dover. Unabridged and unaltered republication of the work first published by Isaac Ware in 1738.

Pappus (1876-1878). *Pappi Alexandrini collectionis quae supersunt*. Berlin: Weidman. Edited by F. Hultsch.

Pasch, M. (1882). *Vorlesungen über Neuere Geometrie*. Leipzig: Verlag von Julius Springer.

Peacocke, C. (1983). *Sense and content : Experience, thought, and their relations*. London: Oxford University Press.

Pecham, J. (1970). Perspectiva communis. In Lindberg (1970).

Peres, Y. (1992). Finite violation of a Bell inequality for arbitrarily large spin. *Physical Review A* 46:4413–4414.

Pinker, S. (1984). *Language learnability and language development*. Cambridge, MA: Harvard University Press.

Pinker, S. (1989). *Learnability and cognition: The acquisition of argument structure*. Cambridge, MA: MIT Press.

Plato (1892). Meno. In *The dialogues of Plato*. London: Macmillian. Translated by B. Jowett. Reprinted in 1937, Random House, New York.

Poincaré, H. C. (1898). The measuring of time. *Revue de Métaphysique et de Morale* 6:1–13.

Poincaré, H. C. (1905). Sur la dynamique de l'électron. *Comptes Rendus de l'Académie des Sciences Paris* 140:1504. Reprinted in Oeuvre de Henry Poincaré, Vol. 9, p. 489, Paris, Gauthier-Villars, 1954.

Poincaré, H. C. (1906). Sur la dynamique de l'électron. *Rendiconti del Circolo Matematico di Palermo* 21:129–175. Reprinted in Oeuvre de Henry Poincaré, Vol. 9, p. 494, Paris, Gauthier-Villars, 1954.

Poincaré, H. C. (1910). *Sechs Vorträge aus der Reinen Mathematik und Mathematishen Physic*. Leipzig: Teubner.

Poincaré, H. C. (1912). *Calcul des probabilités*. Paris: Gauthier-Villars, 2nd edn.

Poisson, S. D. (1833). *Traité de mécanique*. Paris: Bachelier, 2nd edn.

Poisson, S. D. (1837). *Recherches sur la probabilité des jugements en matière criminelle et en matière civile*. Paris.

Poncelet, J. V. (1865). *Traité des proprietés projectives de figures*. Paris: Gauthier-Villars, 2nd edn. First published in 1822.

Popper, K. R. (1957). The propensity interpretation of the calculus of probability and the quantum theory. In S. Körner, ed., *Observation and interpretation in the philosophy of physics*. London: Butterworth.

Popper, K. R. (1959). The propensity interpretation of probability. *British Journal for the Philosophy of Science* 10:26–42.

Popper, K. R. (1974). Replies to my critics. In P. A. Schilpp, ed., *The philosophy of Karl Popper*, vol. 2 of *Library of Living Philosophers, Vol. 14*. Chicago: Open Court Publishing.

Portugal, F. H. and J. S. Cohen (1977). *A century of DNA, a history of the discovery of the structure and function of the genetic substance*. Cambridge, MA: MIT Press.

Post, E. L. (1936). Finite combinatory processes. Formulation I. *Journal of Symbolic Logic* 1:103–5.

Preston, M. G. and P. Baratta (1948). An experimental study of the auction value of an uncertain outcome. *American Journal of Psychology* 61:183–193.

Ptolemy, C. (1984). *The almagest*. New York: Springer. Translated and annotated by G. J. Toomer. Written about 150 A.D.

Rabin, M. O. (1963). Probabilistic automata. *Information and Control* 6:230–245.

Rabin, M. O. and D. Scott (1959). Finite automata and their decision problems. *IBM Journal of Research and Development* 3:114–125. Reprinted in E. F. Moore (Ed.), Sequential machines. Reading MA, Wesley, 1964, 98-114.

Ramsey, F. P. (1931). Truth and probability. In R. B. Braithwaite, ed., *The foundations of mathematics and other logical essays*, pages 156–198. London: Kegan Paul, Trench, Trubner & Co.

Redi, F. (1686). *Esperienze intorno a diverse cose naturali, e particolarmente a quelle, che ci son portate dall'Indie*. Florence: Piero Matini. First published in 1671.

Reichenbach, H. (1924). *Axiomatik der relativistichen Raum-Zeit-Lehre*. Braunschweig: Vieweg and Sons.

Reichenbach, H. (1932). Axiomatik der wahrscheinlichkeitsrechnung. *Math. Zs.* 34:568–619.

Reid, T. (1764/1967). Inquiry into the human mind. In *Philosophical works*, vol. 1. Hildesheim, Germany: George Olms. First published in 1764.

Riemann, B. (1866–1867). Über die Hypothesen, welche der Geometrie zu Grunde liegen. *Gesellschaft der Wissenschaften zu Göttingen: Abhandlungen* 13:133–152. Habilitationsschrift, Göttingen, 1854. Translation: "On the hypotheses which lie at the basis of geometry", Nature VIII, 14-17, 36f (May 1873).

Robb, A. A. (1911). *Optical geometry of motion: A new view of the theory of relativity*. Cambridge, England: Heffer.

Robb, A. A. (1914). *A theory of time and space*. New York: Cambridge University Press.

Robb, A. A. (1921). *The absolute relations of time and space*. New York: Cambridge University Press.

Robb, A. A. (1928). On the connexion of a certain identity with the extension of conical order to n dimensions. *Proceedings of the Cambridge Philosophical Society* 24:357–374.

Robb, A. A. (1930). On a symmetrical analysis of conical order and its relation to time-space. *Proceedings of the Royal Society of London, Series A* 129:549–579.

Robb, A. A. (1936). *Geometry of time and space*. New York: Cambridge University Press.

Roberts, F. S. and P. Suppes (1967). Some problems in the geometry of visual perception. *Synthese* 17:173–201.

Rogers Jr., H. (1967). *Theory of recursive functions and effective computability*. New York: McGraw–Hill.

Rollins, M. (1989). *Mental imagery: On the limits of cognitive science*. New Haven: Yale University Press.

Rosenkrantz, R. D. (1977). *Inference, method and decision*. Dordrecht: Reidel.

Rosenkrantz, R. D., ed. (1983). *E. T. Jaynes: Papers on probability, statistics and statistical physics*. Dordrecht: Reidel.

Rouanet, H., J. M. Bernard, M. C. Bert, B. Lecoutre, M. P. Lecoutre, and B. Le Roux (1998). *New ways in statistical methodology: From significance tests to Bayesian inference*. European University Studies, Series VI-Psychology, Vol. 618. Berne: Peter Lang.

Rubin, G. S. and K. Turano (1992). Reading without saccadic eye movements. *Vision Research* 32(5):895–902.

Rubin, H. and P. Suppes (1954). Transformations of systems of relativistic particle mechanics. *Pacific Journal of Mathematics* 4:563–601.

Rubin, H. and P. Suppes (1955). A note on two-place predicates and fitting sequences of measure functions. *Journal of Symbolic Logic* 20:121–122.

Ruelle, D. (1969). *Statistical mechanics: Rigorous results*. New York: W. A. Benjamin.

Ruelle, D. (1978). *Thermodynamic formalism*. Reading, MA: Addison–Wesley. Encyclopedia of mathematics and its application, Vol. 5.

Rugg, M. D. and M. G. Coles, eds. (1995). *Electrophysiology of mind: Event-related brain potentials and cognition*. New York: Oxford University Press.

Salmon, W. (1979). Propensities: A discussion review. *Erkenntnis* 14:183–216.

Sambursky, S. (1956). On the possible and probable in ancient Greece. *Osiris* 12:35–48.

Savage, L. J. (1954). *The foundations of statistics*. New York: Wiley. Revised edition printed in 1972, Dover, New York.

Schelling, H. (1956). Vision. *Journal of the Optical Society of America* 46:309–315.

Schilpp, P. A., ed. (1963). *The philosophy of Rudolf Carnap*. Library of Living Philosophers, Vol. 11. Chicago: Open Court Publishing.

Schutz, J. W. (1973). *Foundations of special relativity: Kinematik axioms for Minkowski spacetime*. New York: Springer.

Schutz, J. W. (1979). An axiomatic system for Minkowski space-time. Tech. Rep. MPI-PAE/Astro 181, Max-Plank-Institut für Physik und Astrophysik, Munich.

Scott, D. (1964). Measurement structures and linear inequalities. *Journal of Mathematical Psychology* 1:233–247.

Scott, D. and P. Krauss (1966). Assigning probabilities to logical formulas. In Hintikka and Suppes (1966), pages 219–264.

Scott, D. and P. Suppes (1958). Foundational aspects of theories of measurement. *Journal of Symbolic Logic* 23:113–128.

Shepard, R. N. (1966). Learning and recall as organization and search. *Journal of Verbal Learning and Verbal Behavior* 5:201–204.

Shepard, R. N. and L. A. Cooper (1982). *Mental images and their transformations*. Cambridge, MA: MIT Press/Bradford.

Shepard, R. N. and J. Metzler (1971). Mental rotation of three-dimensional objects. *Science* 171:701–703.

Shepherdson, J. C. and H. E. Sturgis (1963). Computability of recursive functions. *Journal of the ACM* 10:217–255.

Shuford, E. H. (1959). A comparison of subjective probabilities for elementary and compound events. Tech. Rep. 20, The Psychometric Lab., University of North Carolina.

Siacci, F. (1878). Del moto per una linea piana. *Atti della Reale Accademia di Torino* 14:750–766.

Sierpenski, W. (1958). *Cardinal and ordinal numbers*. Warsaw: Polska Akademia Nauk, Monografie Matematyczne. Tom. 34 Panstwowe Wydawnictwo Naukowe.

Simon, H. A. (1957). *Models of man*. New York: Wiley.

Sinai, Y. G. (1959). On the concept of entropy of a dynamical system. *Dokl. Akad. Nauk. SSSR* 124:768–771.

Siskind, J. M. (1991). Dispelling myths about language bootstrapping. In *Proceedings of the AAAI Spring Symposium Workshop on Machine Learning of Natural Language and Ontology*.

Siskind, J. M. (1992). *Naive physics, event perception, lexical semantics, and language acquisition*. Ph.D. thesis, Elec. Eng. & Comp. Sci., MIT, Cambridge, MA.

Siskind, J. M. (1994). Lexical acquisition in the presence of noise and homonymy. In *Proceedings of the 12th National Conference on Artificial Intelligence, AAAI-94*, pages 760–766.

Sitnikov, K. (1960). Existence of oscillating motions for the three-body problem. *Doklady Akademii Nauk, USSR* 133(2):303–306.

Smith, C. A. B. (1961). Consistency in statistical inference and decision. *J. R. Statist. Soc., B* 23:1–25.

Smoluchowski, M. (1918). Über den Begriff des Zufalls und den Ursprung der Wahrscheinlichkeitsgesetze in der Physik. *Die Naturwissenschaften* pages 253–263.

Snapper, E. and R. J. Troyer (1971). *Metric affine geometry*. New York: Academic Press.

Sneed, J. D. (1971). *The logical structure of mathematical physics*. Dordrecht: Reidel.

Soare, R. I. (1987). *Recursively enumerable sets and degrees*. New York: Springer.

Solomonoff, R. J. (1964). A formal theory of inductive inference. *Information and Control* 7:1–22, 224–254.

Sommer, R. and P. Suppes (1996). Finite models of elementary recursive nonstandard analysis. *Notas de la Sociedad Matematica de Chile* 15:73–95.

Sommer, R. and P. Suppes (1997). Dispensing with the continuum. *Journal of Mathematical Psychology* 41:3–10.

Stegmüller, W. (1976). *The structure and dynamics of theories*. New York: Springer. Translation of *Theorienstrukturen und Theoriendynamik*, originally published as V. 2, Pt. 2 of the author's *Probleme und Resultate der Wissenschaftstheorie und analytischen Philosophie*, 1973.

Stegmüller, W. (1979). *The structuralist view of theories: A possible analogue of the Bourbaki programme in physical science*. Berlin: Springer.

Stein, W. (1930). Der Begriff des Schwerpunktes bei Archimedes. *Ouellen und Studien zur Geschichte der Mathematik, Physik und Astronomie* 1:221–224.

Stevens, S. S. (1946). On the theory of scales of measurement. *Science* 103:677–680.

Strawson, P. F. (1966). *The bounds of sense: An essay on Kant's Critique of Pure Reason*. London: Methuen.

Suppe, F., ed. (1974). *The structure of scientific theories*. Urbana: University of Illinois Press.

Suppes, P. (1956). The role of subjective probability and utility in decision-making. In *Proceedings of the Third Berkeley Symposium on Mathematical Statistics and Probability, 1954-1955*, vol. 5, pages 61–73.

Suppes, P. (1957/1999). *Introduction to logic*. New York: Van Nostrand. Reprinted in 1999 by Dover, New York.

Suppes, P. (1959). Axioms for relativistic kinematics with or without parity. In Henkin et al. (1959), pages 291–307.

Suppes, P. (1960a). A comparison of the meaning and uses of models in mathematics and the empirical sciences. *Synthese* 12:287–301.

Suppes, P. (1960b). *Axiomatic set theory*. New York: Van Nostrand. Slightly revised edition published by Dover, New York, 1972.

Suppes, P. (1962). Models of data. In E. Nagel, P. Suppes, and A. Tarski, eds., *Logic, methodology and philosophy of science: Proceedings of the 1960 International Congress*, pages 252–261. Stanford: Stanford University Press.

Suppes, P. (1969a). *Studies in the methodology and foundations of science: Selected papers from 1951 to 1969*. Dordrecht: Reidel.

Suppes, P. (1969b). Stimulus-response theory of finite automata. *Journal of Mathematical Psychology* 6:327–355.

Suppes, P. (1969c). Stimulus-response theory of finite automata and TOTE hierarchies: A reply to Arbib. *Psychological Review* 76:511–514.

Suppes, P. (1970a). *A probabilistic theory of causality*. Amsterdam: North-Holland Publishing Co.

Suppes, P. (1970b). Probabilistic grammars for natural languages. *Synthese* 22:95–116.

Suppes, P. (1972). Finite equal-interval measurement structures. *Theoria* 38:45–63.

Suppes, P. (1973a). New foundations of objective probability: Axioms for propensities. In P. Suppes, L. Henkin, G. C. Moisil, and A. Joja, eds., *Logic, methodology, and philosophy of science IV: Proceedings of the Fourth International Congress for Logic, Methodology and Philosophy of Science, Bucharest, 1971*, pages 515–529. Amsterdam: North-Holland Publishing Co.

Suppes, P. (1973b). Congruence of meaning. *Proceedings and Addresses of the American Philosophical Association* 46:21–38.

Suppes, P. (1974a). Popper's analysis of probability in quantum mechanics. In P. A. Schilpp, ed., *The philosophy of Karl Popper*, pages 760–774. Chicago: Open Court Publishing.

Suppes, P. (1974b). The structure of theories and the analysis of data. In F. Suppe, ed., *The structure of scientific theories*, pages 266–283. Urbana, IL: University of Illinois Press, 2nd edn.

Suppes, P. (1974c). Aristotle's concept of matter and its relation to modern concepts of matter. *Synthese* 28:27–50.

Suppes, P. (1974d). The measurement of belief. *Journal of the Royal Statistical Society (Series B)* 36:160–191.

Suppes, P. (1975). From behaviorism to neobehaviorism. *Theory and Decision* 6:269–285.

Suppes, P. (1976). Testing theories and the foundations of statistics. In W. L. Harper and C. A. Hooker, eds., *Foundations of probability theory, statistical inference, and statistical theories of science*, vol. 2, pages 437–455. Dordrecht: Reidel.

Suppes, P. (1977a). Is visual space Euclidean? *Erkenntnis* 36:397–421.

Suppes, P. (1977b). Learning theory for probabilistic automata and register machines. In H. Spada and W. F. Kempf, eds., *Structural models of thinking and learning*, pages 57–79. Bern: Hans Huber Publisher. Proceedings of the 7th IPN-Symposium on formalized theories of thinking and learning and their implications for science instruction.

Suppes, P. (1979). The logic of clinical judgment: Bayesian and other approaches. In H. T. Engelhardt, Jr., S. F. Spicker, and B. Towers, eds., *Clinical judgment: A critical appraisal*, pages 145–159. Dordrecht: Reidel.

Suppes, P. (1980). Limitations of the axiomatic method in ancient Greek mathematical sciences. In J. Hintikka, D. Gruender, and E. Agazzi, eds., *Pisa Conference Proceedings*, vol. 1, pages 197–213. Dordrecht: Reidel.

Suppes, P. (1983). Arguments for randomizing. In P. D. Asquith and T. Nickles, eds., *PSA 1982*, pages 464–475. Lansing, MI: Philosophy of Science Association.

Suppes, P. (1984). *Probabilistic metaphysics*. Oxford: Blackwell.

Suppes, P. (1987). Propensity representations of probability. *Erkenntnis* 26:335–358.

Suppes, P. (1988). Empirical structures. In E. Scheibe, ed., *The role of experience in science, proceedings of 1986 conference of the Académie Internationale de Philosophie des Sciences (Bruxelles)*, pages 23–33. New York: Walter de Gruyter.

Suppes, P. (1989). Current directions in mathematical learning theory. In E. E. Roskam, ed., *Mathematical Psychology in Progress*, pages 3–28. New York: Springer.

Suppes, P. (1991). *Language for humans and robots*. Oxford: Blackwell.

Suppes, P. (1995). Some foundational problems in the theory of visual space. In R. D. Luce, M. D'Zmura, D. Hoffman, G. J. Iverson, and A. K. Romney, eds., *Geometric representations of perceptual phenomena: Papers in honor of Tarow Indow on his 70th birthday*, pages 37–45. Mahwah, NJ: Erlbaum Associates.

Suppes, P. (1998). Pragmatism in physics. In P. Weingartner, G. Schurz, and G. Dorn, eds., *The role of pragmatics in contemporary philosophy*, pages 236–253. Vienna: Holder, Pichler, Tempsky.

Suppes, P. (2000). Quantifier-free axioms for constructive affine plane geometry. *Synthese* 125:263–281.

References

Suppes, P. (2001a). Finitism in geometry. *Erkenntnis* 54:133–144.

Suppes, P. (2001b). Weak and strong reversibility of causal processes. In M. C. Galavotti, P. Suppes, and D. Costantini, eds., *Stochastic causality*, pages 203–220. Stanford, CA: CSLI Publications.

Suppes, P. and R. C. Atkinson (1960). *Markov learning models for multiperson interactions*. Stanford, CA: Stanford University Press.

Suppes, P., M. Böttner, and L. Liang (1995). Comprehension grammars generated from machine learning of natural language. *Machine Learning* 19:133–152.

Suppes, P., M. Böttner, and L. Liang (1996). Machine learning comprehension grammars for ten languages. *Computational Linguistics* 22:329–350.

Suppes, P. and R. Chuaqui (1993). A finitarily consistent free-variable positive fragment of infinitesimal analysis. *Notas de Logica Matematica* 38:1–59. Proceedings of the IX Latin American Symposium on Mathematical Logic, held at Bahia Blanca, Argentina, August 1992.

Suppes, P., A. J. de Barros, and G. Oas (1998). A collection of probabilistic hidden-variable theorems and counterexamples. In R. Pratesi and L. Ronchi, eds., *Waves, information and foundations of physics. Conference proceedings, Vol. 60*, pages 267–291. Bologna: Società Italiana Di Fisica.

Suppes, P., B. Han, J. Epelboim, and Z. L. Lu (1999a). Invariance between subjects of brain wave representations of language. *Proceedings of the United States National Academy of Sciences* 96:12953–12958.

Suppes, P., B. Han, J. Epelboim, and Z. L. Lu (1999b). Invariance of brain-wave representations of simple visual images and their names. *Proceedings of the United States National Academy of Sciences* 96:14658–14663.

Suppes, P., B. Han, and Z. L. Lu (1998). Brain-wave recognition of sentences. *Proceedings of the National Academy of Sciences* 95:15861–15866.

Suppes, P., D. H. Krantz, R. D. Luce, and A. Tversky (1989). *Foundations of measurement volume II: Geometrical, threshold, and probabilistic representations*. San Diego, CA: Academic Press.

Suppes, P. and L. Liang (1996). Probabilistic association and denotation in machine learning of natural language. In A. Gammerman, ed., *Computational learning and probabilistic reasoning*, pages 87–100. New York: Wiley.

Suppes, P. and L. Liang (1998). Concept learning rates and transfer performance of several multivariate neural network models. In C. E. Dowling, F. S. Roberts, and P. Theuns, eds., *Recent progress in mathematical psychology*, pages 227–252. Mahwah, NJ: Erlbaum Associates.

Suppes, P., L. Liang, and M. Böttner (1992). Complexity issues in robotic machine learning of natural language. In L. Lam and V. Naroditsky, eds., *Modelling complex phenomena*, pages 102–127. New York: Springer.

Suppes, P., Z. L. Lu, and B. Han (1997). Brain-wave representations of words. *Proceedings of the United States National Academy of Sciences* 94:14965–14969.

Suppes, P. and W. Rottmayer (1974). Automata. In E. C. Carterette and M. P. Friedman, eds., *Handbook of perception, Vol. 1: Historical and philosophical roots of perception*, pages 335–362. New York: Academic Press.

Suppes, P. and M. Winet (1955). An axiomatization of utility based on the action of utility differences. *Journal of Management Science* 1:259–270.

Suppes, P., D. K. Wong, M. Perreau-Guimaraes, E. T. Uy, and W. Yang (to appear). High statistical recognition rates for some persons' brain-wave representations of sentences.

Suppes, P. and M. Zanotti (1976). Necessary and sufficient conditions for existence of a unique measure strictly agreeing with a qualitative probability ordering. *Journal of Philosophical Logic* 5:431–438.

Suppes, P. and M. Zanotti (1980). A new proof of the impossibility of hidden variables using the principles of exchangeability and identity of conditional distribution. In P. Suppes, ed., *Studies in the foundations of quantum mechanics*, pages 173–191. East Lansing, MI: Philosophy of Science Association.

Suppes, P. and M. Zanotti (1981). When are probabilistic explanations possible? *Synthese* 48:191–199.

Suppes, P. and M. Zanotti (1982). Necessary and sufficient qualitative axioms for conditional probability. *Zeitschrift für Wahrscheinlichkeitstheorie und verwandte Gebiete* 60:163–169.

Suppes, P. and M. Zanotti (1996). Mastery learning of elementary mathematics: Theory and data. In *Foundations of probability with applications: Selected papers, 1974–1995*, pages 149–188. New York: Cambridge University Press.

Suppes, P. and J. L. Zinnes (1963). Basic measurement theory. In R. D. Luce, R. R. Bush, and E. H. Galanter, eds., *Handbook of mathematical psychology*, vol. 1, pages 3–76. New York: Wiley.

Swerdlow, N. M. (1998). *The Babylonian theory of the planets*. Princeton: Princeton University Press.

Szekeres, G. (1968). Kinematic geometry: An axiomatic system for Minkowski space-time. *Journal of Australian Mathematical Society* 8:134–160.

Szmielew, W. (1983). *From affine to Euclidean geometry: An axiomatic approach*. Warszawa, Poland and Dordrecht, Holland: PWN-Polish Scientific Publishers and D. Reidel.

Tait, W. W. (1959). A counterexample to a conjecture of Scott and Suppes. *Journal of Symbolic Logic* 24:15–16.

Tarski, A. (1935). Der Wahrheitsbegriff in den formalisierten Sprachen. *Studia Philosophica* 1:261–405.

Tarski, A. (1953). *Undecidable theories*. Amsterdam: North-Holland Publishing Co.

Tarski, A. and S. Givant (1987). *A formalization of set theory without variables*. Providence, RI: American Mathematical Society.

Toda, M. (1951). Measurement of intuitive-probability by a method of game. *Japan. J. Psychol.* 22:29–40.

Toda, M. (1958). Subjective inference vs. objective inference of sequential dependencies. *Japan. Psychol. Res.* 5:1–20.

Todhunter, I. (1865/1949). *A history of the mathematical theory of probability from the time of Pascal to that of Laplace*. New York: Chelsea Publishing Company. First published in 1865.

Toth, L. F. (1964). *Regular figures*. New York: Macmillan.

Truesdell, C. (1968). *Essays in the history of mechanics*. New York: Springer.

Tulving, E. and F. I. M. Craik, eds. (2000). *The Oxford handbook of memory*. New York: Oxford University Press.

Turing, A. M. (1936). On computable numbers, with an application to the Entscheidungsproblem. *Proc. London Math. Soc. Ser. 2* 42:230–265. A correction, Vol. 43 (1936), 544–546.

Tye, M. (1991). *The imagery debate*. Representation and mind. Cambridge, MA: MIT Press/Bradford.

van der Waerden, B. L., ed. (1968). *Sources of quantum mechanics*. New York: Dover.

van Fraassen, B. C. (1980). *The scientific image*. Oxford: Clarendon Press.

van Lambalgen, M. (1987a). *Random sequences*. Amsterdam: Dutch Foundation for Scientific Research.

van Lambalgen, M. (1987b). Von Mises' definition of random sequences reconsidered. *Journal of Symbolic Logic* 52(3):725–755.

van Lambalgen, M. (1990). The axiomatization of randomness. *Journal of Symbolic Logic* 55(3):1143–1167.

van Lambalgen, M. (1992). Independence, randomness and the axiom of choice. *Journal of Symbolic Logic* 57(4):1274–1304.

van Lambalgen, M. (1996). Randomness and foundations of probability: Von Mises' axiomatisation of random sequences. In T. S. Ferguson, L. S. Shapley, and J. B. MacQueen, eds., *Statistics, probability and game theory. Papers in honor of David Blackwell*, Lecture Notes - Monograph Series, Volume 30, pages 347–367. Hayward, CA: Institute of Mathematical Statistics.

Vaught, R. (1954). Remarks on universal classes of relational systems. *Indagationes Matematicae* 16:589–591.

Veblen, O. (1904). A system of axioms for geometry. *Transaction of the American Mathematical Society* 5:343–384.

Veblen, O. and J. W. Young (1910). *Projective geometry - Vol. 1*. Boston, MA: Ginn. Reprinted by Blaisdell, New York, 1938.

Veblen, O. and J. W. Young (1918). *Projective geometry - Vol. 2*. Boston, MA: Ginn. Reprinted by Blaisdell, New York, 1938.

Ville, J. (1939). *Ètude critique de la notion de collectif*. Paris: Gauthiers-Villars.

Vitruvius (1960). *The ten books on architecture*. New York: Dover. Originally published by Harvard University Press in 1914. Translated by Morris Hicky Morgan.

Voigt, W. (1887). Über das Doppler'sche Prinzip. *Nachr. Ges. Wiss. Göttingen* 41.

Volta, A. (1793/1918). Letter to Tiberius Cavallo, 22 May 1793. In *Le opere di Alessandro Volta*, vol. 1, pages 203–208. Milan: Ulrico Hoepli.

von Mises, R. (1919). Grundlagen der Wahrscheinlichkeitsrechnung. *Mathematische Zeitschrift* 5:52–99.

von Mises, R. (1941). On the foundations of probability and statistics. *Ann. Math. Stat.* 12:191–205, 215–216.

von Neumann, J. (1932/1955). *Mathematical foundations of quantum mechanics*. Princeton, NJ: Princeton University Press. Translated from the German edition of 1932 by R. T. Byer.

von Plato, J. (1983). The method of arbitrary functions. *The British Journal for the Philosophy of Science* 34:37–47.

Wagner, M. (1985). The metric of visual space. *Perceptions & Psychophysics* 38:483–495.

Wald, A. (1936). Sur la notion de collectif dans le calcul des probabilités. *C. R. Acad. Sci. Paris* 202:180–183.

Walker, A. G. (1948). Foundations of relativity. Parts I and II. *Proceedings of the Royal Society of Edimburgh, Section A, Mathematical and Physical Sciences* 62:319–335.

Walker, A. G. (1959). Axioms for cosmology. In Henkin et al. (1959), pages 308–321.

Wexler, K. and P. W. Culicover (1980). *Formal principles of language aquisition*. Cambridge, MA: MIT Press.

Weyl, H. (1928/1931). *The theory of groups and quantum mechanics*. London: Methuen & Co. Ltd. Translation from the second (revised) German edition of *Gruppentheorie und Quantenmechanik*, 1928.

Whittaker, E. T. (1904/1937). *A treatise on the analytical dynamics of particles and rigid bodies*. Cambridge, England: Cambridge University Press, 4th edn. First published in 1904.

Whittaker, E. T. (1910). *A history of the theories of aether and electricity*. London: Longmans, Green and Co.

Wiener, N. (1914). A simplification of the logic of relations. *Proceedings of the Cambridge Philosophical Society* 17:387–390.

Wilson, E. B. and G. N. Lewis (1912). The space-time manifold of relativity. The non-Euclidean geometry of mechanics and electromagnetics. *Proceedings of the American Academy of Arts and Sciences* 48:389–507.

Winnie, J. A. (1977). The causal theory of space-time. In J. Earman, C. Glymour, and J. Stachel, eds., *Foundations of space-time theories*, pages 134–205. Minneapolis: University of Minnesota Press.

Wittgenstein, L. (1922). *Tractatus logico-philosophicus*. New York: Routledge. English and German.

Wolfram, S. (2002). *A new kind of science*. Champaign, IL: Wolfram Media Inc.

Wussing, H. (1984). *The genesis of the abstract group concept*. Cambridge, MA: MIT Press.

Zajaczkowska, A. (1956). Experimental test of Luneburg's theory: Horopter and Alley experiments. *Journal of the Optical Society of America* 46:514–527.

Zeeman, E. C. (1964). Causality implies the Lorentz group. *Journal of Mathematical Physics* 5:490–493.

Zeeman, E. C. (1967). The topology of Minkowski space. *Topology* 6:161–170.

Zvonkin, A. K. and L. A. Levin (1970). The complexity of finite objects and the development of the concepts of information and randomness by means of the theory of algorithms. *Russian Mathematical Surveys* 25:83–124.

Author Index

Abelson, R., 213
Ackermann, W., 75
Aczél, J., 234
Adrian, E. D., 444
Agatharchus, 304
Aho, A. V., 360
Alekseev, V. M., 220
Alexandrov, A. D., 281
Ames Jr., A., 293
Anaxagoras, 304
Andersen, A., 244
Anderson, J. R., 391n
Angell, R. B., 290, 291
Apollonius, 40
Appell, P., 321n
Aquinas, T., 82, 83, 329n
Arbib, M. A., 387–391
Archimedes, 36n, 37–40, 43, 44, 48, 205, 208, 232, 235
Aristotle, 35, 36, 36n, 43, 44, 54n, 81–83, 86, 93, 102, 102n, 103–105, 107, 108, 265, 272, 273, 273n, 328, 329n, 442, 465, 466
Arnold, V. I., 317
Arrow, K. J., 19, 23
Atkinson, R. C., 8n, 381
Attias, H., 286

Bailey, D., 441
Bain, A., 91
Balzer, W., 34
Baratta, P., 246–248
Bayes, T., 140, 140n, 141–144, 159, 166, 183, 193, 199, 223, 226, 242–244
Beck, A., 444
Bell, J. S., 332, 335–338, 342, 343

Bentley, R., 466
Berger, H., 444
Berkeley, G., 289, 290, 293, 302, 452, 453, 464
Berlyne, C., 388
Bernard, J. M., 183, 263
Bernoulli, D., 167
Bernoulli, J., 124, 158
Bernstein, S., 145
Bert, M. C., 183, 263
Bertrand, J. L., 163–165
Bever, T. G., 389
Birkhoff, G. D., 35, 220n
Blank, A. A., 291, 293
Blumenfeld, W., 290–292, 296
Boeder, P., 291, 293
Bohm, D., 273n
Boltzmann, L., 241, 257, 349
Bolyai, J., 45
Borsuk, K., 46, 310
Bose, S., 224
Böttner, M., 419n, 434, 440
Bourbaki, N., 33, 34
Bouwmeester, D., 342
Bowerman, M., 105, 110, 433
Bradley, R. A., 213
Brahe, T., 274n
Breiman, L., 286
Bresnan, J., 361
Brontë, C., 51n
Brouwer, L. E. J., 303
Bruns, H., 328
Brush, S. G., 257
Burks, A., 190
Busemann, H., 285
Bush, R. R., 20, 24, 403

Campbell, N. R., 63
Cantor, G., 61n, 160
Carbone, R., 244
Carbonell, J. G., 441
Carmichael, L., 444
Carnap, R., 129, 190–200
Cartwright, N., 262n
Caton, R., 444
Cauchy, A. L., 12, 149
Cayley, A., 57, 296
Chaitin, G. J., 180
Chasles, M., 47
Choi, S., 433
Chomsky, N., 13, 354, 358, 359, 361, 387, 391
Chuaqui, R., 33, 190, 200, 201, 311
Church, A., 74n, 76, 168, 173, 176, 176n, 177, 178, 220, 360
Clagett, M., 43, 273, 273n
Clauser, J. F., 337
Cohen, J. S., 19
Coles, M. G., 444
Compton, A. H., 328
Condorcet, M. J., 160
Cooley, J. W., 450
Cooper, L. A., 94
Copeland, A. H., 176n
Coren, S., 300
Costantini, D., 224
Cover, T. M., 182
Craik, F. I. M., 86
Crangle, C., 105
Culicover, P. W., 441
Cybulsky, N., 444

da Costa, N. C. A., 33
D'Alembert, J., 160
Daniell, M., 342
Daniels, N., 290, 291
David, F. N., 158
Davidson, D., 28, 239
Davis, H., 444
Davis, M., 81
Davis, R. L., 196
Dawid, A. P., 244
de Barros, A. J., 13, 332n, 340, 341
Dedekind, R., 46

de Finetti, B., 129, 155, 190, 226, 228, 230, 232, 238, 239, 239n, 240–242, 242n, 244, 245, 245n, 248, 250, 253, 254
Dembowski, P., 284, 285, 303
Democritus, 304, 465
de Moivre, A., 158, 187
Dempster, A. P., 253
Desargues, G., 294n
Descartes, R., 44, 52, 53, 65n, 83, 274, 274n, 323n, 327
Diaconis, P., 240n
Dijksterhuis, E. J., 38
Diocles, 40
Dirac, P. A. M., 224, 256, 257
Domotor, Z., 234, 281
Donders, F. C., 302
Doob, J. L., 20, 22, 24, 178
Drösler, J., 296
Du Bois-Reymond, E., 443, 444
Duhem, P., 43

Ehrenfest, P., 349
Einstein, A., 19, 23, 49, 224, 274, 278, 278n, 279–281, 284, 328
Einthoven, V., 444
Epelboim, J., 95n, 451, 452, 459, 460, 460n
Epicurus, 52, 53, 465, 466
Estes, W. K., 20, 24, 377, 381, 403–406, 408–411, 419
Euclid, 10, 35, 36, 36n, 37, 37n, 39–41, 42n, 43–46, 106, 109, 186, 288, 289, 291
Eudoxus, 35, 35n
Euler, L., 159, 323, 323n, 327

Fagin, R., 195
Farah, M. J., 95
Feldman, J. A., 441
Feller, W., 177, 344
Fermat, P., 158
Fermi, E., 224
Fildes, R., 244
Fine, A., 337
Fine, T. L., 262n
Finke, R. A., 94, 95
Fishburn, P. C., 248
Fisher, R. A., 184, 187
Fitelson, B., 198n
FitzGerald, G. F., 278, 279, 281
Fliegel, S. L., 95

Fock, V. A., 257–259, 262
Fodor, J., 93, 389
Foley, J. M., 266, 285, 291, 294, 295, 297, 299, 300, 302
Ford Jr., L. R., 213
Fraenkel, A. A. H., 30
Frank, P., 274n
Freedman, D., 240n
Frege, G., 47, 48, 303
Freudenthal, H., 285
Friedman, J., 263n, 286
Fu, K. S., 286

Gacs, P., 182
Gaifman, H., 190, 200
Galanter, E., 387
Galavotti, M. C., 185n, 220–224
Galen, 42n
Galilei, G., 44, 273
Galison, P., 459
Galvani, L., 443, 444
Garceau, L., 444
Garrett, M., 389
Gauss, J. C. F., 12, 45
Gazdar, G., 361
Geddes, L. A., 444n
Gibbs, E., 444
Gibbs, F. A., 444
Gibbs, W., 22, 187, 241, 257
Giere, R. N., 202, 221
Girgus, J. S., 300
Givant, S., 102, 155
Gödel, K., 49, 360
Gogel, W. C., 291, 293
Goldblatt, R., 282
Goldreich, O., 179
Goldwasser, S., 179
Good, I. J., 129, 245n, 253, 254
Gray, R. M., 182
Green, G., 330
Greenberger, D. M., 332, 339
Greenspan, D. G., 20
Greibach, F. A., 359
Griffith, R. M., 247, 248
Grünbaum, A., 291, 297

Hacking, I., 158, 184n
Hahn, R., 160n
Hamel, G., 322, 323

Han, B., 445, 451, 452, 459, 460, 460n
Hardy, L. H., 291, 293
Harré, R., 22
Harrison, M. A., 391, 392
Hastie, T., 263n, 286
Heath, T. L., 36, 37, 37n
Heisenberg, W., 256, 257
Helmholtz, H., 285, 286, 290, 293, 301
Hermes, H., 319
Hilbert, D., 46–49, 303
Hilbon, M., 244
Hintikka, J., 190, 198, 199
Holland, P. W., 333
Hollbach Weber, S., 441
Holman, M., 220n
Hopcroft, J. E., 360, 371
Hopf, E., 214
Horne, M. A., 332, 337, 339
Huddleston, R., 361
Hudgin, R. H., 281
Hume, D., 83–91, 93, 94, 202n, 238n, 452, 453, 464
Humphreys, P., 223, 224
Huygens, C., 44n

Ictinus, 304
Indow, T., 291, 295, 296
Inoue, E., 291, 296

Jacob, R. L., 174
James, W., 88–92, 94
Jamison, D., 394
Jammer, M., 329n
Jasper, H. H., 444
Jaynes, E. T., 241, 242
Jeffrey, R. C., 226
Jeffreys, H., 129, 185, 185n, 186–188, 188n, 189, 189n, 190, 200, 238
Jones, A., 304n
Jones, O., 98
Jordanus, 43
Joyce, J., 17

Kac, M., 349
Kandel, E. R., 398
Kant, I., 86–88, 91, 265, 297
Kaplan, R. M., 361
Kaufman, L., 445
Kaufmann, S. A., 19

Keller, J. B., 214–216
Kelly, F. P., 346
Kelvin (W. Thomson), 22, 22n, 444
Kemeny, J., 190
Kendler, H. H., 388
Kendler, T. S., 388
Kepler, J., 274n, 322
Keynes, J. M., 131, 184, 185, 185n, 238
Khinchin, A. I., 5, 18, 22
Kieras, D. E., 391
Kirchhoff, G., 257
Kleene, S. C., 13, 74n, 76, 370
Klein, E., 361
Klein, F., 98–100, 102, 156, 290, 296, 303
Knorr, W. R., 35, 35n
Kolmogorov, A. N., 12, 125, 126, 129–131, 144, 165n, 177, 178, 178n, 179, 180, 182, 184, 200, 202, 220
Koopman, B. O., 185n, 230, 234, 248, 250, 253
Kosslyn, S. M., 94, 95
Kraft, C. H., 228, 229
Krantz, D. H., 28, 29n, 63, 69, 71n, 106, 204, 205, 230, 232, 234, 236, 248, 269, 288, 298, 306
Krauss, P. H., 190, 200
Kuratowski, C., 34
Kyburg Jr., H. E., 190, 200, 226

Lagrange, J. L., 12, 159, 330
Lakoff, G., 441
Lamb, H., 302
Lamperti, J., 348
Langley, P., 441
Laplace, P. S., 12, 129, 158–160, 160n, 161–163, 166, 183, 184, 187, 201, 225, 242, 284
Latzer, R. W., 281, 282
Lecoutre, B., 183, 263
Lecoutre, M. P., 183, 263
Leibniz, G. W., 184, 184n, 329, 329n, 331
Lennox, W. G., 444
Le Roux, B., 183, 263
Leucippus, 465
Levelt, W. J. M., 105
Levi, I., 226
Levin, L. A., 182
Levy, P., 177n

Lewandowski, R., 244
Lewes, G. H., 51n
Lewis, G. N., 281
Lhamon, D., 394
Li, M., 182
Liang, L., 419n, 434, 440, 442
Lie, S., 285, 286, 301
Lindberg, D. C., 289
Lindenbaum, A., 102, 106
Lindley, D. V., 129, 245n
Lindsay, R. B., 18, 21
Lindstrom, P., 29
Linsley, D., 444
Lobachevski, N. I., 45
Loève, M., 342
Lorentz, H. A., 49, 104, 278–281
Loschmidt, J. J., 257, 349
Löwenheim, L., 252
Lu, Z. L., 445, 451, 452, 459, 460, 460n
Luce, R. D., 28, 29n, 63, 69, 71n, 106, 204, 205, 212, 227, 230, 232, 234, 236, 246, 248, 254, 269, 288, 298, 306
Lucretius, 52, 53, 466
Luneburg, R. K., 285, 291–297, 299, 301, 302

Mach, E., 40, 274
Macieszyna, S. J., 444
MacLane, S., 35
Makridakis, S., 244
Maltzman, I., 388
Margenau, H., 18
Markov, A. A., 74n, 76, 124, 139, 353
Martin-Löf, P., 177, 179, 182, 220
Matsushima, K., 291, 296
Matteucci, C., 444
Matthews, B. H. C., 444
Maxwell, J. C., 12, 22, 22n, 241, 256, 279–281, 348, 466, 467
Mayo, D. G., 184n
McGlothlin, W. H., 247, 248
McKinsey, J. C. C., 316, 321, 344
Mehlberg, H., 282
Mellor, D. H., 202, 221, 222, 224
Mermin, N. D., 339
Metzler, J., 94
Micali, S., 179
Michelson, A. A., 279

Mill, J., 88
Millenson, J. R., 387
Miller, G. A., 387, 391
Milne, E. A., 281
Minkowski, H., 49, 282
Montgomery, R., 324n
Moody, E. A., 43
Morley, E. W., 279
Moser, J., 219, 220
Mosteller, F., 262, 263, 403
Moulines, C. U., 34
Mundy, B., 282
Murray, N., 220n

Nagel, E., 46, 190, 467
Narayanan, S., 441
Neimark, E. D., 377
Nelson, E., 201
Neugebauer, O., 35, 35n
Newton, I., 12, 37, 44, 44n, 45, 45n, 48, 83, 87, 121, 214, 218, 239, 270n, 274, 289, 316, 322, 322n, 323, 324, 327, 328, 330, 331, 466
Newton, J., 244
Neyman, J., 184, 184n
Nishikawa, Y., 291, 296
Noether, E., 104, 329n
Noguchi, H., 291, 296
Noll, W., 322n
Norman, M. F., 403

Oas, G., 13, 332n
Oates, W. J., 52, 465, 466
Oppenheim, A. V., 447, 450n
Ornstein, D. S., 123n, 126
Osgood, C. E., 388

Padoa, A., 54n
Pais, A., 274n, 278n
Palladio, A., 98, 304
Pan, J. W., 342
Pappus, 304n
Parzen, E., 244
Pascal, B., 158
Pasch, M., 46–48, 267
Peacocke, C., 92, 93
Peano, G., 46, 161
Pearson, E., 184
Pearson, K., 184

Pecham, J., 289
Penfield, W., 444
Peres, Y., 338
Perreau-Guimaraes, M., 452, 461, 462, 464
Philoponus, J., 104
Pieri, M., 46
Pinker, S., 441
Planck, M., 257
Plato, 35, 35n, 42, 54n
Poincaré, H. C., 164, 214, 257, 278–281
Poisson, S. D., 137, 137n, 328
Poncelet, J. V., 45, 47, 47n
Popper, K. R., 202, 221
Portugal, F. H., 19
Post, E. L., 74n, 76
Pratt, J. W., 228, 229
Preston, Malcolm G., 246–248
Pribram, K. H., 387
Proclus, 36, 42
Ptolemy, C., 41, 42, 44, 45n, 107, 108, 272, 273, 304n, 322
Pullum, G., 361
Putnam, H., 190
Pylyshyn, Z., 93

Rabin, M. O., 361, 364, 369, 391, 392
Ramsey, F. P., 239, 239n, 248
Rand, G., 291, 293
Redi, F., 442, 443
Regier, T., 441
Reichenbach, H., 176n, 281
Reid, T., 288–291, 302
Reiser, B. J., 95
Riemann, B., 214, 218, 285
Rittler, M. C., 291, 293
Robb, A. A., 49, 275, 281, 282
Roberts, F. S., 302
Rogers Jr., H., 81
Rollins, M., 93
Rosenbaum, P. R., 333
Rosenkrantz, R. D., 241, 241n, 242
Rottmayer, W., 286, 392, 397
Rouanet, H., 183, 263
Rubin, G. S., 462
Rubin, H., 170, 196, 277, 344
Ruelle, D., 5, 403n, 467
Rugg, M. D., 444
Russell, B., 303, 388

Sag, I., 361
Salmon, W., 224
Sambursky, S., 158
Savage, L. J., 129, 226, 228, 230, 248–252
Schafer, R. W., 447, 450n
Schelling, H., 291, 293
Schutz, J. W., 282
Scott, D., 29, 29n, 67, 118, 190, 200, 229, 230, 234, 235, 252, 254, 361, 364, 391, 392
Seidenberg, A., 228, 229
Shepard, R. N., 93, 94
Shepherdson, J. C., 76, 79, 81
Shimony, A., 337, 339
Shuford, E. H., 247, 248
Siacci, F., 121n
Siegel, S., 28, 239
Sierpenski, W., 61n
Simon, H. A., 19, 23
Simpson, T., 12
Sinai, Y. G., 126
Siskind, J. M., 441
Sitnikov, K., 220
Skolem, A. T., 252
Smith, C. A. B., 253
Smokler, H. E., 226
Smoluchowski, M., 214
Snapper, E., 317
Sneed, J. D., 34
Snir, M., 200
Soare, R. I., 81
Solomonoff, R. J., 180
Sommer, R., 311
Stegmüller, W., 34
Stein, W., 39
Steiner, J., 47
Stevens, S. S., 117
Stolcke, A., 441
Strawson, P. F., 291, 297
Sturgis, H. E., 76, 79, 81
Sugar, A. C., 316, 321
Suppe, F., 30n
Swerdlow, N. M, 37n
Szekeres, G., 282
Szmielew, W., 47, 308, 310

Tait, W. W., 29n

Tarski, A., 18, 20, 21, 25n, 26, 34, 102, 106, 155, 156, 179, 200, 252
Taylor, B., 311
Terry, M. E., 213
Thales, 35
Thomas, J. A., 182
Thomson, W. (Lord Kelvin), 22, 22n, 444
Tibshirani, R., 263n, 286
Toda, M., 247
Todhunter, I., 158
Toth, L. F., 102
Troyer, R. J., 317
Truesdell, C., 322n, 328
Tukey, J. W., 450
Tulving, E., 86
Turano, K., 462
Turing, A. M., 51, 74n, 76, 360, 373
Tversky, A., 28, 29n, 63, 69, 71n, 106, 204, 205, 230, 232, 234, 236, 248, 269, 288, 298, 306
Tye, M., 93

Ullman, J. D., 360, 371
Uy, E. T., 452, 461, 462, 464

van der Waerden, B. L., 257
van Fraassen, B. C., 27, 30n
van Lambalgen, M., 177, 178, 204
Vaught, R., 29
Veblen, O., 49, 284, 285n
Venn, J., 129, 187, 188
Veronese, G., 46
Ville, J., 177, 177n
Vitanyi, P. B. M., 182
Vitruvius, 303, 304
Voigt, W., 278
Volta, A., 443
von Mises, R., 129, 171–173, 177–179, 187, 188
von Neumann, J., 49, 161, 260, 261, 345
von Plato, J., 214
von Staudt, C., 47
Vuillemin, J., 47n

Wagner, M., 266, 296, 297, 299
Wald, A., 176
Walker, A. G., 281, 282
Wallace, D. L., 262, 263
Weinfurter, H., 342

Weiss, B., 123n
Wexler, K., 441
Weyl, H., 259, 260
Whittaker, E. T., 121n, 274, 443
Wiener, N., 34
Williamson, S., 445
Wilson, E. B., 281
Winet, M., 28
Winkler, R., 244
Winnie, J. A., 281
Wittgenstein, L., 193, 197
Wolfram, S., 468
Wong, D. K., 452, 461, 462, 464
Wrinch, D., 238
Wussing, H., 100n

Yang, W., 452, 461, 462, 464
Young, J. W., 285n

Zajaczkowska, A., 291, 293
Zanotti, M., 8n, 230, 234, 236, 254, 333–335
Zeeman, E. C., 281
Zeilinger, A., 332, 339, 342
Zermelo, E., 30, 257, 349
Zinnes, J. L., 28, 61n, 66
Zvonkin, A. K., 182

Index

abstract ideas
- Berkeley, 452–453, 464
- Hume, 452–453, 464

additive
- conjoint measurement, 69
- conjoint structure, 70
- displacement operation, 317
- group, 55, 317
  - vs. multiplicative, 55
- measure, 65, 231
- probability measure, 123
  - countably, 135
- probability space, 134, 135, 139, 161, 168, 169, 212, 227
- property, 131
- representation, 55, 69
- σ-additive, 200

additivity, 69, 191, 227, 232, 298
- axiom, 232
- countable, 139
- de Finetti's axiom, 232
- theorem on finite, 140

adequacy
- of axioms, 63
- of mechanics, 330

affine
- congruence, 269
- geometry, 267n
  - construction, 306
- group of transformations, 268
- parallelogram, 280
- plane, 297
  - bisection construction, 307
  - doubling construction, 307
- space, 268n, 321

- axioms for, 267
- dimensionality, 267
- four-dimensional, 105, 318
- invariance theorem, 268
- ordered, 266
- real, 313, 316–318
- structure, 265
- transformations, *see* transformation

algebra(s), 32, 134, 143, 162, 170, 179, 314
- abstract, 285
- axiomatization, 4
- Boolean, 359
- homomorphism, 58, 58n
- isomorphism, 54
- linear, 213
- of events, 133–134, 162, 225, 230, 233, 234
- of functions, 234
- of indicator functions, 230, 231, 234
  - partial extended, 234
- of operations, 114
- of sets, 64, 133, 134, 136, 165, 253, 387
- qualitative, 231–233
- reduction of geometry to, 52
- σ-algebra, 123, 133–135, 209
- subalgebra, 62, 125, 254, 255

algorithm, 139, 176, 179, 180, 182, 359–362, 367, 374, 441, 450
- complexity of, 180
- for constructing grammars, 358
- learning, 441

alphabet, 80, 177, 361, 362, 364, 365, 374, 376, 380, 383, 384, 396, 402

ambiguity, 357, *see* grammar(s),ambiguous
- theorem on, 360

511

apperception, 88
Archimedean
  – axiom, 205, 208, 232, 235
  – mathematical tradition, 37
association(s), 14, 442
  – and grammatical forms, 422–424, 428
  – and human learning, 425
  – and memory, 95
  – deletion of, 428–429
  – form, 429–430
  – in Hume, 83–84, 86, 87
  – in James, 88–92
  – in Kant, 88
  – incorrect, 420
  – laws of, 88
  – learning principles, 421
  – of ideas, 15
  – of words and internal symbols, 419, 420
  – probabilistic, 421, 423, 426, 429
    – axiom, 429
  – stimulus-response, 13, 389
astronomy, 42, 44, 45n, 284
  – ancient, 42, 42n, 43, 44, 157, 322
automata, 13, see also machine, 392
  – and stimulus-response, see stimulus-response theory
  – connected, 365, 384
  – equivalence, 365
    – weak, 393
  – finite, 361–364, 394
    – and regular grammar, 367–371
    – and stimulus-response models, 380–387
    – deterministic, 361, 367
    – isomorphism, 363–364
    – languages accepted, 364–367, 369
    – languages accepted infinite, 366
    – nondeterministic, 367
    – representation theorem, 370
    – representation theorem by stimulus-response models, 384
    – with output, 363
  – isomorphism, 365
  – linear bounded, 374
    – and context-sensitive grammars, 374
  – probabilistic, 367–370
  – pushdown, 371
  – and context-free grammars, 371–373
  – representation theorem, 373
  – representation
    – in terms of languages, 353
  – theory of, 13, 389
  – transition function, 362
  – two-state, 380
automaton, see automata
automorphism, 99, 122, 303
axiom(s)
  – adequacy, 63
  – and postulates, 36
  – Archimedean, 38, 205, 208, 232, 235
  – associativity, 78
  – bisection, 307
  – completeness, 267
  – congruence, 268
  – construction, 306
  – creative, 53
  – de Finetti's qualitative, 226–230, 238, 250, 252
  – dimensionality, 267
  – doubling, 307
  – dynamical, 320
  – elementary, 252
  – Euclidean, 105, 109, 287
  – first-order, 29
  – for cardinal numbers, Tarski's, 34
  – free, 54
  – Good's, 254
  – grammar-rule generation, 426, 430
  – group, 102
  – impenetrability, 323
  – in Aristotle, 35
  – independence of, 48, 285
  – invariance, see invariance
  – Keynes', 184, 238
  – kinematical, 320
  – Kolmogorov's, 200
  – Koopman's, 250
  – linear learning models, 404–405
  – linearity, 307
  – Luce's choice, 212, 213
  – machine learning, 425–432
    – long-term memory, 427
    – working memory, 426
  – necessity, 238
  – of choice, 40, 305

INDEX

- of free mobility, 286
- of Jeffreys and Wrinch, 238
- of motion, *see* law(s) of, motion
- of set theory, 30
- Pasch's, 46, 267
- Peano's, 161
- principle of sufficient reason, 159
- probability, 134–136, 203
- qualitative, 230–233, 238
- qualitative, of probability, 188
- quantifier-free, 305, 306
- Ramsey's, 239
- rationality, 250–252
- rationality vs. structure, 250
- register learning models, 398–399
    - program construction, 398
    - program execution, 399
    - register structure, 398
    - reinforcement, 399
    - response, 399
    - stimulus encoding, 398
- reinforcement, 405
- Savage's, 249–252
- Schutz's, 282
- Scott's, 235
- stimulus-response theory, 379–380
    - conditioning, 379
    - response, 380
    - sampling, 379
- structure, 250–252
- sufficiency, 238
- waiting-time, 208–210
- Walker's, 282

axiomatizability, 5
- criterion for, 29
- of theory of measurement, 28

axiomatization (of)
- a group, 31
- a theory, 5–7, 10, 17, 26
    - by set-theoretical predicate, 30, 55
- affine plane, 306–308
- affine space, 266–267
    - with restricted congruence, 298–300
- algebra, *see* algebra(s), axiomatization
- and definition, 30–33
- classical particle mechanics, 6, 21–22, 112, 316, 319–323
- classical space-time, 269–272

- decision-making, 250
- Euclidean space, 268
- extrinsic-intrinsic, *see* characterization (of a theory), intrinsic vs. extrinsic
- finite, 28
- geometry, 4, 46, 48, 106
    - verticality, 107
- independent, 33
- measurement
    - bisection, 68
    - conjoint, 69–70
    - difference, 66–67
    - extensive, 64–65
    - hyperordinal, 28–30
    - ordinal, 25–27
- methods of, 10
- partial qualitative expectation structure, 234–235
- probability, 12, 27, 130–157, 203
    - qualitative conditional, 234
    - subjective, 239
- real affine space, 316–318
- register learning models, 397
- restricted relativistic space-time, *see* axiomatization (of), special relativity
- semiorders, 254
- special relativity, 275–276, 281–282
    - qualitative, 280–282
- theory of conditioning, 442
- theory of utility, 239
- vector spaces, 314
- visual space, 292

Bayesian, 184, 185, 189n, 198, 200, 245n, 248, 251, 261, 262
- Bayes' postulate, 141–144
- Bayes' theorem, 140
- inference, 441
- methods, 263
- objective position, 241, 242
- rule of behavior, 141, 143, 144
- subjective position, 242

behavior
- and conditioned reflex, 387
- and stimulus-response theory, 353, 375, 388
- Bayesian rule of, 141, 143, 144

behavior (*continued*)
- equivalence, 393
- experiments, 446
- language, 375
- learning, 13
- rational, 19
- response data, 442
- robot's, 421
- science, 20, 23, 24

behaviorism, 9, 10, 13, 442

behavioristic
- approach to theories, 10
- psychology, 377
- theory of language, 1, 9

belief
- and decision, 245
- and epistemic probability, 183
- and subjective probability, 155, 231, 239, 252
- and utility, 234
- convergence and reasonableness, 243–245
- degrees of reasonable, 185–226
- in Hume, 84
- measurement, 248–252
  - finite approximate, 253
- mistaken, 244
- quantity of, 239

Bell's inequalities, 336–338
- theorem, 337
  - CHSH form, 337
  - three values, 338

Bernoulli process, 124, 125, 127, 182, 241n

betting, *see* gambling

betweenness, 68, 100, 103, 107, 121, 122, 266, 267, 270, 275, 280, 282, 287, 316n

brain
- activity, 88, 90, 287, 442
  - electrical, 442–444, 451
  - observation, 444–446, 463
- and computer, 390
- and language, 354, 442–465
- brain-wave, 8, 10, 14, 448, 453
  - classification, 448, 458
  - prototypes, 453
  - recognition, 445
- discovery of electrical activity, 442–444
- experimental results, 450–453
- censoring data, 463–465
- criticism, 453–461
- extreme statistics, 454–458
- invariance between images and words, 452–453, 464–465
- invariance between subjects, 450–451, 461
- table of results, 460
- test of a timing hypothesis, 461–463
- Fourier analysis of data, 448–450
  - optimal filters, 450
- images, 452, 453
- methods of data analysis, 446–450
- processes, 89
- processing of language, 445
- representation of response, 383

calculus
- formal, *see* calculus, logical
- integral, 42
- logical, 2, 3, 5, 7
  - semantics for, *see* semantics
- of probability, 160

cardinality, 29, 30, 56, 59, 60, 142, 160, 225, 228, 252, 256, 384, 385, 404
- infinite, 305

categorical, 226, 268
- theory, *see* theory, categorical

causal
- account of the mind, 88
- chain, 313
- deterministic stance in classical physics, 260
- dispositional phenomena, 221
- explanation, 83, 170
- law, 322
- processes, 13, 260, 343
  - and noncausal, 260
- reversibility, *see* reversibility

causality, 89, 90, 351
- common cause as hidden variable, 313, 332, 333
- in propensity theory, 221
- probabilistic, 222
- theory of, 222

censoring (statistical), 463–465

chain
- causal chain, 313

INDEX 515

– Markov, *see* Markov chain
– of infinite order, 348
characterization (of a theory), 3
– as theory of measurement, 29
– axiomatic, *see* axiomatization
– intrinsic vs. extrinsic, 5–6, 28, 29
– of the models, 57
– set-theoretical, 48
choice
– axiom of, 40, 212, 249, 305
  – Luce's, 212, 213
– of a unit of measurement, 112
– of coordinate system, 121
– of measure function, 196, 197
– probability, 212, 213
– rational, 19, 23
– set of, 213
– subjective, 199
– theory of, 251
classical particle mechanics, *see* mechanics
closure
– condition, 303, 307
– of a language, 370
– of invariant frames, 270, 276
– property, 29, 79, 100, 133
– under submodels, 29, 29n, 251
coin tossing, 130–131, 136, 137, 147, 390
– mechanics of, 214
– possibly biased, 240
– pre-image of heads, 216
– propensity theory of, 214–218
– representation theorem, 217
collinearity, 266–268, 298, 308
complexity, 221
– and propensity, 221
– Kolmogorov's definition, 129, 177–182
computer, *see also* automata
– and the brain, 390
– learning, *see* machine, learning
– program, 74, 177, 286, 304, 354, 375
– universal, 62, *see also* machine, Turing, universal, *see also* machine, unlimited register, 181
concept(s)
– children's mastery, 401
– formation, 398, 400
– geometric, meaning of, 46
– of a model, 21

– primitive, *see* primitive concepts
– register-machine learning, 398, 401
conditioning, *see also* association(s), 353, 378, 381, 386, 388–391, 393, 399, 401, 442
– and automata, 374, 376
– axioms, 379
– classical, 375–377
– experiments, 376, 442
– function, 377, 414
– of pigeons, 388
– parameters, 381
– pattern, 382
– probability of, *see* probability
– states of, 374, 381, 382, 442
– stimulus-response, 374–377
confirmation
– degree of, 190, 198
– function, 192
– theory, 165, 190–198
  – set-theoretical formulation, 196
congruence, 100, 105, 107, 121, 122, 268, 270, 280, 294, 428, 429, 432
– affine, 269
– axioms for, 268, 300
– class, 428, 433, 435–438
– computation, 427
– meaning, 428, 438
– of parallel segments, 103
– of segments, 275, 283, 287
– perceived, 297
– proper time, *see* time, proper
– restricted
  – axioms for, 298–299
– semantic, 427n
constant(s)
– logical, 24
– nonlogical, 24
continuum mechanics, 22, 322n
– Newton, 322n
correlation, 153, 332–334, 336, 337, 342
– GHZ-type, 341, 342
– matrix, 342, 343
– prediction on, 339
covariant, 122–123
creative definition, *see* definition

criterion (of)
- closure under submodels, 251
- definition, 187
  - eliminability, 53, 78
  - noncreativity, 53, 78
- for a 'pure' state, 259
- for axiomatizability, 29
- for hidden variables, 343
- for joint distribution, 334
- least-squares, 450n, 452, 462
- optimality, 450
- probability spaces generation, 141, 165
- representational consequence, 251
- simultaneity, 343
- Vaught, 29

data
- analysis, 262, 284, 446–450, 459
- and model, 21, 24, 284
- and theory, 7
- auditory, 109
- behavioral response, 442
- brain-wave, 445, 446, 453
- canonical form, 7, 130
- censoring, 463–465
- empirical, 130, 197, 204, 321
- experimental, 144, 198, 284, 287, 377, 461
- haptic, 109
- model of, 66
- observational, 186
- reduction, 144
- relative-frequency, 226, 243
- sensory, 396
- visual, 109

decision
- and subjective probability, 245–248
- behavioral aspects of, 226
- function, 234, 251
- mechanical procedure, 27, 360
- Tarski's procedure, 179
- theory, 226, 250, 251
  - and utility, see utility
  - statistical, 8–10
- under uncertainty, 248–256

definability, 53n

definition
- and axiomatization, 30–33
- as axiom or premise, 53
- as representation, 53–54
- coordinating, 3, 4, 6–8
- creative, 34
- eliminability, see eliminability
- in a theory, 53
- noncreativity, see noncreativity
- recursive, 24, 25
- set-theoretical, 33
- set-theoretical predicate, 32
- theory of, 78

denotation
- denotational-value computation, 427, 431–432
- problem of, 420–421

density, 147, 148
- and distribution, 148, 150n, 210
- asymptotic, 347
- Bernoulli, 136, 150
- binomial, 136, 150
- Cauchy, 149
- conditional, 217
- conditional discrete, 210
- continuous, 149, 216
- discrete, 148, 150, 241
  - properties of, 210
- discrete qualitative, 210–211
- geometric, 138, 150, 151, 211
- of random variable, 148
- piecewise continuous, 148–149
- Poisson, 137, 150, 152
- stationary, 348, 349
- uniform, 150, 210

derivation, see grammar(s), derivation

determinism
- and randomness, 223
- and reversibility, 349–351
- and the three-body problem, 218–220, 223
- and unpredictability, 220
- indeterminism, 90
  - and classical mechanics, 223
  - and propensity, 222–223
- Laplace, 160n

distribution, see also probability, distribution, see also density, 137, 147, 148, 165, 180, 197, 222, 235, 258, 349, 455–457

INDEX 517

- a priori, 193, 197
- and density, 148, 150n, 210
- asymptotic, 125, 346
- beta, 454–458
- binomial, 137, 454, 455
- continuous, 152
- cumulative, 147
- discrete, 369
- exponential, 153, 157
- extreme-statistic, 453–455
- Gaussian, 153, 342, 343
- geometric, 157, 203, 204, 207, 209, 399, 400
- hypergeometric, 406–409, 415n
- joint, 123, 124, 153–155, 332–334, 337, 339–342, 345
  - and simultaneous observation, 343
- maximum-entropy, 242
- mean, 346
- normal, 153, 218
- null-hypothesis, 456, 457
- of random variable, 155, 201, 225
- parameter of, 248, 402
- piecewise continuous, 148–149, 152
- Poisson, 135n, 137
- prior, 188, 198, 241
  - improper, 188n
- random, 180
- rational, 192
- sampling, 414, 415, 415n
- spatial of retinal stimulation, 294
- symmetric, 223
- uniform, 152, 163–165, 201, 400, 402, 421
- unique, 208, 235

Ehrenfest model, 348–349
eliminability
- criterion for subroutines, 78
- criterion of definition, 53, 78
embedding, 62–63, 100, 283, 284, 286, 301, 384
- homomorphic, 62
- isomorphic, 329
- theorem, 62, 329–330
entropy, 123–125, 241n, 349
- and ergodic theory, 14, 97, 123–127
- and isomorphism, 126, 127
- as complete invariant, 97, 123–127, 155, 241
  - for Bernoulli processes, 126
  - for Markov processes, 126
- as measure of uncertainty, 123–127, 241n
- as objective prior distribution, 241
- for a Bernoulli process, 124
- for a Markov process, 124
- for discrete-time processes, 123
- maximum entropy, 242
- of a random variable, 123
- principle of maximum, 241, 242
ergodic
- and stationary, 124, 346
- Markov chain, 125, 346
- process, 97, 124–127, 156, 211, 260
  - birth and death, 348
- theory, 123, 123n
  - and entropy, see entropy
  - fundamental theorem of, 125
- transformation, 124
ergodicity, 124–125
error, 306, 392, 398
- experimental, 340, 341
- in observation, 12
- of approximation, 306
- of measurement, see measurement, error of
event(s), 130
- language of, 131–133
- set-theoretical representation, 130, 131
evidence, 2, 3, 9, 35, 35n, 107, 108, 141, 184, 185, 190, 198, 226, 243, 244, 292, 293, 300, 345
- and hypothesis, 192
- experimental, 186
- theory as organization of, 9
- with zero probability, 202
exchangeability, 156–157, 211
- exchangeable events, 240
- for infinite sequences, 240–241
expectation, 124, 166, 249
- comparison, 234
- conditional, 204n, 236, 334
- function, 232, 233, 235, 236, 238
- mathematical, 166
- moral, see utility, expected

expectation (*continued*)
— of random variable, 149–153, 233, 235, 336–338, 340
— partial qualitative structure, 235–236
expected
— loss, 8
— utility, *see* utility, expected
— value, 149, 230, 407
— of random variable, 149, 150, 230, 231
experiment(s), *see also* data
— and theory, 7
— as infinite sequence of trials, 377
— as random variable, 146
— as set of possible outcomes, 64
— behavioral, 446
— brain, 458–465
— canonical form of data, 130
— classical conditioning, 375, 376, 442
— classical discrimination, 376
— conditioned reflex, 387
— design, 197, 459
— experimenter partition, 405
— Gedanken, 332
— Gestalt, 95
— GHZ, *see* GHZ experiments
— in physics, 23
— in probability, 144–146
— independence of, 144–146
— learning, 183, 403
— MEG, 445
— models, 7
— null-hypothesis, 453
— on hidden variables, 332
— on imagery, 94, 95
— on subjective probability, 246
— on visual space
— Blumenfeld, 290–292
— Foley, 266, 294–295, 299
— Gogel, 293
— Indow, 295–296
— Luneburg, 292–293
— Wagner, 266, 296–297, 299
— quantum entanglement, 458n
— theory of the, 7
— visual-image, 450–451, 459
— with biased mechanism, 162
explanation, 87, 90, 140

— causal, 83, 170
— qualitative, 90
exponential decay, 157
exterior product, *see* vector product

finitism, 266, 303–311
first-order
— logic, *see* logic, first-order
formal method, 35n
— as set-theoretical method, 1
— role of, 1–2
formula(s), 24
— atomic, 24
— molecular, 24
— recursive definition, 25
— validity, 27, 360
— well-formed, 191
Fourier
— analysis, 456
— experimental results, 450–453
— filters, 450
— properties, 448–450
— transformation, 447, 448, 450
frequency
— single-case vs. long-run, 222
function
— clock time, 318
— computable, 74, 76
— conditioning, 377
— confirmation, 192
— decision, 234
— effectively calculable, 74
— extended indicator, 230
— force, 22
— external, 319
— internal, 319
— kinetic energy, 331
— mass, 21
— partial recursive, 74, 74n, 75–76, 394
— over finite alphabet, 80–81
— place-selection, 171
— position, 21, 319
— potential energy, 330, 331
— primitive recursive, 74–75
— utility, *see* utility
— vector-valued, 319

gambling, 157, 158, 174, 246
— and place-selection rule, 171, 172

INDEX 519

- auction game, 246
- betting, 172–174, 245
  - and random sequence, 171
- fair, 177
- martingale, 177
- principle excluding gambling systems, 171, 172
- race track, 246–247
- unfair sequence, 177

game
- auction, 246
- of chance, 158, 161, 167, 178, 242
- problem of points, 158
- two-person, 247

generalization, 426n

geometry, 35, 35n, 36, 39
- absolute, 105, 106, 287, 298
- affine, 103, 105, 267n, 287, 306
- ancient Greek, 38
- and finitism, 266, 303–311
- and use of prepositions, 105
- applications of, 305
- as set-theoretical structure, 98
- axiomatization of, 46, 48
- contextual, 294, 295, 300–301
- demonstrations, 452
- differential, 296
- elliptic, 287, 290
- Euclidean, 40, 43, 48, 103, 105, 106, 121, 287
- foundations of, 45, 48, 285
- group-theoretic, 102
- hyperbolic, 46, 105, 299
- invariants in, 100
- Minkowski, 282
- modern, 45–47
- non-Euclidean, 45
- of visual space, *see* visual space
- ordered, 287
- perceptual, 300
- phenomenal, 297
- projective, 46, 89, 103, 287, 294n
- reduction to algebra, 52
- spherical, 107, 290
- static spatial, 106
- synthetic, 316n
- transformational, 102
- two-dimensional, 289

GHZ experiments, 154n, 338–342
- detector inefficiencies, 339–342
- inequalities, 339–342

grammar(s)
- ambiguous, 360–361
- and automata, 13, 370
- comprehension, 354, 420, 435
  - learning, 419–442
  - rules, 432n
- context-free, 356, 359, 360
  - and pushdown automata, 371–373
  - representation theorem, 372
- context-sensitive, 356, 361
  - and linear bounded automata, 374
  - recursive property, 374
  - representation theorem, 374
- derivation, 355–356
  - leftmost, 360
  - tree, 356, 361, 382
- for internal language, 424, 425
- learning, 376, 441
- lexical-functional, 357, 361
- normal form, 357–359
  - Chomsky, 358
  - Greibach, 359
- of internal language, 430
- phrase-structure, 354, 356, 357, 359, 361
  - and Turing machines, 374
  - generalized, 361
  - representation theorem, 374
- picture, 286–287
- probabilistic, 389
- production, 355
- recursive, 374
- regular, 356, 357, 361
  - and finite automata, 367–371
  - representation theorem, 369
- representation theorems, 361–374
- rule generation, *see* axiom(s), grammar-rule generation
- type-0, *see* grammar, phrase-structure
- types of, 356–357
- weak equivalence, 355

hidden variable(s), 130, 260, 313
- and quantum mechanics, 332–343
- common cause, 313, 332, 333

hidden variable(s) (*continued*)
 – conditional independence, 333
 – deterministic, 333
 – existence, 332, 333
   – Bell's inequalities, 336–338
   – GHZ inequalities, 338–342
 – factorization, 333–334
 – locality, 335
   – Bell, 335
 – representation theorems, 332–343
hierarchy
 – of formal languages, *see* language, formal
 – of geometries, 287–288
 – of grammars, *see* grammar, types of
 – of theories, 8
 – role in learning, 401–403
homogeneity, 259
homomorphism, 28–30, 58–62, 111
 – and isomorphism, 58, 62, 114n
 – for algebras, 58
hypothesis testing
 – authorship of Federalist Papers, 262
 – brain representations, 442–465
 – extreme statistics, 454–458
 – GHZ-type experiments, *see* GHZ-type experiments
 – machine learning, 433–440
 – mental representations, *see* mental, representations
 – visual space, 282–288

impenetrability
 – axiom of, 323
impetus, 273
independence (of)
 – and exchangeability, 240
 – axioms, 48, 68, 284
 – condition, 70
 – conditional, 333
 – distributions, 125, 346
 – events, 166, 168, 178
 – experiments, 144–146
 – irrelevant alternatives, 251
 – path, 310
 – predicates, 191, 196
 – primitive concepts, 320
   – Padoa's principle, 54n
 – random variables, 153–154, 156
 – scale or units, 213
 – the past, 208
 – trials, 154, 455
 – vectors, linear, 316
indeterminism, *see* determinism
induction, *see* inference, inductive
inertial frame(s)
 – classical invariance axioms, 270
 – definition, 269
 – relativistic invariance axioms, 276
inference, 53
 – Bayesian, 441
 – deductive, 192, 193
 – in Hume, 83–85
 – inductive, 129, 185, 242n
 – principles of, 8
   – and theories, 8, 9
   – material, 8
 – statistical, 129, 222, 253
inner product, *see* scalar product
instrumentalism, *see* theory, instrumental view of
intention, 387
intentional, 91
 – component of language, 442
interpretation
 – empirical, 3
   – vs. models, *see* model
 – procedural, 424
 – semantic, 420
invariance, 11, 109, 110, 127, 274
 – absolute convergence, 323
 – as constancy, 97
 – axioms
   – for classical space-time, 270–271
   – for restricted relativistic space-time, 276
 – covariance and, 123, 280n
 – empirical meaning and, 111
 – entropy in ergodic theory, *see* entropy
 – flatness of space, 105
 – geometric, 110, 156
 – groups and, 100
 – in Hume, 84
 – in neural processing, 354
 – in perception, 12, 105, 108, 109
 – in physics, 103, 120

INDEX 521

- in theories of measurement, 110, 112–114
- logic and, 102
- mathematical notion, 114
- meaningfulness and, 110–112
- ordinal transformation, 111
- principle of, 85
- probabilistic, 155–157
  - exchangeability, 156
  - exponential decay, 157
  - independence, 156
- propensity to decay, 208
- proper time, 275, 276
- proper time as a complete invariant, 104
- relativity and, 11, 275
- representation, 97, 111, 112
  - in the brain, 461
  - in the brain, between images and words, 452–453, 464–465
  - in the brain, between subjects, 450–451, 461
  - of language in the brain, 354
- size-distance hypothesis, 285, 295
- spatial preposition, 109
- speed of light, 275, 276
- symmetry and, 11, 97–105
- theorem, *see* invariance theorem (for)
- time and, 313, 343, 344
- under a group of transformations, 6, 104, 105
- under a theory, 104, 104n
- under a time shift, 211

invariance theorem (for), 121
- affine space, 268
- classical space-time, 271
- conservation of energy, 332
- conservation of momentum and angular momentum, 327
- entropy of ergodic Markov processes, 126
- Euclidean space, 269
- finite affine constructions, 310
- Galilean, 271
- measurement
  - bisection, 119
  - conjoint, 119
  - difference, 118
  - extensive, 118
- measurement theories, 112
- qualitative approximate belief measurement, 255
- qualitative conditional expectations, 236
- qualitative conditional probabilities, 236
- qualitative expectations, 232
- qualitative probabilities, 232
- relativistic space-time, 277
- temporal invariance of
  - continuous-time Markov process, 348
  - Markov property, 345
  - random-variable distribution, 347
  - second-order Markov chain, 348
- visual space with restricted congruence, 299

isomorphism, 30, 54–58, 82, 84, 86, 91–93, 364, 391
- absence of, in semantic categories, 433
- and association, 91
- and automata, 363, 365
- and equivalence, 329
- and ergodic processes, 125–127
- and homomorphism, *see* homomorphism
- and mental representation, 81, 86, 89
- and resemblance, 84, 85
- difficulties about, 58
- for algebras, 54
- for groups, 55
- for simple relation structures, 56
- in mental imagery, 95
- in stochastic processes, 126
- of models, 10, 403n
- of models of a theory, 4, 51, 54–57
- of structures, 364
- structural, 74, 111
- types, 28

language
- and automata, 364–367
- and brain, 442–465
- and stimulus-response, *see* stimulus-response theory
- behavior, 375

language (*continued*)
- behavioristic theories of, 2, 9
- brain's processing, 445
- categories of words, 425
- closure, 370
- closure under operations, 359–360
- context-free, 358–360
- context-sensitive, 357, 374
- encoding, 395
- first-order, 200
- for degree of confirmation, 190
- formal, 93, 196, 200, 354, 355, 357, 358, 426n
  - hierarchy, 354–361
- inflected, 438
- internal, 395, 420–422, 424–425, 427n, 430, 432, 434, 438, 440
- learning, *see* learning
- machine, *see* machine
- natural, 13, 110, 355, 361, 419, 420, 427, 428n, 433, 441
- neural processing of, 354
- neural theories of, 2
- of a theory, 24, 25, 30
- of events, 131–133
- phrase-structure, 357, 359
- programming, 177, 375
- recursive/nonrecursive, 361
- recursively enumerable, 361
- regular, 359, 369–371, 387
- representation
  - brain-wave, 8, 10, 446
  - in terms of automata, 353
  - in the brain, 354
- set-theoretical approach to, 2, 357
- set-theoretical operations on, 359–360
- type-0, *see* language, phrase-structure
- unsolvable problems, 360–361

law(s) (of)
- association, *see* association
- causal, 322
- conservation, 104, 122, 327–329, 332
  - Aquinas, 329n
  - Aristotle, 328
  - Bruns, 328
  - Compton effect, 328
  - Descartes, 327
  - Newton, 328
- Donders', 302
- electrodynamics, 279
- expression of, 188
- gravitation, 218
- gravitational attraction, 88
- habit, 90
- inductive, 345
- inertial, 274
- invariant exponential, 157
- linear momentum, 325
- logic, *see* logic, laws of
- mechanics, 345
- motion, 44–45, 121, 321, 323
- nature, 87, 222, 259
- Newton's
  - first, 323
  - second, 323
  - third, 330
- physics, 122, 123, 329
- probability, 123
- reproduction, 87
- the motion of the planets, 322
- vector, 317

learning
- children's mastery of simple concepts, 401
- discrimination, 377
- language, 388, 393, 420, 441
  - and reinforcement, 394
  - by humans, 110, 354, 383, 420, 422, 425, 441
  - by machines, *see* machine learning, language
  - by robots, 419–442
  - natural, 354, 419, 420
  - probabilistic, 421
  - stimulus-response theory, 389
  - without prior knowledge, 441
- linear models, 403, 404, 432
  - and stimulus-sampling models, 403–419
- Luce's alpha model of, 212, 213
- machine, *see* machine learning
- of concepts, 403, 422, 441
- parameter, 404
- register-machine models, 401
  - axiomatization, 397–401
  - primitive concepts, 397

INDEX 523

- program construction axioms, 398
- program execution axioms, 399
- register structure axioms, 398
- reinforcement axioms, 399
- response axioms, 399
- stimulus encoding axiom, 398
- reinforcement models, 419
- role of hierarchies, 401–403
- stochastic models, 130
- theories of, 426n

lexicon
- for internal language, 424

light line, 277, 278

likelihood, 141, 142
- principle of maximum, 143

logic, see also calculus, logical, 10, 200
- and probability, 184–201
- deductive, 185, 186
- elementary, 62
- first-order, 4–6, 17, 25, 62, 252, 360
- formal, 1
- inductive, 185, 198
- laws of, 140
- mathematical, 20, 21, 23
- modal, 155
- of inference, 53
- probability logic, 200
- propositional, 359
- two-valued, 245
- undecidable in, 360

machine, see also automata, 362, 371
- finite, 370
- internal states, 375
- language, 78, 375
- learning, see machine learning
- register, 394–398
- instructions, 398
- single, 80
- unlimited, 51, 74, 76–80, 181, 353, 395
- representation, 14
- of partial recursive functions, 74–81
- sequential, 391
- single register, 395
- transition table, 381
- Turing, 79, 80, 373–374, 394
- universal, 51, 76, 180, 181, 390

- vs. register machine, 394

machine learning, 13, 396–403
- axioms, 425–432
- denotational-value, 427
- form association, 426
- grammatical form generalization, 426
- long-term memory, 427
- using working memory, 426
- congruence computation, 427
- corpora of ten languages, 432–440
- curves, 433–438
- empirical results, 433–440
- language, 354, 421, 425, 434, 441
- by robots, 419–442
- natural, 354, 419, 420
- probabilistic, 420
- problem of denotation, 420–421, 427, 431–432
- unsolved problems, 441–442

Markov chain, 124, 346, 347, 381
- deterministic, 125
- ergodic, 125, 346
- first-order, 139, 344
- nonergodic, 124
- second-order
- and strongly reversible, 348
- states of, 381, 382
- stationary, 346
- strongly reversible, 344, 346
- weakly reversible, 344

Markov process, 126, 241n, 345
- and entropy as a complete invariant, see entropy, as a complete invariant
- aperiodic, 346
- continuous-time, 346, 348
- second-order, 348
- ergodic, 126, 127
- homogeneous, 345
- irreducible, 345

mathematics, 10, 22, 27, 30, 38–41, 46, 48, 53, 57, 102, 120, 161, 305, 375, 389
- and models, 21, 23
- and reduction, 5
- and set-theory, 161
- applied, 20, 33, 311
- finitistic, 303, 311
- in physics, 305, 311

mathematics (*continued*)
  – Babylonian and Egyptian, 35
  – Bourbaki approach, 33–34
  – foundations of, 1, 13, 74, 74n, 186, 303
  – Greek, 35, 36n, 121, 157
  – in Aristotle, 35
  – learning of, 394
  – non-axiomatic, 42
  – of measurement, 63
  – philosophy of, 266
  – pure, 3–6, 33, 37, 40, 58, 180, 186, 303
matrix, 247
  – correlation, 342, 343
  – decision, 249
  – Galilean, 271
  – identity, 271, 278
  – Lorentz, 276–278
  – non-negative definite, 343
  – nonsingular, 271
  – rotation, 268
  – similarity, 269
  – transition, 139
meaning, 430
  – and internal representation, 423, 425, 441
  – and symmetry, 102–103
  – and use, 21
  – closeness of, 427n
  – congruence of, 428, 438
  – constancy of, 21
  – empirical, 45, 112, 321
    – and invariance, 111
  – geometrical, 110
  – intuitive, 39
  – learning, 432
  – meaningfulness
    – and invariance, 110–112
    – of empirical hypothesis, 110
  – multiple, 442
  – objective, in physics, 103–105, 318
  – of geometric concepts, 46
  – of symbols in a theory, 53
  – of the concept of a model, 21
  – physical, 104
measure
  – -theoretic, 177, 178, 182, 201
  – additive, 65, 231
  – Bernoulli, 183
  – complexity of sequence (Kolmogorov's), 129, 177–182
  – degree of belief, 248–252
  – function, 192, 196, 197
  – inner and outer, 253
  – on set of stimuli, 377
  – probability, 136, 155, 203–205, 210, 225, 228, 229, 233, 234, 244, 378, 404
  – a priori, 200
  – additive, 123
  – conditional, 238
  – countably additive, 135
  – for first-order languages, 200
  – measure-theoretic, 178n
  – numerical, 185, 226, 230
  – standard, 227
  – subjective, 231
  – regular (Carnap), 191, 194, 197
  – response strength, 212
  – state-symmetric, 192, 193
  – structure-symmetric, 194
  – symmetric, 193
  – uncertainty, *see* entropy
  – upper and lower, 253
  – Wittgenstein, 193, 197
measurement (of)
  – bisection, 67–68
  – conjoint, 69–70
    – theory of, 39
  – difference, 66–67
  – distance, 55, 115, 275
  – empirical process of, 4
  – error of, 12, 157, 284
  – extensive, 63–66, 232, 236, 247
  – extensive quantities, 55
  – finite approximate for beliefs, 253
  – hardness, 110
  – hyperordinal, 28
  – instruments, 65, 104
  – intelligence, 110
  – intensive, 66–67
  – intensive quantities, 63
  – length, 63, 64
  – mass, 11, 55, 63, 64, 110, 113–115
  – mathematics of, 63
  – methods of, 269
  – of distance, 113
  – ordinal, 25, 26, 110, 114n

INDEX 525

- physical, 68
- precision in, 20
- probability, 110
- procedure, 4, 52, 65, 104, 112, 114, 186, 260
- process, 260, 345
   - in quantum mechanics, 260
- psychological, 68
- racial prejudice, 110
- scales of, see scale
- sensation intensities, 117
- simultaneous of momentum and position, 258
- space-time frame, 269
- speed of light, 276
- structures, 63
- subjective probability, 63, 245, 246, 248
- temperature, 110, 113, 116
- theory of, 4, 5, 11, 28, 55, 58, 59, 62, 95, 112, 120, 226, 403
   - and finite relational structure, 118
   - and weak orderings, 58
   - axiomatizability, 28, 29
   - elementary, 65
   - equal-interval, 97
   - invariance, see invariance, in theories of measurement
- time, 271, 321
- unit of, 11, 110–112, see also scale, 113, 120, 269
- utility, 114, 117, 231
mechanics, 27, 38, 40, 43, 104, 223, 279, 324
- adequacy of, 330
- celestial, 284
- classical particle, 6, 12, 13, 21, 111, 122, 203, 214, 222, 223, 265, 272, 313–332, 334
   - and determinism, 324
   - and indeterminism, 223
   - and reversibility, 344
   - axioms, 319–323
   - primitive concepts, 21, 319–320
   - representation theorems, 329–332
   - system, see systems of particle mechanics
- continuum, 22, 322n

- Newton, 322n
- fluid, 12, 45n, 322n
- medieval, 273, 273n
- of coin tossing, 214–218
- quantum, see quantum mechanics
- relativistic, 111, 343
   - and reversibility, 344
- relativistic (nonquantum), 328
- standard formalization, 27
- statistical, 12, 187, 241, 242, 260, 322n
   - and probability, 256
   - and thermodynamics, see thermodynamics
   - Ehrenfest model, 348–349
   - foundations of, 22
   - model of, 22–23
   - stationarity, 211
- three-body problem, 218–220
mental
- images, 89, 92–95
   - rotation of, 94
- representations, 63, 81–95
   - and isomorphism, 81, 86, 89
method (of)
- asymptotic, 322n
- axiomatic, 10, 35–49
- Bayesian, 263
- constructive, 305, 309
- data analysis, 446–450
- deductive, 48
- finitistic, 311
- formal, 1–2, 35n
- measurement, 269
- model-theoretic, 201
- multidimensional scaling (MDS), 296
- nonconstructive, 176, 178
- nonfinitistic, 305
- reinforcement, 393, 399
- sampling, 243
- set-theoretical, 1, 364
- statistical, 7, 187, 222, 465
model(s), 290
- algebraic, 212
- and data, 284
- and frame of reference, 6
- and reductionism, see reduction in science
- and submodel, 62

model(s) (*continued*)
- as abstraction, 17
- as design standard, 17
- as exemplar, 17
- as linguistic entity, 20
- as set-theoretical entity, 21
- class of, 5, 6, 57, 385, 404
- concept of, 21
  - fundamental, *see* model, in mathematical logic
- construction of, 23
- embedding, *see* embedding
- empirical, 4, 4n, 5, 58
- finite, 252, 311
- homomorphism, *see* homomorphism
- in biology, 23
- in mathematical logic, 21, 23, 24
- in mathematical statistics, 24
- in physics, 23
- in the social sciences, 23
- infinite, 252
- isomorphism, *see* isomorphism
- learning, *see* learning
- meanings of, 17–24
  - as class of models, 20
- multivariate network, 442
- numerical, 4, 5, 11, 58, 114
- of a theory, 3–5, 10, 51, 57, 104, 112, 225, 403
  - and abstraction, 58
- of choice, 212
- of classical particle mechanics, 22
- of electromagnetic phenomena, 22
- of probability, 129
- of rational choice, 19, 23
- of response strength, 211, 214
- of science, 87
- of statistical mechanics (Ehrenfest), 348–349
- of the experiment, 7
  - vs. model of the theory, 7–8
- of theory of measurement, 55
- physical, 17, 21, 22, 24
- physical vs. set-theoretical, 22
- plurality of, 6
- recognition, 462
- related by Galilean transformations, 6
- response strength, 212
- set-theoretical, 21, 22, 24, 54
- stimulus-response, *see* stimulus-response theory
- stimulus-sampling, 403, 407, 414, 418, 419
- submodel, 251
- subset of, 11
- theory of, 5, 20, 190, 200, 228
- vs. empirical interpretations, 3–5

model-theoretic, 200
- approach to probability, 200
- semantics, 426n

momentum, 104, 122, 258, 259, 325–327
- angular, 326
- laws of conservation, *see* law(s), of conservation
- linear, 325
- space, 258

moral expectation, *see* utility, expected

neuroscience, 10, 12, 383, 442–465
noncreativity
- criterion for subroutines, 78
- criterion of definition, 53, 78
normative, 184, 192

observable(s), 144, 332, 333
- and primitive symbols, 3
- as random variables, 332, 333, 339
- data, 321
- expectation of, 338
- Gaussian, 332
- in reinforcement, 392, 394, 405
- in stimulus presentation, 405
- joint distribution of, 334
- mean values of, 257
- normalized, 338
- Peres' definition, 338
- probability, 212
operations, 21, 89, 225, 251, 281, 306, 308, 314, 424
- algebra of, 114
- arithmetical, 4, 52, 111, 362, 363
- binary, 315
- Boolean, 251
- empirical, 4, 111, 113, 114
- set-theoretical, 64, 357
  - on languages, *see* language
optimality, 450

orbital theory, 21
ordering, 5, 6, 33, 59n, 60, 61n, 68, 69, 71,
    73, 231, 234, 253, 323
  – betweenness, 68
  – binary, 68, 251
  – bisection, 68
    – ternary, 68
  – empirical, 111
  – lexicographical, 60, 191, 193
  – of preferences, 239, 250–252
  – partial, 118
  – probability, 230
  – qualitative, 210, 211
  – quaternary, 28, 66, 67, 234, 268, 282,
    283, 308
  – Robb's, 281
  – semiorder, 254
  – separation, 103
  – temporal, 306, 378
  – weak, 39, 55, 56, 58–61, 61n, 62, 64,
    67–70, 73, 111, 211, 227, 253

paradox, 258, 266
  – and set theory, 30
  – Bertrand, 163–164
  – Bertrand's random chord, 164–166
  – Buffon's needle, 164
  – Pascal, 158
parameter, 258, 335, 398, 402, 447, 448,
    450, 453, 455–457, 462, 464
  – conditioning, 381
  – decay, 210
  – distribution, 204
  – estimation of, 7, 8, 24, 183, 203, 401
  – learning, 404, 407
  – of a distribution, 248
  – optimal or unique, 459, 460n
  – physical, 210
  – single, 404, 454
  – value unknown, 188
path, 281
  – independence of, see theorem
  – inertial, 265, 271
  – sample, 123–125, 127, 434
payoff, 247
perception, 86, 89, 92, 109, 286, 421
  – and distinguished points, 301
  – and identity of objects, 84

  – and memory, 86
  – and metric spaces, 288
  – binocular space, 296
  – constancy of, 84
  – distance, 289, 300
    – eye motion, 289, 301–302
  – experiments, 297
  – in Aristotle, 81, 83, 86
  – in Descartes, 83
  – in Hume, 83, 84
  – in Kant, 87
  – invariance in, see invariance
  – minimum visible, 289
  – optical angles, 289
  – spatial, 110
    – semantics of, 110
  – subjective, 261
  – tactile, 289
  – theory of, 286
  – visual, 83, 97, 105, 282, 289, 301, 302
philosophy of science (and)
  – characterization, 1
  – formal method, 1, 2, 28, 49
  – meaningfulness, 104n
  – probability, 14
  – reductionism, 5
  – set-theoretical models, 24
  – structure of scientific theories, 51
pragmatism, 10, 183
  – about probability, 256–263
  – in physics, 261–262
prediction, 222, 244
  – classical or null-hypothesis, 458n
  – on correlation, 339
  – quantum mechanical, 339
  – unpredictability, 220
preference, 66, 249
  – comparing, 234
  – ordering of, 250, 251
  – relation, 239, 249, 252
  – theory of, 28
primitive concepts, 53
  – Euclidean geometry, 106
    – betweenness, see betweenness
    – congruence, see congruence
  – independence, 320
    – Padoa's principle, 54n
  – interpretation, 64

primitive concepts (*continued*)
- of classical particle mechanics, *see* mechanics
- physics, 112
- probability, 130
- projective geometry, 287
- register learning model, 397
- set membership, 54, 78
- stimulus-response theory, 377–378

primitive notions, *see* primitive concepts

probabilistic
- association, 421, 423, 426, 429
  - axiom, 429
- automata, *see* automata
- behavior, 146
- causality, *see* causality
- grammar, *see* grammar
- invariance, *see* invariance
- machine learning, *see* machine, learning
- sampling, 426

probability, 12, 14
- a posteriori, 199
- a priori, 198–200
- and conditioning, 379, 380, 382, 385, 387
- and legal reasoning, 184n
- and modality, 154–155
- axioms of, 134–136, 203
- choice, *see* choice
- classical definition, 157–167, 242, 243
- concepts, 64, 123n, 155, 176, 192, 197, 200, 225, 256
  - in quantum mechanics, 224, 257, 260
  - in statistical mechanics, 256
- conditional, 138, 139, 157, 166, 190, 202, 211, 217, 234, 333, 406, 411, 415
- density, 123, 136, 191, 258, 338, 371
- distribution, 125, 157, 188, 189n, 200, 204, 223, 226, 235, 240, 248, 257, 371
  - in coin tossing, 216
- epistemic, 183
- estimation, 245–248, 250
- foundations of, 14, 129, 130, 159, 161, 177, 184n, 200, 222, 226, 257, 261, 263
- frequentist, 183
- guessing, 386, 398, 399
- interpretations, *see* probability, representations
- inverse, 188, 223
- laws, 123
- logic, 200
- logical theory of, 184–201, 238, 241
- measure or measurement, *see* measure
- model-theoretic approach, 200
- nature of, 11, 49, 129, 183
- objective, 170, 183, 185, 202, 221, 241, 246
- objective priors, 241–242
- of response, 405
- of sampling, 399, 429, 431
- posterior, 141, 141n, 142, 189, 198, 243
- pragmatic view of, 256–263
- prior, 141–143, 164, 185, 188, 189, 243
- propensity representations of, 202–225
- qualitative, 185n, 188, 203–205, 210, 211, 226, 228, 232, 249, 250, 254
  - conditional, 204, 208
  - structure, definition, 227
- quantitative, 250
- random finite sequences, 178–184
- relative-frequency theory, 129, 167–179, 224
- representations, 8, 129, 143, 202
- semantic definition of, 200
- single-case, 202, 203
- space, 123, 129, 130, 131n, 134–135, 141–143, 161, 162, 165, 167, 168, 234, 397
  - additive, 134, 135, 139
  - categoricity, 225
  - finite, 196, 201, 210, 235
  - generation, 140
  - infinite, 165n
  - Laplacean, 161, 162
- standard formal theory of, 12, 27, 200
- statistical, 198
- stopping, 399
- subjective, 63, 64, 155, 184, 198, 204, 208, 210, 211, 224–256, 261
- symmetry principles in, *see* symmetry
- theory, 126, 129, 140, 141, 158, 166, 177, 178, 184, 186, 207, 223–225, 306, 332, 344, 378
- second-order, 334

INDEX

- transition, 345–347, 367, 369
- universal, 182
- upper and lower, 252–256

process
- Bernoulli, *see* Bernoulli process
- birth and death, 348
- brain, 89
- ergodic, *see* ergodic
- stationary, 211
- stochastic, 201
  - continuous-time continuous-state, 201
  - continuous-time discrete-state, 201
  - Laplacean representation, 201

propensity
- and indeterminism, 222–223
- and objective probability, 221
- concept of, 202, 221
- in three-body problem, 218–220
- interpretation, 178
- of coin-tossing, 214–218
- of radioactive decay, 207–210
- of response strength, 211–214
- representation, 202–225

property, 91, 116, 123, 133, 168, 192, 254
- additive, 131
- associative, 233
- closure, *see* closure
- Desarguesian, 294, 294n
- irreflexivity, 196
- measurement, 111
- memoryless, 211
- order, 68, 211
- symmetry, 100
- weakly reversible, 344

proposition, 21, 34, 41, 131, 184, 297, 452
- a priori, 186
- derivation from assumptions, 40
- Desarguesian, 294n
- in geometry, 452
- primitive, 186
- proposition-like account of mental activity, 95
- vs. event in probability, 131, 184
- vs. sentence, 184

psychology, 28, 203, 211, 239, 296, 383, 387
- associationist, 89

- behavioristic, 10, 377
- experiments
  - brain representation of language, 445–453, 458–465
  - on mental imagery, 92–95
  - subjective probability, 246–248
  - visual space, 290–297
- mathematical, 419
- mental representations, *see* mental, representations
- of imagery, 89, 93, 95
- of perception, 286
- of vision, 83, 265, 266, 282–297
  - visual space, *see* visual space
- reduction of, 5
- stimulus-response theory, *see* stimulus-response theory

quantum mechanics, 4, 12, 23, 48, 258, 259, 313, 328
- absence of a converse, 154n
- and causal processes, 343
- and homogeneity, 259
- and measurement, 260
- and probability, 224, 256–261
- and propensity, 202
- and reversibility, 345
- and subjectivity of observation, 260
- and testing, 261
- Bell's inequalities, *see* Bell's inequalities
- GHZ experiments, *see* GHZ experiments
- hidden variables, *see* hidden variable

random
- chord paradox, *see* paradox
- function, 179
- poly-random function, 179
- procedure, 178
- quantity, 155, 240
- sampling, 200, 434, 456
- sequence, 168, 171–173, 176–178, 220, 222, 224
  - definition, *see* randomness, definition
  - exchangeable, 240
  - finite, 178–179, 182, 214

random (*continued*)
- variable, 14, 123, 124, 146–153, 189, 201, 229, 230, 268n, 306, 332–334, 342, 397, 455
  - and experiment, 146
  - continuous, 149
  - correlation, 153
  - covariance, 153
  - discrete, 148
  - expectation, *see* expectation
  - expected value (or mean), *see* expected
  - family of, 155
  - independence, 153–154, 156
  - inequalities, 337
  - observable, *see* observable
  - piecewise continuous, 148
  - with infinite range, 164

randomness, 172
- definition, 168–183, 262
  - Church's, 173–177
  - Kolmogorov's, 179–182
  - Martin-Löf's, 179
  - von Mises', 171–173
- for finite sequences, 168
- in a deterministic system, 223
- propensity for, 218, 222

rational
- behavior, *see* behavior
- choice, *see* choice
- decision, 248, 250, 251
- distribution, *see* distribution

rationality, 248, 251, 252, 442
- axioms, 250–252
  - vs. structure axioms, 250
- conditions of, 243–244

recursive
- definition, 25
- partial recursive functions, 75–81
- primitive recursive functions, 74–75

reduction in science, 52–53, 353, 354, 403n, 419, 465–469
- computational irreducibility, 468
- of geometry to algebra, 52
- pluralism about, 53, 467
- reductionism, 5, 53, 467–469
- representation theorems, 5
- thermodynamics, 467

- to particles or atoms
  - Democritus, 465
  - Epicurus and Lucretius, 52, 465
  - Maxwell, 466
  - Newton, 466

reinforcement, 13, 353, 377, 378, 410
- axiom, 405, 411
- concept of, 394
- determinate, 392–394, 401
- learning, 419
- methods of, 393, 399
- nondeterminate, 392–394
- partial, 393
- schedule of, 374, 376, 381, 385, 403, 414
- set of, 385, 404

relative-frequency theory, 129, 167–179

relativity
- contraction factor, 276, 279
- principle of, 279, 280
- restricted, 282
- space and time functions, 269
- space-time, *see* space-time, restricted relativistic

representation
- additive, 55, 69
- analytic, 121
- and definition, 53–54
- and invariance, 97, 111, 112
- and reduction, 112
- brain-wave, 10, 14, 354, 442–465
- Cartesian, 284
- construction, 119n
- events, *see* event
- homomorphic, 58–62
- in memory, 421
- internal, 420, 421
  - and meaning, 423, 425, 441
- isomorphic, 54–57
- language, *see* language
- Laplacean for stochastic processes, 201
- machine, *see* machine
- mental, *see* mental
- multiplicative, 55
- nature of, 51–54
- numerical, 59, 111, 118n, 281, 284
  - finitistic, 311

INDEX 531

- of probability, *see* probability, representations
- representational consequence, 251
- response, 383
- space, *see* space
- time, *see* time

representation theorem (for), 4, 57
- affine space, 268
- and reductionism, *see* reduction in science
- automata
  - finite, 370
  - finite, by stimulus-response models, 384
  - linear bounded, 374
  - pushdown, 373
- Bell's inequalities, 337
- Cayley's, 57
- center of mass of system, 326
- embedding system of particles, 330
- Euclidean space, 269
- finite affine constructions, 309
- GHZ for perfect measurements, 339
- grammars
  - context-free, 372
  - context-sensitive, 374
  - normal form (Chomsky), 358
  - normal form (Greibach), 359
  - phrase-structure, 374
  - regular, 369
- hidden variable satisfying Locality Condition II, 335
- inequalities for GHZ experiments, 340
- inequalities for three random variables, 335
- inequality for three Gaussian random variables, 343
- joint Gaussian distribution, 342
- joint Gaussian distribution with missing covariances, 342
- linear learning models by stimulus-response models, 418–419
- local hidden variable, 333
- local hidden variable satisfying second-order factorization, 334
- local hidden variable with symmetry, 334
- measurement
  - bisection, 68
  - conjoint, 70
  - difference, 67
  - extensive, 65
- nature of, 223–225, 305
- partial recursive function by register learning model, 400
- partial recursive function by unlimited register machine, 79–80
- perceptual displays, 393
- potential energy function, 332
- probability (by)
  - Carnap's regular measure, 194
  - complex finite sequence in Kolmogorov's sense, 182
  - exchangeable sequences (De Finetti), 240
  - finite Laplacean probability space, 161
  - propensity for heads in coin tossing, 217
  - propensity for randomness in three-body problem, 220
  - propensity in terms of response-strength, 212
  - propensity to decay, 208
  - qualitative approximate belief measurement, 255
  - qualitative conditional expectations, 236
  - qualitative conditional probabilities, 236
  - qualitative expectations, 232
  - qualitative probabilities, 232
  - random infinite sequence in Church's sense, 176
  - Scott's, 229
  - sequences of outcomes finite, 167
  - sequences of outcomes infinite, 169
  - stochastic processes, Chuaqui's, 201
  - stochastic processes, Nelson's, 201
- rate of change of angular momentum of system, 327
- regular languages (Kleene), 370
- regular languages by stimulus-response models, 387
- standard sequences, 205
- tote hierarchy, 387

representation theorem (for) (*continued*)
- visual space with restricted congruence, 299
- weak orderings
  - finite, 59
  - infinite, 61
reversibility, 13, 130, 313, 343–351
- and deterministic systems, 349–351
- of Markov chains, *see* Markov chain
- strong, 344, 346–349
  - harmonic oscillator, 350
- weak, 343–346
risk
- and expected utility, 250
- and truth, 8

sampling, 374, 377, 378, 388, 402, 421
- axioms, 379
- distribution, 414, 415, 415n
- postulate of simple randomness of, 142
- probabilistic, 426
- probability of, *see* probability
- random, 434, 456
- sample space, *see* probability, space
- theorem, 408n
- variance, 415n
scalar product, 315, 319
- bilinearity, 315
- Euclidean metric, 315
- Euclidean norm, 315
- positive definite, 315
- symmetry, 315
scale, 112
- absolute, 112
- classification, 114
- hyperordinal, 117
- interval, 114
- ordinal, 69, 114n, 115–117
- ratio, 112
semantic
- association, 428n
- category, 420, 425, 433, 435, 440
- comparison, 435
- congruence, 427n
- definition of probability, 200
- distinction, 438
- form, 426n
- grammatical rule, 433

- interpretation, 420
- meaning, 425
- operation, 424
- overlapping, 110
- restriction, 419
semantics, 21, 354, 384, 420
- computational, 420
- for logical calculus, 3
- modal, 155
- model-theoretic, 426n
- of spatial perception, 110
- set-theoretical, 420
- vs. syntax, 184, 441
semiorder, *see* ordering
sentential
- connectives, 4, 24, 25
set-theoretical, 161, 172
- approach, 1, 2, 27, 33, 46, 48, 60, 62
  - to language, *see* language
- concepts, 197
- confirmation theory, 196
- definitions, 33, 39
- methods, 1, 364
- model, *see* model
- notation, 132, 133
- operations, 64, 357
- predicate, 10, 17, 30, 32, 34, 55, 130, 320, 357
- representation of events, 131, 133
- semantics, 420
- structure
  - as geometry, 98
  - sample space, 146
  - vector space, 314
- structure of groups, 32
space
- absolute, 270n, 274, 298
- affine, *see* affine
- and finitism, *see* finitism
- constant curvature, 266, 285, 292, 296
- decision, 250, 251
- Euclidean, 268, 318
- flatness of, 105
- geometry of, *see* geometry
- Hilbert, 49
- metric, 103, 288
- momentum, 258
- oriented physical, 105–110

- probability, *see* probability
- projective, 103, 267n
  - separation, 103
- representation of, 265, 266
- sample, 131n, 146, 155, 207, 225, 377
  - finite, 156
  - non-numerical discrete, 146
- transformations, *see* transformations
- vector, *see* vector
- visual, *see* visual space

space-time, 106, 301
- affine, 280
- classical, 265, 269–272
  - axioms for, 270–271
  - Galilean transformations, 271
  - historical background, 272–274
  - invariance theorem, 271
- classical structure, 313, 316, 318–319
- coordinates, 122
- frame of reference, 269
- point, 272
- qualitative approach, 270
- restricted relativistic, 265, 276–282
  - axioms for, 276
  - Galilean matrix, 271
  - historical background, 278–281
  - invariance theorem, 277
  - Lorentz matrix, 276
  - Lorentz transformations, 276
  - proper time, *see* time, proper

standard deviation, 454–458, 458n, 460n
standard formalization, *see* theory, formalization
stationarity, 123–125, 211, 346
statistics
    authorship of Federalist Papers, 262
- extreme, 8, 454–458
- foundations of, 9, 130
- mathematical, 20, 24
- practice, 262–263

stimulus-response theory, 374–380, 386, 388
- and complex behavior, 353, 375, 387–390
- and theory of plans, 387
- association, 13
- connections, 396

- finite automata representation, 353, 380–387, 393
  - response to criticism, 387–394
- hierarchies, 391
- infinite automata representation, 353, 394–403
- insufficiency of, *see* and complex behavior
- language representation, 353, 375, 387, 388, 395
- learning models representation, 353, 403–419
- model, 13, 379, 381, 383–385, 387, 388, 391, 392, 400, 404
- nondeterminate reinforcement, 392, 397
- table, 388
- unlimited register machines representation, 394–403

stimulus-sampling theory, 377, 404
- axioms, 399
- models, 407, 414, 418, 419
  - and linear models, 403–419

stochastic
- approach, 179
- learning model, 130
- process, 13, 123–126, 130, 178, 201, 306, 344–346
  - continuous-time continuous-state, 201
  - continuous-time discrete-state, 201
  - Laplacean representation, 201
- standard concepts, 345

strategy
- minimax, 247

strongly reversible, *see* reversibility
structure(s)
- -symmetric measure, 194
- algebra of sets, 134
- axioms, 250–252
- basic, 33
- biological, 23
- classes of, 33
- concept of, 35
- constituent, 361
- description, Carnap's, 193–195
- event, 162
- finite relational, 118, 283

structure(s) (continued)
 – functional, 361
 – generation, 33
 – group, 32
 – homomorphism of, 58
 – in econometrics, 20
 – information, 394
 – isomorphism of, 4, 10, 54, 55, 95, 364
 – logical, 3
 – measurement, 63
  – additive conjoint, 69
  – bisection, 67
  – difference, 66
  – extensive, 63
 – memoryless waiting-time, 208
 – mother, 33
 – natural language, 355
 – of a theory, 3, 10, 27, 29, 51, 112
 – order, 33
 – representing, 370
 – set-theoretical
  – geometry, 98
  – of measurement, 55
 – simple relation, 56
 – syntactic, 51
 – topological, 33
 – unified of visual space, 300
subjective probability, see probability, subjective
symbolic dynamics, 220, 220n
symmetry, 11, 34, 100, 102, 109, 122, 194, 197–199, 217, 219, 223, 299–301, 337, 450
 – absolute convergence, 323
 – and invariance, see invariance, 98
 – and meaning, see meaning
 – antisymmetric relation, 30
 – bilateral, 98
 – condition, 335
 – in probability, 162–163
 – on conditional expectations, 334
 – principle of indifference or, 163
 – principles of, 105, 165, 167
 – state, 198
 – state-symmetric, 192, 193
 – structure-symmetric, 194
syntax, see grammar(s)
system of particle mechanics
 – conservative, 330–331
systems of particle mechanics, 320
 – axiom of impenetrability, 323
 – constant energy, 332
 – dynamical axioms, 320
 – equivalent, 329
 – kinematical axioms, 320
 – kinetic energy, 331
 – Newton's First Law theorem, 323
 – potential energy, 330
 – single particle, 331
 – subsystem, 329
 – theorem on determinism, 324

theory
 – as linguistic entity, 4
 – as major premises, 8
 – as method of organizing evidence, 9
 – as principles of inference, 8, 9
 – categorical, 10, 57, 225
  – of probability, 225
 – confirmation, 165, 190–198
 – formalization in first-order logic (standard formalization), 4, 24–30
 – instrumental view of, 8–10
 – logical structure of, 3
 – noncategorical, 10, 57, 58
 – noncausal, 322n
 – of categories, 35
 – of groups, 21
 – $(\Omega, \Im)$-categorical, 225
 – possible realization, 21
 – primitive concepts, see primitive concepts
 – quantitative, 24
 – scientific theories, 2–10
  – description of, 2
 – semicategorical, 225
 – set of rules of, 3
thermodynamics, 4, 22, 33
 – and statistical mechanics, 224
 – reduction to statistical mechanics, 5, 403, 467
three-body problem, 222, 223, 225, 324, 328, 331
 – propensity representation, 218–220
time, see also space-time
 – -function, 318

INDEX 535

- and invariance, 313
- beginning of, 104, 105, 265
- continuous or discrete, 306
- coordinate, 275
- direction of, 271, 276, 281
  - reversibility, *see* reversibility
- evolution, 260
- measurement of, 271
- observer-independent, 104
- proper, 104, 275, 275n, 280
  - as a complete invariant, 104
  - invariance, 275, 276
- representation of, 12, 265
- shift, 123
- translation of, 124, 277

transformation(s), 57, 99, 100, 110, 111, 114, 121, 122, 212, 285, 344, 345
- affine, 265, 268, 269, 271, 299
  - Galilean, 6, 111, 122, 265, 269, 271
  - Lorentz, 104, 111, 122, 265, 266, 269, 275, 276, 278–281
  - nonsingular, 277
- directional, 107
- ergodic, 124
- group of, 57, 100, 104, 105, 111
- hypermonotone, 117
- identity, 115
- in grammar, 361
  - learning of, 441
- in physics, 343
- linear, 115, 118, 119, 119n, 213
- metrically isomorphic, 285
- monotone, 116
- monotone-increasing, 111
- nonlinear, 164
- numerical, 114, 118
- of mental images, 94
- ordinal, 111
- projective, 47, 93
- similarity, 114, 115, 118, 232, 269
- space-, 99
- stationary, 124

tree, *see* grammar(s), derivation, tree

trial(s)
- and subtrial, 393
- as ordered pair, 404
- Bernoulli, 124, 136, 138, 154
- conditioning, 386, 393
- discrete, 138
- finite sequence, 168, 180
- in conditioning, 377, 378
- infinite sequence, 187, 188, 240, 377, 404
- learning, 139, 393, 397, 419
- number, 168
- permutation, 211
- randomization, 183
- reinforcement, 377
  - partial, 394
- sampling on, 380
- sequence of, 139

truth, 131, 160, 162, 245, 450
- and probability, 239
- and risk, 8
- general, 452
- L-true, 191
- necessary, 87
- role in evaluation of theories, 8–9
- Tarski's definition, 26
- value, 110, 111
- vs. usefulness, 8

uncertainty
- and decision-making, 248–256
- Heisenberg relations, 257
- measure of, *see* entropy

uniqueness
- of probability measure, 235

unpredictability, *see* prediction

utility, 9, 66, 117, 141n, 239, 249
- and decision theory, 231, 234, 249
- cardinal, 114
- expected, 167, 212, 239, 248, 250
  - maximization, 249
  - theory, 234, 236
  function, 249
- numerical, 252
- of outcomes, 231

utterance, 286, 355, 419, 421, 423, 440
- and internal representation, 422
- production, 420
- semantic interpretation, 420
- superset of, 420

variance, 150
vector product, 316, 319, 323, 326
- bilinearity, 316

vector product (*continued*)
- inner or scalar, *see* scalar product
- Jacobi identity, 316
- orthogonality, 316
- skew symmetry, 316

vector(s), 259, 269, 271, 277
- -velocity, 279
- acceleration, 122
- addition for points, 317
- binary operation on, 315
- in physics, 315
- infinite series, 319
- linearly independent, 316
- of direction of force, 219
- of response strength, 213
- position, 325
- product
  - exterior, *see* vector product
  - inner or scalar, *see* scalar product
- radius, 121n
- spaces, 121, 213, 281, 282, 314–318
  - action on a set of points, 317
  - axioms for, 314
  - basis, 316
  - Cartesian, 315
  - dimensionality, 316
  - real, 315
  - set-theoretical structure, 314
- standard concepts, 316
- translation, 268
- vector-valued function, 319
- velocity, 122

velocity, 104, 214, 222, 258, 259, 279, 324, 344
- as covariant, 122
- constant, 103, 215, 270n
  - relative, 270
- escape, 220
- in coin tossing, 216
- of light, 275, 280
- relative, 276
- uniform, 328
- vector, 122

vision, *see also* visual space
- binocular, 292
- monocular, 288, 290
- psychology of, 83, 265, 266, 282–297
  - visual space, *see* visual space

- theory of
  - Berkeley's, 289
  - Euclidean, 40–41, 288
  - Luneburg's, 293

visual space, 8, 282–302
- as not Euclidean, 292
- constant curvature, 285, 292
- contextual, 300
- contextual effects, 293
- distance perception, *see* perception
- double elliptic, 290
- elliptic, 287
- equidistance tendency, 293
- Euclidean, 283, 288
- experiments, *see* experiments, visual space
- geometry of, 289, 290, 294–296, 300–302
- Helmholtz-Lie problem, 285
- hierarchy of geometries, 287–288
- hyperbolic, 287, 288, 292, 293
- hypothesis testing, 282–287
- kinematics of eye motion, 302
- Luneburg's theory, 292
- nature of, 288–297
  - Berkeley's Essay, 289
  - Euclid's Optics, 288
  - Newton's Opticks, 289
  - Reid's Inquiry, 289
  - Reid's spherical geometry, 290
- objects of, 302
- partial axioms, 297–300
  - affine plane, 297
  - congruence, 298
  - three distinguished points, 297
- spherical, 290

weakly reversible, *see* reversibility